AF849491

Allele Mining for Genomic Designing of Oilseed Crops

This book deliberates on the concept, strategies, tools, and techniques of allele mining in oilseed crops and its application potential in genome elucidation and improvement, including studying allele evolution, discovery of superior alleles, discerning new haplotypes, assessment of intra- and interspecific similarity, and studies of gene expression and gene prediction. Available gene pools in global germplasm collections, specifically those consisting of wild-allied species and local landraces for almost all major crops, have facilitated allele mining. The development of advanced genomic techniques, including PCR-based allele priming and Eco-TILLING-based allele mining, is now widely used for mining superior alleles. Allele's discovery has become more relevant now for employing molecular breeding to develop designed crop varieties matching consumer needs and with genome plasticity to adapt to climate change scenarios. All these concepts and strategies, along with precise success stories, are presented in the chapters dedicated to the major oilseed crops.

1. This is the first book on the novel strategy of allele mining in oilseed crops for precise breeding.
2. This book presents genomic strategies for mining superior alleles underlying agronomic traits from genomic resources.
3. This book depicts case studies of PCR-based allele priming and Eco-TILLING-based allele mining.
4. This book elaborates on gene discovery and gene prediction in major oilseed crops.

This book will be useful to students and faculties in various plant science disciplines, including genetics, genomics, molecular breeding, agronomy, and bioinformatics; scientists in seed industries; and policymakers and funding agencies interested in crop improvement.

Nextgen Agriculture: Novel Concepts and Innovative Strategies
Chittaranjan Kole and Phullara Kole
Series Editor

Allele Mining for Genomic Designing of Fruit Crops
Chittaranjan Kole, Kenta Shirasawa, and Anil Kumar Singh

Allele Mining for Genomic Designing of Cereal Crops
Chittaranjan Kole, Sharat Kumar Pradhan, and Vijay K Tiwari

Allele Mining for Genomic Designing of Vegetable Crops
Chittaranjan Kole, Tusar Kanti Behera, and Prashant Kaushik

Allele Mining for Genomic Designing of Oilseed Crops
Chittaranjan Kole, Manish K. Pandey, and Naveen Puppala

Sustainable Urban Agriculture: New Frontiers
Kheir Al-Kodmany, Madhav Govind, Sharmin Khan, and Chittaranjan Kole

Allele Mining for Genomic Designing of Grain Legume Crops
Chittaranjan Kole, C. Bharadwaj, and Abhimanyu Sarkar

The Vertical Farm: Scientific Advances and Technological Developments
Kheir Al-Kodmany, Andrew Keong Ng, Abel Tablada, and Chittaranjan Kole

For more information, please visit our website: https://www.routledge.com/Nextgen-Agriculture/book-series/NGA

Allele Mining for Genomic Designing of Oilseed Crops

Edited by
Chittaranjan Kole, Manish K. Pandey, and Naveen Puppala

CRC Press
Taylor & Francis Group
Boca Raton London New York

CRC Press is an imprint of the
Taylor & Francis Group, an **informa** business

First edition published 2025
by CRC Press
2385 NW Executive Center Drive, Suite 320, Boca Raton FL 33431

and by CRC Press
4 Park Square, Milton Park, Abingdon, Oxon, OX14 4RN

CRC Press is an imprint of Taylor & Francis Group, LLC

© 2025 selection and editorial matter, Chittaranjan Kole, Manish Kumar Pandey and Naveen Puppala; individual chapters, the contributors Goyary

Reasonable efforts have been made to publish reliable data and information, but the author and publisher cannot assume responsibility for the validity of all materials or the consequences of their use. The authors and publishers have attempted to trace the copyright holders of all material reproduced in this publication and apologize to copyright holders if permission to publish in this form has not been obtained. If any copyright material has not been acknowledged please write and let us know so we may rectify in any future reprint.

Except as permitted under U.S. Copyright Law, no part of this book may be reprinted, reproduced, transmitted, or utilized in any form by any electronic, mechanical, or other means, now known or hereafter invented, including photocopying, microfilming, and recording, or in any information storage or retrieval system, without written permission from the publishers.

For permission to photocopy or use material electronically from this work, access www.copyright.com or contact the Copyright Clearance Center, Inc. (CCC), 222 Rosewood Drive, Danvers, MA 01923, 978-750-8400. For works that are not available on CCC please contact mpkbookspermissions@tandf.co.uk

Trademark notice: Product or corporate names may be trademarks or registered trademarks and are used only for identification and explanation without intent to infringe.

Library of Congress Cataloging-in-Publication Data
Names: Kole, Chittaranjan, editor. | Pandey, Manish K., editor. | Puppala, Naveen, editor.
Title: Allele mining for genomic designing of oilseed crops /
edited by Chittaranjan Kole, Manish Kumar Pandey, and Naveen Puppala
Description: First edition | Boca Raton, FL : CRC Press, 2024 |
Includes bibliographical references
Identifiers: LCCN 2024022096 (print) | LCCN 2024022097 (ebook) |
ISBN 9781032471945 (hardback) | ISBN 9781032471952 (paperback) |
ISBN 9781003385035 (ebook)
Subjects: LCSH: Oilseeds–Genetic engineering. | Plant genomes.
Classification: LCC SB298 .A45 2024 (print) | LCC SB298 (ebook) |
DDC 633.8/5–dc23/eng/20240617
LC record available at https://lccn.loc.gov/2024022096
LC ebook record available at https://lccn.loc.gov/2024022097

ISBN: 978-1-032-47194-5 (hbk)
ISBN: 978-1-032-47195-2 (pbk)
ISBN: 978-1-003-38503-5 (ebk)

DOI: 10.1201/9781003385035

Typeset in Times
by codeMantra

Dedicated to

Dr. Norman Ernest Borlaug

The Father of Green Revolution, the Nobel Peace Prize Winner, and the Founder of the World Food Prize

Who inspired my research and academic works with his generous appreciation and advice.

Series Preface

Agriculture is going through a phase of transformation now. It has both challenges and opportunities. One of the major challenges it is facing is obviously food security. A population of about 9.3 billion projected by 2050 necessitates an increase of crop production by 60–70%. Another 10–15% increase would be needed to counter the reduction of crop production due to global warming and climate change. Nutrition security is another challenge with more than 800 million people suffering from malnutrition. Depletion of fossil fuel and increasing dependence on biofuel from crop residues is another challenge. Loss of biodiversity and land degradation also add to the list of challenges. The cultivable area is also decreasing fast due to housing, transport, and industrialization. Above all, climate change and global warming have a huge impact on agriculture, health, and environment.

Several novel concepts and strategies are emerging to address the above challenges. For example, the practice of precision farming will lead to precise and timely use of inputs resulting in higher yield and better quality of produce on the one hand and reduction of loss of inputs on the other. Similarly, urban agriculture and vertical agriculture will add a lot of cultivable areas in the future. Protected farming will facilitate growing crops under controlled conditions leading to reduction in biotic and abiotic stresses and precise utilization of inputs.

Agriculture requires maintenance of the philosophy farmers have been following since inception over thousands of years which are being sold today in new packages with sophisticatedly coined labels such as traditional agriculture, natural agriculture, sustainable agriculture, conservation agriculture, integrated agriculture, regenerative agriculture, and organic agriculture.

In the changing scenarios, next-generation agriculture will need designed crop varieties and breeds produced through designed practices. This book series aims at deliberations on several novel concepts and innovative strategies developed for next-generation agriculture and will hopefully benefit agricultural education, research, and outreach.

Chittaranjan Kole
Kolkata
January 10, 2024

Preface

The affordability, precision, and high-throughput nature of sequencing technologies in the last two decades have provided an unprecedented understanding of the genomes and target plant traits of economic importance. Furthermore, the advances in third-generation sequencing technologies, including Hi-C, especially in the last decade, have made it possible to develop high-quality reference genomes, pangenomes, and gene annotations, including genome-wide structural variations. These developments have also increased the precision of gene and superior haplotype discovery, followed by mining allelic diversity among breeding lines and genebank collections. The high-throughput and cost-effective genotyping platforms have further boosted the adoption rate of genomic tools in ongoing breeding programs across the world. Deployment of these advances in crops and making them well equipped to scale up the integration of genomic interventions in breeding programs will facilitate the development of future-ready crops.

The above advances have made gene/allele mining more precise and cost-effective for the target traits preferred by farmers, consumers, and industry, as well as efforts toward mitigation of climate change impacts and achieving nutritional security. It is important to accelerate the process of genomic design to develop future-ready crops by bringing new allelic combination(s) to one or more loci underlying desired genes using modern breeding approaches such as marker-assisted breeding, haplotype-based breeding, and gene/allele pyramiding, and also now through gene editing.

These new alleles originate through natural or induced mutations in one or more genomic regions at a time. The traditional plant breeding methods practiced in the 20th century depended solely on phenotypes for the development of new genotypes, primarily through selection and hybridization. The advent of DNA-based molecular markers in the 1980s facilitated indirect identification of the alleles by using linked molecular markers using genetic linkage mapping. Molecular breeding as a supplement to traditional breeding has successfully been employed in almost all major crops for their genetic improvement in a short time. It is worthwhile to mention that transgenic breeding has also contributed significantly to crop improvement, at least for some major input target traits such as herbicide tolerance, insect resistance, etc.; however, this technology is still facing challenges in societal adoption in different parts of the world, especially for food crops. Nevertheless, gene editing has been receiving positive vibes from the policymakers of different countries, creating a conducive environment for accelerating the use of this technology on a large scale.

Despite the huge loss of diversity during domestication, several national and international gene banks have millions of diverse germplasms to explore variability for the traits that are needed to breed climate-resilient and future-ready crops. Genetic diversity played a pivotal role during the entire period of conventional plant breeding, both traditional and molecular. However, plant breeders have been using mainly the primary gene pool constituting the indigenous germplasm lines and local landraces as a genetic resource for new alleles. In the meantime, a drastic dearth of new alleles in the primary gene pool happened due to genetic erosion caused naturally or due to domestication over time, which was aggravated by the injudicious and irrational implementation of the so-called green revolution. Precisely, this highlighted the importance of allied wild crop relatives (WCRs) and thereby a mission of collection, conservation, characterization, and utilization of biodiversity. The emergence of advanced bioinformatics and the availability of cheaper and more innovative genome sequencing technologies have empowered us to explore germplasm and identify a large number of donor alleles, and allele mining has become a reality!

The strategy of association mapping, the most popular trait mapping approach in animal and human sciences, has received huge adoption in plant genomics studies. The availability of large sequenced diverse panels with multilocation phenotyping data has provided much-needed insights into trait genomics through genome-wide association studies (GWAS), leading to the discovery of genes and superior haplotypes in many crops. This has led to the possibility of mining a large number of genes from the available accessions in crops. Mutations occur in the genome as single nucleotide polymorphisms (SNPs) or insertion/deletions (In/Dels), resulting in the generation of new alleles or changes in existing alleles and allele combinations. Using exactly this principle, the approach of Eco-TILLING has facilitated large-scale allele mining in several major crops. On the contrary, sequence-based allele mining has led to the discovery of allelic diversity both at the target gene and in the whole genome. Recent advances in modern genomic tools and technologies such as genome engineering, speed breeding, etc. facilitate the deployment of these new alleles in breeding programs in a relatively short time to accelerate the rate of genetic gains. Truly speaking, allele mining has revolutionized the strategy of utilizing genetic as well as genomic resources and has become an important component of next-generation precision breeding and biotechnology!

Allele discovery has become more relevant now for employing molecular breeding to develop designed crop varieties matching consumer needs and with genome plasticity to adapt to climate change scenarios using targeted and precision breeding tools validated by both genomics and phenomics studies. All these concepts and strategies, along with precise success stories, are presented in over 50 chapters allocated under 5 volumes dedicated to the 5 crop groups, including cereals, oilseeds, grain legumes, fruits, and vegetables.

This volume entitled, "Allele Mining for Genomic Designing of Oilseed Crops" includes 11 chapters focused on Soybean, Oilseed Rape, Peanut, Sunflower, Linseed, Sesame, Mustard and Safflower contributed by 76 scientists from 10 countries including Australia, China, Ghana, Germany, India, Japan, Mali, Serbia, Turkey and USA, We express our thanks for their excellent contributions and timely cooperation.

This book will hopefully be useful to the students, research scholars, and teaching faculties in academia to learn about the fundamentals and applications of allele mining; the scientists practicing plant breeding in public and private institutes in improvising their programs; and also the policy-makers and funding agencies in prioritizing research fields.

Chittaranjan Kole
Kolkata, India
Manish K. Pandey
Hyderabad, India
Naveen Puppala
Clovis, USA

Contents

Contributors

Javed Akhatar
Department of Plant Breeding and Genetics
Punjab Agricultural University
Ludhiana, India

Abhishek Anand
School of Agriculture and Allied Sciences
The Neotia University
Kolkata, India

Chhaya Atri
Department of Plant Breeding and Genetics
Punjab Agricultural University
Ludhiana, India

S.S. Banga
Department of Plant Breeding and Genetics
Punjab Agricultural University
Ludhiana, India

S. K. Bera
Crop Improvement
ICAR-Directorate of Groundnut Research
Jungadh, India

Ramesh S. Bhat
Department of Biotechnology
University of Agricultural Sciences
Dharwad, India

S. Chandra
ICAR-Indian Institute of Soybean Research
Indore, India

Liu Chunyan
College of Agriculture
Northeast Agricultural University
Harbin, People's Republic of China

Haile Desmae
Groundnut Breeding
ICRISAT
Mali, West Africa

Wu Fengjiao
College of Agriculture
Northeast Agricultural University
Harbin, People's Republic of China

Sunil S. Gangurde
Department of Plant Pathology
University of Georgia
Tifton, Georgia

Sri Vidya K Gannavarapu
Department of Molecular Biology
Acharya NG Ranga Agricultural University
Tirupathi, India

Shailendra Goel
Department of Botany
University of Delhi
Delhi, India

M. V. C. Gowda
Department of Genetics and Plant Breeding
University of Agricultural Sciences
Dharwad, India

Anna Goyal
Department of Plant Breeding and Genetics
Punjab Agricultural University
Ludhiana, India

Birgul Guden
Department of Field Crops, Faculty of Agriculture
Akdeniz University
Antalya, Turkey

S. Gupta
ICAR-Indian Institute of Soybean Research
Indore, India

Krishnan Hajella
School of Life Sciences
Devi Ahilya University
Indore, India

Jiang Hongwei
College of Agriculture
Northeast Agricultural University
Harbin, People's Republic of China

Renate Horn
Institute of Biological Sciences, Department of Plant Genetics
University of Rostock
Rostock, Germany

P. Janila
Groundnut Breeding
ICRISAT
Patancheru, India

Jinjin Jiang
Jiangsu Provincial Key Laboratory of Crop Genetics and Physiology
Yangzhou University
Yangzhou, China

Liu Jiarui
College of Agriculture
Northeast Agricultural University
Harbin, People's Republic of China

Yu Kaixin
College of Agriculture
Northeast Agricultural University
Harbin, People's Republic of China

Jignesh H. Kamdar
School of Science, Dept. of Microbiology
R K University
Rajkot, India

M. Karthikeyan
CSIR-National Chemical Laboratory
Pune, India

Harjeevan Kaur
Department of Plant Breeding and Genetics
Punjab Agricultural University
Ludhiana, India

Vikender Kaur
Division of Germplasm Evaluation
Indian Council of Agricultural Research (ICAR) - National Bureau of Plant Genetic Resources
New Delhi, India

R. Kavishwar
ICAR-Indian Institute of Soybean Research
Indore, India

Chittaranjan Kole
Foundation for Science & Society
Kolkata, India

G. Kumawat
ICAR-Indian Institute of Soybean Research
Indore, India

Huan Li
Key Laboratory of Biology and Genetic Improvement of Oil Crops, Ministry of Agriculture and Rural Affairs
Oil Crops Research Institute of the Chinese Academy of Agricultural Sciences
Wuhan, China

Sujan Mamidi
Genome Sequencing Center
Hudson Alpha Institute for Biotechnology
Huntsville, AL

M. Shivakumar
ICAR-Indian Institute of Soybean Research
Indore, India

Sun Minghao
College of Agriculture
Northeast Agricultural University
Harbin, People's Republic of China

B. N. Motagi
AICRP on Groundnut
University of Agricultural Sciences
Dharward, India

Wang Ning
College of Agriculture
Northeast Agricultural University
Harbin, People's Republic of China

Manish K. Pandey
Center of Excellence in Genomics & Systems Biology (CEGSB)
ICRISAT
Patancheru, India

Sarita Pandey
School of Agriculture and Allied Sciences
The Neotia University
Kolkata, India

Shri Hari Prasad
Centre for Plant Biotechnology and Molecular Biology
Kerala Agricultural University
Thrissur, India

Zhaoming Qi
College of Agriculture
Northeast Agricultural University
Harbin, People's Republic of China

Chen Qingshan
College of Agriculture
Northeast Agricultural University
Harbin, People's Republic of China

Moin Qureshi
Department of Field Crops, Faculty of Agriculture
Akdeniz University
Antalya, Turkey

Aleksandra Radanović
Institute of Field and Vegetable Crops
National Institute of Republic of Serbia
Novi Sad, Serbia

Rishiraj Raghuvanshi
ICAR-Indian Institute of Soybean Research
Indore, India

Kirti Rani
Regional Research Station
NBPGR
Jodhpur, India

M. B. Ratnaparkhe
ICAR-Indian Institute of Soybean Research
Indore, India

Supriya Ratnaparkhe
Indore Biotech Inputs and Research Pvt Ltd
Indore

Jayanta Roy
Department of Plant Sciences
North Dakota State University
Fargo, ND

Gyanesh K. Satpute
ICAR-Indian Institute of Soybean Research
Indore, India

N. Shesh
ICAR-Indian Institute of Soybean Research
Indore, India

Kenta Shirasawa
Department of Frontier Research and Development
Kazusa DNA Research Institute
Chiba, Japan

Ruchi Shroti
ICAR-Indian Institute of Soybean Research
Indore, India

Boubacar Sinare
West Africa Centre for Crop Improvement (WACCI)
University of Ghana (UG)
Legon, Accra, Ghana

Ajay Singh
ICAR-National Institute of Abiotic Stress management
Baramati

Devender Singh
Division of Germplasm Evaluation
Indian Council of Agricultural Research (ICAR) - National Bureau of Plant Genetic Resources
New Delhi, India

K.H. Singh
ICAR-Indian Institute of Soybean Research
Indore, India

Mamta Singh
Division of Germplasm Evaluation
Indian Council of Agricultural Research (ICAR) - National Bureau of Plant Genetic Resources
New Delhi, India

Simarjeet Kaur Sra
Department of Plant Breeding and Genetics
Punjab Agricultural University
Ludhiana, India

Subulakshmi Subramanian
ICAR-National Institute of Abiotic Stress management
Baramati, India

Mulpuri Sujatha
ICAR-Indian Institute of Oilseeds Research (IIOR) Rajendranagar
Hyderabad, India

Prashanth Suravajhala
ICAR-National Institute of Abiotic Stress management
Baramati, India

Zhao Tuanjie
MOA Key Laboratory of Biology and Genetic Improvement of Soybean (General), National Centre for Soybean Improvement, State Key Laboratory for Crop Genetics and Germplasm Enhancement, College of Agriculture
Nanjing Agricultural University
Nanjing, China

Bulent Uzun
Department of Field Crops, Faculty of Agriculture
Akdeniz University
Antalya, Turkey

Rajesh V
ICAR-Indian Institute of Soybean Research
Indore, India

Rajeev K. Varshney
Centre for Crop and Food Innovation
Murdoch University
Murdoch, WA

V. Nataraj
ICAR-Indian Institute of Soybean Research
Indore, India

Dalia Vishnudasan
ICAR-National Institute of Abiotic Stress management
Baramati, India

Linhai Wang
Key Laboratory of Biology and Genetic Improvement of Oil Crops, Ministry of Agriculture and Rural Affairs
Oil Crops Research Institute of the Chinese Academy of Agricultural Sciences
Wuhan, China

Youping Wang
Jiangsu Provincial Key Laboratory of Crop Genetics and Physiology
Yangzhou University
Yangzhou, China

Dhammaprakash Pandhari Wankhede
Division of Genomic Resources
ICAR-National Bureau of Plant Genetic Resources
New Delhi, India

Li Xu
College of Agriculture
Northeast Agricultural University
Harbin, People's Republic of China

Neelam Yadav
ICAR-Indian Institute of Soybean Research
Indore, India

Shashank Kumar Yadav
Division of Germplasm Evaluation
Indian Council of Agricultural Research (ICAR) - National Bureau of Plant Genetic Resources
New Delhi, India

Engin Yol
Department of Field Crops, Faculty of Agriculture
Akdeniz University
Antalya, Turkey

Abbreviations

100SDW	100-seed weight
^{13}C	Carbon isotope 13
4-Cl-IAA	4-Chloroindole-3-acetic acid
AA	Ascorbic acid
ABA	Abscisic acid
ABC	ATP-binding cassette transporter
AB-QTL	Advanced backcross QTL
ABRII	Agricultural Biotechnology Research Institute of Iran
ABTS	2,2′-Azinobis-(3-ethylbenzthiazoline-6-sulfonate)
ACC	1-Aminocyclopropane-1-carboxylate
ACE	Angiotensin-converting enzyme
ACL	ATP-citrate lyase
ADF	Acid detergent fiber
ADH	Alcohol dehydrogenase
ADL	Acid detergent lignin
AFLP	Amplified fragment length polymorphism
AGB-IM	Active Germplasm Bank of the Manfredi Experimental Station
AHASL	Acetohydroxyacid synthase large subunit
AhGEA	*Arachis hypogaea* gene expression atlas
AhMITE	*Arachis hypogaea* miniature inverted-repeat transposable element
AhTE	*Arachis hypogaea* transposable element
ALSV	Apple latent spherical virus
AMF	Arbuscular mycorrhizal fungi
AMOVA	Analysis of molecular variance
AOX	Alternate oxidase
APPN	Austrian Plant Phenotyping Network
APX	Ascorbate peroxidase
ARC	Agricultural Corporation Center
ASL	Al-sensitive line
AS-PCR	Allele-specific polymerase chain reaction
ATL	Al-tolerant line
ATP	Adenosine triphosphate
ATPase	Adenosine triphosphatase
B lines	Maintainer lines
BAC	Bacterial artificial chromosome
BC	Back cross
BES	BAC end sequence
BGF	Blue-green fluorescence
BIBAC	Binary BAC
BIH	Branch initiation height
BLAST	Basic local alignment search tool
BLASTn	Nucleotide Basic Local Alignment Search Tool
BM	Biomass
BOAA	b-N-oxalyl-amino-L-alanine
BPMV	Bean pod mottle virus
BSA	Bulked segregant analysis
BSR-Seq	Bulk segregant analysis RNA-Seq

BY	Biomass yield
C16:0	Palmitic acid
C18:0	Stearic acid
C18:1	Oleic acid
C18:2	Linoleic acid
C18:3	Linolenic acid
C20:1	Eicosenoic acid
C22:1	Erucic acid
C5-MTase	Cytosine-5 DNA methyltransferases
CAAS	Chinese Academy of Agricultural Sciences
CAD	Cinnamyl alcohol dehydrogenase
CaMV35S	Cauliflower mosaic virus 35S
CaMV35SDE	Double-enhanced *Cauliflower mosaic virus* 35S
CAPS	Cleaved amplified polymorphic sequence
Cas	CRISPR-associated protein
CAT	Catalase
CBF	C-repeat-binding factor
CBL	Calcineurin B-like protein
CBP	Calmodulin-binding protein
CC	Cellulose content
CcHyPRP	*C. cajan* hybrid proline-rich protein
CcMT1	*Cajanus cajan* metallothionein 1
CC-NBS-LRR	Coiled-coil NBS-LLR
cDNA	Complementary DNA
CDS	Coding sequence
CG	Candidate gene
CGKB	Cowpea Genespace/Genomics Knowledge Base
CGMS	Cytoplasmic genetic male sterility
CGR	Crop growth rate
Chr	Chromosome
CHS8	Chalcone synthase 8
CID	Carbon isotope discrimination
CIPK	CBL interacting protein kinase
CKX4	Cytokinin oxidase
Clp	Chaperone protein
cM	CentiMorgan
CMS	Cytoplasmic male sterility
CMS	Cell membrane stability
CNL	CC-NBS-LRR
CNV	Copy number variation/variant
CO	CONSTANS
COX	Cytochrome oxidase
CPD	Centers of Plant Diversity
cpDNA	Chloroplast DNA
CPO	Centers of Plant Origin
cQTL	Candidate QTL
C-QTLs	Consensus QTLs
Cr	Chromium
CR	Clubroot resistant
CRB	Biological Resource Center for Sunflower
CRF	Cytokinin response factor

CRF-INIA	Centre of Plant Genetic Resources of the National Institute for Agricultural and Food Research and Technology
CRISPR	Clustered regularly interspaced short palindromic repeats
CRK	Cysteine-rich receptor-like protein kinase
cry1AcF	Crystal protein 1AcF
cry1E-C	Crystal protein 1E-C
CS	Clubroot susceptible
CSIC	Institute of Sustainable Agriculture
CWR	Crop wild relative
CWT	Cumulative water transpired
DAF	DNA amplification fingerprinting
DAP	Differentially abundant protein
DArT	Diversity arrays technology
DAS	Days after sowing
dCAPS	Derived CAPS
ddRAD	Double digest restriction-site associated DNA
ddRAD-seq	Double digest restriction-site associated DNA sequencing
DEG	Differentially expressed gene
DF	Days to 50% flowering
DFF	Days to final flowering
DGAT	Diacylgycerol acyltransferase
DGR	Directorate of Groundnut Research
DH	Doubled haploid
DHN	Dehydrin
DI	Disease index
DIF	Days to initial flowering
DII	Drought intensity index
DM	Days to maturity
DMR	Differentially methylated region
DMS	Dimethyl sulfate
DOGMA	Dual organellar genome annotator
DPPN	German Plant Phenotyping Network
DREB	Dehydration responsive element binding
DREB	Dehydration responsive element binding
DSB	Double-stranded break
dsDNA	Double-stranded DNA
DSI	Drought susceptibility index
DSI	Drought stress index
Dt1	Growth habit locus
DTC	Drought tolerance coefficient
DTF	Days to flowering
DTI	Drought tolerance index
DTY	Yield under drought
EBI	European Bioinformatics Institute
EcoTILLING	Ecotype-TILLING
EGS	Early generation selection
EMBO	European Molecular Biology Organization
EMS	Ethylmethane sulfonate
EPPN	European Plant Phenotyping Network
EREBP	Ethylene-responsive element binding protein
ESP	Exchangeable sodium percentage

EST	Expressed sequence tag
ETR	Electron transport rate
F2	Second filial generation
FA	Fatty acid
FAC	Florigen activation complex
FACE	Free air CO2 enrichment
FAD	Fatty acid desaturase
FAD2	Fatty acid desaturase 2
FAO	Food and Agricultural Organization
FAOSTAT	Food and Agriculture Organization Corporate Statistical Database
FAR	Fatty acyl-CoA reductase
FD	Frost damage
FIGS	Focused identification of germplasm strategy
FLA3	Fasciclin-like arabinogalactan proteins
FLC	Flowering locus C
FloCIA	Flower color image analysis
FLS	Frogeye leaf spot
FNR	Ferredoxin-NADP reductase
FOP	Fusarium oxysporum f.sp. phaseoli
FP	Flowering period
FPKM	Fragments per kilobase of transcript per million mapped reads
FRAP	Ferric reducing antioxidant power
FST	Fixation Index
FT	Flowering locus T
FT	Flowering time
GA3	Gibberellic acid
GAB	Genomics-assisted breeding
GATT	General Agreement on Tariffs and Trade
GBS	Genotyping by sequencing
GCIM	Genome-wide composite interval mapping
GC-MS	Gas chromatography-mass spectrometry
GCP	Generation Challenge Program
GE	Genome editing
GE	Genotype environment
GE	Germination energy
GEBV	Genomic estimated breeding value
GEI	Genotype-by-environmental interactions
GIS	Geographic information system
GISH	Genomic in situ hybridization
GM	Genetically modified
GMO	Genetically modified organism
GMS	Genetic male sterility
GMTPS	Soybean terpene synthase genes
GP	Gene pool
GP	Germination percentage
GP	Whole growth period
GP1	Primary gene pool
GP2	Secondary gene pool
GP3	Tertiary gene pool
GPAT5	Glycerol-3-phosphate sn-2-acyltransferase 5
GPX	Glutathione peroxidase

GS	Genomic selection
GSL	Glucosinolate
GSLE	GSL content between eco-environments
GSS	Gene space sequences
GST	Glutathione S-transferase
GWA	Genome-wide association
GWAS	Genome-wide association study/studies
GxE	Genotype x environment interaction
H_2O_2	Hydrogen peroxide
HC	Hemicellulose content
HDR	Homology-dependent repair
HDR	Homology-directed repair
HE	Homologous exchange
HI	Harvest index
HOAC	High oleic acid content
HPT	High-throughput phenotyping
HSF	Heat shock factor
Hsfs	Heat shock protein transcription factors
Hsp	Heat shock protein
IBONE	Instituto de Botánica del Nordeste
ICAR	Indian Council of Agricultural Research
ICARDA	International Center for Agricultural Research in Dry Areas
ICRISAT	International Crops Research Institute for the Semi-Arid Tropics
IFS2	Isoflavone synthase 2
IFVC	Institute of Field and Vegetable Crops – National Institute of the Republic of Serbia
ILs	Introgression lines
Indel	Insertion/deletion
INRA	National Institute for Agricultural Research
INRAE	Institut National de la Recherche Agronomique – National Research Institute for Agriculture, Food and Environment
INTA	Instituto Nacional de Tecnología Agropecuaria – National Agricultural Technology Institute
IP	Intron polymorphism
iPBS	Inter-primer binding site retrotransposon markers
IPGI	Institute of Photogrammetry and Geo Informatics
IPGI	International Peanut Genomic Research Initiative
IPPN	International Plant Phenotyping Network
IPR	Intellectual property right
iProClass	Integrated Protein Classification
IR	Inverted repeat
IRAPs	Inter-Retrotransposon Amplified Polymorphism
IRSTEA	National Research Institute of Science and Technology for the Environment and Agriculture
ISR	Intron spanning region
ISSR	Inter-simple sequence repeat
iTRAQ	Isobaric tag for relative and absolute quantification
KASP	Kompetitive allele-specific PCR
KEGG	Kyoto Encyclopedia of Genes and Genomes
LAI	Leaf area index

LC	Lignin content
LC-MS/MS	Liquid chromatography-mass spectrometry
LD	Linkage disequilibrium
LEA	Late embryogenesis abundant
LER	Leaf expansion rate
LG	Linkage group
LIPM	Laboratory of Plant-Microbe Interactions
LIS	The Legume Information System
LL	Lamina length
LLS	Late leaf spot
LOD	Logarithm of odds
LOX	Lipoxygenases
LP	Low P
LPRA	Root angle at LP
LR	Leaf resistance
LRC	Lateral root cap
LRR	Leucine-rich repeat
LTG	Low-temperature germination
LTL	Leaf total length
LTP	Lipid transfer proteins
LTRs	Long terminal repeats
LW	Lamina width
LWP	Leaf water potential
MAB	Marker-assisted breeding
MABC	Marker-assisted backcrossing
MAGIC	Multi-parent advanced generation intercross
MALDI-TOF	Matrix-assisted laser desorption/ionization time-of-flight mass spectrometry
MAPK	Mitogen-activated protein kinase
MARS	Marker-assisted recurrent selection
MAS	Marker-assisted selection
mcQTLs	Metabolite QTL for complex products
MDA	Malondialdehyde
MIM	Multiple interval mapping
miRNA	MicroRNA
MITE indel	Miniature inverted repeat transposable elements insertion-deletions
MLE	Moringa oleifera leaf extract
MLM	Mixed linear model
MMS	Multi-main stem
MOOB	Multi-objective optimized genomic breeding
MPBQ-MT	2-methyl-6-phytyl-1,4-benzoquinone methyltransferase
MPPs	Multi-parent populations
mQTL	Metabolite QTL
MQTL	Meta-QTL
MRMLM	Multi-locus random-SNP-effect mixed linear model
mRNA	Messenger RNA
MSI	Membrane stability index
msQTL	Metabolite QTL for specific product
MST	Monosaccharide transporter
MT	Maturity time
MTA	Marker-trait association
MTA	Material transfer agreement

NADH	Nicotinamide adenine dinucleotide
NAM	Nested association mapping
NaPPN	North American Plant Phenotyping Network
NAR	Net assimilation rate
NBPGR	National Bureau of Plant Genetic Resources
NBS	Nucleotide-binding site
NBS-LRR	Nucleotide-binding site leucine-rich repeat
NCBI	National Center for Biotechnology Information
NCSU	North Carolina State University
nGBS	Normalized GBS
NGS	Next-generation sequencing
NHEJ	Non-homologous end joining
NIL	Near isogenic lines
NLR	Nucleotide-binding leucine-rich repeat
NPBTs	New plant breeding techniques
NPGS	National Plant Germplasm System
NPs	Nanoparticles
NREF	Non-redundant reference
NUE	Nutrient use efficiency
$\mathbf{O_2^-}$	Superoxide radicals
OA	Osmotic adjustment
OAC	Oleic acid content
OB	Oil body
OC	Oil content
OCRI	Oil Crops Research Institute
ODAP	Oxalyldiaminopropionic acid
ODR	Oxygen diffusion rate
OEE	Oxygen-evolving enhancer
OH	Hydroxyl radical
OilA1	Oil content on linkage group A1
OP	Optimal P
OPV	Open-pollinated variety
ORF	Open reading frame
OsRuvB	Oryza sativa RuvB
OWR	Overwintering rate
P5CR	LΔ1-pyrroline-5-carboxylate reductase
P5CS	Pyrroline-5-carboxylate synthetase
P5CS1	Pyrroline-5-carboxylate synthase
PAGE	Polyacrylamide gel electrophoresis
PAH1	Phosphatidic acid phosphohydrolase 1
PAMSA	PCR amplification of multiple specific alleles
PAV	Presence/absence variants/variations
PAW	Plant available water
PCD	Pulse Crop Database
PCR	Polymerase chain reaction
PDB	Protein Data Bank
PDR	Pleiotropic drug resistance complexes
PDS	Phytoene desaturase
PEBV	Pea early browning virus
PEG	Polyethylene glycol
PEPB	Phosphatidyl ethanolamine-binding protein

PGPR	Plant growth-promoting rhizobacteria
PGR	Plant genetic resources
PH/PHT	Plant height
PHS	Pre-harvest sprouting
pI	Isoelectric point
PIC	Polymorphic information content
PIR	Protein Information Resource
PKAMs	Pigeonpea kompetitive allele-specific PCR assay markers
PL	Petiole length
PN	Lobe number
POD	Pods per plant
POD	Peroxidase
PP	Pea protein
PPB	Participatory plant breeding
PPN	Pea protein nanoparticles
PPR	Pentatricopeptide repeat
PPVFR	Protection of Plant Varieties and Farmer's Right
PRL	Primary root length
Pro	Proline
PRR7	Pseudo response regulator 7
PRSR	Phytophthora root and stem rot
PSD	Protein Sequence Database
PSII	Photosystem II
Psp68	Pisum sativum p68 gene
PSPDB	Plant Stress Protein Database
PUFA	Polyunsaturated fatty acid
PVC	Polyvinyl chloride
PVE	Phenotypic variation explained
PVP	Plant variety protection
QTL	Quantitative trait locus
QTL-HR	QTL hotspot regions
QTLs	Quantitative trait loci
R lines	Restorer lines
RAD	Restriction site-associated DNA
RAD	Root average diameter
RAD-seq	Restriction site-associated DNA sequencing
RAPD	Random amplified polymorphic DNA
Rchit	Rice chitinase gene
RDW	Root dry weight
RE	Restriction enzyme
REMAP	Retrotransposon-microsatellite amplified polymorphism
RF	Root forks
Rf	Fertility restoration gene
RFLP	Restriction fragment length polymorphism
RGA	Rapid generation advance/ advancement
RGA	Resistance gene analog
RGC	Resistance gene candidate
R-gene	Resistance gene
RIL	Recombinant inbred line
RKN	Root-knot nematode
RLD	Root length density

RNA-seq	RNA sequencing
ROS	Reaction oxygen species
RPRV-HN	Hemagglutinin neuraminidase gene of peste des petits ruminants' virus
RPVH	Hemagglutinin gene of render pest virus
RRA	Relative root angle
RSA	Root surface area
RSA	Root system architecture
RSAMPL	Random selective amplification of microsatellite polymorphic loci
RSP	Resting spores per plant
RT	Reverse transcription
RWC	Relative water content
SA	Salicylic acid
SACPD-C	Stearoyl-acyl carrier protein desaturase
SAGE	Serial analysis of gene expression
SAMS	S-adenosyl methionine synthetase
SAT	Semi-arid tropics
SAUR	Small auxin-upregulated RNAs
SB	Speed breeding
SBG	Sequence-based genotyping
SCAR	Sequence characterized amplified region
SCL	Seed coat luster
SCMR	SPAD chlorophyll meter reading
SCN	Soybean cyst nematode
SCoT	Start codon targeted polymorphism
SD	Short day
SD	Stem diameter
SDP1	SUGAR-dependent 1
SDS	Sudden death syndrome
SDS-PAGE	Sodium dodecyl sulfate PAGE
SDW	Shoot dry weight
SFA	Saturated fatty acid
SGC	Seed glucosinolate content
sgRNA	Single guide RNA
SHMT	Serine hydroxymethyltransferase
SIR	Salinity induction response
siRNA	Small interfering RNA
SK	Shikimate kinase
SkimGBS	Skim genotyping by sequencing
SLAF	Specific locus amplified fragment
SLAF-seq	SLAF sequencing
SMD	Sterility mosaic disease
SMG	Selectable marker gene
SMRT	Single-molecule real-time
SMS	Single-molecule sequencing
SMV	Soybean mosaic virus
SNP	Single-nucleotide polymorphism
SNPS	Seed number per silique
SNuPE	Single-nucleotide primer extension
SOC	Seed oil content
SOD	Superoxide dismutase
SOLiD	Supported oligonucleotide ligation and detection

SOS1	Salt overly sensitive 1
SoyOPA-1	Soybean oligo pool all-1
SPAC	Soil–plant–atmosphere continuum
SPD	Seeds per pod
SR	Stem resistance
SRA	Sequence read archives
SRAP	Sequence-related amplified polymorphism
SSCP	Single strand conformational polymorphism
SSD	Single seed descent
SSR	Simple sequence repeat
SSR	Sclerotinia stem rot
STM	Shoot meristemless
STMS	Sequence tagged microsatellite site
STS	Sequence tagged site
SuSy	Sucrose synthase
SV	Structural variant/variation
SW	Seed weight
SWEET	Sugars will eventually be exported transporter
SWNTs	Single-walled carbon nanotubes
SY	Seed yield
TAC	Transcript assembly contig
TAG	Triacylglycerol
TALEN	Transcription activator-like effector nuclease
TAMU	Texas A & M University
TbyS	TILLING by sequencing
TDM	Total dry matter
T-DNA	Transfer DNA
TE	Transposable element
TF	Transcription factor
TFBM	Transcription factor binding motif
TFC	Total flavonoid content
TFs	Transcription factors
TILLING	Targeting induced local lesions in genomes
TIR-NBS-LRR	Toll-interleukin-like receptor–NBS-LRR
TN	Tapidor and Ningyou7
TNL	TIR-NBS-LRR
TOL	Tolerance index
TPC	Total phenolic content
TRAP	Target region amplification polymorphism
TRIPS	Trade-related Aspects of Intellectual Property Rights
TRL	Total root length
TRT	Total root tips
TRV	Tobacco rattle virus
TRV	Total root volume
Trx	Thioredoxin
TS	Training set
TSA	Total root surface area
TSA	Transcriptome shotgun assembly
TSW	Thousand-seed weight
TTC	Triphenyl tetrazolium chloride
TuYV	Turnip yellows virus

TWAS	Transcriptome-wide association studies
UNEAK	Universal Network Enabled Analysis Kit
UPGMA	Unweighted pair group method with arithmetic mean
URGI	National Research Institute for Agriculture, Food and Environment research unit in genomics and bioinformatics
USDA	United States Department of Agriculture
USDA-ARS	USDA – Agricultural Research Service
USDA-ARS-NCRPIS	USDA-ARS North Central Regional Plant Introduction Station
USDA-NASS	USDA – National Agricultural Statistics
USFA	Unsaturated fatty acid
USLP1.0	Universal Soy Linkage Panel 1.0
V. f.	Vicia faba
VCF	Variant call format file
VIGS	Virus-induced gene silencing
VIR	Federal Research Center N. I. Vavilov All-Russian Institute of Plant Genetic Resources
WAP	Weeks after planting
WBC	White-brown complexes
WFD	Winter frost damage
WGRS	Whole genome re-sequencing
WGS	Whole genome sequencing
WL	Waterlogging
WTO	World Trade Organization
WUE	Water use efficiency
XTH	Xyloglucan endotransglucosylases/hydrolases
YLD	Yield
YRR	Yield reduction rate
YST	Yellow seed trait
ZFN	Zinc finger nuclease
ZIP	Iron-regulated transporter-like
α-ALA	Alpha-linolenic acid
γ-TMT	Gamma-tocopherol methyltransferase

Editor Biography

Prof. Chittaranjan Kole is an internationally reputed scientist with an illustrious professional career spanning over 40 years and is globally recognized for his original contributions in the fields of plant genomics, biotechnology, and molecular breeding, leading to the publication of more than 160 quality research articles. He has edited over 190 books for the leading publishers in the world, including Springer, Wiley-Blackwell, and Taylor & Francis Group. His scientific contributions and editing acumen have been appreciated by seven Nobel Laureates, including Profs. Norman Borlaug, Arthur Kornberg, Werner Arber, Phillip Sharp, Günter Blobel, Lee Hartwell, and Roger Kornberg. He has been honored with a number of fellowships, honorary fellowships, and national and international awards, including the 'Outstanding Crop Scientist Award' conferred by the International Crop Science Society. He has served in all prestigious positions in academia, including as Vice-Chancellor of the BC Agricultural University and Project Coordinator of Indo-Russian Center of Biotechnology in India, and Director of Research of the Institute of Nutraceutical Research of Clemson University in the United States. He also worked at the Pennsylvania State University and Clemson University as Visiting Professor in the United States. He was awarded with the Raja Ramanna Fellowship by the Department of Energy, Government of India. He is also heading the International Climate-Resilient Crop Genomics Consortium, International Phytomedomics and Nutriomics Consortium, and Genome India International as their Founding President. He has recently established the Prof. Chittaranjan Kole Foundation for Science and Society and is working for his Founding Chairman.

Dr. Manish K. Pandey is currently leading the Groundnut and Pigeonpea Genomics, Prebreeding & Bioinformatics Group at the International Crops Research Institute for the Semi-Arid Tropics (ICRISAT), Hyderabad, India. His key contributions include the reference genome sequence for diploid progenitor and both the subspecies of cultivated tetraploid groundnut, gene expression atlas, low- to high-density SNP genotyping assays, diagnostic markers for key traits, quality control panel, marker-assisted pyramiding, and genomic selection in groundnut. Through genomics-based breeding, his team developed five high oleic groundnut varieties, which are released for cultivation in six states of India. He has published more than 200 scientific articles (>10,000 citations) in various international journals, including *Nature Genetics*, *PNAS-USA*, and *Molecular Plant*. His efforts have been recognized at the international level, as he has been invited to deliver presentations at several international conferences, review proposals for international funding agencies, and collaborate with a large number of scientists at the international level. He is also Adjunct Associate Professor at the University of Southern Queensland (USQ), Australia, and Distinguished Professor at the Institute of Crop Genetic Resources of Shandong Academy of Agricultural Sciences (SAAS), China. He has now been inducted into many scientific academies in India, namely, as Fellow of the National Academy of Agricultural Sciences (NAAS)-New Delhi and Telangana Academy of Agricultural Sciences (TAAS)-Telangana, Associate Fellow/Member of Andhra Pradesh Academy of Agricultural Sciences (APAS)-Andhra Pradesh, and Member of The National Academy of Sciences-India (NASI). He was also among the top 2% highly cited researchers in 2019–2022 compiled by Stanford University, USA.

Naveen Puppala is a Professor at New Mexico State University, USA. His research focuses on the genetic characterization and improvement of Valencia peanuts through molecular techniques and conventional breeding approaches. His Valencia breeding emphasizes variety development for high yield, high oleic, four seeds per pod, good taste, and disease resistance. He has published 86 peer-reviewed journal articles, 6 book chapters, and 170 meeting abstracts as author or co-author.

He has given numerous invited presentations at national and international conferences. Seventeen graduate students have been supported with grants secured from USAID funds, USDA-NIFA, New Mexico Peanut Research Board, and National Peanut Research, where he was the Principal Investigator or co-PI of the project. He has served as an advisor, co-advisor, and on a graduate student committee. He has mentored national peanut breeders from ten different countries to develop high-yielding, disease-resistant Valencia peanuts to increase the livelihood of economically poor farmers. He served as President-Elect for the American Peanut Research and Education Society from 2014 to 2015. He has released 14 peanut cultivars. He served as a peanut crop germplasm committee chair from 2009 to 2011. He served on grant review panels for USDA-NIFA. He was an associate editor for *BMC Plant Biology*, *Journal of Plant Registration*, and *Peanut Science*.

1 Soybean Allele Mining for Molecular Breeding
A Review

Zhaoming Qi, Yu Kaixin, Sun Minghao, Wang Ning, Wu Fengjiao, Li Xu, Liu Jiarui, Jiang Hongwei, Zhao Tuanjie, Liu Chunyan, and Chen Qingshan

1.1 INTRODUCTION

Soybean, known as "菽" in ancient times, originated and domesticated in China. Soybean is not only rich in plant protein and edible oil but also contains beneficial nutrients such as isoflavones and lecithin. It is an important edible, feed and oil crop. China used to be the largest soybean producer and exporter in the world, but since 1996, due to the impact of imported soybeans and low comparative benefits, the self-sufficiency rate of domestic soybeans has gradually declined, and the degree of dependence on soybean imports has been maintained at a high level of more than 80% for a long time. China's soybean industry has gradually shrunk, which is also related to the development of transgenic technology. In 2019, China put forward a soybean revitalization plan, which aims to gradually improve the self-sufficiency level of domestic soybeans by means of expanding planting areas, increasing yield and improving quality. Given that it is difficult to increase soybean planting area significantly, it is of great significance to use modern biotechnology to cultivate new genetically modified soybean varieties with high yields and high quality.

The geographical environment of China is very complex, which determines that soybean varieties need to have the ability to adapt to different regional production, which requires soybean breeding to enhance wide adaptability. At present, the functional mining of related genes in soybeans is mainly focused on yield, biostress, abiotic stress, flowering and quality, especially in order to improve yield. However, if the breeding goal is only high yield, which appears to be relatively single, and the quality and resistance to extensive stress such as insect resistance and disease resistance are not prominent, it will seriously restrict the stability of soybean production and lead to very high production costs. The research progress of soybeans was reviewed mainly from the aspects of soybean quality (fatty acids and proteins), important concentrated metabolites (isoflavones and terpenoids), stress resistance, florescence and so on. It has played a positive role in promoting the development of the soybean industry.

1.2 RESEARCH PROGRESS ON SYNTHESIS AND ACCUMULATION OF FATTY ACIDS IN SOYBEAN

Soybean is one of the most important sources of vegetable oil in the world. The fatty acid composition of soybean oil has a great influence on its nutritional value, storage resistance and processing technology. The composition and accumulation of oils are affected by a variety of enzyme activities in the fatty acid synthesis pathway; the expression of these genes is also regulated at the pre-transcriptional, transcriptional and post-transcriptional levels, and many related genes are involved in this process.

DOI: 10.1201/9781003385035-1

1.2.1 Regulation of Transcription Factors on Fatty Acid Content in Soybean

LEC1, LEC2, WRI1, ABI3 and FUS3 are all classic transcription factors regulating lipid content. The *GmWRI1* protein can bind to aw-box, which is involved in various steps of oil biosynthesis, and positively regulate oil accumulation in soybean seeds through a gene expression network related to fatty acid biosynthesis. Moreover, the oil content of transgenic seeds increases steadily, and no significant effect on protein content or yield is observed (Chen et al., 2018; Wang et al., 2022). *GmABI3* is responsible for the biosynthesis of seed-specific TAG and long-chain fatty acids in soybeans (Manan & Zhao, 2020). FUS3 plays a positive role in the regulation of oil accumulation and also plays an upregulation role in the expression of fatty acid synthesis. The oil content of abi3-deficient mutants decreased, so ABI3 may be related to oil accumulation. In fus3-deficient mutants, the accumulation of triacylglycerol was three times as high, suggesting that ABI3 could synergically and independently regulate the accumulation of seed oil (Yang et al., 2022). *GmLEC1* and *GmLEC2* are upstream positive regulators of *GmFUS3, GMABI3, GmDof11* and *GmWRI1* (Manan et al., 2017). Some transcription factors also regulate lipid synthesis. *GmDREBL* is involved in the regulation of fatty acid accumulation by controlling WRI1 and its downstream gene expression (Zhang et al., 2016). *GmMYB73* may decrease GL2 and then release GL2-inhibited PLDα1 expression for lipid accumulation, which may improve the oil yield of leguminous plants (J. Liu et al., 2014). *GmDOF4* and *GmDOF11* positively regulate genes related to fatty acid biosynthesis to improve the oil content of soybean seeds (H. Wang et al., 2007).

1.2.2 Regulation of Soybean Fatty Acids by Various Enzymes

Various enzymes and genes that affect fatty acid synthesis also regulate lipid synthesis. Fatty acid desaturase (FAD) is the key enzyme to introduce double bonds into the fatty acid chain in the biosynthetic pathway of glycerol in plants. Deletion of alleles in the FAD family leads to a decrease in linoleic acid and linolenic acid content and a significant increase in oleic acid content in soybean (Flores et al., 2008; Haun et al., 2014; Pham et al., 2011). Moreover, studies have shown that the decrease in linolenic acid content depends on the number of mutated FAD3 and fAD2-1 genes, the severity of the fAD2-1A gene mutation and the testing environment (Pham et al., 2012). Stearoyl-acyl carrier protein desaturase (SACPD-C) can control stearic acid accumulation in seeds. It encodes a unique δ 9 isomer, stearoyl-ACP-desaturase, which is responsible for converting stearic acid to oleic acid. High levels of SACPD-C transcripts were detected only in developing seed tissue, which indicates that the encoded desaturase has the function of enhancing oleic acid biosynthesis ability (Zhang et al., 2008). At the same time, the loss of SACPD-C resulted in a moderate increase in stearic acid content in seeds, while overall tuberculous accumulation and nitrogen fixation were not affected (Gillman et al., 2014). There are two kinds of fatty acid thioesterase (Fat) in higher plants, respectively, FatA and FatB. *GmFATA* and *GmFATB* mutated alleles increase oleic acid content and decrease stearic acid and palmitic acid content, thus regulating saturated fatty acid content (Ma et al., 2021; Zhou et al., 2021). Acyl CoA (diacylgycerol acyltransferase, DGAT) is a very important enzyme in organisms. Its main mechanism is to add diacylglycerol to fatty acid acylglycerol to form triacylglycerol. Oleic acid increased and linoleic acid decreased in the soybean seeds of overexpressed DGAT transgenic plants (Jing et al., 2021; Xu et al., 2021). The composition of fatty acids was also regulated by *GmSDP1* (SUGAR Dependent1-1), *GmGPDHp1* (Kanai et al., 2019; Zhao et al., 2021) and *GmPLDγ* (Bai et al., 2020).

1.2.3 Mapping of Fatty Acid-Related Genes in Soybean

Some genes or loci closely associated with oil were also identified by association, functional analysis and population genetic analysis. The sucrose efflux transporter gene, *GmSWEET39*, controls soybean seed oil content and is in a selective state. Its sequence variation affects its relative expression and oil content in soybean seeds (Miao et al., 2020). The CCT gene POWR1 is the basis of large effector protein/oil quantitative trait loci (QTLs). Pathogenic TE insertion truncates the CCT domain and significantly increases seed oil content, weight and yield while reducing the protein content (Goettel et al., 2022). Transcriptomic results showed that *GmOLEO1* positively enhanced oil accumulation by affecting triglyceride metabolism (Zhang et al., 2019).

1.3 RESEARCH PROGRESS ON SYNTHESIS AND METABOLISM OF SOYBEAN ISOFLAVONES AND TERPENOIDS

Soy isoflavones are flavonoids, a class of secondary metabolites formed during soybean growth, and a biologically active substance. They determine the economic value of soybeans, and the quality of soybeans is directly related to the health of eaters. Therefore, the majority of researchers are more and more concerned. The genes affecting soybean isoflavone content mainly encode enzymes and transcription factors.

1.3.1 Regulation of Isoflavones in Soybean by Various Enzymes

The current research found that the main regulators of isoflavone content in soybeans are chalcone synthase and isoflavone synthase. In 2000, Woosuk Jung et al. used yeast expression analysis to screen soybean ESTs encoding cytochrome P450 proteins and identified soybean isoflavone synthase genes *IFS1* and *IFS2* (Jung et al., 2000). In 2008, Hao Cheng et al. cloned, sequenced and analyzed the *IFS1* and *IFS2* genomic regions of 33 Chinese soybean germplasms. The results showed that three SNPs in the *IFS1* gene and two SNPs in the *IFS2* gene ($P < 0.05$) were closely associated with all individual and total isoflavone levels in seeds, regardless of population structure (Cheng et al., 2008).

The chalcone synthase genes are *CHS6, CHS7* and *CHS8*. Transcriptome analysis by Sangeeta Dhaubhadel et al. (2007) showed that the expression of multiple genes for isoflavone biosynthesis, including phenylalanine ammonia lyase and chalcone synthase (*CHS8*, *CHS8* and isoflavone synthase 2), coincided with the beginning of the accumulation of isoflavone compounds in embryos. Comparative analysis of genes involved in isoflavone biosynthesis in RCAT Angora (high seed isoflavone) and Harovinton (low seed isoflavone) showed that the expression level of *CHS7* and *CHS8* in RCAT Angora (high seed isoflavone) was significantly higher than that in Harovinton. The difference in seed isoflavone levels between these two varieties may be caused by the differential expression of *CHS7* and *CHS8* in the later stage of seed development (Dhaubhadel et al., 2007). In 2009, Zernova et al. used three DNA boxes containing *CHS6* chalcone synthase and IFS2 isoflavone synthase genes and a PAL5 phenylalanine ammonia lyase gene to transform soybean into embryo culture in order to change the isoflavone content and composition of soybean seeds. It was found that two lines of T4 homozygous seeds had reduced isoflavone concentrations (about 70%) (Zernova et al., 2009).

Glycosyltransferases also affect isoflavone content.

1.3.2 Effects of Transcription Factors on Metabolism of Soybean Isoflavones

CHS8 expression is regulated by the transcription factors *GmMYB176*, *GmbZIP5* and *GmMYB29*. In 2010, Jinxin Yi et al. proved that *GmMYB176* regulates CHS8 expression and affects isoflavone synthesis in soybean. *GmMYB176* encodes R1 MYB protein and is expressed in soybean seeds during maturation. Transient expression of *GmMYB176* in soybean germ protoplasts increased endogenous CHS8 transcription by 169 times after 48 h. RNAi silencing *GmMYB176* in hair roots leads to a decrease in isoflavone compound levels, suggesting that *GmMYB176* is necessary for isoflavone biosynthesis (Yi et al., 2010). In 2017, Shanshan Chu et al. found that isoflavone concentration-related SNP was located in the 5′-untranslated region (5′-UTR) of R2R3 MYB transcription factor *GmMYB29*. A transient reporter gene assay showed that *GmMYB29* activates *IFS2* (isoflavone synthase 2) and *CHS8* (Chalcone synthase 8) gene promoters, and *GmMYB29* overexpression and RNAi-mediated silencing in soybean hair roots lead to increased and decreased isoflavone content, respectively (Chu et al., 2017). In 2021, Arun Kumaran Anguraj Vadivel et al. identified the *GmMYB176* interactor that regulates soybean isoflavone biosynthesis, demonstrated the co-immunoprecipitation of THE bZIP transcription factors, *GmbZIP5* and *GmMYB176*, and demonstrated protein-protein interactions in plants. RNAi silencing *GmbZIP5* can reduce isoflavone content in the hairy roots of soybeans (Anguraj Vadivel et al., 2021).

1.3.3 Published Important Genes of Soybean Isoflavones

In addition to enzymes and transcription factors, other genes were found to be related to isoflavones. In 2021, Ming-Sin Ng et al. found that *GmMATE1* and *GmMATE2* localized to the vacuolar membrane and promoted isoflavone accumulation, which may be the real isoflavone transporters promoting isoflavone accumulation in soybean seeds. *GmMATE1* defective soybean mutants significantly reduced the content of total isoflavones in seeds, while the overexpression of *GmMATE1* in transgenic soybeans promoted the accumulation of isoflavones in seeds (Ng et al., 2021).

1.3.4 Research in Synthesis and Metabolism Progress of Soybean Terpenes

Terpenes are important defense compounds against herbivores and pathogens. In 2013, Man Zhang et al. reported the identification of a new monoterpene synthase gene, *GmNES*, from soybean. Transcription of *GmNES* is upregulated in soybeans infected with cotton leaf worm, mechanically damaged or treated with salicylic acid (SA). Gas chromatography-mass spectrometry (GC-MS) analysis showed that the recombinant *GmNES* enzyme specifically produced NEROL. This finding suggests that *GmNES* is the nerol synthase gene in soybeans. Subcellular localization using GFP fusion indicated that *GmNES* was localized in chloroplasts (Zhang et al., 2013). In addition to *GmNES*, *GmTPS3* is also a gene related to terpene synthesis. In 2014, Jianyu Liu et al. conducted genome analysis of 23 soybean terpene synthase genes (GMTPS) distributed on 10 out of 20 chromosomes and constructed transgenic tobacco plants with exogenous expression of *GmTPS3*. Transgenic tobacco lines expressing *GmTPS3* showed stronger resistance to cotton leaf worm and higher geraniol levels in dual-choice feeding preference and force-feeding tests (Liu et al., 2014).

1.4 RESEARCH PROGRESS ON THE FUNCTION OF 7s GLOBULIN (B-COGLYCININ) AND 11s GLOBULIN (GLYCININ) AND EDIBLE RELATED QUALITY GENES IN SOYBEAN

As an important source of nutrition, soybean protein accounts for about 40% of soybean seeds. 7s globulin (b-coglycinin) and 11s globulin (glycinin) account for about 70% of the total seed protein. The functional properties of 7s globulin and 11s globulin are different due to different amino acid compositions and structures. And the interaction between them has an important effect on the functional properties of soybean protein.

1.4.1 Functional Introduction of 7s Globulin (b-Coglycinin) and 11s Globulin (Glycinin)

Soybean protein is composed of various components, which are divided into 2S (25 kDa), 7S (180 kDa), 11S (350 kDa) and 15S (600 kDa) globulins according to different centrifugal sedimentation coefficients (Sui et al., 2021). β-Coglycinin (7S) and glycinin (11S) are the main components of soybean protein, accounting for about 70%–80% of soybean protein, and their contents are significantly negatively correlated, among which 7S globulin accounts for about 30% and 11S globulin accounts for about 40%, but the ratio of 7S–11S also varies with soybean varieties, about 0.6–4.1 (John et al., 2021). 11S globulin is a hexameric protein with each of the six subunits composed of acidic (A1a, A1b, A2, A3, A4 and A5) and basic (B1a, B1b, B2, B3 and B4) polypeptides, which are linked by disulfide bonds (Ma et al., 2010). To date, several genes encoding 11S have been identified, such as *Gy1* to *Gy5* (encoding the glycine subunits of A1aB2, A2B1a, A1bB1b, A5A4B3 and A3B4, respectively), *Gy6* (pseudogene) and *Gy7* (weakly expressed) (Beilinson et al., 2002; Fischer & Goldberg, 1982; Nielsen et al., 1989; Scallon et al., 1985). Reported 15 genes (*CG1-G15*) encode the 7S subunit, where *CG-1* encodes the α′-subunit and *CG-4* encodes the β-subunit (Harada et al., 1989). Soybean 7S globulin is one of the main allergenic soybean proteins, and its α-subunit can be recognized by 25% of the serum of soybean-sensitive patients with atopic dermatitis (Ogawa et al., 1995). They have shown that the main factor causing allergy in soybean storage protein is the α'-subunit in 7S globulin. Due to the existence of a large number of α'-subunits, some people are allergic to soybean protein. Studies have shown that α-subunit-deficient soybean varieties can not only improve the nutritional quality and processing characteristics of soybeans, but also reduce the necessary technological procedures to remove the sensitization effect of soybean protein and improve the utilization rate of soybean protein. They found that IgE binds specifically and preferentially to the 7S globulin beta subunit. And it is the first report to identify transdermal sensitized soybean allergens in a mouse model (Murakami et al., 2018). This study found that 7S protein was the active component of soybean in lowering plasma cholesterol. The enhancing effect of isoflavones on the blood cholesterol lowering activity of 7S protein is achieved by inhibiting cholesterol absorption and de novo synthesis in hypercholesterolemic hamsters (Liu et al., 2019).

Therefore, changing the 7S globulin content and cultivating α-subunit-deficient varieties have also become important research directions in soybean quality breeding. Since there is currently no effective food allergy treatment, preventing food allergies becomes an effective measure. With the increasing application of plant protein in food, even if allergy sufferers can avoid direct consumption of soy products, it is more and more difficult to avoid processed foods with added soy protein, so the demand for hypoallergenic soy products is also increasing. The commonly used methods in the field of food processing to remove the allergenicity of soybean protein mainly include desensitization processes such as heat denaturation (Ohishi et al., 1994), chemical reagent treatment (Tsuji et al., 1997) and enzymatic hydrolysis (Lee, 2004). However, soybean antigen protein is heat stable and exhibits anti-nutritional properties. The above methods not only increase the processing cost to varying degrees but also affect the functional properties of soybean protein in subsequent processing to varying degrees. In animal feed processing, hypoallergenic soybeans can effectively prevent animal allergies. Using hypoallergenic soybeans as raw materials in food processing can provide nutritional support for allergic people and improve product quality. Therefore, cultivating hypoallergenic soybean varieties can improve the nutritional quality and functional properties of soybean protein while removing soybean allergenic protein. Hypoallergenic soybeans and their products have great market potential.

1.4.2 Research Progress of Related Genes in Soybean Food Processing

Soybean is rich in nutritional value and contains a variety of nutrients, including protein, fat, carbohydrates and other energy-producing nutrients, as well as calcium, phosphorus, iron and other mineral elements (Song et al., 2015). Soybean is also rich in dietary fiber, which can lower plasma cholesterol and regulate intestinal function (Kim et al., 2016; Shin et al., 2015). Soybeans are often processed into tofu, yuba, dried bean curd, tempeh and other soy products. Tofu is widely consumed, inexpensive and nutritious. In the process of traditional tofu processing, cooked soybean milk is obtained by refining, cooking and filtering soybeans soaked for a long time, and then adding a coagulant to make it solidify. Finally, traditional tofu is usually dehydrated and shaped by physical pressure (Kim & Wicker, 2005; Liu, 2012; Obatolu, 2008). It was found that fermentation for 18 h was the most suitable due to the content of reducing sugars, total peptides and isoflavone aglycones (daidzein and genistein), as well as antioxidant activity and the effect of freeze-thawed tofu. Digestibility was improved (Lee et al., 2018). It was found that angiotensin-converting enzyme (ACE) bacterially fermented soybean food had higher ACE inhibitory activity than fungal fermented soybean food, and 3-day sprouted bean sprouts were better than 5-day and 5-day sprouted soybean food. Seven-day bean sprouts had higher ACE inhibitory activity. The ACE inhibitory ability of 11S hydrolyzate was higher than that of 7S (Handa et al., 2020).

Scholars at home and abroad have done a lot of research on the relationship between soybean protein components and processing quality, but the results are slightly different. . There is a negative correlation between soy protein and the firmness and breakability of tofu (Murphy et al., 1997). Soybean varieties with a high 11S/7S ratio produced lower yields of tofu but greater firmness and cohesion (Ji et al., 1999). The ratio of 11S/7S has little correlation with the quality of tofu (Skurry et al., 2010; Taira et al., 1990). . Studies suggest that the total protein content and water-soluble protein content in soybeans have a very significant positive or negative correlation with the texture of tofu, and the elasticity has a very significant positive correlation with the 7S component. The effect of soybean protein subunits on tofu quality was studied, and the results showed that the A3 subunit of 11S globulin had the greatest impact on tofu quality, while the A4 subunit had a negative impact on tofu quality (Poysa et al., 2006). The firmness of tofu made from soybeans with 11S globulin A5 subunit deletion was high (Nishinari et al., 2014). The subunit in 11S globulin is closely related to the gelation speed and transparency of soybean protein (Fukushima, 1991). Believes that the gel formed by heating 11S globulin is rich in water retention and has high tensile stress and shear force, and the gel hardness and elasticity are better when it is made into tofu (Saio et al., 2014). Showed that soybean 11S globulin content was positively correlated with the firmness of tofu (Mujoo et al., 2003).

1.4.3 Research Progress of Genes Related to Soybean Food Processing

Food scientists have been working hard to improve the taste, quality and function of traditional soy foods, and with the help of emerging processing technologies, today's soy protein development is making great progress. Modification of soybean proteins through interactions with polyphenols, Maillard reactions and even non-thermal treatment is a promising method that can minimize heat-induced denaturation of soybean proteins (Jan et al., 2017; Oliver et al., 2006; Zhang et al., 2018). Utilizing fibrillation and electrospinning, it may provide a new method for designing and manufacturing soybean protein with different applications (Cao & Mezzenga, 2019; Xu et al., 2012). Furthermore, three-dimensional (3D) printing, one of the most important pillars of Industry 4.0, has been introduced into food manufacturing (Sun et al., 2015). With this technology, food can be efficiently produced locally to meet individual needs by using advanced 3D printing and inkjet technology. Over the past few years, there has been a growing interest in plant-based meats and unprecedented success in high-moisture extrusion of plant-based proteins.

1.5 RESEARCH PROGRESS ON FUNCTIONAL GENES RELATED TO STRESS RESISTANCE IN SOYBEAN

1.5.1 Research Progress on Salt Stress in Soybean

The area of saline-alkali land in the world is increasing year by year. Aiming at the large area of underutilized saline-alkali land, improving salt-alkali tolerance of soybeans and cultivating salt-tolerant varieties are some of the important measures to make use of abandoned land resources, increase soybean planting area and increase soybean yield. Soybean response to salt stress is a complex biological process that depends on the co-expression of multiple genes in a complex network. At present, some progress has been made in the mechanism of salt tolerance in soybeans.

The damage of soil salinization to plants is mainly the toxic effect of Na^+. SOS1 protein (salt overly sensitive 1) is a Na^+/H^+ reverse transporter on the cell membrane, which mainly mediates the efflux of Na^+ to reduce the effect of Na^+ on the damage to cells (Yue et al., 2012). The SOS1 gene in soybean (*Glyma.08G092000*) is expressed in the true leaves, epicotyl, hypocotyl and root tissues of the plant, and the highest expression level is in the root (Wang-xing et al., 2015). Transcription factors are also key links in regulating salt tolerance in soybeans. There are 156 genes in soybean belonging to the MYB family members, 43 of which are involved in abiotic stress response, and the pathways in response to salt stress are also diverse (Mochida et al., 2009). Additionally, when plants receive salt stress signals, stress-related hormones such as abscisic acid (ABA), gibberellin and auxin, which are hormones related to growth and development, play an important role in balancing growth and stress responses (Yu et al., 2020).

1.5.2 Research Progress on Disease Resistance of Soybean

Phytophthora root rot, soybean gray spot, cyst nematode disease and mosaic virus disease are several popular soybean diseases in China, which will affect the normal growth of soybean plants and lead to a reduction in soybean yield and even no harvest. How to improve the disease resistance of soybeans has always been the focus of soybean breeders.

Phytophthora root and stem rot (PRR) is one of the devastating diseases affecting soybean (Glycine max) production, which can cause 10%–50% of soybean yield loss (Bernard et al., 1957).

Existing studies have shown that the introduction of disease resistance genes such as *hrpZm, hrf2* and *CaAMP1* by transgenic technology can enhance soybean tolerance to PRR, and overexpression of the disease course-related protein gene *Glym41*, harpin-like protein genes *hrf2* (Niu et al., 2019) and *hrpZm* (Du et al., 2018), *Capsicum annuum* anti-microbial protein 1, *CaAMP1* (Niu et al., 2020), etc., can significantly enhance soybean resistance to epidemics of root rot resistance.

Frogeye leaf spot of soybean (FLS) is a fungal disease caused by *Cercospora sojina* Hara. The disease affects soybean leaves, stems, pods and kernels.

At present, for the research on the location of soybean gray spot resistance genes, three dominant genes, Rcs1, Rcs2 and Rcs3, have been identified in the United States, which control the resistance to American gray leaf spot races 1, 2 and 5 respectively. Rcs3, at the same time, is resistant to American soybean cinerea race 2 but not at the same locus as Rcs2 (Boerma & Phillips, 1983). At this stage, the resistance of soybean varieties that have been bred and used in production in the southern United States is mostly determined by the Rcs3 gene. Two dominant genes, Hrcs1 and Hrcs2, have been identified in my country, which control the resistance of soybean gray spot 1 in China, respectively. The resistance of races No. 1 and 7 and the two genes had an interaction effect (Mian et al., 2009).

Soybean cyst nematode (Heterodera glycines, SCN) is also called Verticillium wilt, commonly known as "fire dragon seedlings". The resistance of SCN is a quantitative trait controlled by multiple loci, and five resistance loci have been identified so far: *rhg1, rhg2, rhg3* (Yuan et al., 2012), *rhg4* (X Liu et al., 2011) and *cqSCN-003* (Afzal et al., 2012). Among them, *rhg1* and *rhg4* are the two most concerned sites.

1.5.3 Effects of Soybean Flowering Related Genes on Yield and Quality of Soybean

The yield level of soybeans in China is not high, and an important influencing factor is the regional environment. The difference between regional environments is largely attributed to the difference in photoperiod. Soybeans are short-day crops, and their growth and development are very sensitive to the photoperiod response. Short-day light can promote flowering, while long-day light can inhibit flower bud growth. Hence, soybean is considered as an important model plant in the early study of the photoperiod pathway. As a result, high-quality varieties of soybean cannot adapt to a wide range of planting areas, and yields cannot be increased over the long term. Therefore, the photoperiod regulation mechanism of soybeans must be thoroughly explored to lay the foundation for cultivating high-yield soybeans.

The current research has found a number of major genes controlling the flowering and maturity of soybean: *E1* and *E2*(Bernard., 1971), *E3* (Buzzell, 1971), *E4* (Buzzell & Voldeng, 1980), *E6* (Bonato & Vello, 1999), *E7* (Cober & Voldeng, 2001), *E8* (Cober et al., 2010), *E9* (Kong et al., 2014). *E10* (Ray et al., 1995; Samanfar et al., 2017), *FTs* (Zhao et al., 2016), and *Tof genes* (Lu et al., 2020), among which, *E2, E3, E4, E7, E8, E10* are late-flowering genes, and *E6, E9* and J are early-flowering genes.

1.5.4 Research Progress on Heavy Metal Stress in Soybean

Heavy metal refers to a class of metal elements whose atomic density is greater than 5 g/cm^3. There are 45 kinds of metal elements, including lead (Pb), cadmium (Cd), chromium (Cr), mercury (Hg), copper (Cu) and so on. With the rapid development of modern industry, heavy metals gradually accumulate in soil, air and water and accumulate in plants through the metabolism of plant roots and leaves. Among them, crops such as rice, wheat, corn and soybeans bear the brunt of heavy metal pollution. At the same time, heavy metals also enter the human body through these food products, causing a variety of diseases and causing harm to the human body. Recent research results from the Ecological Center of the Chinese Academy of Sciences show that at present, the area of cultivated land polluted by heavy metals such as cadmium, arsenic, chromium, lead, copper and so on is nearly 20 million hectares, and food security is closely related to human health. Heavy metal pollution in the soil has posed a great threat to the yield and quality of crops.

At present, several major genes that enhance soybean's tolerance to heavy metals have been found: L2, GmHMA3, SALI3-2, GMGLYI-7, GMGLYiI-8, GMGLYI-6, GMGLyi-9, GMGLyi-20, GMGLYiI-5, GMGLYII-10 and GmNRAMP genes. On the other hand, several genes with potential functions in the soybean heavy metal response have been identified: P1B-ATpases, GmABCB23, GmABCB25, GmABCB48, GmABCB52, GmABCI1, GmABCI5, GmABCI13 and GmGIF genes.

At the same time, some genes in soybeans were confirmed to have potential functions in response to heavy metals. Genome-wide features of the P1B-ATPases gene family have potential functions in the reaction of heavy metal cadmium (Fang et al., 2016). In 2021, Huang et al. found that the soybean genes *GmABCB23, GmABCB25, GmABCB48, GmABCB52, GmABCI1, GmABCI5* and *GmABCI13* were toxic to Al. These genes were confirmed to have potential functions in response to heavy metals (Huang et al., 2021).

REFERENCES

Afzal A J, Srour A, Saini N, et al. Recombination suppression at the dominant Rhg1 /Rfs2 locus underlying soybean resistance to the cyst nematode [J]. *Theor Appl Genet*, 2012, 124(6): 1027–1039.

Anguraj Vadivel AK, McDowell T, Renaud JB, Dhaubhadel S. A combinatorial action of GmMYB176 and GmbZIP5 controls isoflavonoid biosynthesis in soybean (Glycine max). *Commun Biol*, 2021, 4(1): 356.

Bai Y, Jing G, Zhou J, et al. Overexpression of soybean GmPLDγ enhances seed oil content and modulates fatty acid composition in transgenic Arabidopsis[J]. *Plant Sci*, 2020, 290: 110298.

Beilinson V, Chen Z, Shoemaker R, et al. Genomic organization of glycinin genes in soybean[J]. *Theor Appl Genet*, 2002, 104(6): 1132–1140.

Bernard R L. Two major genes for time of flowering and maturity in soybeans. *Crop Sci*, 1971, 11(2): 242–244.

Bernard R L, Smith P E, Kaufmann M J, et al. Inheritance of resistance to Phytophthora root and stem rot in soybean[J]. *Agron J*, 1957, 49: 391.

Boerma H R, Phillips D V. Genetic implications of the susceptibility of Kent soybean to Cercospora sojina. *Phytopathology*, 1983, 73(12): 1666–1668.

Bonato ER, Vello NA. E6, A dominant gene conditioning early flowering and maturity in soybeans. *Genet Mol Biol*, 1999, 22(2): 229–232.

Buzzell RI. Inheritance of a soybean flowering response to fluorescent-daylength conditions. *Can J Genet Cytol*, 1971, 13(4): 703–707.

Buzzell RI, Voldeng HD. Inheritance of insensitivity to long day length. *Soybean Genet Newsl*, 1980, 7: 26–29.

Cao Y, Mezzenga R. Food protein amyloid fibrils: Origin, structure, formation, characterization, applications and health implications[J]. *Adv Colloid Interface Sci*, 2019, 269: 334–356.

Chen L, Zheng Y, Dong Z, et al. Soybean (*Glycine max*) WRINKLED1 transcription factor, GmWRI1a, positively regulates seed oil accumulation[J]. *Mol Genet Genom*, 2018, 293(2): 401–415.

Cheng H, Yu O, Yu D. Polymorphisms of IFS1 and IFS2 gene are associated with isoflavone concentrations in soybean seeds. *Plant Sci*, 2008, 175(4): 505–512.

Chu S, Wang J, Zhu Y, et al. An R2R3-type MYB transcription factor, GmMYB29, regulates isoflavone biosynthesis in soybean. *PLoS Genet*, 2017, 13(5): e1006770.

Cober ER, Molnar SJ, Charette M, et al. A new locus for early maturity in soybean. *Crop Sci*, 2010, 50(2): 524–527.

Cober ER, Voldeng HD. Low R:FR light quality delays flowering of E7E7 soybean lines. *Crop Sci*, 2001, 41(6): 1823–1826.

Dhaubhadel S, Gijzen M, Moy P, Farhangkhoee M. Transcriptome analysis reveals a critical role of CHS7 and CHS8 genes for isoflavonoid synthesis in soybean seeds. *Plant Physiol*, 2007, 143(1): 326–338. doi: 10.1104/pp.106.086306.

Du Q, Yang X D, Zhang J H, et al. Over-expression of the Pseudomonas syringae harpin-encoding gene hrpZm confers enhanced tolerance to Phytophthora root and stem rot in transgenic soybean[J]. *Transgen Res*, 2018, 27(3): 277–288.

Fang X, Wang L, Deng X, et al. Genome-wide characterization of soybean P 1B-ATPases gene family provides functional implications in cadmium responses[J]. *BMC Genomics*, 2016, 17(1): 1–15.

Fischer R L, Goldberg R B. Structure and flanking regions of soybean seed protein genes[J]. *Cell*, 1982, 29(2): 651–660.

Flores T, Karpova O, Su X, et al. Silencing of GmFAD3 gene by siRNA leads to low α-linolenic acids (18: 3) of fad3-mutant phenotype in soybean [Glycine max (Merr.)][J]. *Transgen Res*, 2008, 17(5): 839–850.

Fukushima D. Structures of plant storage proteins and their functions[J]. *Food Rev Int*, 1991, 7(3): 353–381.

Gillman J D, Stacey M G, Cui Y, et al. Deletions of the SACPD-C locus elevate seed stearic acid levels but also result in fatty acid and morphological alterations in nitrogen fixing nodules[J]. *BMC Plant Biol*, 2014, 14(1): 1–19.

Goettel W, Zhang H, Li Y, et al. POWR1 is a domestication gene pleiotropically regulating seed quality and yield in soybean[J]. *Nat Commun*, 2022, 13(1): 1–11.

Handa C L, Zhang Y, Kumari S, et al. Comparative study of angiotensin I-converting enzyme (ACE) inhibition of soy foods as affected by processing methods and protein isolation[J]. *Processes*, 2020, 8(8): 978.

Harada J J, Barker S J, Goldberg R B. Soybean beta-conglycinin genes are clustered in several DNA regions and are regulated by transcriptional and posttranscriptional processes[J]. *Plant Cell*, 1989, 1(4): 415–425.

Haun W, Coffman A, Clasen B M, et al. Improved soybean oil quality by targeted mutagenesis of the fatty acid desaturase 2 gene family[J]. *Plant Biotechnol J*, 2014, 12(7): 934–940.

Huang J, Li X, Chen X, et al. Genome-wide identification of soybean ABC transporters relate to aluminum toxicity[J]. *Int J Mol Sci*, 2021, 22(12): 6556.

Jan A, Sood M, Sofi S A, et al. Non-thermal processing in food applications: A review[J]. *Int J Food Sci Nutr*, 2017, 2(6): 171–180.

Ji M P, Cai T D, Chang K C. Tofu yield and textural properties from three soybean cultivars as affected by ratios of 7S and 11S proteins[J]. *J Food Sci*, 1999, 64(5): 763–767.

Jing G, Tang D, Yao Y, et al. Seed specifically over-expressing DGAT2A enhances oil and linoleic acid contents in soybean seeds[J]. *Biochem Biophys Res Commun*, 2021, 568: 143–150.

John H, Chandra P, Giri S K, et al. Effect of Processing Methods on 11S/7S Protein and Nitrogen Solubility Index of Soy Protein Isolate[J]. *J Inst Engs (India): Sr A*, 2021, 102(4): 989–994.

Jung W, Yu O, Lau SM, O'Keefe DP, Odell J, Fader G, McGonigle B. Identification and expression of isoflavone synthase, the key enzyme for biosynthesis of isoflavones in legumes. *Nat Biotechnol*, 2000, 18(2): 208–212. doi: 10.1038/72671. Erratum in: Nat Biotechnol 2000 May;18(5):559. PMID: 10657130.

Kanai M, Yamada T, Hayashi M, et al. Soybean (Glycine max L.) triacylglycerol lipase GmSDP1 regulates the quality and quantity of seed oil[J]. *Sci Rep*, 2019, 9(1): 1–10.

Kim H S, Yu O K, Byun M S, et al. Okara, a soybean by-product, prevents high fat diet-induced obesity and improves serum lipid profiles in C57BL/6J mice[J]. *Food Sci Biotechnol*, 2016, 25(2): 607–613.

Kim Y, Wicker L. Soybean cultivars impact quality and function of soymilk and tofu[J]. *J Sci Food Agric*, 2005, 85(15): 2514–2518.

Kong FJ, Nan HY, Cao D, et al. A new dominant gene E9 conditions early flowering and maturity in soybean. *Crop Sci*, 2014, 54(6): 2529–2535.

Lee J. A new technique to produce hypoallergenic soybean proteins using three different fermenting microorganism[J]. *J Allerg Clin Immunol*, 2004, 113(2): S239.

Lee M K, Kim J K, Lee S Y. Effects of fermentation on SDS-PAGE patterns, total peptide, isoflavone contents and antioxidant activity of freeze-thawed tofu fermented with Bacillus subtilis[J]. *Food Chem*, 2018, 249: 60–65.

Liu J, Huang F, Wang X, Zhang M, Zheng R, Wang J, Yu D. Genome-wide analysis of terpene synthases in soybean: functional characterization of GmTPS3. *Gene*, 2014, 544(1): 83–92.

Liu K S. *Soybeans: Chemistry, Technology, and Utilization[M]*. Springer, 2012.

Liu X, Liu S, Jamai A, et al. Soybean cyst nematode resistance in soybean is independent of the Rhg4 locus LRR-RLK gene [J]. *Funct Integr Genom*, 2011, 11(4): 539–549.

Liu Y, Yang J, Lei L, et al. Isoflavones enhance the plasma cholesterol-lowering activity of 7S protein in hyper-cholesterolemic hamsters[J]. *Food Funct*, 2019, 10(11): 7378–7386.

Liu Y F, Li Q T, Lu X, et al. Soybean GmMYB73 promotes lipid accumulation in transgenic plants[J]. *BMC Plant Biol*, 2014, 14: 73.

Lu S, Dong L, Fang C. Stepwise selection on homeologous PRR genes controlling flowering and maturity during soybean domestication. *Nat Genet*. 2020, 52(4): 428–436.

Ma J, Sun S, Whelan J, et al. CRISPR/Cas9-mediated knockout of GmFATB1 significantly reduced the amount of saturated fatty acids in soybean seeds[J]. *Int J Mol Sci*, 2021, 22(8): 3877.

Ma X, Sun P, He P, Han P, Wang J, Qiao S, Li D. Develop-ment of monoclonal antibodies and a competitive ELISA detection method for glycinin, an allergen in soybean. *Food Chem*, 2010, 121: 546–551

Manan S, Ahmad M Z, Zhang G, et al. Soybean LEC2 regulates subsets of genes involved in controlling the biosynthesis and catabolism of seed storage substances and seed development[J]. *Front Plant Sci,* 2017, 8: 1604.

Manan S, Zhao J. Role of glycine max ABSCISIC ACID INSENSITIVE 3 (GmABI3) in lipid biosynthesis and stress tolerance in soybean[J]. *Funct Plant Biol*, 2020, 48(2): 171–179.

Mian R, Bond J, Joobeur T, et al. Identification of soybeangenotypes resistant to Cercospora sojina by field screening and molecular markers. *Plant Dis*, 2009, 93(4): 408–411.

Miao L, Yang S, Zhang K, et al. Natural variation and selection in GmSWEET39 affect soybean seed oil content[J]. *New Phytol*, 2020, 225(4): 1651–1666.

Mochida K, Yoshida T, Sakurai T, et al. In silico analysis of transcription factor repertoire and prediction of stress responsive transcription factors in soybean [J]. *DNA Res*, 2009, 16(6): 353–369.

Mujoo R, Trinh D T, Ng P. Characterization of storage proteins in different soybean varieties and their relation-ship to tofu yield and texture[J]. *Food Chem*, 2003, 82(2): 265–273.

Murakami H, Ogawa T, Takafuta A, et al. Identification of the 7S and 11S globulins as percutaneously sensitiz-ing soybean allergens as demonstrated through epidermal application of crude soybean extract[J]. *Biosci Biotechnol Biochem*, 2018, 82(8): 1408–1416.

Murphy P A, Chen H P, Hauck C C, et al. Soybean protein composition and tofu quality[J]. *Food Technol (USA)*, 1997, 51(1): 86–110.

Ng MS, Ku YS, Yung WS, Cheng SS, Man CK, Yang L, Song S, Chung G, Lam HM. MATE-type proteins are responsible for isoflavone transportation and accumulation in soybean seeds. *Int J Mol Sci*, 2021, 22(21): 12017.

Nielsen N C, Dickinson C D, Cho T J, et al. Characterization of the glycinin gene family in soybean[J]. *Plant Cell*, 1989, 1(3): 313–328.

Nishinari K, Kohyama K, Zhang Y, Kitamura K, Sugimoto T, Saio K, Kawamura Y. Rheological study on the effect of the A5 subunit on the gelation characteristics of soybean proteins[J]. *Agricultural and Biological Chemistry*, 2014, 55(2): 351–355.

Niu L, Yang J, Zhang J H, et al. 2019. Introduction of the harpin Xooc-encoding gene hrf2 in soybean enhances resistance against the oomycete pathogen Phytophthora sojae[J]. *Transgen Res*, 28(2): 257–266.

Niu L, Zhong X F, Zhang Y Y, et al. Enhanced tolerance to Phytophthora root and stem rot by over-expression of the plant antimicrobial peptide CaAMP1 gene in soybean [J]. *BMC Genetics*, 2020, 21(1): 68.

Obatolu V A. Effect of different coagulants on yield and quality of tofu from soymilk[J]. *Eur Food Res Technol*, 2008, 226(3): 467–472.
Ogawa T, Bando N, Tsuji H, et al. Alpha-subunit of beta-conglycinin, an allergenic protein recognized by IgE antibodies of soybean-sensitive patients with atopic dermatitis[J]. *J Agric Chem Soc Jpn*, 1995, 59(5): 831–833.
Ohishi A, Watanabe K, Urushibata M, et al. Detection of soybean antigenicity and reduction by twin-screw extrusion[J]. *J Am Oil Chem Soc*, 1994, 71(12): 1391–1396.
Oliver C M, Melton L D, Stanley R A. Creating proteins with novel functionality via the Maillard reaction: A review[J]. *Crit Rev Food Sci Nutr*, 2006, 46(4): 337–350.
Pham A T, Lee J D, Shannon J G, et al. A novel FAD2-1 A allele in a soybean plant introduction offers an alternate means to produce soybean seed oil with 85% oleic acid content[J]. *Theor Appl Genet*, 2011, 123(5): 793–802.
Pham A T, Shannon J G, Bilyeu K D. Combinations of mutant FAD2 and FAD3 genes to produce high oleic acid and low linolenic acid soybean oil[J]. *Theor Appl Genet*, 2012, 125(3): 503–515.
Poysa V, Woodrow L, Yu K. Effect of soy protein subunit composition on tofu quality[J]. *Food Res Int*, 2006, 39(3): 309–317.
Ray JD, Hinson K, Mankono JEB, et al. Genetic control of a long-juvenile trait in soybean. *Crop Sci*, 1995, 35(4): 1001–1006.
Saio K, Kamiya M, Watanabe T. Food processing characteristics of soybean 11S and 7S proteins[J]. *Agric Biol Chem*, 2014, 33(9): 1301–1308.
Samanfar B, Molnar ST, Charette M, et al. Mapping and identification of a potential candidate gene for a novel maturity locus, EI0, in soybean. *Theor Appl Genet*, 2017, 130(2): 377–390.
Scallon B, Thanh V H, Floener L A, et al. Identification and characterization of DNA clones encoding group-II glycinin subunits[J]. *Theor Appl Genet*, 1985, 70(5): 510–519.
Shih M C, Hou H J, Chang K C. Process optimization for soft tofu[J]. *J Food Sci*, 2010, 62(4): 833–837.
Shin J. Preparation and characterization of fine powdered whole soybean curd[J]. *J Exer Nutr Biochem*, 2015, 19(4): 297.
Song J F, Liu C Q, Jiang X Q, et al. Quality evaluation of vacuum microwave-dried immature vegetable soybean (G. lycine Max [L.] Merr.)[J]. *J Food Qual*, 2015, 38(5): 337–346.
Sui X, Zhang T, Jiang L. Soy protein: Molecular structure revisited and recent advances in processing technologies[J]. *Annu Rev Food Sci Technol*, 2021, 12: 119–147.
Sun J, Zhou W, Huang D, et al. An overview of 3D printing technologies for food fabrication[J]. *Food Bioprocess Technol*, 2015, 8(8): 1605–1615.
Taira H. Quality of soybeans for processed foods in Japan. *J Agric Res Quart*, 1990, 24: 224–230.
Thanh VH, Shibasaki K. Major proteins of soybean seeds. Subunit structure of β-conglycinin. *J Agric Food Chem*, 1978, 26: 692–695.
Tsuji H, Okada N, Yamanishi R, et al. Fate of a major soybean allergen, Gly m Bd 30K, in rice, barley and soybean-koji miso (fermented soybean paste) during fermentation[J]. *Food Sci Technol Intl Tokyo*, 1997, 3(2): 145–149.
Wang H, Guo J, Lambert K N, et al. Developmental control of Arabidopsis seed oil biosynthesis[J]. *Planta*, 2007, 226(3): 773–783.
Wang H W, Zhang B, Hao Y J, et al. The soybean Dof-type transcription factor genes, GmDof4 and GmDof11, enhance lipid content in the seeds of transgenic Arabidopsis plants[J]. *Plant J*, 2007, 52(4): 716–729.
Wang Z, Wang Y, Shang P, et al. Overexpression of soybean GmWRI1a stably increases the seed oil content in soybean[J]. *Int J Mol Sci*, 2022, 23(9): 5084.
Wang-xing N, Xu L, Yu B J. A putative soybean GmsSOS1 confers enhanced salt tolerance to transgenic Arabidopsis sos1-1 mutant[J]. *Protoplasma*, 2015, 252: 127–134.
Xu X, Jiang L, Zhou Z, Wu X, Wang Y. Preparation and properties of electrospun soy protein isolate/polyethylene oxide nanofiber membranes. *ACS Appl Mater Interfaces*, 2012, 4(8): 4331–4337.
Xu Y, Yan F, Liu Y, et al. Quantitative proteomic and lipidomics analyses of high oil content GmDGAT1-2 transgenic soybean illustrate the regulatory mechanism of lipoxygenase and oleosin[J]. *Plant Cell Rep*, 2021, 40(12): 2303–2323.
Yang Z, Liu X, Wang K, et al. ABA-INSENSITIVE 3 with or without FUSCA3 highly up-regulates lipid droplet proteins and activates oil accumulation[J]. *J Exp Bot*, 2022, 73(7): 2077–2092.
Yi J, Derynck MR, Li X, Telmer P, Marsolais F, Dhaubhadel S. A single-repeat MYB transcription factor, GmMYB176, regulates CHS8 gene expression and affects isoflavonoid biosynthesis in soybean. *Plant J*, 2010, 62(6): 1019–1034.

Yu Z, Duan X, Luo L, et al. How plant hormones mediate salt stress responses[J]. *Trends Plant Sci*, 2020, 25(11): 1117–1130.

Yuan C P, Li Y H, Liu Z X, et al. DNA sequence polymorphism of the Rhg4 candidate gene conferring resistance to soybean cyst nematode in Chinese domesticated and wild soybeans[J]. *Mol Breed*, 2012, 30(2): 1155–1162.

Yue Y, Zhang M, Zhang J, et al. SOS1 gene overexpression increased salt tolerance in transgenic tobacco by maintaining a higher K+/Na+ ratio[J]. *J Plant Physiol*, 2012, 169(3): 255–261.

Zernova OV, Lygin AV, Widholm JM, Lozovaya VV. Modification of isoflavones in soybean seeds via expression of multiple phenolic biosynthetic genes. *Plant PhysiolBiochem*, 2009, 47(9): 769–77. doi: 10.1016/j.plaphy.2009.05.006.

Zhang D, Zhang H, Hu Z, et al. Artificial selection on GmOLEO1 contributes to the increase in seed oil during soybean domestication[J]. *PLoS Genetics*, 2019, 15(7): e1008267.

Zhang M, Cao X, Jia Q, et al. FUSCA 3 activates triacylglycerol accumulation in Arabidopsis seedlings and tobacco BY 2 cells[J]. *Plant J*, 2016, 88(1): 95–107.

Zhang M, Liu J, Li K, Yu D. Identification and characterization of a novel monoterpene synthase from soybean restricted to neryl diphosphate precursor. *PLoS One*, 2013, 8(10): e75972.

Zhang P, Burton J W, Upchurch R G, et al. Mutations in a Δ9-stearoyl-ACP-desaturase gene are associated with enhanced stearic acid levels in soybean seeds[J]. *Crop Sci*, 2008, 48(6): 2305–2310.

Zhang Y, Chen S, Qi B, et al. Complexation of thermally-denatured soybean protein isolate with anthocyanins and its effect on the protein structure and in vitro digestibility[J]. *Food Res Int*, 2018, 106: 619–625.

Zhang Y Q, Lu X, Zhao F Y, et al. Soybean GmDREBL increases lipid content in seeds of transgenic Arabidopsis[J]. *Sci Rep*, 2016, 6(1): 1–13.

Zhao C, Takeshima R, Zhu J, et al. A recessive allele for delayed flowering at the soybean maturity locus E9 is a leaky allele of FT2a, a FLOWERING LOCUS T ortholog. *BMC Plant Biol*. 2016, 16: Article 20.

Zhao Y, Cao P, Cui Y, et al. Enhanced production of seed oil with improved fatty acid composition by overexpressing NAD+-dependent glycerol-3-phosphate dehydrogenase in soybean[J]. *J Integr Plant Biol*, 2021, 63(6): 1036–1053.

Zhou Z, Lakhssassi N, Knizia D, et al. Genome-wide identification and analysis of soybean acyl-ACP thioesterase gene family reveals the role of GmFAT to improve fatty acid composition in soybean seed[J]. *Theor Appl Genet*, 2021, 134(11): 3611–3623.

2 Allele Mining in Soybean

M.B. Ratnaparkhe, Rishiraj Raghuvanshi, Ruchi Shroti, V. Nataraj, M. Shivakumar, S. Chandra, G. Kumawat, R. Kavishwar, N. Shesh, Neelam Yadav, Rajesh V, Ajay K. Singh, Supriya Ratnaparkhe, Shri Hari Prasad, Subulakshmi Subramanian, Dalia Vishnudasan, Prashanth Suravajhala, Gyanesh K. Satpute, S. Gupta, M. Karthikeyan, Krishnan Hajella, and K.H. Singh

2.1 INTRODUCTION

Soybean [*Glycine max* (L.) Merr] is a major oilseed crop in the world and an important source of protein and oil for both humans and animals. It is also used as a raw product for many human health and industrial applications. Other than edible oil (18%–22%), it contains protein (38%–45%), ash, carbohydrate, minerals, nutritional elements and antioxidants that are largely beneficial for humans. Therefore, it has gained popularity in the food, feed, health, and pharmaceutical industries worldwide. Soybean also improves soil fertility by fixing free atmospheric nitrogen through symbiotic association with soil microorganisms. Soybean also contains several minerals, and useful nutraceuticals like isoflavones and tocopherols, which have several health benefits. Therefore, increasing soybean production remains crucial for global food security.

2.2 GENOMIC AND TRANSCRIPTOMIC RESEARCH

2.2.1 Genomes

The construction of reference genomes for individual species has played an important role in genomics studies and has boosted the discovery of candidate genes (Xie et al. 2024). Single-nucleotide polymorphisms (SNPs), copy-number variations (CNVs), and insertion-deletions (InDels) were identified using the reference genomes of several plant species. This significant development became the basis for genome-wide association studies (GWAS) and genomic selection (GS), where diverse sequencing datasets are compared with reference genomes and the identified variations are then statistically associated with phenotypes (Varshney et al. 2005; Crossa et al. 2017). The soybean genome sequencing project was accomplished by DOE-JGI-Community Sequencing Program. The genome sequence assembly was termed as Glyma-1.0. This reveals an approximately 950-Mb genome of the expected 1,115-Mb assembled on 20 chromosomes (Schmutz et al. 2010). The protein-coding regions were predicted to be 66,153, of which over 46,000 genes were predicted with a high confidence level. The genome sequence data and gene annotation of soybeans are housed in the Phytozome database (http://www.phytozome.net/). Genomic technologies and approaches were further utilized, which contributed to the increased production and productivity of soybeans under different conditions and regions. The identification of new genes and loci that control yield, tolerance to biotic and abiotic stresses, and other agronomic traits in soybean plants has significantly increased. Several research projects were further conducted, which have provided large scale datasets for comparative and functional genomics studies (Lam et al. 2010; Valliyodan et al. 2016, 2019; Kim et al. 2019; Kajiya-Kanegae et al. 2021, Sun et al. 2024). These studies identified a number of

DOI: 10.1201/9781003385035-2

important loci associated with higher yield, biotic and abiotic stresses, and seed composition traits. The comparative genomic analysis also provided evolutionary information, polyploidization, copy number variation, and presence-absent variations (PAV) in soybean. Ha et al. (2019) developed a database called Soybean-VCF2Genomes to identify the closest accession in soybean germplasm collection. Genomic studies have been applied extensively in soybean for identification of genes associated with key agronomic and physiological traits and for understanding the genome structure (Innes et al. 2008; Wawrzynski et al. 2008; Kim et al. 2010; Livingstone et al. 2010; Ma et al. 2010; Grant et al. 2010; Ha et al. 2012; Chan et al 2012; Ashfield et al. 2012; Deshmukh et al. 2014; Valliyodan et al. 2016; Li et al. 2017; Chaudhary et al. 2019; Ferreira-Neto et al. 2019; Kim et al. 2019; Lin et al. 2019; Liu et al. 2020a; Pagano et al. 2020; Satpute et al. 2021; Valliyodan et al. 2021; Bayer et al. 2022; Li et al. 2024a). The other interesting technology is the use of genome editing (GE), which, for example, resulted in the production of a soybean plant with a high rate of oleic acid in 2019. Although soybeans have undergone CRISPR/Cas-based targeted genome edits for imparting altered seed protein or oil composition and herbicide tolerance, it is anticipated that the technology will be explored for other agronomic and physiological traits (Bao et al. 2021). Besides developing various products of interest, GE could be used in the functional genomics studies of soybean, wherein unequivocal assignment of gene-function relationships was possible.

2.2.2 Transcriptomes

Expression studies on single and global gene expression patterns are an integral part of any crop improvement program. New short-read sequencing technologies have the potential to rapidly change the transcriptome research landscape of soybeans and other crops. The gene expression patterns in soybeans have been investigated using global expression analysis techniques such as high-density expression arrays, Serial Analysis of Gene Expression (SAGE), and other functional genomics approaches. The use of microarrays in soybean gene expression studies was conducted for functional studies of key genes associated with biotic and abiotic stresses (Maguire et al. 2002; Thibaud-Nissen et al. 2003; Vodkin et al. 2004; Hao et al 2023). In the characterization of genetic elements defining tolerance to biotic and abiotic stresses, seed composition and increasing yield have gained greater interest (Thao et al. 2013). Initial exploration for drought tolerance in soybeans showed strong upregulation of around 3,000 root-derived genes and the metabolite coumestrol (Tripathi et al. 2016). In addition, the early transcriptional response of soybean roots to drought stress has been studied in great detail by Neto et al. (2013). Also, differential expression of genes involved in osmo-protectant biosynthesis also conferred drought tolerance (Ha et al. 2015). Comparative expression and protein-protein interaction analysis of AQPs in cultivated and wild soybeans helped in identifying *GmTIP2;1* as a novel candidate gene that would confer salt and water stress tolerance to the plants (Zhang et al. 2017). A comprehensive list of investigations exploring the drought tolerance mechanism in soybeans utilizing transcriptomic approaches is presented in Table 2.1. Functional genomics studies were also conducted to identify the role of microRNAs (miRNAs). miRNAs are key regulators of gene expression and play important roles in many aspects of plant biology. Turner et al. (2012) identified a number of novel miRNAs and previously unknown family members for conserved miRNAs in the recently released soybean genome sequence. They classified all known soybean miRNAs based on their phylogenetic conservation (conserved, legume-specific, and soybean-specific miRNAs) and examined their genome organization, family characteristics, and target diversity. Currently, there are several ongoing projects, including genome resequencing, digital gene expression, and whole transcript profiling in soybeans for various traits (Chu et al. 2021, Huang et al. 2024). RNA sequencing can provide a more accurate estimate of the number of genes and genomic regions involved in adaptation and be used to detect potential candidate genes. Expression analysis indicating differences may reflect the early process of adaptive divergence at the population level. These sequence variations and expression variations may be involved in the adaptation of a species to its natural habitat (Niu et al. 2024).

TABLE 2.1
Potential Genetic Resources for Abiotic Stress Tolerance

Tolerant Genetic Resource/Genotype	Source
Drought Tolerance	
PI 416937	USDA, ARS
Young	USDA, ARS
Jackson	–
PI 407155	–
R01-416F and R01-581F	USA
93705-36 and PI471938	USDA
PI 468917	–
C12 and W05	CUHK, China
PI 567690 and PI 567731	China
EC 538828, JS 97-52 and EC 602288	India
NTCPR94-5157, N09-13890, NC-Raleigh, and SC07-1518RR	USA
PK 1180 and SL 46	India
TGX 709-7E and EC 389174	India
Water Logging Tolerance	
Edison, GR 8836, CX 415	USA
Archer	USA
PI 408105A	Korea
Misuzudaizu	Japan
91210-350, 91210-316	USA
Kefeng No. 1	China
Peking	China
Nam Vang, VND2, ATF15-1	Cambodia
AGS 313	Bangladesh
JS 97-52 and JS 20–38	India
Iyodaizu	Japan
Kokubu 7, Maetsuezairai 90B, Yahagi	Japan
UA 5615C, R10–4892, R13–12552, R07–6669, Walters, R04–342, S11–25108, S12–1362, S11–25615	USA
JS 20–38, C 2797, Hardee, IS 128 and JS 71–05	India
PI 561271, PI 567651, PI 567343, PI 407184,PI 603910C, PI 567394B, PI 467162, PI 479751, PI 407229, PI 597459C, PI 424082, PI 378699A, PI 424107A, PI 366124	USA
Salt Tolerance	
S-100, Centennial, Cook, D49–2491, Dillon, Forrest, Gordon, Haskell, Hill, Hutton, Johnston, Lee, Manokin, Wright	USA
Nannong 1138-2	China
Jindou no. 6 (PI574484)	China
FT-Abyara (PI628838)	Brazil
JWS156-1	Japan
PI483463	China
Tiefeng 8	China
W05	China
Fiskeby III	Sweden
Osage	USA

(Continued)

TABLE 2.1 (*Continued*)
Potential Genetic Resources for Abiotic Stress Tolerance

Tolerant Genetic Resource/Genotype	Source
Jidou 12	China
SL 1226, SL 1258	India
PI597458B, PI342434, PI548198, PI561389B, PI407202, PI407220, PI424107A, PI479752, PI407083, PI468908, PI080837, PI417500, PI424116, PI483460B, PI562551	USDA Soybean Germplasm Collection

Source: Ratnaparkhe et al. (2022a).

2.2.3 Epigenomes

Epigenetic variation plays an important role in determining various biological processes (Ramesh et al. 2014, Zhang et al. 2023). The importance of natural genetic variation for crop domestication and improvement has been widely investigated. However, the contribution of epigenetic variation in crop domestication at the population level has rarely been explored. To understand the impact of epigenetics on crop domestication, Shen et al. (2018a) investigate the variation of DNA methylation during soybean domestication and improvement by whole-genome bisulfite sequencing of 45 soybean accessions, including wild soybeans, landraces, and cultivars. Through methylomic analysis, we identified 5,412 differentially methylated regions (DMRs). These DMRs exhibit characters distinct from those of genetically selected regions. In particular, they have significantly higher genetic diversity. Association analyses suggest only 22.54% of DMRs can be explained by local genetic variations. Intriguingly, genes in the DMRs that are not associated with any genetic variation are enriched in carbohydrate metabolism pathways. Recently, Zhang et al (2023) provided important insights into the epigenomic landscape of soybean cultivar ZH13 assembled genome.

2.2.4 Genetic Maps

Molecular markers have been extensively used for the development of genetic maps in soybeans. These markers include protein markers, isozymes, RFLP, AFLP, RAPD, SSR, and SNP markers. The first report of restriction fragment length polymorphism (RFLP) on soybean was an assessment of the molecular genetic diversity of the nuclear genome (Apuya et al. 1988). Subsequently, RFLP markers were extensively used for genetic diversity analysis for soybean cultivars, landraces, and germplasm lines and for genome mapping (Keim et al. 1989 Keim et al. 1990, 1997; Diers et al. 1992; Skorupska et al. 1993; Akkaya et al. 1995; Lorenzen et al. 1995; Shoemaker and Specht 1995; Mansur et al. 1996; Cregan et al. 1999; Ferreira et al. 2000; Yamanaka et al. 2001; Lightfoot et al. 2005; Chan et al. 2012; Ratnaparkhe et al. 2013) until SSR and SNP markers became popular. High-density linkage maps were developed using a combination of SSR and/or SNP markers. Comparative mapping was conducted to join maps developed through different mapping populations (Frary et al. 2008).

2.2.5 Genetic Markers

Different types of molecular markers have been used in soybeans to map major genes and QTLs for different traits. QTLs representing more than 100 agronomically important traits have been mapped in soybean. Current information on all mapped QTLs in soybean is available on the USDA-ARS soybean genetic database *SoyBase* (http://soybase.org). Since SNP markers are abundant, they have become more popular for QTL analysis of various agronomic traits in soybean (https://soybase.org, http://soykb.org). With the availability of whole genome sequences, Gene/QTL mapping in soybean has become more common (Schmutz et al. 2010). Genome sequencing has greatly assisted in the development of thousands of molecular markers for genetic mapping studies. QTL analysis played a significant role in identifying genomic regions associated with various traits. QTL mapping and marker development have progressed for agronomic and physiological traits, insect and pest resistance, and also for resistance to several climatic stresses with improved yield. Hyten et al. (2008) developed a multiplex assay of 384 SNPs designated as soybean oligo pool all-1 (SoyOPA-1) for genotyping in the complex genome of soybean. In order to develop more SNPs, Hyten et al. (2010a) sequenced a total of 3,268 SNP-containing robust STS in six diverse genotypes, resulting in the identification of 13,042 SNPs with an average of 3.5 SNPs per polymorphic STS. These SNPs along with 5,551 SNPs discovered by Choi et al. (2007) were used to design two Illumina custom 1,536 SNP GoldenGate assays designated as SoyOPA-2 and SoyOPA-3. A set of 1,536 SNPs from the 3,456 SNPs present in three SoyOPAs was selected to include sufficient polymorphic SNP markers distributed throughout the genome that could be used for most of the QTL mapping applications. This set of 1,536 SNPs GoldenGate assay was designated as Universal Soy Linkage Panel 1.0 (USLP1.0). Hyten et al. (2010b) sequenced a reduced representation library of soybeans to identify SNPs using high-throughput sequencing methods. A total of 7,108 to 25,047 SNPs were detected by using multiple SNP detection methods. A total of 1,536 SNPs were selected from this pool of 7,108 SNPs to create an Illumina GoldenGate assay (SoyOPA-4). The SoyOPA-4 produced 1,254 successful GoldenGate assays, indicating a validation and assay conversion rate of 81.6% for the predicted SNPs. Chaisan et al. (2010) used 335,857 publicly available ESTs derived from 18 genotypes for EST clustering and *in silico* SNP identification. A total of 3,219 EST contigs were established based on three to nine genotypes, and a total of 26,735 SNPs were identified. The confirmation of *in silico* identified SNPs by Sanger sequencing yielded a 15.7% accuracy rate between two cultivars, Williams 82 and Harosoy. These studies resulted in the development of a large number of SNP markers in soybeans, which could be utilized for the mapping of complex traits as well as molecular breeding applications (Liu et al 2020c; Du et al 2023, Han et al 2024; Gao et al 2024).

2.2.6 Plant Genetic Resources

Soybean germplasm, landraces, and wild species have been collected extensively and maintained at various institutes. Currently, more than 170,000 *G. max* accessions are maintained by more than 160 institutions in nearly 70 countries. The USDA Soybean Germplasm Collection is one of the most intensely used germplasm collections in the world and the most intensely used in the NPGS. The collection includes more than 1,100 wild soybeans from China, Korea, Japan, and Russia, and more than 18,000 cultivated soybeans from China, Korea, Japan, and 84 other countries (Song et al. 2015). Most of the cultivated soybeans from China, Korea, and Japan are landraces that are not the product of modern plant breeding (Carter et al. 2004). Potential germplasm resources used for the improvement of abiotic stress tolerance are listed in Tables 2.1 and 2.2.

2.3 PROGRESS IN SOYBEAN GENOMICS

2.3.1 Genome Sequencing

The use of a single reference genome has several limitations for investigating the enormous genetic diversity present across diverse accessions within the species. Mapping of sequencing reads on the single reference genome frequently misses highly polymorphic regions and regions that are absent from the reference genome, and therefore a more robust and thorough methodology is necessary. A pangenome is one such strategy that appears to have the potential to capture a species' whole genetic repertoire. Tettelin et al. (2005) introduced the term "pangenome" to characterize the variety of

TABLE 2.2
List of Soybean Cultivars/Germplasm Used for QTLs Mapping Studies for Drought Tolerance Traits

Parents Involved	Population/ Size	Traits Related to Drought Tolerance	Linked Marker/s	Type of Marker	LG/Ch No.	QTLs
Young/ PI416937	F_4 (120)	Water use efficiency (WUE), leaf ash	cr497-1	RFLP	J	3
S100/ Tokyo	F_2 (116)	WUE	A489H	RFLP	L	2
Minsoy/ Noir 1	RIL (236)	Transpiration efficiency (TE), carbon isotope discrimination (CID), yield	Satt205, Satt489	SSR	C2	3
Kefeng 1/ Nannong 1138-2	RIL	Dry root weight, total root length and root volume, canopy-wilting	–	–	N6-C2 N8-D1b + WN11-E, N18-K	13
Jackson/ KS4895	RIL (81)	Leaf wilting	Sat_044	SSR	K	1
Hutcheson/ PI471938	F_4 (140)	Yield, and wilting	Satt226, Sat_375, Sat_074	SSR	D2, F1, F2	6
KS 4895/ Jackson	RIL (92)	Canopy wilting	Satt177, Satt362	SSR	A2, B2, D2, F	4
Kefeng1/ Nannong1138-2	RIL (184)	Seed yield per plant, drought susceptibility index	–	–	C2, H	10
Kefeng1/ Nannong1138-2	RIL (184)	Leaf water, seed yield	–	–	C2, D1b, H, A2	17
PI 416937/ Benning	RIL (128)	Root traits	Satt383, Satt339, Sct_191, Satt429 Sat_299	SSR	1, 3, 4, 8, 20	5

(Continued)

TABLE 2.2 (*Continued*)
List of Soybean Cultivars/Germplasm Used for QTLs Mapping Studies for Drought Tolerance Traits

Parents Involved	Population/ Size	Traits Related to Drought Tolerance	Linked Marker/s	Type of Marker	LG/Ch No.	QTLs
Hongfeng 11/ Clark	BC	Relative water content (RWC), water holding capacity, root traits, canopy wilting	Sat_136, Satt167, Satt398, Satt694, GMSL514	SSR	A1,A2, B1, B2, D1a,C2, E, F, K, G, I, L, M, N, O	40
Hongfeng 11/ Clark	BC (46)	Germination rate and seedling stage drought	Satt449, Satt499, Satt440, Sat_180	SSR	A1, K I, H	31
PI 416937/ Benning	RIL (150)	Canopy wilting	Satt 302	SSR	2, 4, 5, 12, 14, 17,19	7
Hongfeng 11/ Harosoy	BC2F3 (95)	–	Satt253, Satt513, Satt693, Satt240, Satt323, Satt255	SSR	–	18
SNWS 0048/ Jinda73	BIL	Physiological trait	–	–	D2, G, M, N	9
Jingdou 23/ ZDD 2315	RIL (447)	Root and shoot traits of seedlings	Satt333, Satt327, Satt519, Satt597	SSR	–	24
373 Genotypes	–	WUE	39 MTAs	SNP	–	21
93705 KS4895/ Jackson, 08705 KS4895/ Jackson KS4895/ PI 424140, A5959/ PI 416937, Benning/ PI 416937	–	Canopy-wilting trait	–	–	2, 5, 11, 14,17, 19	13
Dunbar/ PI 326582A,	BIL (251)	Root and shoot traits	Satt315, Satt253, Satt142	SSR	8, 12	4
CPI 26671/G 2120; Valder/ G 2120	RIL	Epidermal conductance, RWC, plant survival after stress	soPt-856602	DArT	6	106 34
M8206/ TongShan, ZhengYang/ M8206	–	Drought tolerance	–	SNP	chromosomes from 01 to 20	111
PK 1180/ UPSL 298	F_{2}/ BSA	Seedling survivability	Satt277	SSR	6	One gene
PI 416997/ PI 567201D	RIL (196)	CID	–	SNP	6,7,10, 11, 15, 17, 20	16

Source: Ratnaparkhe et al. (2022a).

genes in *Streptococcus agalactiae*. A pan-genome consists of a core genome, encompassing the genes present in all accessions, and a dispensable genome, which contains the accessory genes that are shared by some but not all strains.

As the cost of sequencing reduced and new sequencing technologies developed, a sudden increase in plant genome sequencing and resequencing projects was seen. Resequencing, however, displayed its limitations when it came to the collection of a variety of structural variations, particularly for wild relatives of the crop that have a greater genetic diversity than their domesticated equivalents. To map resequencing short reads to a reference genome, high levels of sequence similarity are required. This came at the risk of losing information from more divergent genomic regions. Thus, the first plant pan-genome was developed in 2014 in the wild soybean *Glycine soja* (Li et al. 2014). Subsequently, numerous pan-genomes were developed for various crop species and their wild relatives. Kim et al. (2010) assembled the genome sequence of a wild soybean (*G. soja* var. IT182932) using Illumina-GA and GS-FLX. Shen et al. (2018b) de novo assembled a high-quality genome for cultivar 'Zhonghuang 13' (Gmax_ZH13) using single-molecule real-time (SMRT) sequencing, optical mapping, chromosome conformation capture sequencing (Hi-C), and next-generation sequencing (HiSeq) Shen et al. (2018b). Xie et al. (2019) then assembled a high-quality reference genome for wild soybean W05. Valliyodan et al. (2019) de novo assembled references for another two cultivars and one wild soybean using a combination of short- and long-read technologies. In addition to the de novo assembly of reference genomes, the pan-genome also progressed significantly in soybean. In 2014, Li et al. reported the first soybean pan-genome built by the assembly of seven wild soybeans decoded using second-generation sequencing technology (Li et al. 2014). Later, Liu et al. (2020a) constructed a soybean pan-genome by de novo assembly of 26 representative wild and cultivated soybeans using long-read sequencing. This assembly produced not only golden-grade genomes for each accession but also, for the first time, a graph-based genome in plants, which provides a promising platform for future in-depth soybean functional genomic studies. Liu et al. (2020a) individually *de novo* assembled 26 soybean genomes and constructed a graph-based genome using these assembled genomes. Pan-genome analyses of these genomes identified numerous genetic variations, including large structure variations and gene fusion events that were not detected by previous studies. It also facilitated the identification of candidate genes associated with agronomic traits. In addition, Liu et al. sequenced a total of 2,898 soybean accessions using Illumina technology, with an average coverage depth of more than 13× for each accession. These accessions included 103 wild soybeans, 1,048 landraces, and 1,747 cultivars, which represented a full range of soybean geographic distributions. Liu et al. (2022) re-sequenced 60 vegetable soybean accessions and performed comparative analysis with other re-sequenced genomes. Other 103 previously resequenced wild soybean and 155 previously resequenced grain soybean accessions were included for comparative genomics studies, population structure analysis, and selective sweeps of vegetable, grain, and wild soybean. Results indicated that total 1,112 and 1,047 genes are under selection in the vegetable soybean and grain soybean populations as compared with the wild soybean population, respectively. A total of 134 selected genes were shared between the vegetable soybean and grain soybean populations. Additionally, four sucrose synthase genes, one sucrose-phosphate synthase gene, and four sugar transport genes were identified as candidate genes related to important traits such as seed sweetness and seed size in vegetable soybeans.

2.3.2 Genetic Diversity and Population Structure of Wild and Domesticate Populations

The cultivated soybean and related wild species are rich sources of resistance genes for various biotic and abiotic stresses. Wild species are extremely diverse morphologically, cytologically, and genetically; they grow in very diverse climates and have a wide geographical distribution. They are a rich source of agronomically useful genes and alleles for biotic and abiotic stress tolerance. Recently, Maranna et al. (2023) developed improved soybean genotypes for extra early maturity, higher yield, and Mungbean Yellow Mosaic India Virus (MYMIV) resistance using *Glycine soja* as a resistance source for MYMIV. The annual and perennial soybean species are significantly distantly related. Wild perennial *Glycine* species have great potential for soybean improvement. The genus *Glycine* is divided into two subgenera, *Glycine* Willd. (perennial) and *Soja* (Moench) F.J. Herm (annual). The subgenus *Soja* includes two species: the soybean [(*G. max* (L.) Merr.)] and its wild annual progenitor, *G. soja* Sieb. & Zucc. (Ratnaparkhe et al. 2010). The subgenus *Glycine* contains 30 wild perennial species. Soybean genetic resources may be categorized into four plausible gene pools (GP). Soybean gene pool -1 (GP-1) consists of biological species that can be crossed to produce vigorous hybrids that exhibit normal meiotic chromosome pairing and possess total seed fertility. Based on this definition, all soybean (*G. max*) germplasm and the wild soybean, *G. soja*, are included in GP-1 with the qualification that seed sterility can be associated with chromosomal structural changes such as inversions and translocations. Gene segregation is normal, and gene exchange is generally easy. GP-2 species can hybridize with GP-1 easily, and F_1 plants exhibit at least some seed fertility (Harlan and de Wet 1971). *Glycine max* is without GP-2 because no known species has such a relationship with soybean. It is possible that species of the soybean GP-2 do exist in Southeast Asia, where the *Glycine* genus may have originated. However, extensive plant exploration in this part of the world is required to validate this assumption. GP-3 is the third outer limit of potential genetic resources. Hybrids between GP-1 and GP-3 are lethal, and gene transfer is not possible or requires radical techniques such as embryo rescue and so on (Harlan and deWet 1971). Based on this definition, GP-3 includes the 26 wild perennial species of the subgenus *Glycine*. These species are indigenous to Australia and are geographically isolated from *G. max* and *G. soja*. Only three species (*G. argyrea*, *G. canescens*, and *G. tomentella*) have been successfully hybridized with soybean; the F_1 hybrids were rescued by embryo culture, were sterile, and most researchers could not proceed beyond the amphidiploid stage, with the exception of Singh et al. (1998a, b). This suggests that only three species belong to the GP-3. GP-4 is the extreme outer limit of potential genetic resources. Pre- and post-hybridization barriers inhibit embryonic development, and premature embryic abortion occurs. Rarely can hybrid seedling lethality, hybrid seed inviability, and inviable F_1 plants be circumvented by bridge crosses within the genus *Glycine* (Singh et al. 2007a, b). Only a few wild perennial *Glycine* species have been hybridized with soybeans. Thus, the majority of species belong to soybean GP-4, as they have not been hybridized with GP-1 and, when hybridized, did not produce viable F_1 plants (Singh and Hymowitz 1987). Although the wild perennial species exhibits resistance traits to several biotic and abiotic stresses, the transfer of useful genes into soybean has not been accomplished. Thus, currently, breeders/geneticists really only have access to the primary gene pool for expanding the germplasm base.

2.3.3 Evolution, Demography, and Phylogeography

Patterns of molecular diversity provide insights into the location and number of domestication events. Molecular diversity in soybean and its wild species has been examined using various markers, including random amplified polymorphic DNA (RAPD) markers, amplified fragment length polymorphisms (AFLPs), simple-sequence repeats (SSRs), and SNPs. The level of genetic diversity and geographic differentiation in Chinese cultivated soybean have been extensively studied using the coefficient of parentage (Cui et al. 2000a, b), morphological traits (Dong et al. 2004), and SSR markers (Li et al. 2008), showing a clear geographic effect on genetic structure. Comparisons between the diversity of different samples of Asian soybean landraces and that of North American cultivars have demonstrated a lower level of diversity in the American pools than in the Asian pools, using either phenotypic characterization (Cui et al. 2000a, 2001) or the coefficient of parentage (Cui et al. 2000b). This reduced diversity was confirmed using sequence analyses to show successive genetic bottlenecks between wild and cultivated soybeans and between Asian landraces and North American cultivars (Hyten et al. 2006). Genetic diversity in soybeans is covered in great detail by Carter et al. (2004). Molecular studies have also highlighted the pattern of genetic diversity in soybeans and various *Glycine* species. Typically, genetic diversity clusters by taxon, with a clear differentiation between wild and domesticated taxa (Powell et al. 1996). The genetic structure of *G. max* and *G. soja* typically agrees with geographic location (Abe et al. 2003; Xu and Gai 2003; Dong et al. 2004; Li et al. 2008). Molecular studies have indicated that Chinese and Japanese *G. soja* populations form distinct germplasm pools (Kuroda et al. 2006) and Asian accessions of the *G. max* group in general accordance with the planting region and also the sowing season (Abe et al. 2003).

There has been a comprehensive study of the genetic relationships of all species in the genus *Glycine*. Zhuang et al 2022). The annual (subgenus *Soja*) and perennial (subgenus *Glycine*) soybean species are significantly distantly related and have diverged from a common ancestor around 5 MYA (Innes et al. 2008). Several attempts to hybridize species between the subgenus *Soja* and the subgenus *Glycine* were undertaken, but they were unsuccessful. In these studies, the pods resulted from interspecific hybridization and eventually aborted and abscised, although pod development could be initiated (Ladizinsky et al. 1979a, b). Later, the intersubgeneric F_1 hybrids of *G. max*×*G. clandestina*, *G. max*×*G. tomentella*, and *G. max*×*G. canescens* were obtained *in vitro* either through embryo rescue or using transplanted endosperm as a nurse layer (Broué et al. 1979; Broué et al. 1982; Newell and Hymowitz 1982). In general, the cultivated soybean could only hybridize with members of the subgenus *Glycine* imperfectly. The progeny of such intersubgeneric hybrids were completely sterile and obtained with great difficulty. Research also demonstrated that the cultivated soybean does not hybridize with any of the wild relatives in other genera of the tribe (Hymowitz 2004). The wild soybean (*G. soja*) has accumulated tremendously rich genetic diversity (Kofsky et al. 2018), and this adaptive evolutionary process has eventually led to the formation of diversification patterns in the wild soybean. This includes diversity for many obvious morphological traits such as flower, pubescence, seed and hilum color, disease and insect resistance traits, physiological and biochemical traits, as well as the content of protein, oil, and carbohydrates and their constituents (Boerma and Specht 2004).

2.3.4 Phylogenomics

The first sequenced soybean genome was cv. Williams 82, a cultivar developed in the 1980s in the US (Schmutz et al. 2010). The availability of this high-quality soybean genome has opened a new chapter in soybean structure and functional genomics research. However, soybean yield improvement is changing at a slower pace, and this major challenge needs to be addressed to meet the global demand for food and feed. Natural genetic variation analysis of soybean germplasm using genome sequence information will greatly help identify genes that are associated with major traits and components for soybean improvement. Deeper sequencing, high-quality SNP data from representative soybean lines provide an unprecedented opportunity to understand the landscape of soybean diversity. Previous studies have indicated that cultivated soybean was domesticated from its wild relative [*Glycine soja* (Sieb. and Zucc.)] approximately 5,000 years ago in temperate regions of China (Zhou et al. 2015). Chung et al. (2014) catalogued genome variation in an annual soybean population by high-depth resequencing of 10 cultivated and 6 wild accessions and obtained 3.87 million high-quality SNPs. Nuclear genome phylogeny supported a single origin for cultivated soybeans. In this study, 10-fold longer linkage disequilibrium (LD) in the wild soybean relative was identified as compared to wild maize and rice. Despite the small population size, the long LD and large SNP data allowed identification of 206 candidate domestication regions with significantly lower diversity in the cultivated but not in the wild soybeans. Valliyodan et al. (2016) reported genomic sequences of 106 soybean accessions representing a wide variety of geographic origins and with desirable traits used in the soybean research program for crop improvement. Results indicated that US soybean cultivars were grouped into the elite group, and most Chinese cultivars were grouped into the landrace group. Individuals from the same geographical region tended to cluster together, which reflected isolation by distance during evolution and/or parallel selections in similar ecological habitats accompanied by gene flow. The genetic diversity analysis using whole-genome SNPs showed a lower level of genetic diversity in cultivated soybeans as compared to wild soybeans, and the genetic diversity decreased from wild to landraces to elite cultivars. The total number of SNPs and the number of non-synonymous SNPs were 15% higher in wild soybeans than in the landrace and elite groups. Also, the number of large-effect SNPs was nearly the same between landraces and elite groups (Valliyodan et al. 2016). Kim et al. (2021) analyzed the genomic variation of 781 soybean individuals consisting of *G. max*, *G. soja*, and hybrid (*G. max*×*G. soja*) accessions using whole genome sequence data and identified putative deleterious mutations in soybean populations. The results indicated that there is a 7.1% reduction in overall deleterious mutations in domesticated soybeans relative to wild soybeans and a further 1.4% reduction from landrace to improved accessions. The detected domestication-selective sweeps also show reduced levels of deleterious alleles. In domesticated soybean, LD increased to 97 kb, similar to that of predominantly selfing cultivated rice (*indica* and *japonica*), but was much higher than outcrossing cultivated maize (Kim et al. 2021). Zhuang et al. (2022) assembled chromosome-level genomes of representative perennial species across the genus *Glycine*, including five diploids and a young allopolyploid, and constructed a *Glycine* super-pangenome framework by integrating 26 annual soybean genomes. These perennial diploids exhibit greater genome stability and possess fewer centromere repeats than the annuals. Biased subgenomic fractionation occurred in the allopolyploid, primarily by the accumulation of small deletions in gene clusters through illegitimate recombination, which was associated with pre-existing local subgenomic differentiation. The super-pangenome framework of the *Glycine* genus includes 109,827 non-redundant protein-coding genes from perennials. The study indicated that ~70% of protein-coding genes were absent in the annual soybean pan-genome, representing a huge repertoire of genetic potential for improvement of the annual crop. This study also unveils the propensities and consequences of polyploid genome evolution, genetic determinants of the life-history strategy transition, and the causes and mechanisms for subgenome fractionation.

2.3.5 Domestication

A single origin of domesticated soybean has been suggested by several researchers and also supported by recent genome resequencing studies (Zhou et al. 2015; Sedivy et al. 2017). Resequencing of 302 wild, landrace, or improved soybeans suggests that all domesticated soybeans are derived from a single cluster of *G. soja* wild soybeans. This supports the single origin hypothesis that all cultivated soybeans originated from a single domestication event (Zhou et al. 2015). Several genetic bottlenecks occurred during soybean domestication and diversification, mainly in the domestication of Asian landraces and in the introduction of a few landraces to North America (Hyten et al. 2006). Genomic regions associated with soybean domestication are sought based on test statistics using the levels and pattern of nucleotide diversity, and the extent of LD and haplotype extension for complete or partial selective sweeps. Population differentiation analysis has also been employed in the identification of genomic regions that potentially underwent selection in geographic differentiation or modern breeding programs. Loss of pod shattering is a key agronomic trait that was targeted by human selection and is regarded as a milestone of crop domestication. Loss of pod shattering in soybean lines is promoted by a NAC transcription factor, SHAT1–5 (Dong et al. 2014). The domesticated allele of this gene is expressed 15-fold higher than the wild allele. *Pdh1*, encoding a dirigent-like protein involved in lignification, is another gene that affects soybean's pod shattering phenotype (Funatsuki et al. 2014). Shattering-resistant varieties carry a single nucleotide substitution at the beginning of the coding sequence that produces a stop codon. The shattering-resistant allele of *Pdh1* is observed at low frequencies in Japanese and Korean landraces and cultivars and at moderate frequency in China, while about 75% of South Asian landraces carry the resistant allele. Most modern North American cultivars possess the resistant allele, indicating that the *Pdh1* gene was utilized as an additional shattering-resistance locus in modern breeding programs in North America.

Seed hardness is another important trait for soybean domestication and improvement. The gene associated with seed harness has been identified as *GmHs1-1*, which encodes a calcineurin-like metallo-phosphoesterase transmembrane protein (Sun et al. 2015). Determinacy is another important trait associated with the domestication process of soybeans. Results demonstrated that soybean stem growth habit is regulated by an epistatic interaction between two major loci, *Dt1* and *Dt2* (Bernard 1972). A major focus of the soybean domestication and diversification process was selection for adaptation to a particular latitudinal photoperiod (Cober and Morrison 2010a). As a short-day flowering plant, its latitudinal expansion requires the loss of photoperiod sensitivities. Photoinsensitivity has played a great role in the adaptation of soybeans by helping it enter the reproductive phase in long-day conditions at higher latitudes and rapidly complete the life cycle in the frost-free period. When genotypes adapted to higher latitudes are introduced to lower latitudes, they enter the reproductive phase without attaining sufficient biomass (Islam et al. 2018). Several genes underlying soybean's latitudinal adaptation have been identified. Analyses of genetic regions that control photoperiod sensitivities of cultivated soybeans, known as maturity loci (*E* loci), indicated variation in the DNA sequence of flowering-associated genes. Among 180 cultivated soybeans surveyed, the percentages of recessive alleles at the major maturity loci *E1*, *E2*, *E3*, and *E4* are 38.3%, 84.5%, 36.3%, and 7.2%, respectively (Zhai et al. 2014), These maturity loci have significantly contributed to diversification or local adaptation. Among these *E* loci, *E1* shows a predominant effect on photoperiodic control of flowering and maturation. Another maturity locus *E2* encodes GmGIa, a homolog of *Arabidopsis* GIGANTEA (GI) that is a component of the circadian clock and a regulator of photoperiodic flowering (Watanabe et al. 2011). The dominant *E2* allele delays flowering and maturity, while the homozygous *e2* allele elevates expression of *GmFT2a*, leading to early flowering. The *E2* allele is more prevalent in wild soybeans than the *e2* allele in cultivated soybeans (Langewisch et al. 2014). The recessive *e2* haplotypes, H1, H2, and H3, display unique geographic patterns (Wang et al. 2016). H1 is widely distributed among cultivated soybeans, while H2 is present in southern China. H3 is assumed to have been later introgressed from wild soybeans independently and is restricted to the Northeast region of China. Several loci that are associated with the photoperiod response in soybean have been identified (Cober et al. 1996a; Cober and Voldeng 2001; Liu et al. 2008; Watanabe et al. 2009, 2011; Cober et al. 2010b; Kong et al. 2010; Xia et al. 2012; Ha et al. 2012;Zhai et al. 2014; Wang et al. 2019a; Lu et al. 2020; Gupta et al. 2021; Lin et al. 2021; Dong et al. 2022; Kou et al. 2022). Three loci, viz., E2, Tof5, and Tof12, were mainly associated with domestication (Wang et al. 2016; Lu et al. 2020; Kou et al. 2022) and showed sequence diversity among cultivated soybeans and its annual relative *G. soja*. Parallel selection was observed for Tof5 loci in cultivated soybean and its annual wild relatives, both leading to adaptation to high latitudes (Dong et al. 2022).

2.4 ALLELE MINING IN SOYBEAN

Allele mining is a promising approach to dissecting naturally occurring allelic variation at candidate genes controlling key agronomic traits, which has potential applications in crop improvement programs. Allele mining can be effectively used for the discovery of superior alleles by 'mining' the gene of interest from diverse genetic resources (Figure 2.1). It can help in accessing and determining the change in nucleotide sequence linked with superior alleles, as well as developing an understanding of phenotypic changes associated with novel traits at the molecular level.

The discovery of sequence polymorphisms for a specific gene in naturally occurring populations can be done using one of two main methods: (i) a modified Targeting Induced Local Lesions in Genomes (TILLING) strategy or (ii) sequencing-based allele mining. Furthermore, accessing gene banks effectively requires an understanding of genetic variety in general and allelic diversity at candidate genes of interest in particular. Therefore, allele mining appears to be promising in terms of its value for enhancing the desired feature in the characterization of genetic diversity or allelic/genic diversity among the accessions of the collection. The use of genetic resources for crop improvement depends critically on the availability of sequence and sequence variation that impacts plant phenotypics. Allele mining aids in the analysis of allele evolution. Additionally, it supports the discovery

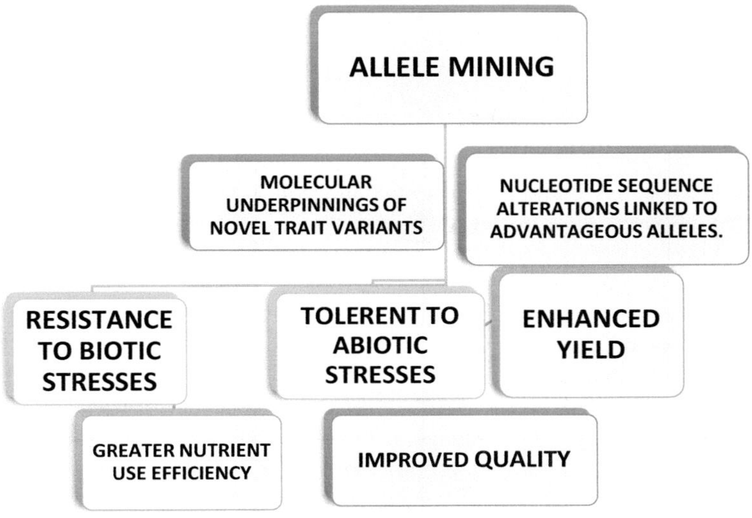

FIGURE 2.1 Allele mining for improving yield, tolerance to biotic and abiotic stresses, and improved seed composition traits in soybean.

of novel haplotypes and the establishment of allele-specific markers for marker-assisted selection. This ability will provide breeders with direct access to the necessary alleles conferring biotic stress resistance, abiotic stress tolerance, greater nutrient usage efficiency, increased yield, and quality improvements. Additionally, it can shed light on the molecular underpinnings of novel trait variants and pinpoint the nucleotide sequence alterations linked to advantageous alleles.

The identification of advantageous alleles is challenging when it comes to a multi-genetic quantitatively inherited characteristic (like yield), as many more harmful and undesirable alleles will also segregate in populations descended from *G. max* × *G. Soja* reproduces. It has been suggested that the use of sophisticated backcross populations could be used to address this issue (Tanksley and Nelson 1996). Backcrossing decreases the number of wild species alleles received from donors in any particular backcross-derived line, allowing the numerous undesired or harmful alleles to be easily identified. Exposed, but more critically, making it simpler to identify the few advantageous donor alleles. Then, with little or no concomitant introgression of any unfavorable alleles, those few favorable alleles, now present in a more elite background, can be easily introgressed into the elite germplasm pool. The method involves using molecular linkage maps and a breeding procedure known as the advanced backcross QTL method, which enables an elite cultivar's genetic background to be analyzed for a subset of alleles from a wild or exotic plant. The molecular linkage map is used to purify lines so that they only contain a particular "wild QTL" in an elite genetic background and to determine the chromosomal location of "wild" alleles that have been passed down to the progeny. It is also used to identify which of the wild species introgressions are associated with the superior performance of lines.

Any crop species that presents allelic variety results from mutations, which are the impetus of evolution. Mutations alter the current allele and allelic combinations or produce new alleles. They occur as insertions and deletions or SNPs in both coding and non-coding areas of the genome (InDel). By changing the encoded protein structure and/or function, it may have a significant impact on the phenotype in the coding area. For instance, the AtAHASL protein encoded by csrl–2 varies from the native AtAHASL protein by one amino acid, the substitution of serine for asparagine at position 653 (S653N), resulting in tolerance to herbicides containing imidazolinone. The protein still serves its biological purpose in the plant in addition to the modified herbicide binding.

Herbicides containing imidazolinone are tolerated by the soybean variety CV127. Another example is the drop in linolenic acid concentration in soybean seeds caused by mutations in the genes for soybean microsomal omega-3 fatty acid desaturase (Bilyeu et al. 2005).

2.4.1 Approaches for Allele Mining

There are three methods for finding alleles, including the eco-tilling-based allele mining method, association mapping-based allele mining, and sequencing-based allele mining.

2.4.1.1 Eco-tilling

While eco-tilling detects natural mutations, modified tilling processes known as TILLING (Targeting Induced Local Lesions IN Genomes) can find single base-pair allelic variation in target genes (more precisely, induced point mutations). Only one genome needs to be sequenced for each haplotype, making it cost-effective and enabling the quick discovery of variance in a large number of individuals. The development of a mutagenized population, DNA preparation and pooling, and mutation identification are the three main phases of tilling. Initially, chemical mutagens are used to cause random mutations in genomes. By using chemicals such as ethylmethanesulfonate (EMS), seeds can be made mutagenic. The self-fertilized M1 plants that come from this process are utilized to prepare DNA samples for mutational screening from M2 individuals. Using gene-specific infrared dye-labeled primers, PCR amplification of a target fragment is the first step in the mutation screening process. PCR-amplified products are digested by a single-strand nuclease to remove mismatched base pairs. Several enzymes, including S1 nuclease 5, T4 endonuclease VII6, and Cel-17, have been employed for mismatch-specific cleavage. It is possible to see cleaved bands that represent mutations or polymorphisms using a denaturing polyacrylamide gel.

2.4.1.2 Sequence Based Allele Mining

Alleles in various genotypes are amplified by PCR in sequencing-based allele mining, which is followed by DNA sequencing to identify nucleotide variation. (Figure 2.2). In other words, by multiplying a particular section of deoxyribose nucleotides as well as by using sequence-based allele mining to identify different alleles among the cultivars.

2.4.1.3 Next-Generation Sequencing Technology for Allele Mining

The method of sequencing makes it possible for us to comprehend how this nucleotide is arranged in DNA molecules. In the past few years, additionally, "massively parallel" techniques have been developed, resulting in "next-generation" sequencing platforms with higher throughput and accuracy. Re-sequencing, alignment of the sequence data, and comparison with the reference genome all involve these techniques. The first of its kind, which relied on pyrosequencing and did away with the necessity for cloning, was commercialized by 454 Life Sciences. It is possible to generate 100 Mb of sequence with 99.5% accuracy and increase read length by averaging over 250 bases using this 454 sequencing process.

2.4.2 Allele Mining for Different Abiotic Stresses in Soybean

2.4.2.1 Unraveling Flowering Genes: Insights from Allele Mining

Understanding of the molecular mechanisms involved in photoperiodic induction of flowering has significantly advanced since the discovery of FLOWERING LOCUS T (FT) in *Arabidopsis* (*Arabidopsis thaliana*) and its equivalent, Heading date 3a (*Hd3a*), in rice (*Oryza sativa*). These genes encode florigens, which act as leaf-derived long-distance signals promoting the transition to flowering (Corbesier et al. 2007; Itoh and Izawa 2013).

The FT/Hd3a protein is conserved across various plant species and plays a crucial role in floral induction (Pin and Nilsson 2012). Soybean possesses at least 10 *FT* homologs, organized as

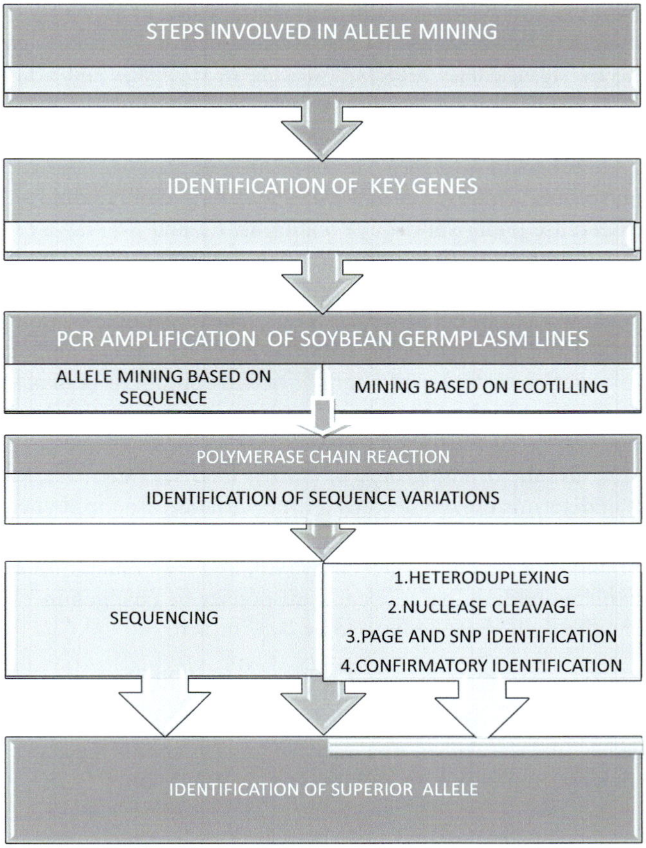

FIGURE 2.2 Steps involved in allele mining.

FIGURE 2.3 Screening for charcoal rot (a) and anthracnose (b) resistance.

five pairs of linked genes in different chromosomal regions (Kong et al. 2010). Among these, six genes stimulate early flowering when expressed in *Arabidopsis* (Kong et al. 2010), while *FT4* has an inhibitory effect (Zhai et al. 2014). Two of these genes, *FT2a* and *FT5a*, demonstrate transcriptional patterns closely associated with changes in photoperiod (Kong et al. 2010). The expression of *FT2a* and *FT5a* is regulated by *E3* and *E4*, which encode the phytochrome A (PHYA) proteins

GmPHYA3 and GmPHYA2, respectively (Liu et al. 2008; Kong et al. 2010). Overexpressing *FT2a* and *FT5a* promotes soybean flowering even under non-inductive conditions (Sun et al. 2011).

The maturity gene *E1* significantly affects flowering in soybeans and acts as a floral repressor. It encodes a putative transcription factor containing a bipartite nuclear localization signal and a DNA-binding B3 domain. *E1* overexpression suppresses *FT2a* and *FT5a* expression and inhibits floral induction in a photoperiod-insensitive cultivar (Xia et al. 2012). *E1* expression is regulated by photoperiodic control mediated by *E3* and *E4*; under long-day conditions, *E1* transcript levels increase, while they decrease under short-day conditions. *E3* and *E4* play a crucial role in the photoperiodic control of *E1* expression (Xia et al. 2012). Even in the *e3* and *e4* backgrounds, *E1* still inhibits flowering, particularly under far-red light-enriched long-day conditions with red:far-red light quantum ratios below 1 (Cober et al. 1996). The soybean genome contains at least three functional *PHYA* genes, *E3*, *E4*, and *GmPHYA1*, with the latter two being homologs and redundantly affecting hypocotyl elongation under far-red light-enriched conditions (Liu et al. 2008). These three PHYA genes may be involved in the photoperiodic control of *E1* expression, which, in turn, regulates the photoperiodic control of *FT2a* and *FT5a* expression in soybean.

The loci *E1, E3, E4*, and the ortholog of *Arabidopsis* GIGANTEA (*GI*), known as *E2*, together play significant roles in determining the flowering time variation among soybean cultivars (Xu et al. 2013). Photoperiod insensitivity in soybean cultivars adapted to high-latitude environments is the result of independent and repeated mutations at *E1, E3*, and *E4* (Tsubokura et al. 2013). However, the molecular mechanisms underlying soybeans' responses to changes in photoperiod and short periods of light during the dark phase remain to be determined.

2.4.2.2 Unraveling Drought Genes: Insights from Allele Mining

The response to drought stress is primarily governed by the regulation of gene expression, which involves numerous genes. In *Arabidopsis*, this intricate process entails the expression of transcription factors that regulate downstream genes, including RD20A, RD22, and RD29B in the ABA-dependent pathway, and ERD1 in the ABA-independent pathway (Yamaguchi-Shinozaki and Shinozaki K. 2007). RD20A, RD22, RD29B, and ERD1 are widely recognized as water-deficit markers in *Arabidopsis* and other plant species. As a result, identifying these genes in soybean would be crucial for future research in this crop (Pellegrineschi et al. 2004; Huang et al. 2011; Stolf-Moreira et al. 2011).

Studies have examined the expression profiles of drought-stress response genes under water-deficit stress conditions using hydroponic systems (HSys) (Fujita et al. 2004). However, some researchers prefer to assess drought responses in pot-based systems (PSys) because they better simulate field conditions (Casagrande et al. 2001). In PSys, drought induces a slower water deprivation process, allowing the plant to adapt to the water deficit gradually. In contrast, in HSys, the water deficit occurs abruptly due to the removal of the plant from the nutrient solution or the addition of osmolytes to the solution (Cowan 1965). The instantaneous water deficit in HSys leads to severe consequences for soybean plants, hindering a continuous acclimation process. Therefore, the water-deficit stress responses in PSys and HSys may differ significantly, leading to the expression of distinct sets of genes.

2.4.2.3 Unraveling Oil Percentage Genes: Insights from Allele Mining

The composition of higher plant seeds includes the embryo, endosperm, and seed coat. The embryo and endosperm originate from the fertilized egg cell and central cell, respectively, while the seed coat develops from the sporophytic integument. As a result, the size of seeds is determined by a combination of signals from maternal and zygotic tissues, which coordinate the growth of the embryo, endosperm, and seed coat (Li et al. 2019). In *Arabidopsis* and rice, several signaling pathways controlled by maternal factors have been identified to regulate seed size. These pathways involve G-protein signaling, ubiquitin-proteasome signaling, mitogen-activated protein kinase (MAPK) signaling, phytohormone signaling, and various transcriptional regulators. Additionally, the HAIKU (IKU) pathway and certain phytohormones also play a partial role in regulating the growth of zygotic tissues (Li et al. 2019). However, compared to *Arabidopsis* and rice, the understanding of molecular networks that regulate seed size in soybeans is currently less developed.

Several transcriptional factors (TFs) have been identified as critical regulators of seed size in soybeans. For example, BIG SEEDS1 (BS1), belonging to the group II TIFY TF family, plays a crucial role in controlling seed, pod, and leaf size by regulating cell proliferation in *Medicago truncatula*, a model legume (Ge et al. 2016). In soybeans, the downregulation of BS1 orthologous genes (*GmBS1* and *GmBS2*) led to increased seed size and amino acid content. Another TF, *SLB1*, encodes an F-box protein that is part of the SKP1/Cullin/F-box E3 ubiquitin ligase complex. *SLB1* interacts with *BS1*, controlling lateral branching and organ growth by modulating BS1 protein stability in *Medicago truncatula*. Overexpression of *SLB1* resulted in increased leaf and seed size in both *Medicago truncatula* and soybean, suggesting the conservation of *SLB1's* function (Yin et al. 2020).

Plant WRKY TFs are involved in various biological processes, including embryogenesis and seed development (Luo et al. 2005). WRKY15a, a TF in soybeans, showed differential expression during pod development in cultivated and wild soybeans. The WRKY15a gene had four haplotypes (H1–H4) with varying CT-core microsatellite repeats at the 5′-untranslated region (5′-UTR). The H1 haplotype with six CT repeats was found exclusively in cultivated soybeans, while the H3 haplotype with five CT repeats was predominant in wild soybeans. The seed weight of haplotype H1 was higher than that of wild soybeans carrying haplotypes H2, H3, and H4, and this weight was positively correlated with WRKY15a expression, indicating a positive impact of WRKY15a on seed size (Gu et al. 2017).

2.5 SOYBEAN IMPROVEMENT THROUGH ALLELE MINING AND GENOMICS TECHNOLOGIES

New sequencing technologies have the potential to rapidly change the molecular research landscape in soybeans. There is a significant increase in the identification of novel alleles through large-scale sequencing, transcriptome sequencing, genome-wide association studies, and further characterization of key alleles through GE strategies. Recently, the Allele Catalog Tool was designed with soybean genomic data and other resources. The Allele Catalog datasets were generated using variant calling pipeline (SnakyVC) and the Allele Catalog pipeline (AlleleCatalog) Chen et al. (2023). Using the Allele Catalog Tool, it is possible to discover genes with different types of mutations along with the functional effect changes and amino acid changes (Figure 2.4). Allele mining can

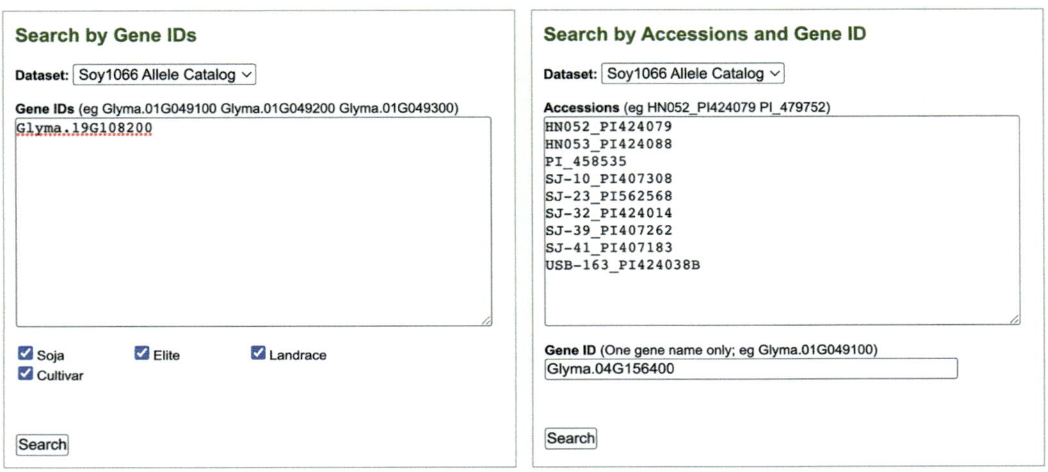

FIGURE 2.4 The Soybean Allele Catalog Tool: a web-based interactive tool for allele discovery and analysis (Chan et al. 2023)

make significant progress in soybean improvement for biotic and abiotic *stresses* seed composition traits and higher yields. Liu et al. (2020) selected 26 accessions and performed de novo genome assembly for soybean accession. Through a comparative genome analysis, a total of 14,604,953 SNPs and 12,716,823 Indels, 27,531 copy number variations, and 723,862 present and absent variations were identified. These structural variations contribute to agronomic trait variations and also play important roles in driving genome evolution and gene structure variation. Soybean population genetics has also greatly advanced due to the availability of a soybean reference genome and the development of next-generation sequencing technology (Valliyodan et al. (2019). Niu et al. (2024) developed an SNP dataset for efficiently evaluating soybean germplasm resources using the genome sequencing data of 3,661 soybean accession (Figure 2.5). Yang et al. (2022) sequenced 250 soybean accessions, which were collected from the core Northeast China soybean germplasm pool, consisting of 134 accessions of landraces and cultivars from Northeast and Northwest China and 116 accessions from European and North American cultivars. The high-depth whole-genome resequencing

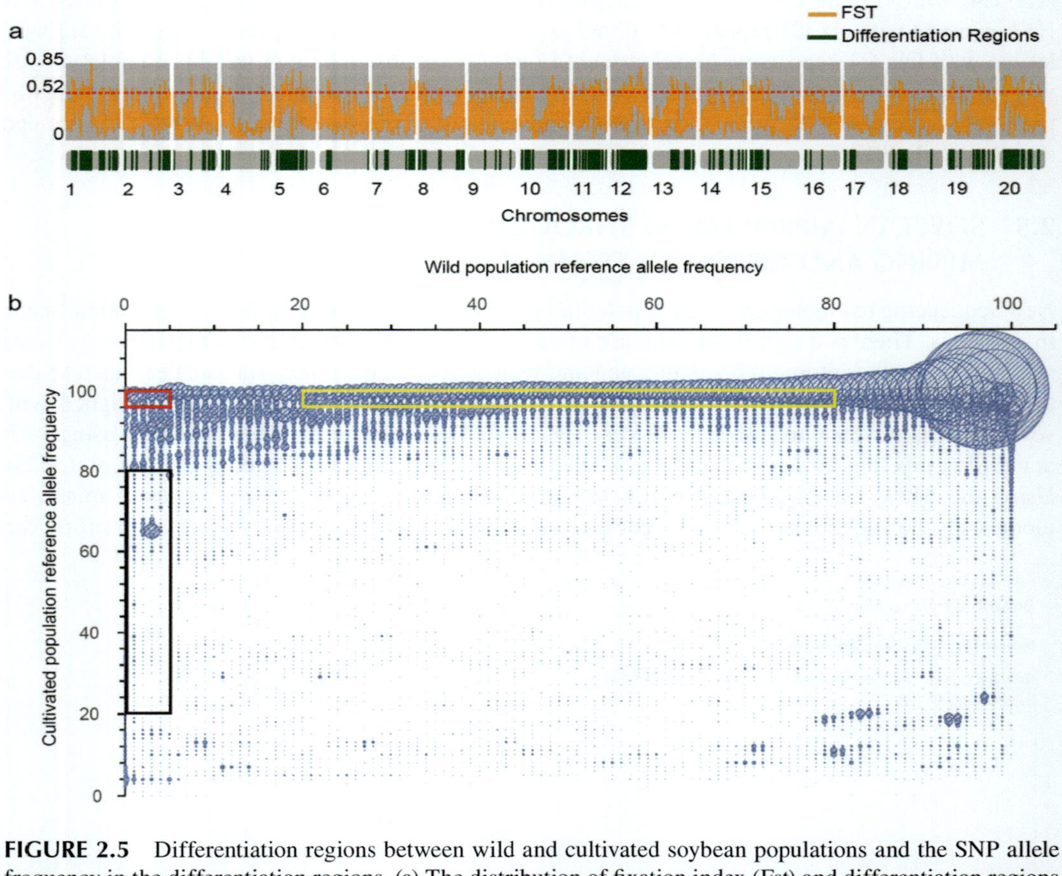

FIGURE 2.5 Differentiation regions between wild and cultivated soybean populations and the SNP allele frequency in the differentiation regions. (a) The distribution of fixation index (Fst) and differentiation regions along the chromosomes. Yellow lines represent Fst values, and Fst values above the red dashed line represent the top 5% Fst values of the genome. The green bars indicate the genomic differentiation regions between the wild and cultivated soybean populations. (b) Bubble chart of SNP allele frequency in the differentiation regions. The ordinate represents the frequency of reference alleles in the wild population, the abscissa represents the frequency of reference alleles in the cultivated soybean population, and the circle size represents the number of SNPs at that frequency. The red, black, and yellow boxes represent the allele frequency distributions of type 1, type 2, and type 3 SNPs, respectively.

Source: Niu et al. (2024)

and comprehensive analyses of these 250 soybean accessions revealed valuable information on soybean genome structure, novel genes associated with important agronomic traits, and genetic networks. These genetic resources provide unique references for further exploring molecular breeding and evolution in soybeans. The resequencing of wild and cultivated soybean accessions revealed the consequences of the artificial selection accompanying domestication and showed that genetic diversity was significantly decreased after domestication (Lam et al. 2010; Li et al. 2013; Valliyodan et al. 2021). Integration of genome-wide association studies (GWAS) and selection sweeps on domestication traits using the data from a resequencing of 302 wild soybeans, landraces, and improved cultivars disclosed the genetic loci responsible for traits related to soybean domestication, improvement, and local breeding (Zhou et al. 2015). Moreover, it was found that naturally occurring introgression was widespread and counteracted genetic bottlenecks during soybean domestication (Wang et al. 2019b). Valliyodan et al. (2016) examined the natural variation in 106 PI lines for known maturity (*E1–E4*) and plant architecture (*Dt*) genes. The maturity of plants depends upon the complex and coordinated regulation of photoreceptors, floral meristems, and flowering time genes, as well as geographical distribution. The functional analysis of mutant alleles for maturity genes showed an early flowering time phenotype. The haplogroup analysis of maturity and plant architecture genes showed a strong correlation, and distinct clusters were associated with growth habits and geographic origin. The analysis of *E1–E4* genes (gene plus 3.5 kb upstream/downstream regions) found that the lower maturity group (MG) lines (MG-0, –I, –II) retain one or more mutant alleles. Soybean accession Fiskeby-III from Sweden belonged to MG-0, and showed an entire gene deletion for the E1 locus, leading to the *e1-nl* allele. In addition, this line showed all of the major mutant alleles for other maturity (*E2* and *E3*) and *Dt* genes. Recently, Li et al (2024) identified domestication gene GmCYP82C4 underlying the major quantitative trait locus for the seed weight in soybean.

2.5.1 Transcriptomics

With the development of sequencing technology, next-generation sequencing (NGS) has made it possible to obtain genome-wide expression across populations, greatly promoting the investigation of population transcriptomics and molecular ecology. RNA sequencing (RNA-Seq) based on NGS can provide a more accurate estimate of the number of independent loci and detect potential candidate genes. Expression variation may occur before genetic variation and may be heritable; therefore, expression differences may reflect the early process of adaptive divergence at the population level. In addition to identifying gene expression variations, RNA-Seq data can also allow the development of SNPs on a large scale, which can capture potential sequence variations. These sequence variations and expression variations may be involved in the adaptation of a species to its natural habitat. There are several online databases for studying soybean gene expression and transcriptome studies (Table 2.3).

2.5.2 Epigenomics

DNA methylation profiling analyses revealed that hypomethylation could affect the expression of neighboring genes (Song et al. 2013). Kim et al. (2015) found that CG body-methylated genes were abundant in duplicated genes that exhibited a higher expression level than single-copy genes. Moreover, methylation changes proximal to the transposable elements (TEs) Shen et al. (2018) investigated the variation of DNA methylation during soybean domestication and improvement and found that DMRs that are not associated with any genetic variation were enriched in carbohydrate metabolism pathways. It was found that DNA demethylation/methylation also plays critical roles in stress responses, such as continuous cropping (Liang et al. 2019), salinity stress (Song et al. 2013), and cyst nematode infection (Rambani et al. 2015). In addition, hundreds of small RNAs had been identified in soybean (Arikit et al. 2014; Lin et al. 2020), some of which showed tissue- or time-specific transcriptional patterns, indicating their biological relevance (Arikit et al. 2014; Razzak

TABLE 2.3
List of Online Databases Used for Soybean Genomics Studies

Databases	Features	Tools	Website
SoyBase and the SoybeanBr-eeder's Toolbox	Genetic and physical maps, QTL, genome sequence, transposable elements, annotations, graphical chromosome visualizer	BLAST search, ESTs search, SoyChip Annotation Search, Potential Haplotype (pHap) and Contig Search, Soybean Metabolic Pathways, Fast Neutron Mutants Search, RNA-Seq Atlas	http://soybase.org/
SGMD The Soybean Genomics and Microarray Database	Integrated view genomic, EST and microarray data	Analytical tools allowing correlation of soybean ESTs with their gene expression profiles	http://bioinformatics.towson.edu/SGMD/
SoyKB-Soybean Knowledge Base,	Multi-omics datasets, genes/proteins, miRNAs/sRNAs metabolite profiling, molecular markers, information about plant introduction lines and traits,	Germplasm browser, QTL and trait browser, fast neutron mutant data, differential expression analysis, phosphorylation data, phylogeny	http://soykb.org/
SoyDB-Soybean transcription factors database	Protein sequences, predicted tertiary structures, putative DNA binding sites, Protein Data Bank (PDB), Protein family classifications	PSI-BLAST, browse database, family prediction by HMM, FTP data retrieve	http://casp.rnet.missouri.edu/soydb/
SoyMetDB-The soybean metabolome database	Soybean metabolomic data	Pathway viewer	http://soymetdb.org
soyTEdb-Soybean transposable elements database,	Williams 82 transposable element database	Browse for repetitive elements, transposable element and map position, data retrieval tools	www.soybase.org/soytedb/
SoyProDB-Soybean proteins database,	Several 2D gel images showing isolated soybean seed proteins	Search tool for 2D spots, navigation tools for protein data	http://bioinformatics.towson.edu
DaizuBase-An integrated soybean genome database including BAC-based physical maps	BAC-based physical map, linkage map and DNA markers, BACend, BAC contigs, ESTs, full-length cDNAs	Gbrowse, Unified Map, Gene viewer, BLAST	http://daizu.dna.affrc.go.jp/
Soybean network (SoyNet)			www.inetbio.org/soynet
A knowledge database of soybean functional networks(SoyFN)		Functional gene network, microRNA functional network, gene annotation, genome browser	http://nclab.hit.edu.cn/SoyFN

(Continued)

TABLE 2.3 (*Continued*)
List of Online Databases Used for Soybean Genomics Studies

Databases	Features	Tools	Website
SoyXpress-Soybean transcriptome database,	Soybean ESTs, metabolic pathways, gene ontology terms, Swiss-prot Identifiers and Affymetrix gene expression data	BLAST search, microarray experiments, pathway search, etc.	http://soyxpress2.agrenv.mcgill.ca
SoyGD-The Soybean GBrowse Database, Southern Illinois	Physical map and genetic map, bacterial artificial chromosome (BAC) fingerprint database, associated genomic data	Sequence data retrieval tools, navigation tool for sequence information of different builds	http://soybeangenome.siu.edu/
Deltasoy-An Internet-Based Soybean Database for Official Variety Trials	Official variety trial (OVT) information in soybean, Mississippi OVT data, including yield, location, and disease information	Comparison tools for variety trail data, phenotypic data and disease related data	http://msucares.com/deltasoy/testlocationmap.htm
Soybean Cyst Nematode proteins database (SCNProDB)		SCN protein identification, 2D gel images data	http://bioinformatics.towson.edu/
Soybean Functional Genomics Database (SFGD)		Gbrowse, microarray expression profiling, transcriptome data, gene co-expression regulatory network, acyl-lipid metabolism pathways, cis-element significance analysis	http://bioinformatics.cau.edu.cn/SFGD/
Soybean Proteome Database		Proteome, metabolome, transcriptome datasets, 2D-PAGE and proteomics information, comparative proteomics under flooding, drought and salt stress	http://proteome.dc.affrc.go.jp/Soybean/
Soybean-VCF2Genomes	to map single sample variant call format (VCF) file against known soybean germplasm collection for identification of the closest soybean accession		http://pgl.gnu.ac.kr/soy_vcf2genome/

Source: Kavishwar et al. (2021)

et al. 2022). Population genetic analyses suggested a co-evolution of MIRNA and miRNA targets during soybean domestication.

2.5.3 Genome-Wide Association Studies

The genome-wide association study (GWAS) approach provides opportunities to explore the tremendous allelic diversity existing in natural soybean accessions. Recent advances in genome sequencing have played a key role in genome-wide association studies. The term linkage disequilibrium (LD) was used to describe changes in the genetic variation within a population over time. Variation in LD in a particular genomic region is affected by mutation, domestication, population admixture, population substructure, level of inbreeding and selection, and confounding. The extent of LD is also dependent on the recombination rate. Soybeans usually show less decay of LD (a longer region is in LD) because recombination is ineffective to cause LD decay in a homozygous genetic background, whereas high LD decay (a shorter region in LD) is common in outcrossing crop species. SNPs provide an abundant source of DNA polymorphism compared to SSRs, thereby improving the success rate in a diversity of applications and the determination of genetic relatedness among individuals. Zhu et al. (2003) sequenced 25 diverse soybean genotypes to determine SNP frequency in coding and noncoding regions of the soybean genome. The nucleotide diversity (θ) observed was 0.00053 and 0.00111 in coding and in nocoding perigenic DNA, respectively, whereas the combined nucleotide diversity of the whole sequence analyzed was 0.00097. Squared allele frequency correlations (r^2) among haplotypes at 54 loci with two or more SNPs indicated low genome-wide LD. A worldwide haplotype map for soybeans (GmHapMap) was constructed using whole-genome sequence data for 1,007 *Glycine max* accessions (Torkamaneh et al. 2021). A lower level of genome-wide genetic diversity was observed in soybeans as compared to other major crops. Genome-wide LD studies in soybeans have facilitated the identification of markers and key genes associated with various abiotic stresses.

2.5.3.1 GWAS for abiotic stresses and Root Trait Architecture

GWAS is routinely being used in soybeans and other plant species, for abiotic and abiotic stresses and other agronomic traits. GWAS for quantitative traits like drought tolerance are predictable to be affected by a confounding population. Dhanapal et al. (2015) analyzed the carbon isotope ratio (δ13C) in a population of 373 genotypes and found an association of 39 SNPs, which are tagged to 21 different loci. Similarly, Kaler et al. (2017) reported 54 SNPs associated with $\delta^{13}C$ and 47 SNPs associated with $\delta^{18}O$. These SNPs were tagged with 46 putative loci and 21 putative loci for $\delta^{13}C$ and $\delta^{18}O$, respectively. Several markers and loci have been reported for various drought-related traits, viz., chlorophyll fluorescence (Hao et al. 2012; Herritt et al. 2018), canopy temperature (Kaler et al. 2017), delayed canopy wilting (Steketee et al. 2020), and drought susceptibility index (Chen et al. 2020). GWAS studies in soybeans for drought tolerance were reported in a germplasm association panel containing 259 Chinese cultivars for drought-related traits. The investigation was based on a total of 4,616 SNPs, and 15 SNP trait associations were identified by GWAS, among which three SNPs were associated with two of the drought tolerance indices (Liu et al. 2020b). Yu et al. (2019) conducted GWAS on a panel of 347 soybean genotypes to identify SNPs associated with seed-flooding tolerance-related traits. In this study, three major QTNs, viz., QTN13, qNSR-10, and qEC-7-2, were identified. Furthermore, QTN13 was consistently identified in all three studied traits and in multiple environments. Wu et al. (2020) conducted GWAS on a panel of 384 soybean lines for flooding tolerance, and a total of 14 SNPs were identified across all environments and models. Kan et al. (2015) conducted an association mapping study in soybeans by analyzing seed germination under salt stress and identified three loci on chromosomes Gm08, Gm09, and Gm18. Using 283 diverse lines of worldwide soybean accessions, Zeng et al. (2017) identified eight loci on chromosomes Gm02, Gm7, Gm08, Gm10, Gm13, Gm14, Gm16, and Gm20 that are significantly associated with leaf chloride concentrations and leaf chlorophyll concentrations and also confirmed a major locus on Gm03. Huang et al. (2018) used 192 soybean germplasm lines and identified six genomic regions associated with salt tolerance on soybean chromosome Gm02, Gm03, Gm05, Gm06, Gm08, and Gm18. Do et al. (2019) used two GWAS populations for GWAS studies and identified a major locus on chromosome Gm03 and three new loci on Gm01, Gm08, and Gm18 associated with salt tolerance. Several SNPs were significantly associated with four traits: leaf scorching score, chlorophyll content ratio, leaf sodium content, and leaf chloride content on chromosomes Gm08 and Gm18, of which two adjacent peak SNPs were present in the significant region on Gm08 (Do et al. 2019). Zhang et al. (2019b) identified 18 significant SNPs that were located on chromosomes Gm08 and Gm18 and are associated with salt tolerance at the germination stage. Seventeen of the eighteen significant SNPs were located in a major QTL*qST-8*, which was identified by linkage mapping in recombinant inbred lines (RILs) (Zhang et al. 2019b).

Zhang et al. (2022) identified several QTLs for drought conditions by using a RIL population comprising 234 F6:10 lines and a GWAS panel of 259 soybean accessions. They constructed a genetic map by using SNP markers, and 18 QTLS were identified on 7 soybean chromosomes in two environments. Beside this, 53 QTLs were also identified in the GWAS panel on 19 soybean chromosomes. A combination of the two populations showed that two SNPs fell within two of the QTL (qPH7-4 and qPH7-6) confidence intervals. They re-located several previously reported drought tolerance genes in soybeans and some other crops but also identified a number of non-synonymous stress-related mutation site differences between the two parents, involving Glyma.07g093000, Glyma.07g093200, Glyma.07g094100, and Glyma.07g094200. One previously unreported new gene related to drought stress, Glyma.07g094200, was found by regional association analysis. The significant SNP CHR7–17619 (G/T) was within an exon of the Glyma.07g094200 gene. In the RIL population, the DSP value of the "T" allele of CHR7–17619 was significantly ($P<0.05$) larger than the "G" allele in different environments.

Drought stress is a complex polygenic trait that comprises several morpho-physiological sub-traits (Valliyodan et al. 2016; Valliyodan et al. 2017; Kavishwar et al. 2021; Satpute et al. 2022). Canopy wilting is one of the most important traits that directly justify drought stress. Steketee et al. (2020) conducted a GWAS study on the canopy wilting trait in 162 diverse maturity group VI-VIII genotypes and identified 45 SNPs tagging 44 loci that are significantly associated with the canopy wilting trait; among them, some are found near the previously reported QTLs. The locus identified in this study could be useful in improving canopy wilting in soybeans. The genomic regions discovered across environments, in addition to the new slow wilting germplasm identified with favorable alleles, can be exploited by breeders to improve soybean drought tolerance. Another GWAS study was conducted with two contrasting models, the Mixed Linear Model (MLM) and Multi-Locus Random-SNP-Effect Mixed Linear Model (mrMLM), for seed flooding tolerance. (Sharmin et al. 2020). Several common SNPs in between these two models were identified; among these SNPs, two key SNPs, Gm_08_11971416 and Gm_08_46239716, were consistently connected with seed-flooding tolerance-related traits such as electrical conductivity and germination rate.

Root system architecture (RSA) is the most important trait that influences several biotic and abiotic stresses, as well as nutrient use efficiency. Genomic-assisted breeding for new varieties with efficient root system architecture has great potential for increasing nutrient use efficiency and adaptation to challenging climates. An important GWAS study was carried out for RSA by Mandozai et al. (2021) with the help of specific locus amplified fragment sequencing (SLAF-seq). They efficiently applied five different GWAS models (GLM, MLM, CMLM, FaST-LMM, and EMMAX) to 260 spring soybean genotypes and identified 27 significant SNPs distributed on chromosomes 2, 6, 8, 9, 13, 16, and 18; two of them were found to be associated with multiple root-related traits. Eleven promising root and shoot regulating genes, viz., Glyma.09G179600, Glyma.08G060300, Glyma.16G138900, Glyma.16G208400, Glyma.09G153400, Glyma.13G303800, Glyma.08G060600, Glyma.16G173300, Glyma.02G113900, Glyma.06G148800, and Glyma.18G094200, overlapping with significant SNPs, were identified and also verified by expression analysis. These genes showed a role in positive regulation of root branching number in the spring soybean panel, which could be useful in genomic-assisted breeding programs.

2.5.3.2 GWAS for Seed Composition Traits

Over the past two decades, numerous studies have revealed numerous alleles and QTLs for seed protein and oil (Fliege et al. 2022; Jiang et al. 2023), SoyBase, the USDA, and the ARS Soybean Genetics and Genomics Database). Hwang et al. (2014) carried out a GWAS study for seed protein and oil content using 298 genetically diverse soybean genotypes. A significant association with oil content was found for 25 SNPs, and these were physically located in 13 regions on 12 of the 20 chromosomes, while the seed protein QTL on chromosome 20 has been reported seven times (Bolon et al. 2010). A key locus in the 8.4 Mb region on chromosome 20 for seed protein content was extensively discussed and depicts 12 potential candidate genes. Furthermore, it was narrowed to a 2.4 Mb region based on a GWAS study (Hwang et al. 2014) that harbors only six genes, viz., Glyma20g19620, Glyma20g19630, Glyma20g19680, Glyma20g21030, Glyma20g21040, and Glyma20g21080. These genes were also confirmed by expression analysis. The narrower GWAS-defined locus will help to design more precise genomic-assisted allele selection and expedite positional cloning of the genes associated with protein content. Recently, Liu et al. (2024) conducted De novo genome assembly of a high-protein soybean variety HJ117. The sequence dataset is a valuable resources for further soybean quality improvement research, and will aid in the elucidation of regulatory mechanisms underlying soybean protein content.

2.5.4 GS in Soybean

GWAS and genome-wide prediction are key tools in genomic-assisted varietal development (Figure 2.6). GWAS is more commonly used to assist molecular breeding programs by identifying significantly associated markers (Chaudhary et al. 2019; Niu et al 2024; Wang et al 2024). Markers

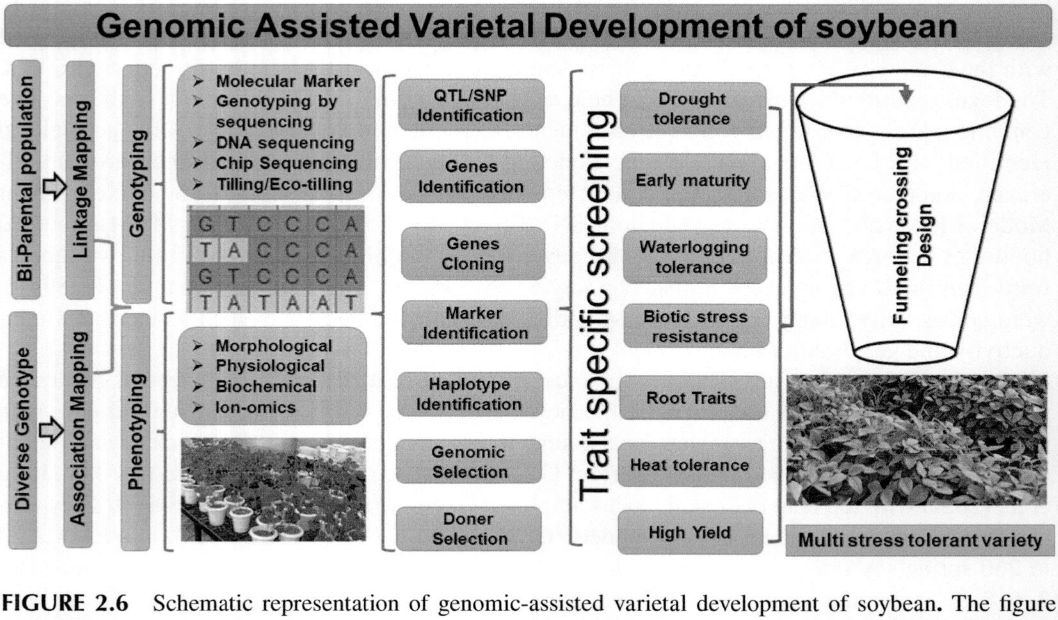

FIGURE 2.6 Schematic representation of genomic-assisted varietal development of soybean. The figure shows the process by which phenomics, in combination with genomics approaches, will lead to the development of a multiple stress-tolerant variety.

developed from GWAS provide key information for producing SNP arrays for different crops, while genomic prediction calculates the estimated breeding values in order to rank the selection candidates in molecular breeding. The major difference between GWAS and genomic prediction is that in GWAS, usually a single SNP is associated with the phenotype, while in genomic prediction, all SNPs are simultaneously fitted into the model, associating with the phenotype (Veerkamp et al. 2016). The genomic model takes all SNPs into consideration, while the GWAS is more focused on the association between a single SNP and different phenotypes. GS helps in crop improvement programs by shortening the breeding cycle. GS predicts the breeding values of lines using genome-wide marker profiling and allows the selection of lines prior to field-phenotyping. It helps in the faster and more accurate selection of superior individuals. Genomic models recreate the framework utilized for pedigree analysis, but using genomic relationships instead.

Studies involving GS in plants have been mostly focused on prediction for advancement purposes, hence restricted to the evaluation of genetic gain within a single generation. Studies of long-term gains based on GS are expensive and time-consuming (Gorjanc et al. 2018; Allier et al. 2019). GS can potentially accelerate the genetic improvement of soybeans by reducing the time to complete breeding cycles. GS utilizes similar statistical models as GWAS, but it seeks to exploit larger genomic variations than individual genomic regions (Meuwissen et al. 2001). GS has been shown to reduce selection time in soybean breeding (Matei et al. 2018), and the United States soybean germplasm collection has proven to be a valuable resource for creating GS models (Jarquin et al. 2016). Promising results have displayed successful prediction of grain yield, protein and oil content, plant height, maturity, and seed weight (Ma et al. 2016; Duhnen et al. 2017; Stewart-Brown et al. 2019; Ravelombola et al. 2021), as well as soybean cyst nematode (SCN) resistance (Ravelombola et al. 2019, 2020).

With the advent of new high-throughput platforms, genotyping has made GS more affordable and efficient. The basic concept behind GS is that a set of breeding materials is used as a training set (TS). The TS is both genotyped and phenotyped for traits of interest in order to calculate marker effects, which then predict the performance of a test set that has been genotyped but not phenotyped. These phenotypic predictions are often referred to as genomic estimated breeding values. To evaluate the effectiveness of GS, cross-validation is often implemented. Cross-validation involves bisecting a set of lines that have been both genotyped and phenotyped into a TS and a validation set. The TS is used to estimate marker effects and calculate genomic estimated breeding values for the validation set. The genomic estimated breeding values are correlated with the observed phenotypic values of the validation set, and this determines predictive ability (Combs and Bernardo 2013; Jacobson et al. 2014). The higher the correlation coefficient, the more predictive. There are several factors that often influence the accuracy of GS. These factors include trait architecture and heritability, TS size and composition, marker density, and a statistical model for the estimation of marker effects (Jannink et al. 2010). Estimation of prediction accuracy using simulated data has suggested up to 40% better efficiency than marker-assisted selection (Bernardo and Yu 2007). There have been several studies examining the potential for GS in soybeans. Jarquin et al. (2014) was one of the first studies examining the potential for GS in soybeans for seed yield prediction. They reported a prediction accuracy of 0.64 for seed yield across 301 experimental lines from the soybean breeding program and found little improvement in accuracy when TS size (N_P) exceeded 100 breeding lines. Predicted success when performing GS tends to be higher in studies reporting results with prediction accuracy *vs.* predictive ability, especially for lower heritability traits. The BARCSoySNP6K iSelect BeadChip was used to genotype a mixed population of 235 soybean cultivars by Ma et al. (2016), and the potential for GS was examined for plant height and seed yield. The potential to utilize GS has also been investigated within larger populations, such as the SoyNAM population, which is composed of over 5,500 lines across 40 bi-parental populations (Xavier et al. 2016). Traits investigated were seed yield, days to maturity, plant height, pod number, node number, and pods per node.

2.6 APPLICATION OF ALLELE MINING FOR BIOTIC STRESSES

Recently, several new alleles/genes have been identified associated with key soybean diseases (Kumawat et al. 2016; Chandra et al. 2022; Ratnaparkhe et al. 2022b; Chicowski et al. 2024). Genetic diversity studies have also been conducted to identify variations in the pathogen. These studies will help in the molecular breeding program for improving biotic stresses.

2.6.1 Nematode Resistance

Soybean Cyst Nematode (SCN), *Heterodera glycines*, is a major pest of soybeans worldwide, causing significant yield losses (Mahmood et al. 2024). Genetic diversity in SCN populations has been extensively studied, as it plays a crucial role in the evolution of the nematode and its interactions with host plants (Vuong et al. 2010; Kim et al. 2016). Several genes have been identified in soybeans that confer resistance to SCN, including *Rhg1*, *Rhg4*, and *Rhg5*. These genes are located on different chromosomes and encode different proteins that are involved in the plant's defense response against the nematode. The Rhg1 locus is the most studied SCN resistance locus and has been mapped to soybean chromosome 18 (Vuong et al. 2010). It is a complex locus that encodes multiple proteins, including a serine hydroxymethyltransferase (SHMT), a calmodulin-binding protein (CBP), and a resistance gene analog (RGA). The *Rhg4* and *Rhg5* loci are also important for SCN resistance and have been mapped to soybean chromosomes 8 and 10, respectively. In summary, genetic diversity in SCN populations is influenced by factors such as mutation, recombination, and gene flow. The genes and loci that control resistance to SCN, such as *Rhg1*, *Rhg4*, and *Rhg5*, are important for understanding the evolution of the nematode and its interactions with soybean.

2.6.2 Fungal Diseases

Fungal diseases are a significant threat to soybean production worldwide and can cause significant yield losses if not managed properly. Some common fungal diseases affecting soybeans include soybean rust, phytophthora root and stem rot (PRSR), pod and stem blight, anthracnose, fusarium wilt, sudden death syndrome (SDS), sclerotinia stem rot (white mold), (Ellis et al. 2013; Marmat et al. 2017; Nataraj et al. 2019; Roth et al. 2020; Kumar et al. 2021; Ramteke et al. 2024). Every fungal pathogen has a unique way of spreading. While some are carried by wind, others rely on rain droplets or water movement in the soil. Depending on the local climate, certain diseases caused by fungi may not survive the winter season, but their spores can still travel through the wind from regions where the disease can survive the winter. Genetic resistance is a promising approach to controlling these diseases, and there has been considerable research in this area. Soybean breeding programs have been successful in developing soybean varieties with resistance to oomycete diseases, and molecular markers have been used to identify and select for resistance genes (Zhong et al. 2018).

PRSR is a devastating disease that affects soybean crops worldwide, causing significant yield losses. The disease is caused by the soil-borne oomycete, *Phytophthora sojae*. To manage the disease, soybean breeders have been developing cultivars with resistance to PRSR. The resistance to PRSR in soybeans is conferred by several genes that act in a complex and interacting manner. In the 1950s, soybean researchers discovered the first Rps gene (*Rps1a*). Since then, more than 20 additional Rps genes have been identified and located on 9 chromosomes, specifically chromosomes 2, 3, 7, 10, 13, 14, and 16–19. These discoveries were reported in various studies conducted by different research groups, including Sugimoto et al. (2012), Lin et al. (2013), Zhang et al. (2013), Sun et al. (2014), Ping et al. (2015), Li et al. (2016), Cheng et al. (2017), and Zhong et al. (2017). The best-characterized R-gene is *Rps1k*, which confers resistance to most *P. sojae* isolates in North America. Other R-genes, such as *Rps2*, *Rps3*, and *Rps8*, have been identified in soybean cultivars from different regions of the world (Cheng et al. 2017; Zhong et al. 2017).

Recent advances in genomics and molecular breeding have facilitated the identification and cloning of several R-genes in soybean. For instance, the *Rps1k* gene was cloned using a map-based cloning approach, which allowed the identification of the candidate gene and its functional validation. Other R-genes, such as *Rps11* and *Rps12*, were identified using genome-wide association studies (GWAS) and quantitative trait loci (QTL) mapping (Lin et al. 2022).

Rhizoctonia solani Kuhn is a fungal pathogen that causes aerial blight disease in soybean plants. This disease can lead to significant yield losses in soybean crops, making it a major concern for farmers (Ajayi-Oyetunde and Bradley 2017). The resistance to *Rhizoctonia solani* Kuhn in soybean is complex, involving multiple genes and pathways. RAB resistance is still unexplored; however, one good study has been reported (Surbhi et al. 2021). They evaluated some SSR markers to identify soybean genotypes that are resistant and susceptible to RAB. The markers Satt 119, Sat_076, Satt 433, Satt 281, Satt 277, Satt 245, and Satt 520 were tested, and they were able to clearly distinguish between the two types of genotypes by amplifying different banding patterns. In particular, Satt 433 and Satt 520 showed patterns that were highly similar to the results obtained from field screening, which is a common method for evaluating plant disease resistance. Moreover, the study found that 18 soybean genotypes that exhibited moderately resistant reactions to the disease in field conditions over 2 years had a banding pattern that was similar to the pattern produced by the resistant check PS-1583 under the SSR marker amplification. This suggests that these soybean genotypes have some level of resistance to the disease.

Charcoal rot is another important disease in soybeans (Rajput et al. 2023; Maranna et al. 2024). For the identification of charcoal rot resistance genotypes, a core set of 100 diverse soybean genotypes was subjected to screening for resistance (Talukdar et al. 2009). Details of the work done in finding the resistance source against charcoal rot are presented in Table 2.4. None of the genotypes were immune, but seven genotypes (viz., DS 9712, DS 9814, JS 335, PK 564, EC 439618, EC 439619, and DS61) were identified (Talukdar et al. 2009) as resistant. Expression of the disease reaction is continuous, that is, a wide variety was observed, starting from highly susceptible through moderately resistant to highly resistant, suggesting the involvement of multiple genetic loci in controlling the resistance of the disease. Advancements have been made to map QTL for charcoal rot resistance

TABLE 2.4
Soybean Germplasm Lines/Varieties for Charcoal Rot Resistance

S.N	Genotypes with Resistant Reaction		Screening Method
1.	Resistant	DS9712, DS9814, JS335, PK564, EC439618, EC439619, and DS61	Paper towel method
	Moderately resistant	BR11, DS-201-A, NRC67, NRC37, and NRC7	
2.	Resistant	DS9712, DS9814, PK564, EC439619, JS335, EC439618, and EC44303	Field screening
	Resistant	JS335, NRC7, NRC37, PK564, EC44303, EC439618, C439619, DS61, DS9712, and DS9814	Pot inoculation
	Resistant	DS9712, DS9814, and JS335	Blotter paper
	Resistant	DS9814 and JS335	In-vitro screening
3.	Moderately resistant	DT97–4290	Field screening
4.	Moderately resistant	DG3905, Manokin, DT99–16864, DT99–17483, DT98–7553, and DT99–17554	Field screening
5.	Partially resistant	PI 548302 and PI 548414	Cut-stem inoculation method
6.	-	DT97–4290, DT98–7553, DT99–17554, and DT99–16864	Cut-stem inoculation method
7.	Absolute resistance (AR)	JS 20–69, RVS 2001–4, and MACS 1336	Field screening at Amravati

Source: Ratnaparkhe et al. (2022b)

TABLE 2.5
Genes/QTL Identified Against Charcoal rot

S.N	Gene/QTL	Chromosome	Phenotyping	Population	Reference
1	Glyma.04g053100	4	Field screening	A collection of 459 PI lines spanning MG I to MG III from USDA	Coser et al. (2017)
	Glyma.14g002000	14			
	Glyma.18g248100	18			
	Glyma.18g228600	18			
	Glyma.06g176100/ Glyma.06g176200	6	Cut stem inoculation method		
	Glyma.08g306800/ Glyma.08g306900	8			
	Glyma.08g315900/ Glyma.08g316500	8			
	Glyma.09g230300	9			
	Glyma.12g216200	12			
	Glyma.12g006300	12			
	Glyma.18g262800	18			
	Glyma.20g197000	20			

in soybeans and identify linked molecular markers (Talukdar et al. 2009). Coser et al. (2017) identified a set of candidate genes governing charcoal rot resistance in a collection of 459 PI lines through association studies (Table 2.5). QTL mapping of charcoal rot resistance in the PI 567562A soybean accession was done by da Silva et al. (2019), and genomic regions governing resistance to charcoal rot in soybeans were identified on chromosome 16. Later, da Silva et al. (2020) conducted bulked segregant analysis using next-generation sequencing for the identification of genetic loci for charcoal rot resistance in soybean. Three genomic regions on chromosomes 5, 8, and 14 were identified as associated with charcoal rot resistance in soybean.

2.6.3 Viral Diseases

Soybean is vulnerable to viral diseases, which can result in significant yield losses (Usovsky et al. 2022). There are several viral diseases that affect soybeans, including soybean mosaic virus (SMV), bean pod mottle virus (BPMV), and soybean vein necrosis virus (SVNV). These viruses have different modes of transmission, symptoms, and severity, but they can all impact soybean yields. Genetic diversity plays a crucial role in the development of resistance to viral diseases in soybeans. Different soybean varieties have unique resistance levels to specific viral diseases, and this diversity can be exploited to develop new, resistant soybean varieties. Genetic diversity is also essential in maintaining resistance over time, as viruses can evolve and overcome previously resistant varieties. Cho and Goodman (1979) developed the SMV isolate classification system in the United States based on the responses of two susceptible soybean cultivars, "Clark" (Plant Introduction (PI) 548533) and "Rampage," and six resistant cultivars, "Davis" (PI 553039), "York" (PI 553038), "Kwanggyo," "Marshall" (PI 548693), and "Ogden" (PI 548477) and Buffalo. There are several viral resistance genes identified in soybean, i.e., *Rsv1* is a dominant gene that provides resistance against SMV, one of the most common viral diseases of soybean; *Rsv3* confers resistance against SMV and BPMV. It has been shown to be effective against multiple strains of SMV. *Rsv4* provides resistance against soybean dwarf virus (SbDV), which is transmitted by aphids. *Rsv5* provides resistance against SMV and SbDV; *Rsv6* provides resistance against SMV and BPMV (Usovsky et al. 2022).

Overall, allele mining, genetic markers, and genes are powerful tools for studying genetic diversity and identifying resistance to various diseases in soybeans. By understanding the genetic basis of disease resistance and the novel allele associated with disease resistance, researchers and breeders can develop more robust and resilient soybean varieties that are better able to withstand soybean diseases (Ratnaparkhe et al. 2020; Li et al. 2023).

2.6.3.1 Genome Editing for Soybean Improvement

Application of GE technologies utilizing CRISPR/Cas (clustered regularly interspaced short palindromic repeat/CRISPR-associated) proteins has opened up novel avenues for the development of commercially important crops, either through transgenic or non-transgenic approaches. The combination of CRISPR/Cas and developments in the field of plant regeneration has offered an opportunity to introduce this commercially important trait into the cultivated genotypes. Although soybeans have undergone CRISPR/Cas-based targeted genome edits for imparting altered seed protein or oil composition and herbicide tolerance, it is anticipated that the technology will be explored for defining biotic stress tolerance (Bao et al. 2021). Besides developing various products of interest, GE could be used in the functional genomics studies of soybean, wherein unequivocal assignment of gene-function relationships was possible. Although GE technologies are powerful, they may not replace the genetic modification technologies that are in vogue due to various challenges, including their regulatory regimen.

2.7 CONCLUSIONS

Allele mining is a promising approach and plays important role in soybean improvement programs. The availability of high-quality genome assemblies and the use of conventional and molecular breeding techniques have greatly helped in allele mining, population genomic studies, and understanding genome evolution. With the advent of omics technologies, the application of forward and reverse genetics approaches has gained momentum in the identification and characterization of rare alleles. Although much advancement has been made in the application of novel technologies to improve soybeans, much remains in the field of population genomics. A combination of multiple omics approaches involving proteomic, transcriptomic, metabolomics, and ionomic, along with genomic science, is required to comprehensively uncover the genomic regions associated with domestication, biotic and abiotic stresses, seed composition and other agronomics and physiological traits.

REFERENCES

Abe J, Xu DH, Suzuki Y, Kanazawa A, Shimamoto Y (2003) Soybean germplasm pools in Asia revealed by nuclear SSRs. *Theor. Appl. Genet* 106: 445–453.

Ajayi-Oyetunde OO, Bradley CA (2017) Identification and characterization of *Rhizoctonia* species associated with soybean seedling disease. *Plant Dis* 101: 520–533

Apuya NR, Frazier BL, Keim P, Roth EJ, Lark KG (1988) Restriction fragment length polymorphisms as genetic markers in soybean, *Glycine max (L.)* Merrill. *Theor Appl Genet* 75: 889–901

Allier A, Lehermeier C, Charcosset A, Moreau L, Teyssèdre S (2019) Improving short- and long-term genetic gain by accounting for within-family variance in optimal cross-selection. *Front. Genet* 10: 1006. 10.3389/fgene.2019.01006

Akkaya MS, Shoemaker RC, Specht JE, Bhagwat AA, Cregan PB (1995) Integration of simple sequence repeat (SSR) DNA markers into a soybean linkage map. *Crop Sci* 35: 1439–1445.

Arikit S, Xia R, Kakrana A, Huang K, Zhai J, Yan Z, et al. (2014) An atlas of soybean small RNAs identifies phased siRNAs from hundreds of coding genes. *Plant Cell*. 26: 4584–601.

Ashfield T, Egan A, Pfeil BE, Chen NW, et al. (2012) Evolution of a complex disease resistance gene cluster in diploid *Phaseolus* and tetraploid *Glycine*, *Plant Physiol* 159: 336–354.

Bao A, Zhang C, Huang Y, Chen H, Zhou X, Cao D (2021) Genome editing technology and application in soybean improvement. *Oil Crop Science* 5: 31–40. doi: 10.1016/j.ocsci.2020.03.001. 20/21

Bayer PE, Valliyodan B, Hu H, Marsh JI, Yuan Y, Vuong TD, Patil G, Song Q, Batley J, Varshney RK, Lam H-M, Edwards D, Nguyen HT (2022) Sequencing the USDA core soybean collection reveals gene loss during domestication and breeding. Plant Genome 15(1):e20109. https://doi.org/10.1002/tpg2.20109

Bernard R. (1972) Two genes affecting stem termination in soybeans. *Crop Sci* 12: 235–239.

Bernardo, R., Yu J. (2007) Prospects for genome-wide selection for quantitative traits in maize. *Crop Sci* 47: 1082–1090.

Bilyeu K, Palavalli L, Sleper D, Beuselinck P (2005) Mutations in soybean microsomal omega-3 fatty acid desaturase genes reduce linolenic acid concentration in soybean seeds. *Crop Sci* 45: 1830–1836.

Bolon YT, Joseph B, Cannon SB, Graham MA, Diers BW, Farmer AD. et al. (2010) Complementary genetic and genomic approaches help characterize the linkage group I seed protein QTL. *BMC Plant Biology*, 10, 41.

Boerma HR, Specht JE (eds) (2004) *Soybeans: Improvement, production, and uses*, vol 16, 3rd edn, Agronomy monograph. ASA, CSSA, SSSA, Madison, WI.

Broue P, Douglass J, Grace JP, Marshall DR (1982) Interspecific hybridization of soybeans and perennial glycine species indigenous to Australia via embryo culture. *Euphytica* 31: 715–724.

Broue P, Marshall DR, Grace JP (1979) Hybridization among the Australian wild relatives of the soybean. *J Aust Inst of Agric Sci* 45: 256–257.

Carter TE Jr, Nelson RL, Sneller CH, Cui Z (2004) Genetic diversity in soybean. In: Boerma HR, Specht JE (eds) *Soybeans: Improvement, production, and uses*, vol 16, 3rd edn, Agronomy monograph. ASA, CSSA, SSSA, Madison, WI, pp. 303–416.

Casagrande EC, Farias JRB, Neumaier N, et al. (2001) Expressão gênica diferencial durante déficit hídrico em soja. *Rev Bras Fisiol Veg* 13: 168–184

Cheng Y, Ma Q, Ren H, et al. (2017) Fine mapping of a Phytophthora-resistance gene *RpsWY* in soybean (*Glycine max* L.) by high-throughput genome-wide sequencing. *Theor. Appl. Genet.* 130: 1041–1051. doi: 10.1007/s00122-017-2869-5

Chaisan T, Van K, Kim MY, Kim KD, Choi BS, Lee SH (2010) *In silico* single nucleotide polymorphism discovery and application to marker-assisted selection in soybean. *Mol Breed* doi:10.1007/s11032-010-9541-y

Chan C, Qi XP, Li MW et al. (2012) Recent developments of genomic research in soybean. *J Genet Genom.* 39: 317–324.

Chan YO, Dietz N, Zeng S, Wang J, Flint-Garcia S, Salazar-Vidal MN, Škrabišová M, Bilyeu K, Joshi T (2023) The Allele Catalog Tool: a web-based interactive tool for allele discovery and analysis. *BMC Genomics*.10;24(1):107.

Chaudhary J, Shivaraj S, Khatri P et al (2019) Approaches, applicability, and challenges for development of climate-smart soybean. *Genomic designing of climate-smart oilseed crops*. Kole C. Eds, Springer, Berlin, pp. 1–74.

Chen L, Fang Y, Li X, Zeng K, Chen H, Zhang H, Yang H, Cao D, Hao Q, Yuan S, Zhang C, Guo W, Chen S, Yang Z, Shan Z, Zhang X, Qiu D, Zhan Y, Zhou XA (2020) Identification of soybean drought-tolerant genotypes and loci correlated with agronomic traits contributes new candidate genes for breeding. *Plant Mol Biol* 102(1–2):109–122

Chandra S, Choudhary M, Bagaria PK, Nataraj V, Kumawat G, Choudhary JR, et al. (2022). Progress and prospectus in genetics and genomics of Phytophthora root and stem rot resistance in soybean (*Glycine max* L.). *Front. Genet.* Nov 14, 13.

Chaudhary J, Shivaraj S, Khatri P, Ye H, Zhou L, Klepadlo M, Dhakate P, Kumawat G, Patil G, Sonah H, Ratnaparkhe MB et al. (2019) Approaches, applicability, and challenges for development of climate-smart soybean. *Genomic designing of climate-smart oilseed crops*. Springer, Berlin, pp 1–74.

Chicowski AS, Bredow M, Utiyama AS, Marcelino-Guimarães FC, Whitham SA (2024) Soybean-Phakopsora pachyrhizi interactions: towards the development of next-generation disease-resistant plants. *Plant Biotechnol J.* 22(2):296-315.

Cho EK, Goodman RM (1979) Strains of soybean mosaic virus: Classification based on virulence in resistant soybean cultivars. *Phytopathology* 69: 467. doi: 10.1094/Phyto-69-467.

Choi IY, Hyten DL, Lakshmi KM et al (2007) A soybean transcript map: Gene distribution, haplotype and SNP analysis. *Genetics* 176: 685–696.

Chu, J.SC., Peng, B., Tang, K. et al. (2021)Eight soybean reference genome resources from varying latitudes and agronomic traits. *Sci Data* 8, 164

Chung WH, Jeong N, Kim J et al (2014) Population structure and domestication revealed by high-depth resequencing of Korean cultivated and wild soybean genomes. *DNA Res* 21(2): 153–167.

Cober ER, Molnar SJ, Charette M, Voldeng HD (2010a) A new locus for early in soybean. *Crop Sci* 50: 524–527. doi: 10.2135/cropsci2009.04.0174

Cober ER, Tanner JW, Voldeng HD (1996a) Genetic control of photoperiod response in early-maturing near-isogenic soybean lines. *Crop Sci* 36: 601–605

Cober ER, Tanner JW, Voldeng HD (1996b) Soybean photoperiod-sensitivity loci respond differentially to light quality. *Crop Sci* 36(3): 606–610.

Cober ER, Voldeng HD (2001) A new soybean and photoperiod-sensitivity locus linked to E1 and T. *Crop Sci* 41: 698–701.

Cober ER, Morrison MJ (2010b) Regulation of seed yield and agronomic characters by photoperiod sensitivity and growth habit genes in soybean. *Theor Appl Genet* 120: 1005–1012.

Cober ER, Voldeng HD (2001) Low R: FR light quality delays flowering of E7E7 soybean lines. *Crop Sci* 41: 1823–1826.

Combs E, Bernardo R (2013) Accuracy of genome-wide selection for different traits with constant population size, heritability, and numbers of markers. *Plant Gene* 6: 1–7.

Corbesier L, Vincent C, Jang S et al (2007) FT protein movement contributes to long-distance signaling in floral induction of Arabidopsis. *Science* 316(5827): 1030–1033.

Coser SM, Chowda Reddy RV, Zhang J, Mueller DS, Mengistu A, Wise KA, Allen TW, Singh A, Singh AK (2017) Genetic architecture of charcoal rot (*Macrophomina phaseolina*) resistance in soybean revealed using a diverse panel. *Front Plant Sci* 8:1626

Cowan IR (1965) Transport of water in the soil-plant-atmosphere system. *J Appl Ecol* 2: 221–239.

Cregan PB, Jarvik T, Bush AL et al (1999) An integrated genetic linkage map of the soybean. *Crop Sci* 39: 1464–1490.

Crossa J, Pérez-Rodríguez P, Cuevas J et al (2017) Genomic selection in plant breeding: Methods, models, and perspectives. *Trends in Plant Sci* 22(11): 961–975.

Cui Z, Carter Jr TE, Burton JW (2000a) Genetic base of 651 Chinese soybean cultivars released during 1923 to 1995. *Crop Sci* 40: 1470–1481.

Cui Z, Carter Jr TE, Burton JW (2000b) Genetic diversity patterns in Chinese soybean cultivars based on coefficient of parentage. *Crop Sci* 40: 1780–1793.

Cui Z, Carter TE Jr, Burton JW, Wells R (2001) Phenotypic diversity of modern Chinese and North American soybean cultivars. *Crop Sci* 41: 1954–1967.

da Silva MP, Klepadlo M, Gbur EE et al (2019) QTL mapping of charcoal rot resistance in PI 567562A soybean accession. *Crop Sci*. 59: 474–479.

da Silva MP, Zaccaron AZ, Bluhm BH et al (2020) Bulked segregant analysis using next-generation sequencing for identification of genetic loci for charcoal rot resistance in soybean. *Physiol Mol Plant Pathol* 109: 101440.

Deshmukh R, Sonah H, Patil G et al (2014) Integrating omic approaches for abiotic stress tolerance in soybean. *Front Plant Sci* 5: 244.

Dhanapal AP, Ray JD, Singh SKet al (2015) Genome-wide association study (GWAS) of carbon isotope ratio ($\delta^{13}C$) in diverse soybean [*Glycine max* (L.) Merr.] genotypes. *Theor. Appl. Genet* 128*:* 73–91.

Diers BW, Keim P, Fehr WR, Shoemaker RC (1992) RFLP analysis of soybean seed protein and oil content. *Theor Appl Genet* 83: 608–612.

Do TD, Vuong TD, Dunn D et al (2019) Identification of new loci for salt tolerance in soybean by high-resolution genome-wide association mapping. *BMC Genom…ics* 20: 318.

Dong Y, Yang X, Liu J, Wang B-H, Liu B-L, Wang Y-Z. (2014) Pod shattering resistance associated with domestication is mediated by a NAC gene in soybean. *Nat Commun* 5: 3352.

Dong LD, Cheng Q, Fang C et al (2022) Parallel selection of distinct Tof5 alleles drove the adaptation of cultivated and wild soybean to high latitudes. *Mol Plant* 15: 308–321.

Dong YS, Zhao LM, Liu B, Wang ZW, Jin ZQ, Sun H (2004) The genetic diversity of cultivated soybean grown in China. *Theor Appl Genet* 108: 931–936.

Du H, Fang C, Li Y, Kong F, Liu B (2023) Understandings and future challenges in soybean functional genomics and molecular breeding. *J Integr Plant Biol*. 65(2):468–495.

Duhnen A, Gras A, Teyssèdre S, Romestant M, Claustres B (2017) Genomic selection for yield and seed protein content in soybean: A study of breeding program data and assessment of prediction accuracy. *Crop Sci*. 57: 1325–1337.

Ellis, M. L., M. M. Díaz Arias, D. R. Cruz Jimenez, G. P. Munkvold, and L. F. Leandro. (2013). First report of Fusarium commune causing damping-off, seed rot, and seedling root rot on soybean (Glycine max) in the United States. *Plant Dis*. 97: 284.

Ferreira AR, Foutz KR, Keim P (2000) Soybean genetic map of RAPD markers assigned to an existing scaffold RFLP map. *J Hered* 91: 392–396.

Ferreira-Neto JRC, da Silva MD, Benko-Iseppon AM, Pandolfi V, Binneck E, Nepomuceno, AL, Abdelnoor RV, Kido EA (2019) Inositol phosphates and Raffinose family oligosaccharides pathways: Structural genomics and transcriptomics in soybean under root dehydration. *Plant Gene* 20: 100202

Fliege CE, Ward RA, Vogel P, Nguyen H, Quach T, Guo M, Viana JPG, Dos Santos LB, Specht JE, Clemente TE, Hudson ME, Diers BW (2022) Fine mapping and cloning of the major seed protein quantitative trait loci on soybean chromosome 20. *Plant J.* 110(1):114–128.

Frary A., Doganlar S., Ratnaparkhe M.B. (2008) Compara–tive mapping. In: Kole C., Abbott A.G. (Eds.), Princi–ples and practices of plant genomics, vol. 1. Genome mapping. Science Publishers, USA, pp. 205–228.

Fujita M, Fujita Y, Maruyama K et al (2004) A dehydration-induced NAC protein, RD26, is involved in a novel ABA-dependent stress-signaling pathway. *Plant J* 39: 863–876.

Funatsuki H, Suzuki M, Hirose A, Inaba H, Yamada T, Hajika M, Komatsu K, Katayama T, Sayama T, Ishimoto M et al. (2014) Molecular basis of a shattering resistance boosting global dissemination of soybean. *Proc Natl Acad Sci U. S. A.* 111: 17797–17802.

Gao W, Ma R, Li X, Liu J, Jiang A, Tan P, Xiong G, Du C, Zhang J, Zhang X, Fang X, Yi Z, Zhang J (2024) Construction of Genetic Map and QTL Mapping for Seed Size and Quality Traits in Soybean (Glycine max L.). *Int J Mol Sci.* 1;25(5):2857.

Ge L, Yu J, Wang H et al (2016) Increasing seed size and quality by manipulating BIG SEEDS1 in legume species. *Proc Natl Acad Sci U. S. A.* 113(44): 12414–12419. doi: 10.1073/pnas.1611763113

Gorjanc G, Gaynor RC, Hickey JM (2018) Optimal cross selection for long-term genetic gain in two-part programs with rapid recurrent genomic selection. *Theor. Appl. Genet* 131: 1953–1966.

Grant D, Nelson RT, Cannon SB, Shoemaker RC (2010) SoyBase, the USDA-ARS soybean genetics and genomics database. *Nucleic Acids Res* 38: D843–D846.

Gu Y, Li W, Jiang H, Wang Y, Gao H, Liu M, et al. (2017). Differential expression of a WRKY gene between wild and cultivated soybeans correlates to seed size. *J Exp Bot* 68 (11), 2717–2729.

Gupta S, Kumawat G, Agrawal N, Tripathi R, Rajesh V, Nataraj V, et al. (2022). Photoperiod trait: Insight in molecular mechanism for growth and maturity adaptation of soybean (*Glycine max*) to different latitudes. *Plant Breed.* 141: 483–500. doi: 10.1111/pbr.13041

Gupta, S. et al. Long juvenility trait (2021) A vehicle for commercial utilization of soybean (Glycine max) in lower latitudes. Plant Breed. 140, 543–560

Ha, CV, Watanabe Y, Tran UT et al (2015) Comparative analysis of root transcriptomes from two contrasting drought-responsive Williams 82 and DT2008 soybean cultivars under normal and dehydration conditions. *Front Plant Sci 6*: 551.

Ha J, Abernathy B, Nelson W et al (2012) Integration of the Draft Sequence and Physical Map as a Framework for Genomic Research in Soybean (*Glycine max* (L.) Merr.) and Wild Soybean (*Glycine soja* Sieb. and Zucc.). *G3 (Bethesda, Md.)* 2(3): 321–329.

Ha J, Jeon HH, Woo DU et al (2019) Soybean-VCF2Genomes: A database to identify the closest accession in soybean germplasm collection. *BMC Bioinformatics* 24(20)(Suppl 13): 384.

Han D, Zhao X, Zhang D, Wang Z, Zhu Z, Sun H, Qu Z, Wang L, Liu Z, Zhu X, Yuan M (2024) Genome-wide association studies reveal novel QTLs for agronomic traits in soybean. *Front Plant Sci.* 2024 May 14;15:1375646. doi: 10.3389/fpls.2024.

Hao D, Chao M, Yin Z, Yu D (2012) Genome-wide association analysis detecting significant single nucleotide polymorphisms for chlorophyll and chlorophyll fluorescence parameters in soybean (*Glycine max*) landraces. *Euphytica* 186: 919–931

Hao Q, Yang H, Chen S, Qu Y, Zhang C, Chen L, Cao D, Yuan S, Guo W, Yang Z, Huang Y, Shan Z, Chen H, Zhou X (2023) RNA-Seq and Comparative Transcriptomic Analyses of Asian Soybean Rust Resistant and Susceptible Soybean Genotypes Provide Insights into Identifying Disease Resistance Genes. *Int J Mol Sci.* 30;24(17):13450.

Harlan JR, de Wet JMJ (1971) Towards a rational classification of cultivated plants. *Taxon* 20: 509–517.

Herritt M, Dhanapal AP, Purcell LC, Fritschi FB (2018) Identification of genomic loci associated with 21 chlorophyll fluorescence phenotypes by genome-wide association analysis in soybean. *BMC Plant Biol* 18(1): 312.

Huang G-T, Ma S-L, Bai L-P. (2011) Signal transduction during cold, salt, and drought stresses in plants. *Mol Biol Rep* 39: 969–987.

Huang L, Zeng A, Chen P, Wu C, Wang D, Wen Z (2018) Genome-wide association analysis of salt tolerance in soybean [*Glycine max* (L) Merr]. *Plant Breed* 137: 714–720.

Huang Y, Koo DH, Mao Y, Herman EM, Zhang J, Schmidt MA (2024) A complete reference genome for the soybean cv. *Jack. Plant Commun.* 12;5(2):100765

Hwang, EY., Song, Q., Jia, G. et al. (2014) A genome-wide association study of seed protein and oil content in soybean. *BMC Genom* 15(1). doi: 10.1186/1471-2164-15-1

Hymowitz T (2004) Speciation and cytogenetics. In: Caldwell BE (ed.) *Soybeans: Improvement Production and Uses*, vol 16, 3rd edn, Agron Monogr. ASA, CSSA, SSSA, Madison, WI, pp. 97–136.

Hyten DL, Choi I-Y, Song Q et al (2010a) A high density integrated genetic linkage map of soybean and the development of a 1,536 Universal Soy Linkage Panel for QTL mapping. *Crop Sci* 50: 960–968.

Hyten DL, Cannon SB, Song Q et al (2010b) High-throughput SNP discovery through deep resequencing of a reduced representation library to anchor and orient scaffolds in the soybean whole genome sequence. *BMC Genom* 11: 38.

Hyten DL, Song Q, Choi IY et al (2008) High-throughput genotyping with the GoldenGate assay in the complex genome of soybean. *Theor Appl Genet* 116: 945–952.

Hyten DL, Song Q, Zhu Y et al (2006) Impacts of genetic bottlenecks on soybean genome diversity. *Proc Nat Acad Sci* 103: 16666–16671.

Innes RW, Ameline-Torregrosa C, Ashfield T et al (2008) Differential accumulation of retroelements and diversification of NB-LRR disease resistance genes in duplicated regions following polyploidy in the ancestor of soybean. *Plant Physiol* 148: 1740–1759.

Islam MR, Fujita D, Watanabe S, Zheng SH (2018) Variation in photosensitivity of flowering in the world soybean minicore collections (GmWMC). *Plant Prod Sci.* doi: 10.1080/1343943X.2018.1561197

Itoh H, Izawa T. (2013) The coincidence of critical day length recognition for florigen gene expression and floral transition under long-day conditions in rice. *Molecular Plant* 6(3): 635–649.

Jacobson A, Lian L, Zhong S, Bernardo R (2014) General combining ability model for genomewide selection in a biparental cross. *Crop Sci.* 54: 895–905.

Jannink JL, Lorenz AJ, Iwata H (2010) Genomic selection in plant breeding: From theory to practice. *Brief. Funct. Genomics* 9: 166–177.

Jarquin D, Specht J, Lorenz A (2016) Prospects of genomic prediction in the USDA soybean germplasm collection: Historical data creates robust models for enhancing selection of accessions. G3 6: 2329–2341.

Jiang G-L, Rajcan I, Zhang Y-M, Han T, Mian R (2023) Editorial: Soybean molecular breeding and genetics. Front. Plant Sci 14: 1157632. doi: 10.3389/fpls.2023.1157632

Kajiya-Kanegae H, Nagasaki H, Kaga A et al (2021) Whole-genome sequence diversity and association analysis of 198 soybean accessions in mini-core collections. *DNA Res* 19(1): 28.

Kaler AS, Dhanapal AP, Ray JD, King CA, Fritsch FB, Purcell LC (2017) Genome-wide association mapping of carbon isotope and oxygen isotope ratios in diverse soybean genotypes. *Crop Sci* 57: 3085–3100.

Kan G, Zhang W, Yang W, Ma D, Zhang D, Hao D, Hu Z, Yu D (2015) Association mapping of soybean seed germination under salt stress. *Mol Gen Genom* 290(6): 2147–2162.

Kavishwar R, Chandra S, Satpute G et al (2021) Role of omics technologies in understanding drought stress tolerance in Soybean. *Soybean Res.* 19(1) 90–97.

Keim P, Diers BW, Olson TC, Shoemaker RC (1990) RFLP mapping in soybean: Association between marker loci and variation in quantitative traits. *Genetics* 126: 735–742.

Keim P, Schupp JM, Travis SE et al (1997) A high-density soybean genetic map based on AFLP markers. *Crop Sci* 37: 537–543.

Keim P, Shoemaker RC, Palmer RG (1989) Restriction fragment length polymorphism diversity in soybean. *Theo. App. Gene* 77: 786–792.

Kim MY, Lee S, Van K, Kim TH, Jeong SC, Choi IY, Kim DS, Lee YS, Park D, Ma J, Kim WY, Kim BC, Park S, Lee KA, Kim DH, Kim KH, Shin JH, Jang YE, Kim KD, Liu WX, Chaisan T, Kang YJ, Lee YH, Kim KH, Moon JK, Schmutz J, Jackson SA, Bhak J, Lee SH (2010) Whole-genome sequencing and intensive analysis of the undomesticated soybean (*Glycine soja* Sieb and Zucc) genome. *Proc Natl Acad Sci U. S. A.* 21;107(51): 22032–22037

Kim KD, El Baidouri M, Abernathy B, Iwata-Otsubo A, Chavarro C, Gonzales M, et al. (2015) A comparative epigenomic analysis of polyploidy-derived genes in soybean and common bean. *Plant Physiol* 168: 1433–1447.

Kim JY, Jeong S, Kim KH et al (2019) Dissection of soybean populations according to selection signatures based on whole-genome sequences. *Gigascience* 8(12): giz151.

Kim KS, Vuong TD, Qiu D et al (2016) Advancements in breeding, genetics, and genomics for resistance to three nematode species in soybean. *Theor Appl Genet* 129: 2295–2311.

Kim MS, Lozano R, Kim JH et al (2021) The patterns of deleterious mutations during the domestication of soybean. *Nat Commun* 12(1): 97.

Kofsky J, Zhang H, Song BH. (2018) The untapped genetic reservoir: The past, current, and future applications of the wild soybean (*Glycine soja*). *Front Plant Sci* 9: 949.

Kong F, Liu B, Xia Z et al (2010) Two co-ordinately regulated homologs of FLOWERING LOCUS T are involved in the control of photoperiodic flowering in soybean. *Plant Physiol* 154: 1220–1231.

Kou K, Yang H, Li HY et al (2022) A functionally divergent SOC1 homolog improves soybean yield and latitudinal adaptation. *Curr Biol* 32: 1–15.

Kumawat G, Maranna S, Gupta S, Tripathi R, Agarwal N, Singh V, Rajesh V, Chandra S, Kamble V, Nataraj V, Bharti A, Sharma M P, Jadhav P V, Ratnaparkhe M B, Satpute G K, Bhatia VS (2020) Identification of novel genetic sources for agronomic and quality traits in soybean using multi-trait allele specific genic marker assays. *J Plant Biochem Biotechnol* 30: 160–171.

Kumar, S., Rajput, L.S., Ramteke, R.H. et al. (2021). First report of root rot and damping off disease in soybean (Glycine max) caused by Pythium deliense in India. *Plant Disease,* 105, https://doi.org/10.1094/PDIS-01-21-0059-PDN.

Ladizinsky G, Newell CA, Hymowitz T (1979a). Giemsa staining of soybean chromosomes. *J Hered* 70: 415–416.

Ladizinsky G, Newell CA, Hymowitz T (1979b). Wide crosses in soybean: Prospects and limitations. *Euphytica* 28: 421–423.

Lam HM, Xu X, Liu X et al (2010) Resequencing of 31 wild and cultivated soybean genomes identifies patterns of genetic diversity and selection. *Nat Genet.* 42(12): 1053–1059.

Langewisch T, Zhang H, Vincent R, Joshi T, Xu D, Bilyeu K (2014) Major soybean maturity gene haplotypes revealed by SNPViz analysis of 72 sequenced soybean genomes. *PLoS ONE* 9(4): e94150.

Li L, Lin F, Wang W et al (2016) Fine mapping and candidate gene analysis of two loci conferring resistance to *Phytophthorasojae* in soybean. *Theor. Appl. Genet.* 129: 2379–2386.

Li Y, Zhao W, Tang J, Yue X, Gu J, Zhao B, Li C, Chen Y, Yuan J, Lin Y, Li Y, Kong F, He J, Wang D, Zhao TJ, Wang ZY (2024b) Identification of the domestication gene GmCYP82C4 underlying the major quantitative trait locus for the seed weight in soybean. *Theor Appl Genet.* 28;137(3):62.

Li N, Xu R, Li Y (2019) Molecular networks of seed size control in plants. *Ann. Rev. Plant Biol.* 70: 435–463. doi: 10.1146/annurev-arplant-050718-095851

Li D, Wang Q, Tian Y, Lyv X, Zhang H, Hong H, Gao H, Li YF, Zhao C, Wang J, Wang R, Yang J, Liu B, Schnable PS, Schnable JC, Li YH, Qiu LJ (2024a) TWAS facilitates gene-scale trait genetic dissection through gene expression, structural variations, and alternative splicing in soybean. *Plant Commun.* 2024 Jun 25:101010. doi: 10.1016/j.xplc.2024.101010

Li, D., Zhang, Z., Gao, X. et al. (2023) The elite variations in germplasms for soybean breeding. *Mol Breeding* 43, 37 (2023). https://doi.org/10.1007/s11032-023-01378-0

Li W, Han Y, Zhang D et al (2008) Genetic diversity in soybean genotypes from north-eastern China and identification of candidate markers associated with maturity rating. *Plant Breed* 127: 56–61.

Li Y, Liu X, Deng W et al (2022) Fine mapping the soybean mosaic virus resistance gene in soybean cultivar heinong 84 and development of CAPS markers for rapid identification. *Viruses* 14(11): 2533.

Li YH, Li W, Zhang C et al. (2010) Genetic diversity in domesticated soybean (*Glycine max*) and its wild progenitor (*Glycine soja*) for simple sequence repeat and single-nucleotide polymorphism loci. *New Phytol* 188: 242–253.

Li YH, Zhao SC, Ma JX et al (2013) Molecular footprints of domestication and improvement in soybean revealed by whole genome re-sequencing. *BMC Genom.* 28(14): 579.

Li YH, Zhou G, Ma J et al. (2014) De novo assembly of soybean wild relatives for pan-genome analysis of diversity and agronomic traits. *Nat Biotechnol* 32: 1045.

Liang X, Hou X, Li J et al (2019) High-resolution DNA methylome reveals that demethylation enhances adaptability to continuous cropping comprehensive stress in soybean. *BMC Plant Biol.* 19: 79.

Lightfoot DA, Njiti VN, Gibson PT, Kassem MA, Iqbal JM, Meksem K (2005) Registration of the Essex by Forrest recombinant inbred line mapping population. *Crop Sci* 45: 1678–1681.

Lin F, Chhapekar SS, Vieira CC et al (2022) Breeding for disease resistance in soybean: A global perspective. *Theor. Appl. Genet* 135: 3773–3872.

Lin F, Zhao M, Ping J et al (2013) Molecular mapping of two genes conferring resistance to *Phytophthora sojae* in a soybean landrace PI 567139B. *Theor. Appl. Genet.* 126: 2177–2185.

Lin Y, Li W, Zhang Y et al (2019) Identification of genes/proteins related to submergence tolerance by transcriptome and proteome analyses in soybean. *Sci Rep* 9: 14688

Lin XY, Liu BH, Weller JL, Abe J, Kong FJ (2021) Molecular mechanisms for the photoperiodic regulation of flowering in soybean. *J Integr Plant Biol* 63: 981–994.

Lin X, Lin W, Ku YS, Wong FL, Li MW, Lam HM, Ngai SM, Chan TF (2020) Analysis of Soybean Long Non-Coding RNAs Reveals a Subset of Small Peptide-Coding Transcripts. *Plant Physiol.* 182(3):1359-1374.

Liu B, Kanazawa A, Matsumura H, Takahashi R, Harada K, Abe J. (2008) Genetic redundancy in soybean photoresponses associated with duplication of the phytochrome A gene. *Genetics* 180(2): 995–1007.

Liu, Z., Li, H., Gou, Z. et al. (2020b) Genome-wide association study of soybean seed germination under drought stress. *Mol Genet Genomics* 295: 661–673

Liu N, Niu Y, Zhang G, Feng Z (2022) Genome sequencing and population resequencing provide insights into the genetic basis of domestication and diversity of vegetable soybean. *Horticulture Res.* 9: uhab052.

Liu Y, Du H, Li P et al (2020a) Pan-genome of wild and cultivated soybeans. *Cell* 182(1): 162–176.

Liu S, Zhong H, Meng X, Sun T, Li Y, Pinson SRM, Chang SKC, Peng Z. (2020c) Genome-wide association studies of ionomic and agronomic traits in USDA mini core collection of rice and comparative analyses of different mapping methods. *BMC Plant Biol.* 24;20(1):441.

Livingstone JM, Cheng KC, Strömvik MV (2010) Bioinformatics as a tool. In: Bilyeu K, Ratnaparkhe MB, Kole C (eds) *Genetics, genomics, and breeding of soybean*. CRC, New Hampshire, UK.Pp 263-290

Lorenzen L, Boutin S, Young N, Specht J E, Shoemaker R C (1995) Soybean pedigree analysis using map-based molecular markers: I. Tracking RFLP markers in cultivars. *Crop Sci* 35: 1326–1336.

Lu SJ, Dong LD, Fang C et al (2020) Stepwise selection on homeologous PRR genes controlling flowering and maturity during soybean domestication. *Nat Genet* 52: 428–436.

Luo M, Dennis ES, Berger F, Peacock WJ, Chaudhury A (2005) MINISEED3 (MINI3), a WRKY family gene, and HAIKU2 (IKU2), a leucine-rich repeat (LRR) KINASE gene, are regulators of seed size in Arabidopsis. *Proc. Natl. Acad. Sci. U. S. A* 102(48): 17531–17536. doi: 10.1073/pnas.0508418102

Ma J, Shoemaker R, Jackson S, Cannon S (2010) Comparative genomics. In: Bilyeu K, Ratnaparkhe MB, Kole C (eds) *Genetics, genomics, and breeding of soybean*. CRC, New Hampshire, UK, pp. 245–262

Ma Y, Reif JC, Jiang Y, Wen Z, Wang D, Liu Z, et al. (2016) Potential of marker selection to increase prediction accuracy of genomic selection in soybean (*Glycine max* L.). *Mol Breed.* 36: 113. https://doi.org/10.1007/s11032-016-0504-9.

Maguire TL, Grimmond S, Forrest A, Iturbe-Ormaetxe I, Meksem K, Gresshoff P (2002) Tissue-specific gene expression in soybean (*Glycine max*) detected by cDNA microarray analysis. *J. Plant Phys* 159: 1361–1364.

Mahmood A, Bilyeu KD, Škrabišová M, Biová J, De Meyer EJ, Meinhardt CG, Usovsky M, Song Q, Lorenz AJ, Mitchum MG, Shannon G, Scaboo AM (2023) Cataloging SCN resistance loci in North American public soybean breeding programs. *Front Plant Sci.* 20;14:1270546.

Mandozai A, Moussa AA, Zhang Q, Qu J, Du Y, Anwari G, et al. (2021) Genome-wide association study of root and shoot related traits in spring soybean (*Glycine max* L.) at seedling stages using SLAF-seq. *Front Plant Sci* 12: 568995.

Mansur LM, Orf JH, Chase K, Jarvik T, Cregan PB, Lark KG (1996) Genetic mapping of agronomic traits using recombinant inbred lines of soybean. *Crop Sci* 36: 1327–1336.

Maranna S, G Kumawat, V Nataraj et al (2023) Development of improved genotypes for extra early maturity, higher yield and Mungbean Yellow Mosaic India Virus (MYMIV) resistance in soybean (*Glycine max* L.). *Crop and Pasture Sci* 11(1): 22853.

Maranna, S, Kumawat, G, Nataraj, V, Gireesh, C, Gupta, S, Satpute, GK, Ratnaparkhe, MB and Yadav DP (2019). NAM population–a novel genetic resource for soybean improvement: development and characterization for yield and attributing traits. *Plant Genet Resour-C* 17 (6): 545–553.

Maranna S, Nataraj V, Kumawat G, Mehetre SP, Reddy R, Jaybhay SPGS, Rathod S, Agrawal N, et al. (2024). Understanding of G×E interactions of yield attributes in soybean MAGIC population and characterization for charcoal rot resistance. *Agronomy J* 1–12. https://doi.org/10.1002/agj2.21572

Marmat, N., Ansari MM, Ratnaparkhe, M. et al (2017). Molecular and phylogenetic studies of Colletotrichum truncatum associated with soybean anthracnose in India. *Plant Pathol. Q.* 7, 146–152.

Matei G, Woyann LG, Milioli AS et al (2018) Genomic selection in soybean: Accuracy and time gain in relation to phenotypic selection. *Mol Breed* 38: 117.

Meuwissen THE, Hayes BJ, Goddard ME (2001) Prediction of total genetic value using genome-wide dense marker maps. *Genetics* 157: 1819–1829.

Nataraj V, Kumar S, Kumawat G et al. (2019) Charcoal rot resistance in Soybean: current understanding and future perspectives. In: Wani S (ed) Disease resistance in crop plants. Springer, Cham. https://doi.org/10.1007/978-3-030-20728-1_10

Neto JRCF, Pandolfi V, Guimaraes FCM et al (2013) Early transcriptional response of soybean contrasting accessions to root dehydration. *PLoS One 8*(12): e83466.

Newell, CA and Hymowitz, T (1980) A taxonomic revision in the genus *Glycine* subgenus *Glycine* (Leguminosae). *Brittonia* 32: 63–69

Niu Y, Yung WS, Sze CC, Wong FL, Li MW, Chung G, Lam HM (2024) Developing an SNP dataset for efficiently evaluating soybean germplasm resources using the genome sequencing data of 3,661 soybean accessions. *BMC Genomics*. 14;25(1):475.

Niu M, Tian K, Chen Q, Yang C, Zhang M, Sun S, Wang X (2024) A multi-trait GWAS-based genetic association network controlling soybean architecture and seed traits. *Front Plant Sci*. 14:1302359.

Pagano MC, Miransari M, Corrêa EJA, Duarte NF, Yelikbayev BK. (2020) Genomic research favoring higher soybean production. *Curr Genomics* 21(7): 481–490.

Pellegrineschi A, Reynolds M, Pacheco M et al (2004) Stress-induced expression in wheat of the *Arabidopsis thaliana* DREB1A gene delays water stress symptoms under greenhouse conditions. *Genome* 47: 493–500.

Pin PA, Nilsson O. (2012) The multifaceted roles of FLOWERING LOCUS T in plant development. *Plant, Cell & Environment* 35(10): 1742–1755.

Ping J, Fitzgerald JC, Zhang C et al (2015) Identification and molecular mapping of *Rps11*, a novel gene conferring resistance to *Phytophthora sojae* in soybean. *Theor. Appl. Genet* 129: 445–451.

Powell W, Morgante M, Doyle JJ, McNicol JW, Tingey SV, Rafalske AJ (1996) Gene pool variation in genus *Glycine* subgenus *soja* revealed by polymorphic nuclear and chloroplast micro-satellites. *Genetics* 144: 793–803.

Rajput LS, Kumar S, Nataraj V, Shivakumar M, Pathak K, Jaiswal S, Pandey V. (2023) Recent advancement in management of soybean charcoal rot caused by *Macrophomina phaseolina*. In *Macrophomina Phaseolina*. Academic Press, Cambridge, MA, pp. 55–74.

Rambani A, Rice JH, Liu J, Lane T, Ranjan P, Mazarei M, et al. (2015) The methylome of soybean roots during the compatible interaction with the soybean cyst nematode. *Plant Physiol* 168: 1364–1377.

Ramesh SV, Ratnaparkhe MB, Kumawat G, Gupta GK, Husain SM (2014) Plant miRNAome and antiviral resistance: a retrospective view and prospective challenges. *Virus Genes* 48: 1–14

Ramteke, R., Rajput, L.S., Nataraj, V. et al. Evaluation of soybean genotypes for resistance to collar rot (Sclerotium rolfsii) under field and glasshouse condition. *Trop. plant pathol.* (2024). https://doi.org/10.1007/s40858-024-00652-6

Ratnaparkhe MB, Marmat N, Kumawat G et al (2020) Whole genome re-sequencing of soybean accession EC241780 providing genomic landscape of candidate genes involved in rust resistance. *Curr Genomics* 21(7): 504–511.

Ratnaparkhe MB, Satpute GK, Kumawat G et al (2022a). Genomic designing for abiotic stress tolerant soybean. In: C. Kole (ed) *Genomic designing for abiotic stress resistant oilseed crops*, Springer, Cham, pp. 1–73.

Ratnaparkhe MB, Nataraj V, Shivakumar M, Chandra S (2022b) Genomic design for biotic stresses in soybean. In: C. Kole (ed) *Genomic designing for biotic stress resistant oilseed*, Springer, Cham, pp. 1–54.

Ratnaparkhe MB, Ramesh SV, Giriraj K, Husain SM, Gupta S (2013) Advances in soybean genomics. In: Gupta (ed) *Legumes in the omic era*, Springer, pp. 41–72.

Ratnaparkhe MB, Singh RJ, Doyle JJ (2010) Glycine. In: C. Kole (ed) *Wild Crop Relatives*, Springer: Genomic and Breeding Resources, pp. 83–116.

Mitchell G Roth, Richard W Webster, Daren S Mueller, Martin I Chilvers, Travis R Faske, Febina M Mathew, Carl A Bradley, John P Damicone, Mehdi Kabbage, Damon L Smith, Integrated Management of Important Soybean Pathogens of the United States in Changing Climate, *Journal of Integrated Pest Management*, Volume 11, Issue 1, 2020, 17, https://doi.org/10.1093/jipm/pmaa013

Ravelombola WS, Qin J, Shi A, Nice L, Bao Y (2019) Genome-wide association study and genomic selection for soybean chlorophyll content associated with soybean cyst nematode tolerance. *BMC Genom*. 20: 904. doi: 10.1186/s12864-019-6275-z

Ravelombola WS, Qin J, Shi A, Nice L, Bao Y (2020) Genome-wide association study and genomic selection for tolerance of soybean biomass to soybean cyst nematode infestation. *PLoS One* 15: e0235089. doi: 10.1371/journal.pone.0235089

Ravelombola WS, Qin J, Shi A, Song Q, Yuan J (2021) Genome-wide association study and genomic selection for yield and related traits in soybean. *PLoS One* 16: e0255761. doi: 10.1371/journal.pone.0255761

Razzaq, A, et al. (2022). miRNAs in Soybean Improvement. In: Wani SH, Sofi NuR, Bhat MA, Lin F (eds) *Soybean Improvement*. Springer, Cham. https://doi.org/10.1007/978-3-031-12232-3_12

Saindon G, Voldeng HD, Beversdorf WD, Buzzell RI (1989) Genetic control of long daylength response in soybean. *Crop Sci* 29: 1436–1439

Satpute, G.K. et al. (2022). Dissection of Physiological and Biochemical Bases of Drought Tolerance in Soybean (Glycine max) Using Recent Phenomics Approach. In: Wani, S.H., Sofi, N.u.R., Bhat, M.A., Lin, F. (eds) Soybean Improvement. Springer, Cham. https://doi.org/10.1007/978-3-031-12232-3_2

Satpute GK, Ratnaparkhe MB, Chandra S et al (2021) Breeding and molecular approaches for evolving drought-tolerant soybeans In: Giri B, Sharma MP, (eds) *Plant stress biology*, Springer, Singapore, pp. 83–130.

Schmutz J, Cannon SB, Schlueter J et al (2010) Genome sequence of the paleopolyploid soybean. *Nature* 463: 178–183.

Sedivy EJ, Wu F, Hanzawa Y (2017) Soybean domestication: the origin, genetic architecture and molecular bases. *New Phytol*. 214(2): 539–553.

Sharmin RA, Bhuiyan MR, Lv W et al (2020) RNA-Seq based transcriptomic analysis revealed genes associated with seed-flooding tolerance in wild soybean (*Glycine soja* Sieb &Zucc). *Environ Exp Bot* 171: 103906.

Shen Y, Zhang J, Liu Y et al (2018a) DNA methylation footprints during soybean domestication and improvement. *Genome Biol* 19(1): 128.

Shen et al. (2018b) De novo assembly of a Chinese soybean genome. *Sci China Life Sci*. DOI: 10.1007/s11427-018-9360-0

Shoemaker RC, Specht JE (1995) Integration of the soybean molecular and classical genetic linkage groups. *Crop Sci* 35: 436–446.

Singh RJ, Chung GH, Nelson RL (2007a) Landmark research in Legumes. *Genome* 50: 525–537.

Singh RJ, Hymowitz T (1987) Intersubgeneric crossability in the genus Glycine Willd. *Plant Breed* 98: 171–173.

Singh RJ, Klein TM, Mauvais CJ, Knowlton S, Hymowitz T, Kostow CM (1998a) Cytological characterization of the transgenic soybean. *Theor Appl Genet* 96: 319–324.

Singh RJ, Kollipara KP, Hymowitz T (1998b) The genomes of *Glycine canescens* F J Herm and *G. tomentella* Hayata of Western Australia and their phylogenetic relationships in the genus *Glycine* Willd. *Genome* 41: 669–679.

Singh RJ, Nelson RL, Chung GH (2007b) Soybean (*Glycine max* (L) Merr). In: Singh RJ (ed) *Genetic resources, chromosome engineering and crop improvement, vol 4, Oilseed crops*. CRC, Boca Raton, FL, pp. 13–50.

Singh RK, Prasad A, Muthamilarasan M, Parida SK, Prasad M (2020) Breeding and biotechnological interventions for trait improvement: Status and prospects. *Planta* 252(4): 1–18.

Skorupska HT, Shoemaker RC, Warner A, Shipe ER, Bridges WC (1993) Restriction fragment length polymorphism in soybean germplasm of the southern USA. *Crop Sci* 33: 1169–1176

Song QX, Lu X, Li QT et al (2013) Genome-wide analysis of DNA methylation in soybean. *Mol Plant*. 6(6): 1961–1974.

Song Q, Hyten DL, Jia G, Quigley CV, Fickus EW, Nelson RL, Cregan PB (2015) Fingerprinting soybean germplasm and its utility in genomic research. *G3 (Bethesda)* 28; 5(10): 1999–2006

Steketee CJ, Schapaugh WT, Carter TE, Li Z (2020) Genome-wide association analyses reveal genomic regions controlling canopy wilting in soybean. *G3 Genes Genomes Genetics* 10: 1413–1425.

Stewart-Brown, BB, Song Q, Vaughn JN, Li Z (2019) Genomic selection for yield and seed composition traits within an applied soybean breeding program. *G3* 9: 2253–2265.

Stolf-Moreira R, Lemos E, Carareto-Alves L et al (2011) Transcriptional profiles of roots of different soybean genotypes subjected to drought stress. *Plant Mol Biol Rep*. 29: 19–34.

Sugimoto T, Kato M, Yoshida S et al (2012) Pathogenic diversity of *Phytophthora sojae* and breeding strategies to develop Phytophthora-resistant soybeans. *Breed. Sci* 61: 511–522.

Sun H, Jia Z, Cao D et al (2011) GmFT2a, a soybean homolog of FLOWERING LOCUS T, is involved in flowering transition and maintenance. *PLoS ONE* 6(11): e29238.

Sun Z, Lam HM, Lee SH, Li X, Kong F (2024) Soybean functional genomics: bridging theory and application. *Mol Breed*. 12;44(1):2.

Sun J, Li L, Zhao J et al (2014) Genetic analysis and fine mapping of *RpsJS*, a novel resistance gene to Phytophthora sojae in soybean [*Glycine max* (L.) Merr.]. *Theor. Appl. Genet* 127: 913–919.

Sun L, Miao Z, Cai C, Zhang D, Zhao M, Wu Y, Zhang X, Swarm SA, Zhou L, Zhang ZJ et al. (2015) *GmHs1-1*, encoding a calcineurin-like protein, controls hard-seededness in soybean. *Nat Genet* 47: 939–943.

Surbhi K, Singh KP, Singh NK, Aravind T (2021) Assessment of genetic diversity among soybean genotypes differing in response to aerial blight (*Rhizoctonia solani* Kuhn) using SSR markers. *J. Phytopathol.* 169(1): 37–44.

Talukdar A, Verma K, Gowda DSS, Sinha P (2009) Molecular breeding for charcoal rot resistance in soybean I. Screening and mapping population development. *Ind. J. Genet. Plant Breeding* 69(4): 367–370.

Tanksley SD, Nelson JC (1996) Advanced backcross QTL analysis: A method for the simultaneous discovery and transfer of valuable QTLs from unadapted germplasm into elite breeding lines. *Theor. Appl Genet* 92(2): 191–203.

Thao NP, Thu NB, Hoang XL, Van Ha C, Tran LS (2013) Differential expression analysis of a subset of drought-responsive GmNAC genes in two soybean cultivars differing in drought tolerance. *Intl J Mol Sci* 14: 23828–23841.

Thibaud-Nissen F, Shealy RT, Khanna A, Vodkin LO (2003) Clustering of microarray data reveals transcript patterns associated with somatic embryogenesis in soybean. *Plant Phys* 132(1): 118–136.

Torkamaneh D, Laroche J, Valliyodan B, O'Donoughue L, Cober E, Rajcan I, et al. (2021) Soybean (*Glycine max*) Haplotype Map (GmHapMap): a universal resource for soybean translational and functional genomics. *Plant Biotechnol J.* 19(2): 324–334.

Tripathi P, Rabara RC, Reese RN et al (2016) A toolbox of genes, proteins, metabolites and promoters for improving drought tolerance in soybean includes the metabolite coumestrol and stomatal development genes. *BMC Genom* 17: 102.

Tsubokura Y, Matsumura H, Xu M et al (2013) Genetic variation in soybean at the maturity locus E4 is involved in adaptation to long days at high latitudes. *Agronomy* 3(1): 117–134.

Turner M, Yu O, Subramanian S (2012) Genome organization and characteristics of soybean microRNAs. *BMC Genom* 13: 169.

Usovsky M, Chen P, Li D et al (2022) Decades of genetic research on *Soybean mosaic virus* resistance in soybean. *Viruses* 14(6): 1122.

Valliyodan B, Brown AV, Wang J et al (2021) Genetic variation among 481 diverse soybean accessions, inferred from genomic re-sequencing. *Sci Data* 8: 50. doi: 101038/s41597-021-00834-w

Valliyodan B, Cannon SB, Bayer PE et al (2019) Construction and comparison of three reference-quality genome assemblies for soybean. *Plant J* 100(5): 1066–1082.

Valliyodan B, Qiu D, Patil G et al (2016) Landscape of genomic diversity and trait discovery in soybean. *Sci Rep* 6: 23598.

Valliyodan B, Ye H, Song L, Murphy M, Shannon JG, Nguyen HT (2017) Genetic diversity and genomic strategies for improving drought and waterlogging tolerance in soybeans. *J Exp Bot* 68(8): 1835–1849. doi: 101093/jxb/erw433 PMID: 27927997

Varshney RK, Graner A, Sorrells ME. (2005) Genomics-assisted breeding for crop improvement. *Trends in Plant Sci* 10(12): 621–630.

Veerkamp RF, Bouwman AC, Schrooten C, Calus MP (2016) Genomic prediction using preselected DNA variants from a GWAS with whole-genome sequence data in Holstein-Friesian cattle. *Genet Sel E* 48(1): 95.

Vodkin LO, Khanna A, Shealy Ret al (2004) Microarray analysis for global expression constructed with a low redundancy set of 27,500 sequenced cDNAs representing an array of developmental stages and physiological conditions of the soybean plant. *BMC Genom* 5: 73.

Vuong TD, Sleper DA, Shannon JG, Nguyen HT (2010) Novel quantitative trait loci for broad-based resistance to soybean cyst nematode (*Heterodera glycines* Ichinohe) in soybean PI 567516C. *Theor Appl Genet* 121: 1253–1266.

Wang F, Nan H, Chen L et al (2019a) A new dominant locus, E11, controls early flowering time and maturity in soybean. *Mol Breed* 39: 70.

Wang X, Chen L, Ma J (2019b) Genomic introgression through interspecific hybridization counteracts genetic bottleneck during soybean domestication. *Genome Biol* 20: 22.

Wang Y, Gu YZ, Gao HH et al (2016) Molecular and geographic evolutionary support for the essential role of GIGANTEAa in soybean domestication of flowering time. *BMC Evol Biol* 16: 79.

Wang J, Yang Q, Chen Y, Liu K, Zhang Z, Xiong Y, Yu H, Yu Y, Wang J, Song J, Qiu L (2024) QTL mapping and genomic selection of stem and branch diameter in soybean (Glycine max L.). *Front Plant Sci.*;15:1388365. doi: 10.3389/fpls.2024.1388365.

Watanabe S, Hideshima R, Xia Z et al (2009) Map-based cloning of the gene associated with the soybean maturity locus *E3*. *Genetics* 182: 1251–1262.

Watanabe S, Xia Z, Hideshima R, Tsubokura Y, Sato S, Harada K (2011) A map-based cloning strategy employing a residual heterozygous line reveals that the GIGANTEA gene is involved in soybean maturity and flowering. *Genetics* 188: 395–407. doi: 10.1534/genetics.110.125062

Wu C, Mozzoni LA, Moseley D et al (2020) Genome-wide association mapping of flooding tolerance in soybean. *Molecular Breeding* 40(1): 1–4.

Xia Z, Watanabe S, Yamada T et al (2012) Positional cloning and characterization reveal the molecular basis for soybean maturity locus E1 that regulates photoperiodic flowering. *Proc Natl Acad. Sci USA* 109(32): E2155–E2164.

Xie, M., Chung, C.YL., Li, MW. et al. (2019) A reference-grade wild soybean genome. *Nat Commun* 10, 1216.

Xie, L., Gong, X., Yang, K. et al. (2024) Technology-enabled great leap in deciphering plant genomes. *Nat. Plants* 10, 551–566.

Xu DH, Gai, JY (2003) Genetic diversity of wild and cultivated soybeans growing in China revealed by RAPD analysis. *Plant Breed* 122: 503–506.

Xu M, Xu Z, Liu B et al (2013) Genetic variation in four maturity genes affects photoperiod insensitivity and PHYA-regulated post-flowering responses of soybean. *BMC Plant Biol* 13: 91–104.

Yamaguchi-Shinozaki K, Shinozaki K. (2006) Transcriptional regulatory networks in cellular responses and tolerance to dehydration and cold stresses. *Annu Rev Plant Biol.* 57:781-803.

Yamanaka N, Ninomiya S, Hoshi M et al (2001) An informative linkage map of soybean reveals QTLs for flowering time, leaflet morphology and regions of segregation distortion, *DNA Res.* 8: 61–72.

Yang C, Jun Y, Shuqin J et al (2022) Resequencing 250 soybean accessions: New insights into genes associated with agronomic traits and genetic networks. *Genom. Proteom.* Bioinform. 20(1): 29–41.

Yin P, Ma Q, Wang H et al (2020) SMALL LEAF AND BUSHY1 controls organ size and lateral branching by modulating the stability of BIG SEEDS1 in *Medicago truncatula. New Phytologist* 226(5): 1399–1412. doi: 10.1111/nph.16449

Yu Z, Chang F, Lv W, Sharmin RA, Wang Z, Kong J, Bhat JA, Zhao T (2019) Identification of QTN and candidate gene for seed-flooding tolerance in soybean [Glycine max (L.) Merr.] using genome-wide association study (GWAS). *Genes* 10:957

Zhai H, Lü S, Liang S et al (2014) GmFT4, a homolog of flowering locus T, Is positively regulated by E1 and functions as a flowering repressor in soybean. *PLoS ONE* 9(2): e89030.

Zhang D, Wang X, Li S et al (2019) A post-domestication mutation, Dt2, triggers systemic modification of divergent and convergent pathways modulating multiple agronomic traits in soybean. *Molecular Plant* 12(10): 1366–1382. doi: 10.1016/j.molp.2019.05.010

Zhang, DY, Kumar, M, Xu, L et al (2017) Genome-wide identification of major intrinsic proteins in *Glycine soja* and characterization of GmTIP2;1 function under salt and water stress. *Sci Rep* 7: 4106. doi: 10.1038/s41598-017-04253-z

Zhang J, Xia C, Wang X et al (2013) Genetic characterization and fine mapping of the novel Phytophthora resistance gene in a Chinese soybean cultivar. *Theor. Appl. Genet* 126: 1555–1561

Zhang W, Liao X, Cui Y, Ma W, Zhang X, Du H, et al. (2019) A cation diffusion facilitator, *GmCDF1*, negatively regulates salt tolerance in soybean. *PLoS Genet* 15(1): e1007798.

Zhang Y, Liu Z, Wang X, Li Y, Li Y,Gou Z,et al.(2022) Identification of genes for drought resistance and prediction of gene candidates in soybean seedlings based on linkage and association mapping. *Crop J.* 10: 830–839.

Zhang C, Xie L, Yu H, Wang J, Chen Q, Wang H (2023) The T2T genome assembly of soybean cultivar ZH13 and its epigenetic landscapes. *Mol Plant*. 6;16(11):1715-1718.

Zeng A, Chen P, Korth K, et al. (2017) Genome-wide association study (GWAS) of salt tolerance in worldwide soybean germplasm lines. *Mol Breeding* 37: 30.

Zhong C, Sun S, Li Y, Duan C, Zhu Z (2017) Next-generation sequencing to identify candidate genes and develop diagnostic markers for a novel Phytophthora resistance gene, RpsHC18, in soybean. *Theor. Appl. Genet.* doi: 10.1007/s00122-017-3016-z

Zhong C, Sun S, Yao L, Ding J, Duan C, Zhu Z (2018) Fine mapping and identification of a novel phytophthora root rot resistance locus RpsZS18 on chromosome 2 in soybean. *Front Plant Sci* 9: 44. doi: 10.3389/fpls.2018.00044.

Zhou, Z, Jiang, Y, Wang, Z et al. (2015) Resequencing 302 wild and cultivated accessions identifies genes related to domestication and improvement in soybean. *Nat Biotechnol* 33: 408–414. doi: 10.1038/nbt.3096

Zhu YL, Song QJ, Hyten DL, Van Tassell CP, Matukumalli LK, Grimm DR, et al. (2003) Singlenucleotide polymorphisms in soybean. *Genetics*. 163: 1123–34.

Zhuang Y, Wang X, Li X et al. (2022) Phylogenomics of the genus *Glycine* sheds light on polyploid evolution and life-strategy transition. *Nat. Plants* 8: 233–244.

3 Allele Mining and Its Importance in Rapeseed

Sri Vidya K Gannavarapu, Jayanta Roy, and Sujan Mamidi

3.1 INTRODUCTION

Rapeseed (*Brassica napus* L., *AACC*, $2n = 4x = 38$) is a recent allopolyploid species formed by the hybridization event of two diploid species, *Brassica rapa* (Asian cabbage, AA, $2n = 2x = 20$) and *Brassica oleracea* (Mediterranean cabbage, CC, $2n = 2x = 18$) (Nagaharu 1935). It is estimated that the hybridization event took place about 7,500 years ago (Chalhoub et al. 2014). It is the second most important oil seed crop in the world, after soybean (Rahman et al. 2022), with an average of 72 million tons per year (FAO 2020). This is extensively grown in Canada, Europe, China, and Australia, with an increase in acreage over the last 15 years due to its increased demand for human consumption and biodiesel (Kirkegaard et al. 2020). It is also a rich source of protein feed next to soybeans in the world (Gatrell et al. 2014, Chao et al. 2017, Wang et al. 2018). In addition to providing vegetable oil and feed for animals, this crop has a major role in contributing to the diversification of rotations, improved ground cover, improving soil organic status, reducing weeds and pathogens, and improving soil microbiome conditions in wheat-based cropping systems (Chan and Heenan 1996, Hansen et al. 2018, Nelson et al. 2022).

B. napus after its formation ~7,500 years ago was initially cultivated for human consumption and animal fodder, *B. napus* subsp. *rapifera* (swede and rutabaga), and *B. napus* subsp. *pabularia* (fodder rape and Siberian kale), as illustrated in (Figure 3.1). Rapeseed was also domesticated as an oilseed crop (*B. napus* subsp. *oleifera*) after the industrial revolution (Allender and King 2010, An et al. 2019). Local adaptation for early maturity and heterosis and the development of 'double-low' (low glucosinolate (GSL) and erucic acid content in seeds) cultivars through breeding performed since the 1950s (Krzymanski 1978) have established rapeseed as a major global oil crop and greatly improved its industrial value and status.

Production systems of canola generally fall into three major development or "season" types or ecotypes globally.

a. The winter type is grown during the winter and is mainly cultivated in Europe, the Southern Plains of the United States of America (USA), and China. These winter-type canolas are seeded in the fall, vernalized over the winter to induce flowers, and harvested in the summer (Rahman and McClean 2013). This is the original type that started to be cultivated in Europe in the late Middle Ages (Hu et al. 2022).
b. Spring types of canola are mainly grown in Canada, the Northern Plains of the USA, and Australia. The spring canola is planted in the early spring and harvested in the late spring of the same growing season (Rahman and McClean 2013). This was developed around the year 1,700, with a short life-cycle period. No vernalization is required to induce flowering in these ecotypes (Lu et al. 2019, Hu et al. 2021).
c. Semi-winter-type canola is grown in winter in Australia, China, South Africa, and South America. The semi-winter type has a shorter period of vernalization to induce flowers (Wang et al. 2011, Kirkegaard et al. 2020). Canola when introduced to China in the 1930s–1940s produced this new ecotype, with moderate vernalization requirements after environmental adaptation and selection (Wei et al. 2017, Song et al. 2020).

DOI: 10.1201/9781003385035-3

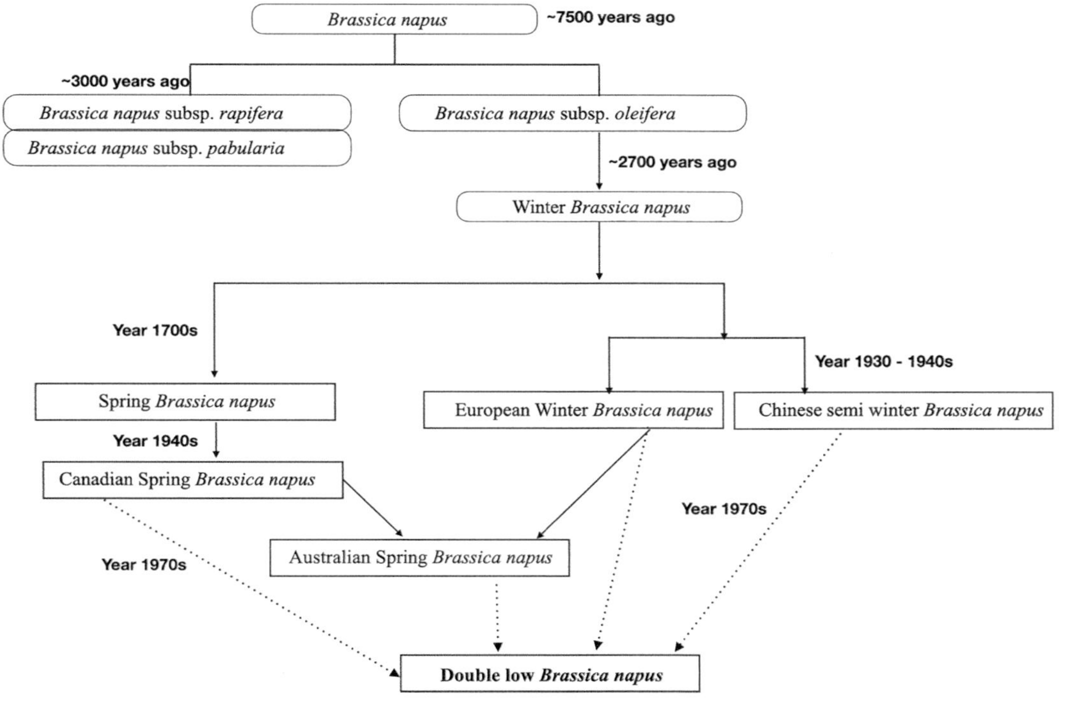

FIGURE 3.1 History of rapeseed (Cited from Hu et al. 2021).

Canada is the largest producer of rapeseed. More than 20 million metric tons of rapeseed were produced in 2018, with Saskatchewan province accounting for over half of production. Rapeseed production increased by 32.1% nationally to 18.2 million tons (Mts) in 2022. Yields increased by 37.6% to 37.7 bushels per acre, while harvested area fell 3.9% to 21.2 million acres. Rapeseed yield in Saskatchewan increased by 48.0% to 37.0 bushels per acre, offsetting the lower harvested area (–4.8% to 11.4 million acres), resulting in a 41.0% production increase to 9.5 Mts. In Alberta, rapeseed production was up by 28.8% to 5.6 Mts, due to higher yields (+31.8% to 38.1 bushels per acre) offsetting the lower harvested area (–2.2% to 6.5 million acres). In Manitoba, rapeseed production rose 14.4% to 2.9 Mts. Harvested area decreased by 5.0% to 3.2 million acres, while yields rose 20.5% to 39.4 bushels per acre (Statistics Canada 2022).

In the USA, rapeseed production increased 13.5-fold from the five-year average of 0.11 Mts (1991–1995) to the five-year average of 1.49 Mts (2015–2019; NASS 2022). At the same time, rapeseed oil consumption has increased rapidly in the last few years. Statistics shows that though rapeseed production increased, it was not enough to meet the growing demand. For this reason, every year, the USA imports a huge amount of rapeseed oil (2.50 Mts in 2019) from other countries (USDA 2020). In the USA, rapeseed production is restricted to the north-central region, and North Dakota (ND) is the leading canola-growing state, where 83% of US rapeseed is grown. The North Dakota State University (NDSU) rapeseed-breeding program could play a vital role in the rapeseed economy by developing high-yielding varieties, shortening the breeding cycle, and expanding canola-growing acreage.

Australia is the fifth-largest producer of rapeseed. In Australia, rapeseed is the third-largest crop, after wheat and barley. The country's rapeseed production has jumped nearly three-fold since 2000. In 2022–23, Australian rapeseed farmers are expected to harvest a record 3.4 million hectares, up 12% from last year. This is due to high rapeseed prices, which replaced acres planted with barley and oats. The crop is harvested in November and December. Rapeseed futures surged 59% in 2021 as soaring demand for vegetable oils amid lower supply sparked a global oilseed price rally.

Rapeseed, popularly known as mustard in India, has the highest oil content (35%–40%) among the country's nine main oilseeds, including groundnut, sunflower, and soybeans. Rapeseed output in India, the world's biggest importer of vegetable oils, is likely to rise by as much as 29.4% this year as farmers plant more area with the winter-sown oilseed. Rapeseed production in India was 8.5 Mts in the 2020–21 crop year and is expected to stand at 10–11 Mts in the 2022 crop year (Central Organization for Oil Industry and Trade; https://cooit.in/). This increase is necessary given that India spent a record 1.17 trillion rupees ($15.71 billion) on vegetable oil imports in 2020/21. Early maturing rapeseed crops are sown from mid-September to October, while the sowing of late-maturing varieties continues until mid-November. It is harvested from February onward.

Rapeseed is mainly cultivated for vegetable oil and animal feed, and the primary focus of the breeder is seed quality, seed oil content (SOC), and GSL content. Many of the rapeseed cultivars contain high GSL concentrations that reduce the nutritional value of the seed. Given the high importance of rapeseed as an oil seed crop, it is necessary to develop new varieties with low GSL content and durable resistance to stresses. This chapter deals with allele mining and its prospects in polyploid crops like rapeseed, various conventional and recent approaches to allele mining, and functional validation of potential alleles under study.

3.2 ALLELE MINING

The systematic research, identification, and characterization of new allelic variants for one or multiple genes in large and diverse collections are referred to as "***allele mining***" (Guerra et al. 2022). Allele mining can be used in the discovery of superior alleles, candidate genes, and transcripts that alter phenotypic trait expressions. With its origin and limited domestication/selection history, it is difficult to express and maintain variations for many important traits. To overcome this problem, allele mining can be visualized as a potential technique in rapeseed breeding to exploit and maintain diversity, provide broad spectrum resistance to various abiotic and biotic stresses, and identify various agronomically important traits.

3.2.1 Formation of New Alleles

Mutations are the fundamental cause of the formation and discovery of new alleles. Single nucleotide polymorphism (SNP), insertions and deletions (Indels), transversions, transitions, and other structural variations (SVs) lead to huge allelic diversity among various cultivars, landraces, and wild types. Allelic diversity in the coding or exonic areas is considered to be efficient as it alters the protein structure and expression of the phenotype. Many recent discoveries in the field of genomics led to the conclusion that mutations in the coding regions alone do not contribute to allelic expression and diversity. Variations in the intronic and non-coding regions, promoter sites, and others are also involved in the trait expression and allelic diversity studies, which are denoted as true allele mining. The discovery of these alleles and their functional annotation and validation would provide opportunities for breeders and geneticists to take advantage of true allele mining by implementing marker-assisted selection (MAS), thereby shortening the breeding cycle for the trait of interest.

3.2.2 Private Allele

Alleles that are found in only one of several populations in a larger collection are referred to as private alleles. The frequency of private alleles has been used to determine the population structure and the rate of gene flow. When one marker is located at more than one locus, alleles are regarded as belonging to the same locus if they show an obvious co-dominant relationship from their segregation patterns. When amplicons of alleles were very close in molecular size and belonged to distinctive stuttering patterns, they were considered allelic. As most multi-locus markers detected two homoeologous loci in the A and C genomes of *B. napus*, it was often possible to infer from the other species of Brassica, including *B. rapa* (AA genome) and *B. oleracea* (CC genome), whether a particular locus belonged to the A or C genome. Private alleles are indicators of both genetic diversity and genetic distinctiveness.

3.2.3 Role of Allele Mining in Rapeseed Breeding

Plant breeders have until recent times, mostly focused on the selection of only a few candidate genes and cultivars, which led to the loss of genetic variation. There might always be some useful potential alleles and genes in the unexplored regions of the rapeseed genome that are very useful in the development of multiple traits with desirable qualities (deVincente and Tanksley 1993, Xiao et al. 1996, McCouch et al. 2007). Therefore, the discovery and mining of potential alleles improve the genetic variation and development of new cultivars.

There is an exacerbating effect of abiotic stresses on the yield of rapeseed. Drought, heat, carbon assimilation, salinity, cadmium toxicity, and high temperatures are the major abiotic stresses in rapeseed. There is a divergence among these stresses that leads to defective seed yield and oil quality. Drought affects carbon assimilation by limiting diffusion by stomata. Similarly heat impairs carbon dioxide assimilation. High temperatures cause protein denaturation and enzyme inactivation. Breeding for the improvement of these characters by molecular approaches may lead to the development of tolerant rapeseed against stresses. To achieve this, allele mining can be considered one of the best techniques for the identification of true alleles and enhancing tolerance. However, allele mining has its challenges in rapeseed, given the extensive population structure of rapeseed based on growth habits (also linked geographically) and limited population history that also leads to limited diversity.

3.3 ALLELE MINING APPROACHES

For the use of allele mining in breeding and genetics, there are three major steps (Figure 3.2):

1. Identification in populations
2. Validation in additional populations
3. Deployment into breeding programs

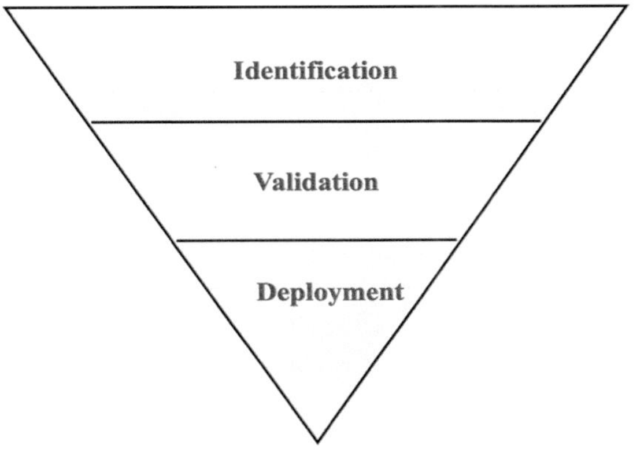

FIGURE 3.2 Three major steps for allele mining in plant breeding.

A typical polyploid crop like rapeseed contains a gene in its multiple copies, which makes it difficult to identify the loss or gain of function by a single mutation at one locus. Considering the mentioned aspects, novel allele mining approaches and their validation need to be practiced for the discovery of functional alleles. Conventional and sequence-based methods are the major approaches in allele mining.

3.3.1 Conventional Allele Mining Approaches

Breeders used two traditional allele mining approaches, *viz.*, targeting induced local lesions in genomes (TILLING) and ecotilling, which cause mutations and lead to the identification of novel alleles in biparental populations and natural populations, respectively.

3.3.1.1 Tilling

Tilling is one of the reverse genetic approaches that screens the large, mutagenized population for the discovery of key alleles. The main advantage of tilling is that it can be applicable to any plant species and can be used to analyze large genomes that are often difficult using insertional mutagenesis (Kurowska et al. 2011). It involves the pooling up of deoxyribose nucleic acid (DNA) from the mutated population and the identification of multiple variations of the gene using CEL1 endonuclease, which recognizes the mismatches in double-stranded DNA (dsDNA). The nonsense/splice junction mutations, or SNPs, result in the loss of function of alleles that alter gene function, whereas missense mutations lead to a partial loss of gene activity by altering a range of alleles. Thus, the tilling approach has found enormous applications in the field of functional genomics. Raghavan et al. (2006) described agarose-based gel detection of SNP haplotypes in tilling. Here, they scanned amplicons of 3kb in agarose gels, which helps the breeders with less equipment like polyacrylamide gel electrophoresis (PAGE) for SNP haplotyping. Though the resolution is low, this is an effective method to map candidate genes from parental lines that show low levels of polymorphism.

3.3.1.2 Ecotilling

A modified tilling approach identifies the allelic variations by detecting the variations from naturally occurring alleles in primary (GP1) and secondary gene pools (GP2) (Comai et al. 2004, Comai and Heinkoff 2006). This method generally relies on the pooling of DNA from the natural variant population, followed by the enzymatic cleavage of heteroduplex DNA with a single-strand-specific nuclease, and its detection through Li-cor genotypes. Modifications in the EcoTILLING led to the identification of SNPs and novel allelic variations in the natural population by sequencing the amplicons.

Though these conventional methods are cost-effective and involve SNP discovery, the development of GP, and cleavage by endonuclease, detection using PAGE gels is cumbersome and a time-consuming process that can be replaced by following sequence-based allele mining approaches.

3.3.2 Sequence-Based Allele Mining

Sequence-based approaches use multiple variant types (SNP, indels, copy number variation (CNV), and other SVs), which are obtained by using next-generation sequencing techniques such as genotyping-by-sequencing (GBS), next-generation sequencing (NGS), arrays, chips, etc,, as mentioned in Figure 3.3. These are utilized in a variety of populations that have recombination accumulated over time. Sequence-based approaches help to analyze the variants and lead to the discovery of candidate genes. It is a highly sophisticated and precise technique over conventional approaches that include sequence information for pinpointed identification of allelic variations and key allele discovery without creating any mutations.

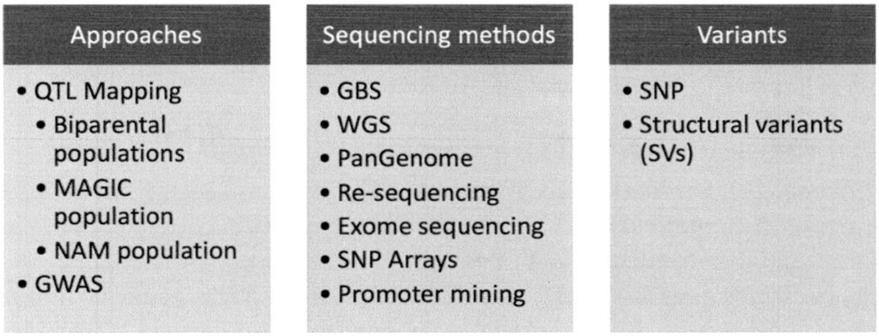

FIGURE 3.3 Outline of sequence-based allele mining.

3.3.2.1 Quantitative Trait Loci Mapping

3.3.2.1.1 Biparental Mapping

Linkage mapping involves the identification of desirable candidate genes and alleles using biparental populations like recombinant inbred lines (RILs), near isogenic lines (NILs), backcross (BCs), etc. In biparental populations, two inbred lines exhibiting contrasting phenotypes are usually crossed to produce one or more segregating progenies (Figure 3.4). Parents are selected based on their genetic and phenotypic diversity for a trait of interest, allowing for the detection of genomic regions associated with the target trait. Desirable quantitative trait loci (QTLs) and alleles are identified by studying the polymorphism between parents and also within the population by using specific markers for a trait. Populations derived from biparental crosses only capture a small picture of the genetic factors that affect target traits in the species. This method involves limited genetic diversity, which makes it vulnerable to discovering diverse and de novo alleles for various important traits.

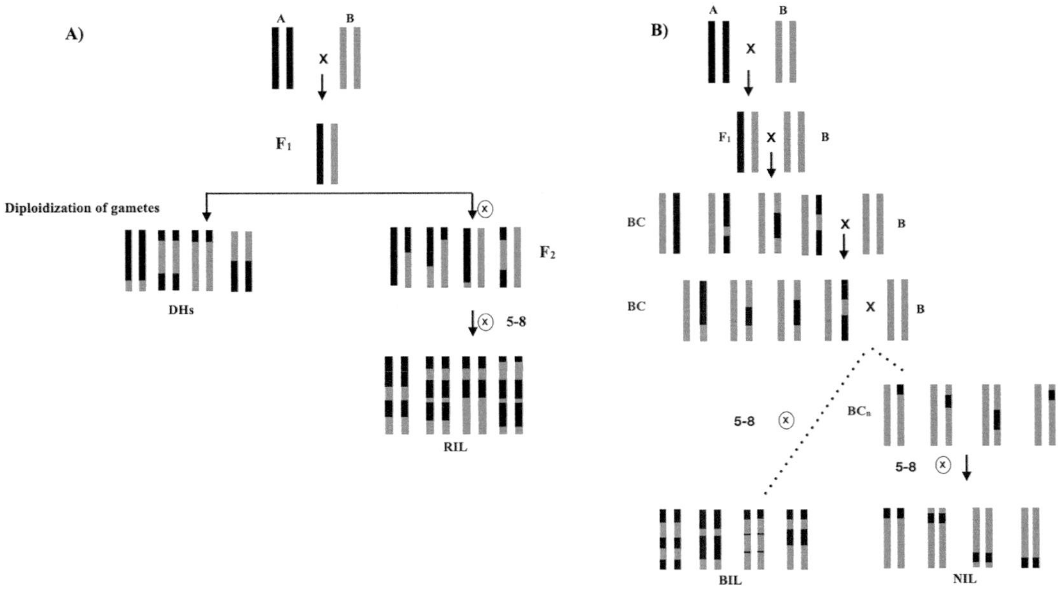

FIGURE 3.4 Parental founders P_1 and P_2 are contrasting for the trait/s of interest, and F_1 is the simple hybrid derived from founder cross: (a) doubled haploid (DHs), F_2, and RILs developed by successive generations of selfing. (b) BC, backcross inbred lines (BILs), and NILs obtained by successive generations of selfing (Cited from Arrones et al. 2020).

This shortage of diversity that is limited to both parents QTLs detected in a biparental population might not be expressed in other genetic backgrounds. Complex mapping populations can be developed as an alternative to this method in allele discovery studies.

3.3.2.1.2 Multi-Parent Advanced Generation Intercross Population

Multi-parent advanced generation intercross (MAGIC) populations are panels of RILs that are fine-scale mosaics with theoretically equal proportions of the multiple founder genomes (Figure 3.5). They have a larger number of recombinant events and are an intermediate standing between naturally occurring accessions and the simple, minimal recombining RILs. Multiple founders in MAGIC populations have higher allelic diversity when compared to those derived from biparental crosses. Also, multiple cycles of parental inter-crossing result in a set of rearranged genomes with a high level of recombination. This increases the power of QTL detection. However, the development and maintenance of these populations require significant time and are also highly laborious to maintain. There are three different ways of getting MAGIC populations.

3.3.2.1.3 Nested Association Mapping Populations

Nested association mapping (NAM) is a genome-wide trait dissection strategy that integrates the advantages of both linkage analysis and association mapping into a single population (Figure 3.6). Multiple RIL populations are developed by crossing one single male parent with a set of other parents. This leads to the identification of QTL with great precision and accuracy. These populations serve as a reservoir for genetic diversity.

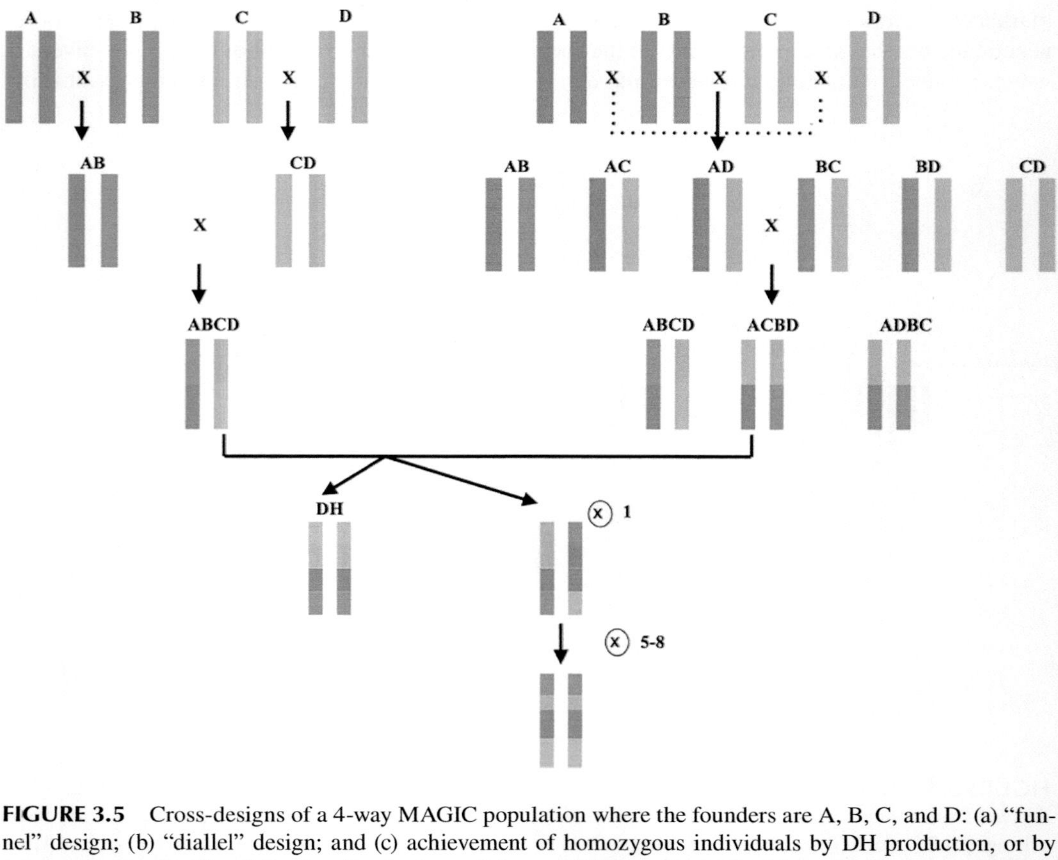

FIGURE 3.5 Cross-designs of a 4-way MAGIC population where the founders are A, B, C, and D: (a) "funnel" design; (b) "diallel" design; and (c) achievement of homozygous individuals by DH production, or by several rounds of selfing following the single-seed descent (SSD) method (Cited from Arrones et al. 2020).

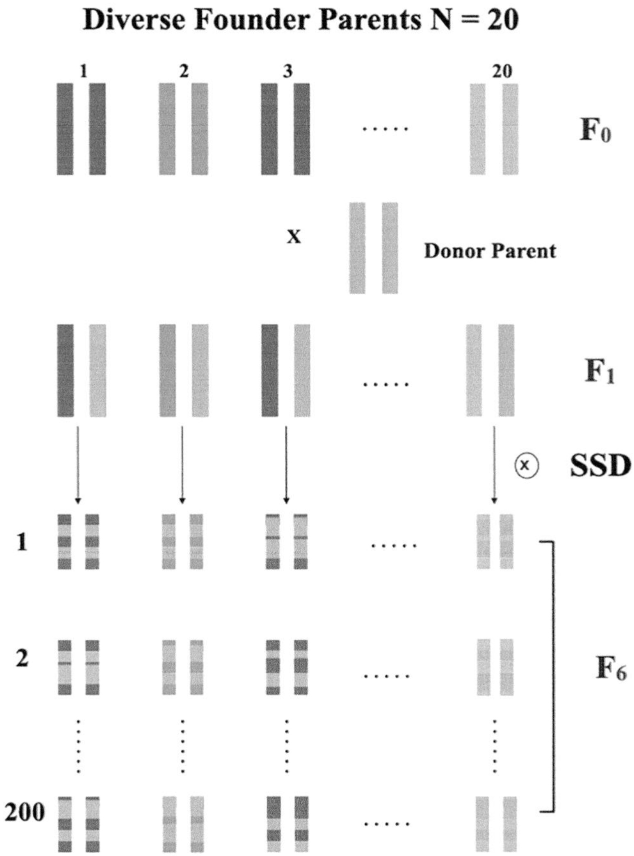

FIGURE 3.6 NAM population development

(Source: https://reworkhow.github.io/XSim.jl/case/NAM/).

3.3.2.2 Genome-Wide Association Studies

Genome-wide association studies (GWAS), also termed as linkage disequilibrium (LD) mapping, is useful to identify the genetic variations of a trait using the sequence information of large natural populations. Association mapping takes advantage of high-genetic diversity panels, including collections of selected individuals with unknown kinship and the historical accumulation of recombination events during thousands of generations. It is useful to study QTLs contributing to abiotic and biotic stresses and various agronomically important traits. The application of GWAS for the more precise mapping of novel genes related to complex traits in crop plants has been demonstrated in many studies. Phenotypic expression depends upon the alleles and gene-to-gene interactions, which are narrow in biparental populations such as RILs, NILs, and other crosses. Association mapping considers various recombinant events in the population under study, which is a major advantage in identifying rare and diverse alleles in recombinant populations (Figure 3.7). Greater genetic diversity can be observed over here, and many major and minor QTLs across the genome can be identified for various traits. It is performed to identify favorable alleles that promote high allelic diversities among various cultivars. However, the main limitations of GWAS are LD and population substructure. In addition, GWAS requires very large samples and many markers to have enough

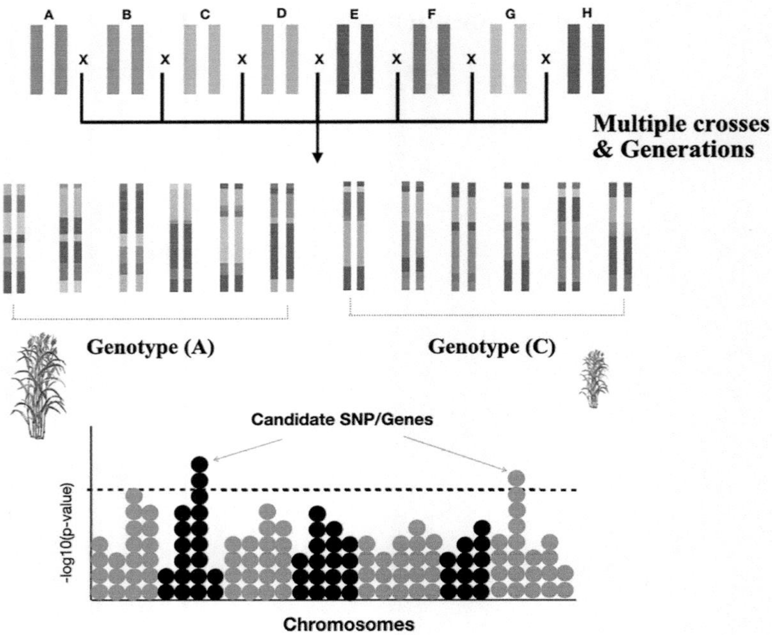

FIGURE 3.7 Historical recombination events and natural genetic diversity of germplasm collections used for GWAS (Cited from Arrones et al. 2020).

power to detect genomic regions of interest, and its efficiency is limited by unknown pedigrees and missing parental information.

Rapeseed has multiple biparental populations that are specifically developed for the trait of interest using contrasting parents. QTLs identified here are good only for the restricted germplasm. However, the QTL identified can be used for MAS in that breeding program. Association mapping uses diverse collections (natural collections or cultivars in the case of rapeseed). However, GWAS studies in rapeseed are limited given the significant population structure (by ecotype type). NAM has been recently developed for the spring type in Canada. This NAM population has about 2,500 lines (RILs). These are derived from crossing or recombining 50 diverse parental lines with one reference line (https://www.canolacouncil.org/). In China, a NAM population was developed that includes 2,425 F_6 RILs derived from crosses between the common parent Zhongshuang11 (ZS11) and 15 other diverse founder lines (semi-winter and spring inbred lines; Hu et al. 2018).

3.4 SEQUENCING METHODS FOR ALLELE MINING

3.4.1 GBS-Based Mining

Polyploid species contain multiple alleles with different dosages. Single-dose (SD) SNPs are the crucial molecular markers in polyploids that are useful in linkage analysis and QTL mapping because of their simple inheritance pattern. The percentage of SD SNPs decreases consistently with improved levels of ploidy because of the high heterozygosity in polyploid hybrids (Gracia et al. 2013). These SD SNPs and SVs can be identified easily by the GBS technique, which in turn helps polyploids like *Brassica* in the determination of novel alleles.

GBS is a sequencing method that reduces genome complexity by using restriction enzymes (REs). It is highly reproducible and can detect genome-wide SNPs without relying on previous sequence information (Elshire et al. 2011). The discovery of novel alleles using the GBS technique will be of great use in mining studies.

3.4.2 Whole Genome Sequencing

Next generation sequencing approaches help researchers explore unthinkable possibilities in allele mining and other studies. Advances in NGS have led to the development of whole genome assemblies using a combination of multiple long reads (Pacbio, Oxford Nanopore) and short read sequences. Whole genome sequencing (WGS) provides the DNA sequences for all the chromosomes. Additionally, annotation identifies the positions of genes and various DNA elements.

Sequencing a complete reference genome yields an exceptional resource of genetic information. The first reference genome sequence of the European winter-type cultivar "*Darmor-bzh*" was released in 2014 (Chalhoub et al. 2014). A whole genome shot gun approach using a short-read NGS strategy was adapted to sequence the *B. napus* genome. Recently, another high-confidence genome sequence of canola, 921.5 Mb, was derived from the cultivar ZS11 using PacBio Single Molecule Real Time (SMRT) sequencing data. The advent of high-throughput sequencing platforms and the genomic sequences help to identify various QTLs, markers, and candidate gene(s) that are part of selection during post-domestication and improvement of various resistant cultivars with low erucic acid content of canola (Chen et al. 2021).

3.4.3 Pan-Genome

A single reference genome sequence is not sufficient for identifying the best candidate genes for molecular breeding or for understanding the genomic background of a population due to the prevalence of genomic SVs. A pan-genome is the summary of the genomes of a species obtained by comparing many resequenced genomes of a species or, occasionally, a genus. A pan-genome can help us understand the size of a core genome (defined as the conserved part among the related genomes), the size of a pan-genome, and the amount and nature of variations within a species or genus, which improves our understanding of the evolution of a species/genus as well as of agronomic traits. Comparative genomic analysis using pan genome data helps in identifying the presence/absence variations (PAVs) and promoter regions of functionally important genes (Gao et al. 2019). In addition, aligning multiple whole genome sequences leads to SV identification. In addition, aligning resequencing libraries to a graph-based pan genome or a close progenitor will reduce false positives.

3.4.4 Resequencing

Resequencing is an NGS method where a sample is sequenced to a good depth, generally larger than GBS coverage. This data can be used in a variety of studies, like linkage mapping and association mapping, to identify variants close to casual variants. With higher depth and advanced software tools, we can identify several variants, like SNPs, indels, and other structural variants. Resequencing of the genome helps identify variants with high precision. Resequencing also helps identify haplotypes around beneficial alleles.

3.4.5 Exome sequencing

Exome sequencing is a sequencing technique where all the protein-coding regions of genes in a genome are sequenced. The first step in this technique is to extract DNA that encodes proteins.

These regions are known as exons and contribute to about 1% of the human genome. The second step is to sequence using the NGS. The main advantage of this sequencing is that the coding variants have a greater impact on the trait variation. Also, with limited sequencing, this method can cover all genic regions, whereas whole genome sequencing needs much more depth.

3.4.6 SNP Arrays

An SNP array is a type of DNA microarray that is used to detect polymorphisms within a population. SNP calls can be obtained directly without the need for bioinformatics work. These have a set of imprinted SNPs, which were developed based on previous SNP studies. These can easily be tested and deployed to large gene banks for a quick and efficient way of classification.

3.4.7 Promoter Mining

There is an increase in the number of studies that study gene regions, but the detection of promoter (regulatory) sites has not been given much importance so far. Promoters are the *cis* acting elements that interact with transcriptional factors and promote the transcription and expression of a gene. They are also involved in the upregulation and downregulation of gene expression. Targeting the discovery of associated alleles in these areas helps provide good insights in expression studies.

3.5 VARIANTS

3.5.1 SNPs

SNPs are the most common type of genetic variation among organisms. Each SNP represents a difference in a single nucleotide. For example, a SNP may replace the nucleotide cytosine (C) with the nucleotide thymine (T) in a certain stretch of DNA. Most commonly, SNPs are found throughout the genome, with the majority of them residing in between the genes. They act as biological markers to locate genes that are associated with disease. SNPs occurring in a gene or in a regulatory region play a direct role in trait variation. These can be obtained using GBS, NGS, SNP array, and exome sequencing.

3.5.2 SVs

SVs are large genomic alterations that can have a large effect and are considered more deleterious than SNPs. SVs in the genome can be in the form of indels, duplication, transposition, or the inversion of sequences (Figure 3.8). SVs are generally rare and occur only in a few individuals in a population. SVs form a great resource for GWAS (Spencer et al. 2009). Several studies indicate that they have a greater impact on phenotype than non-synonymous SNPs. SV can best be identified with long-read sequencing and synteny analysis.

3.6 ALLELE MINING FOR OIL CONTENT AND MAJOR DISEASES IN RAPESEED

3.6.1 SOC in *Brassica napus*

One of the major objectives of oilseed rape breeding is to increase the SOC and quality. SOC in *B. napus* is a quantitative trait controlled by many QTL/genes, each with a minor effect. In *B. napus*, an increase in seed yield of 2.3%–2.5% is equal to a 1% increase in oil content (Wang et al. 2013). Numerous linkage mapping studies and association mapping studies have been used to identify

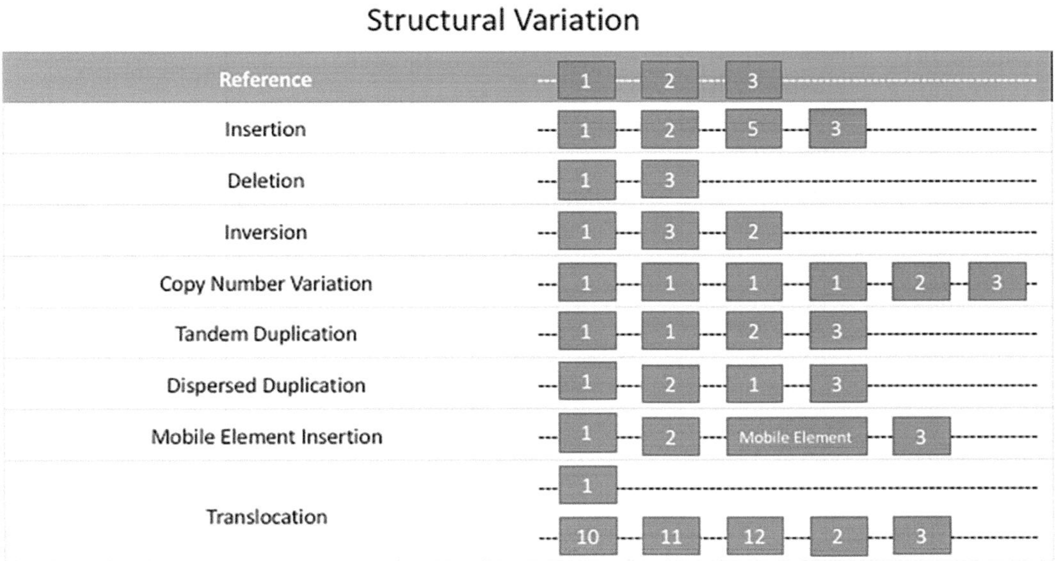

FIGURE 3.8 Depiction of different types of SVs compared to the reference genome. Each different number represents a different gene (Source: Wikipedia).

QTLs controlling SOC in *B. napus*. These identified QTL are distributed throughout the genome (Delourme et al. 2006, Wang et al. 2013, Jiang et al. 2014, Liu et al. 2016, Sun et al. 2016, Tang et al. 2021). Delourme et al. (2006) discovered 14 and 10 SOC-related QTLs in two DH populations, DY and RNSL, respectively, but only one common QTL was detected in both populations. Wang et al. (2013) identified 47 oil-related meta-QTLs on A1, A3, A6, A7, and C2 linkage groups, and Jiang et al. (2014) reported 46 distinct loci that control oil content, with loci on A1, A3, A8, C2, and C3 linkage groups. Sun et al. (2016) detected a total of 40 QTLs and 29 marker-trait associations by linkage and association mapping. Among these seven common QTLs located on chromosomes A02, A05, A08, A10, C02, and C05, they were identified using different populations and different environments. Liu et al. (2016) used GWAS with 521 diverse *B. napus* accessions and identified 50 significant loci associated with SOC. A novel locus located on A05 was validated using independent biparental linkage populations. This increases oil content by 1.5%–1.7%. In another GWAS study with 588 *B. napus* accessions, 17 QTL were identified. Of these, 12 SNPs were distributed on A3, five on C5, and C7 chromosomes, respectively (Xiao et al. 2019). Recently, Tang et al. (2021) used 505 diverse inbred lines and combined GWAS and transcriptome-wide association studies (TWAS). They detected 27 QTLs appearing in multiple environments, of which 13 QTLs are not identified earlier. A pair of homologous genes, *BnPMT6*, and two alleles were identified that have a negative regulation of SOC using multiple omics studies. The QTLs/genomic regions that were detected in multiple populations across diverse environments are good targets for fine mapping. This would facilitate opportunities for map-based cloning of the gene of interest.

3.6.2 Diseases in *Brassica napus*

Intense cultivation of *B. napus* to meet the increasing demands for oil and other uses has resulted in increased incidence and disease severity (Van de Wouw et al. 2016, Roy et al. 2021). However, the type of disease incidence varies from one geographic region to another. There are several pathogens that cause major diseases in *B. napus*, leading to substantial losses in seed yield and oil quality worldwide. These diseases include blackleg (*Leptosphaeria maculans*), Sclerotinia stem rot (SST) (*Sclerotinia sclerotiorum*), clubroot (CR) (*Plasmodiophora brassicae*), Alternaria blight (*Alternaria brassicae, A. brassicicola*), downy mildew (*Hyaloperonospora parasitica*), etc.

3.6.2.1 Blackleg

Blackleg or stem canker is a destructive fungal disease caused by the hemibiotrophic pathogen *Leptosphaeria maculans*. The nature of *L. maculans* resistance is both qualitatively and quantitatively inherited (Raman et al. 2012). Qualitative or R gene(s)-mediated resistance in *B. napus* is controlled by major gene effects, which are usually tested efficiently and accurately at the cotyledon stage. Conversely, quantitative resistance is controlled by several genes of small allelic effect, providing durable resistance at the adult plant stage under field environments (Amas et al. 2021). However, rapeseed breeders prefer quantitative resistance due to its durability and lower chance of resistance breakdown. Genetic mapping through QTL linkage mapping and GWAS approaches was used to identify the marker-trait associations associated with blackleg resistance. These studies revealed multiple QTLs distributed throughout 19 chromosomes of the *B. napus* genome (Raman et al. 2020). Jestin et al. (2015) used multiple-connected populations derived from crossing resistant parent Darmor and the other three resistant parents (Aviso, Canberra, and Grizzly) with a susceptible cultivar, Bristol, to compare the effects of different resistant alleles and to validate and know the physical location of QTL precisely. They detected a total of 17 QTL-spanning chromosomes A01, A02, A04, A06, A08, C01, C02, C04, C05, and C08 on conferring field resistance and four QTLs on A02, C02, C04, and C08 were found to be overlapped with the physical interval reported by Pilet et al. (1998). A combination of both QTL mapping using three segregating populations and GWAS analyses utilizing 166 winter-type diverse lines were performed to identify marker-trait associations for blackleg resistance (Kumar et al. 2018). QTL mapping resulted in the identification of 16 common QTLs with at least two biparental populations. Among these, one QTL on C04 was found to be overlapped between the populations, whereas nine QTLs were common to the previously reported QTL in the connected multi-parental design by Jestin et al. (2015). Moreover, five significant genomic regions detected by GWAS were found to be located within the support intervals of QTLs on chromosomes A06, A06, A09, C01, and C04. Most recently, Raman et al. (2020) performed a GWAS analysis using 421 diverse accessions of *Brassica* species and found a total of 59 significant SNP associations across 17 chromosomes of *B. napus*, and many of them were found to be overlapped within the vicinity of the physical positions of the previously identified marker-trait associations conferring blackleg resistance (Jestin et al. 2011, FopaFomeju et al. 2014, Larkan et al. 2016, Rahman et al. 2016, Kumar et al. 2018, Fikere et al. 2020). Furthermore, they also identified chromosomes, A04 and A08, as genomic hotspots for blackleg resistance based on the abundance of significant SNP detection.

3.6.2.2 Sclerotinia stem rot

SSR, caused by *Sclerotinia sclerotiorum* (Lib.) de Bary, is one of the most destructive fungal diseases affecting canola production worldwide (Bolton et al. 2006, Roy et al. 2022). The yield losses due to this pathogen vary from 10% to 80% from year to year depending on the disease severity, and every unit increase in SSR incidence imposes a 0.5%–0.7% reduction in canola seed yield (Del Río et al. 2007, Wu et al. 2016). Classical genetic analyses have shown that the nature of *S. sclerotiorum* resistance in *B. napus* is quantitatively inherited, controlled by polygenes with additive effect, and displays medium to high heritability (Zhao et al. 2006, Wu et al. 2013, Wei et al. 2016, Roy et al. 2021, 2022). Classic QTL mapping studies using biparental populations derived from resistant and susceptible parents have detected several QTLs on chromosomes A01, A02, A03, A06, A07, A08, A09, A10, C01, C02, C03, C04, C05, C06, C07, C08, and C09

(Zhao et al. 2006, Yin et al. 2010, Wu et al. 2013, Wei et al. 2014, Qasim et al. 2020, Shahoveisi et al. 2021). However, these QTL were specific to each study and varied with mapping populations, environments, inoculation methods, and developmental stages (Roy 2021). In only a few instances, common QTL were mapped in different mapping populations (Behla et al. 2017, Shahoveisi et al. 2021). To achieve high mapping resolution and to map the genetic markers/QTLs linked to the gene of interest, GWAS analyses utilizing diverse populations were performed in *B. napus* for *S. sclerotiorum* resistance. (Gyawali et al. 2016) used 84 simple sequence repeat markers to conduct a GWAS study using 152 *B. napus* accessions at the flowering stage in a controlled environment and identified 34 significantly associated loci, of which 21 alleles contributed to the SSR resistance. Recently, Roy et al. (2021) identified a total of 123 significant loci conferring adult plant stem resistance in multiyear field environments. They found a cluster of significant SNPs located on chromosome C06 (33.2–34.1 Mb) regions that were also reported as conserved QTL regions for SSR resistance (Li et al. 2015). These genomic regions show overlapping confidence intervals with the genomic regions/QTLs for stem resistance detected from GWAS (Wei et al. 2016, Wu et al. 2016) and QTL mapping (Zhao et al. 2006, Wu et al. 2013, Qasim et al. 2020, Shahoveisi et al. 2021). Moreover, they also reported a new genomic region on chromosome A09 (35.6–45.8 Mb) consisting of numerous significant SNPs, which was also identified for seedling stage resistance from a GWAS mapping study by (Roy et al. 2022). Recent identification of the overlapping genomic regions could provide confidence to narrow down the genomic regions and identify the putative candidate genes conferring SSR resistance that could guide the map-based cloning of the SSR resistant genes to facilitate future MAS.

3.6.2.3 Clubroot

CR is one of the most important soilborne diseases caused by a biotrophic pathogen. *Plasmodiophora brassicae* possesses a serious threat to cruciferous plants, especially *B. rapa, B. oleracea*, and *B. napus*. The yield loss in severely CR-infested rapeseed crops was estimated to range from 10% to 100% depending on the geographical location (Pageau et al. 2006, Strelkov et al. 2007). The inheritance of CR resistance mechanisms is complex in nature. Sources of *P. brassicae* resistance in A-genome (*B. rapa*) have been particularly promising as they are race-specific, controlled by a major dominant gene (Voorrips and Visser 1993, Hirani et al. 2018), while C-genome (*B. oleracea*) resistance is broad-spectrum, often controlled by various combinations of major and minor genes (Diederichsen et al. 2009, Hasan et al. 2021). To date, a total of 24 loci conferring CR resistance have been mapped on the seven chromosomes A01, A02, A03, A05, A06, A07, and A08 of A-genome of *B. rapa* using different pathotypes/races of *P. brassicae* (Lee et al. 2002, Suwabe et al. 2006, Sakamoto et al. 2008, Yu et al. 2017, Nguyen et al. 2018, Pang et al. 2018, Zhu et al. 2019, Karim et al. 2020). However, the majority (11 of 24) of these loci were mapped on A03 and five on A08 chromosome. Two major clusters of CR-resistant genes were found on A03-first cluster spanning 15.5–16.3 Mb regions containing three resistant CR loci, CRd, CRk, and Crr3, and the second cluster at about 23.8–26.8 Mb regions harboring seven CR loci, CRa, CRb, CRq, Rcr1, Rcr2, Rcr4, and Rcr5 (Hasan et al. 2021). On the other hand, multiple loci with major or minor effects were reported for clubroot resistance in all C-genome chromosomes. Breeding efforts were made to introgress CR-resistant alleles from the secondary gene pool (*B. rapa* and *B. oleracea*) to the oilseed rape/canola (*B. napus*). However, incorporating resistance into the crop germplasm from the secondary pool was complex due to the difficulties of producing viable hybrids, which are more frequent in *B. oleracea* than *B. rapa* (Rahman 2013, Rahman et al. 2014). Therefore, the use of GP1 as a source of resistant alleles would be more effective in developing CR-resistant oilseed rapeseed cultivars. Several QTLs/SNPs associated with CR resistance have been mapped to the A01, A02, A03, A08, A10, C02, C03, C05, C06, C07, and C09 in *B. napus*. Manzanares-Dauleux et al. (2000) mapped a major dominant gene, Pb-Bn1 on A04, and two small-effect QTLs on A4 and C5 in a winter type dwarf Darmor-bzh*B. napus* line. Hasan and Rahman (2016) used rutabaga-derived mapping populations and identified a genomic region of chromosome A08 conferring resistance to five *P. brassicae* pathotypes. Fredua-Agyeman et al. (2020) evaluated 124 rutabaga-type *B. napus* accessions against 13 *P. brassicae* pathotypes to perform a GWAS analysis and revealed chromosomes A03 and A08 as genomic hotspots associated with resistance to different pathotypes. Dakouri et al. (2021) performed a GWAS using 177 diverse *B. napus* accessions and identified 13 significant SNP loci, with 9 SNPs mapped to the A-genome and 4 SNPs to the C-genome. However, none of the significant SNPs were found to be overlapped with the reported genomic regions by Fredua-Agyeman et al. (2020). Even though in most of the cases, A-genome appeared to carry more QTL than C-genome for CR resistance, resistant alleles harbored in C-genome provide opportunities for the genetic improvement of CR resistance in rapeseed.

3.7 FUNCTIONAL VALIDATION OF IDENTIFIED ALLELES IN CANOLA AND OTHER CROPS

Validation of the discovered alleles or genes plays a key role in describing their functionality and identifying the link between the phenotypic and genetic architecture of the concerned traits governed by these alleles/genes.

3.7.1 Tilling and EcoTILLING in Canola

A polyploid crop like rapeseed contains a complex genome, and each gene is present in multiple copies. A mutation in a single locus cannot identify the loss of function of the gene. In this scenario, a reverse genetic approach like ecotilling can independently identify the mutations in the homologous genes and identify the function of the gene.

The *Brassica napus* TILLING project (*B. napus* TILLING data) was initiated and funded by Genome Canada/Genome Alberta (https://www3.botany.ubc.ca/can-till/Projects.html#napus), focusing on identifying various mutations in the canola population of approximately 3,000 EMS lines. The rate of mutation is 1 mutation per 100 kb. It is also useful in the identification of null and deleterious alleles for each gene.

An ethyl methane sulfonate (EMS) population of 3,158 lines of *B. napus* cultivar DH12075 has been generated and screened for mutations in 1,920 lines by Gilchrist et al. (2013). A total of 432 unique mutations in 26 genes have been reported, thus providing multiple mutant alleles for each gene, which can be utilized in the development of various agronomically important traits.

3.7.2 QTL Mapping

Zhou et al. (2021) conducted a detailed experiment in an elite inbred cultivar (ZS11) for allele mining of alleles contributing to five agronomic traits like high SOC, low seed glucosinolate content (SGC), long siliques, and stable yield. The DH population was developed from ZS11 into 195-14A, a cytoplasmic male sterile line. A high-density genetic map has been constructed using the Brassica 60k SNP array and microsatellite repeats. When tested in seven different environments, 71 consensus QTLs were identified by meta-analysis, of which 8 were assigned as major QTLs. A total of 72 candidate genes for 5 different traits have been reported, which helps in the utilization of favorable alleles for the development of rapeseed cultivars.

3.7.3 GWAS in Canola

GWAS in rapeseed breeding enhanced resistance to SSR by identifying alleles linked to SSR resistance in diverse lines. Gyawali et al. (2016) performed GWAS for SSR resistance in 152 *B. napus* accessions and were able to identify 21 alleles contributing to resistance.

Recently, Li et al. (2014) reported marker trait associations for the genetic architecture of seed weight and seed quality in canola using Illumina Brassica 60k SNP array. Xu et al. (2016) identified 41 SNPs correlated with flowering time. Wang et al. (2016) conducted GWAS for 448 canola-inbred lines and identified 40 QTLs associated with flowering time, 117 genomic regions related to divergent growth habits, and 224 flowering genes. The reported elite alleles can be utilized by researchers in rapeseed breeding for the development of various traits.

GWAS methods have a lot of obstacles that lower the false positive rate, which affects the identification accuracy of alleles. The restricted two-stage multi-locus, multi-allele GWAS (RTM-GWAS) was developed to explore whole genome QTL alleles controlling various traits in plants. This procedure is applied in cotton to identify the QTL alleles controlling fiber content yield traits (Su et al. 2020).

Jan et al. (2019) conducted haplotype-based GWAS to capture haplotypes for improved efficiency. He suggested that epistatic interactions and additive effects between alleles at different marker loci contribute to diversity for flowering and seed GSL traits. Qian et al. (2017) mentioned that 3097 conserved haplotype blocks have been identified in rapeseed, which contain major and minor genes contributing to agronomically important traits.

3.7.4 Whole Genome and Pan Genome

Brassica napus is an allotetraploid with a complex genome with a genome size of 1130 Mb (Chalhoub et al. 2014). Since the first genome assembly of B. napus was released, the high-quality genomes of five cultivars have also been revealed using advanced sequencing technologies. This led to improvements in the size of the assembly from 634.19 Mb to 1,008 Mb and an average N50 scaffold size from 777.3 Kb to 57.88 Mb, respectively (Hu et al. 2021). Song et al. (2020) de novo assembled and annotated eight *B. napus* accessions. Using comparative analysis, millions of small variations and 77.2–149.6 Mbp presence and absence variations (PAVs) were identified. More than 9.4% of the genes contained large-effect mutations or structural variations. Recently, a PAV-based genome-wide association study (PAV-GWAS) was performed to identify causal structural variations for silique length, seed weight, and flowering time. These genomic analyses showed that the A and C genomes of *B. napus* are highly homologous, and homoeologous exchange events (HEs) are ongoing processes that have greatly promoted the generation of novel, large-scale structural variation in *B. napus* (Chalhoub et al. 2014).

The pan genomes of Brassica provide insights on the allele variations leading to differential gene expression for a particular trait within species. It also provides high coverage for precise characterization and prediction of gene content within species. Also, pangenomes provide a huge repertoire of SVs. These pan genome resources help for a better understanding of the genome architecture and have the ability to accelerate the genetic improvement of *B. napus* (Table 3.1).

3.7.5 Resequencing

Hu et al. (2022) used whole-genome resequencing of 418 diverse rapeseed accessions to identify the genomic basis for the selection of adaptation and agronomic traits. GWAS identified 628 associated loci-related causative candidate genes for 56 agronomically important traits, including plant architecture and yield traits. The identified nonsynonymous mutations in candidate genes for agronomic traits have significant differences in allele frequency distributions across the improvement process. Lu et al. (2019) resequenced 588 *B. napus* accessions. By integrating genome-wide association studies, selection signals, and transcriptome analyses, they identified genes associated with improved stress tolerance, oil content, seed quality, and ecotype improvement.

TABLE 3.1
List of Pangenomes Available

Cultivars	Assembly Genome (Mb)	Anchored Genome (Mb)	Scaffold 50 (Mb)	Gene Number	Reference
Darmor-*bzh*v4.1	849.7	646	0.764	101,040	Chalhoub et al. (2014)
Darmor-*bzh*v8.1	850.28	798.95		80,382	Bayer et al. (2017)
Darmor-*bzh*v10	924	867	50.2	108,190	Rousseau-Gueutin et al. (2020)
Tapidor v6.3	636.4	627.9	0.197	70,162	Bayer et al. (2017)
ZS11	976	855	0.602	101,942	Sun et al. (2017)
NY7	994	892	6.91	104,179	Zou et al. (2019)
Express 617	925	765	4.882	89,857	Lee et al. (2020)
ZS11	1,160	921.5	53.2	106,059	Chen et al. (2021)
ZS11	1,008	960.8	57.88	100,919	Song et al. (2020)
Westar	1,007	942	55.24	97,514	Song et al. (2020)
No2127	1,011	932	53.9	95,385	Song et al. (2020)
Zheyou7	1,015	924	50.9	96,209	Song et al. (2020)
Gangan	1,033	934	46.68	96,843	Song et al. (2020)
Shengli	1,001	930	52.53	94,586	Song et al. (2020)
Tapidor	1,014	932	52.41	96,117	Song et al. (2020)
Quinta	1,003	933	55.9	95,492	Song et al. (2020)

These data were stored in public platforms like the Brassica Database (http://39.100.233.196), the *Brassica napus* Genome Browser (https://wwwdev.genoscope.cns.fr/brassicanapus/) (Chalhoub et al. 2014), BnPedigome (http://ibi.zju.edu.cn/bnpedigome/index.php) (Zou et al. 2019), BnaSNPDB (http://rapeed.zju.edu.cn:3838/bnasnpdb) (Yan et al. 2020), and BnPIR (http://cbi.hzau.edu.cn/bnapus/) (Song et al. 2021). High-throughput genotyping by whole-genome sequencing, target segment sequencing and SNP-chip array are ongoing for Brassica germplasm. With the advancements in technology, these data resources are becoming increasingly inexpensive and convenient for exploring Brassica pan genomic variation, and especially for identifying and introducing favorable genes and alleles into the *B. napus* GP.

3.8 ALLELE MODIFICATIONS

Allelic variations and polymorphisms are the causal variations for genetic diversity and gene expression. Major allelic modifications have been described in detail in the following sections.

3.8.1 Allele Replacement

Complex traits in higher plants are conferred by single nucleotide substitutions, changes in gene structure, and an altered level of gene expression (Hua et al. 2019). SNPs, transposon insertions, and other genetic modifications are the causal agents of mutations in plants. It is considered that many of the complex traits and crop improvement traits are governed by a large number of minor effect QTL alleles, which are likely organized in an epistatic network. Thus, targeted replacements with new alleles and editing of alleles could facilitate breeding without linkage drag and the development of allelic variations that do not occur naturally, which are required to address the complexity of the plant genome.

3.8.2 Prime Editing

A recently developed technique that enables all possible 12 transitions and transversions of DNA bases, DNA insertions, and deletions (Anzalone et al. 2019). Prime editings (PEs) are composed of a cas9 nickase (H840A), which fuses with the reverse transcriptase protein, and a PE guide RNA (pegRNA) that contains a reverse transcription (RT) template encoding desired edits and a prime binding site (PBS) for hybridization of 3^1 end of the nicked DNA strand to initiate RT.

A surrogate plant prime editor 2 (pPE2) system was tested in rice and led to the generation of T0 lines of up to 31.3% frequencies (Xu et al. 2020). In a different study based on the PE3 platform, N-terminal fusion of the M-MLV-RTR to Cas9 nickase (H840A) yielded average editing efficiencies up to 24.3% across the 13 target sites in rice when the pegRNAs contained RT sequences with multiple-nucleotide substitutions (Butt et al. 2020). Improvements in REs like paired pegRNA strategy, use of engineered pegRNAs with 3^1 extensions to minimize pegRNA degradation (Liu et al. 2021, Nelson et al. 2021), manipulation of cellular DNA repair pathway through activation of strategic silent mutations near the intend-ended edit regions.

3.8.3 Base Editing

A precise modification of genomes at a single base level without the requirement of double-stranded breaks (DSBs). Majorly available base editors in plants are categorized into cytosine BEs (CBEs), adenine BEs (ABEs), C-to-G BEs (CGBEs), dual-BE, and organelle BEs. CBEs and ABEs are fusions of catalytically impaired cas9 nucleases with single-stranded DNA (ssDNA)-specific deaminases that facilitate the C:G to T:A and A:T to G:C base transitions, respectively (Hess et al. 2016, Komor et al. 2016, Ma et al. 2016, Nishida et al. 2016). To improve the efficiency, precision, and specificity of BEs utilization of engineered deaminase, alter the spatial location of the deaminase, reduce the indwelling of CBEs by increasing the copy of UGI, reduce off-target effects by employing deaminase with a lower affinity for ssDNA, and prevent epigenetic modification by utilizing engineered deaminase.

3.9 FUTURE PROSPECTS

Allele mining has been instrumental in handling genomic resources and the exploitation of allelic variations, which is useful in the development of elite cultivars. Sequence-based allele mining relies mostly on approaches like haplotype-based mining, exome-based mining, and pan genome-based mining, which hastens the discovery of target-based allelic variations for improving crops with novel allelic diversity. Deployment of alleles in gene banks after validation serves as a repository to maintain allelic diversity. Developing molecular profiles of the alleles in gene banks serves as a guide in the identification of the evolutionary potential of key alleles and associated target candidate genes. Targeting and manipulating these genes helps in the maintenance of the diversity among cultivars, which is lost through rapid selection, genetic drift, spontaneous mutations, and domestication. Deleterious alleles may affect the expression of targeted traits and thus cause variation at the phenotype level. Conventional breeding approaches are not that efficient at handling these deleterious alleles. Rapid improvements in genome editing provide a way to handle these deleterious alleles. Clustered regularly interspaced palindromic repeats (CRISPR)-based multiplex editing in the targeted regions serves as a potent solution to get rid of deleterious alleles.

3.10 CONCLUSION

Increased demand for rapeseed oil production requires continuous improvement in rapeseed breeding for elite cultivars, despite changes in climate and growth conditions. Exploitation of variations among cultivars requires an in-depth understanding of phenotype architecture and the genetic makeup of the crop. In polyploids like rapeseed, multiple alleles govern the major agronomically important traits. Getting insights into allelic diversity patterns in these crops plays an important role in handling the variations that affect phenotypic expression. Hence, widespread knowledge on allele mining, various strategies of mining, and innovative technologies like genome editing optimize research efforts in developing cultivars with enhanced trait qualities.

REFERENCES

Allender CJ, King GJ (2010) Origins of the amphidiploid species *Brassica napus* L. investigated by chloroplast and nuclear molecular markers. *BMC Plant Biol* 10:54

Amas, JR, Anderson, DE, Cowling W, Batley J (2021) Status and advances in mining for blackleg (*Leptosphaeria maculans*) Quantitative Resistance (QR) in oilseed rape (*Brassica napus*). *Theor Appl Genet* 134(10):3123–3145

An H, Qi X, Gaynor ML, Hao Y, Gebken SC, Mabry ME, McAlvay AC, Teakle GR, Conant GC, Barker MS, Fu T, Yi B, Chris Pires J (2019) Transcriptome and organellar sequencing highlights the complex origin and diversification of allotetraploid *Brassica napus*. *Nat Commun* 28(10):2878

Anzalone AV, Randolph, PB, Davis JR, Sousa AA, Koblan LW, Levy JM, Chen PJ, Wilson C, Newby GA, Raguram A, Liu DR (2019) Search-and-replace genome editing without double-strand breaks or donor DNA. *Nature* 576(7785):149–157

Arrones A, Vilanova S, Plazas M, Mangino G, Pascual L, Diez MJ, Prohens J, Gramazio P (2020) The dawn of the age of multi-parent MAGIC populations in plant breeding: Novel powerful next generation resources for genetic analysis and selection of recombinant elite material. *Biology* 9(8):229

Bayer PE, Hurgobin B, Golicz AA, Chan CK, Yuan Y, Lee H, Renton M, Meng J, Li R, Long Y, Zou J (2017) Assembly and comparison of two closely related *Brassica napus* genomes. *Plant Biotechnol J* 15(12):1602–1610

Behla R, Hirani AH, Zelmer CD, Yu F, Fernando WGD, McVetty PBE, Li G (2017) Identification of common QTL for resistance to *Sclerotinia sclerotiorum* in three doubled haploid populations of *Brassica napus* (L.). *Euphytica* 213:260

Bolton MD, Thomma BPHJ, Nelson BD (2006) *Sclerotinia sclerotiorum* (Lib.) de Bary: Biology and molecular traits of a cosmopolitan pathogen. *Mol Plant Pathol* 7(1):1–16

Butt H, Rao GS, Sedeek K, Aman R, Kamel R, Mahfouz M (2020) Engineering Herbicide Resistance via Prime Editing in Rice. *Plant Biotechnol J* 18(12):2370–2372

Canola Council of Canada (https://www.canolacouncil.org/)

Central Organization for Oil Industry and Trade: (https://cooit.in/)

Chalhoub B, Denoeud F, Liu S, Parkin IAP, Tang H, Wang X, Chiquet J, Belcram H, Tong C, Samans B, Corréa M, Silva CD, Just J, Falentin C, Koh CS, Clainche IL, Bernard M, Bento P, Noel B, Labadie K, Alberti A, Charles M, Arnaud D, Guo H, Daviaud C, Alamery S, Jabbari K, Zhao M, Edger PP, Chelaifa H, Tack D, Lassalle G, Mestiri I, Schnel N, Le Paslier MC, Fan G, Renault V, Bayer PE, Golicz AA, Manoli S, Lee TH, Thi VHD, Chalabi S, Hu Q, Fan C, Tollenaere R, Lu R, Battail C, Shen J, Sidebottom CHD, Wang X, Canaguier A, Chauveau A, Bérard A, Deniot G, Guan M, Liu Z, Sun F, Lim YP, Lyons E, Town CD, Bancroft I, Wang X, Meng J, Ma J, Pires JC, King GJ, Brunel D, Delourme R, Renard M, Aury JM, Adams KL, Batley J, Snowdon RJ, Tost J, Edwards D, Zhou Y, Hua W, Sharpe AG, Paterson AH, Guan C, Wincker P (2014) Early allopolyploid evolution in the post-Neolithic *Brassica napus* oilseed genome. *Science* 345(6199):950–953

Chan KY, Heenan DP (1996) The influence of crop rotation on soil structure and soil physical properties under conventional tillage. *Science Direct* 37(2-3):113–125

Chao H, Wang H, Wang X, Guo L, Gu J, Zhao W, Li, B, Chen D, Rabonatahiry N, Li M (2017) Genetic dissection of seed oil and protein content and identification of networks associated with oil content in *Brassica napus*. *Sci Rep* 7:46295

Chen X, Tong C, Zhang X, Song A, Dong W, Chen F, Wang Y, Tu J, Liu S, Tang H, Zhang L (2021) A high-quality *Brassica napus* genome reveals expansion of transposable elements, subgenome evolution and disease resistance. *Plant Biotechnol J* 19(3):615–630

Comai L, Heinkoff S (2006) TILLING: practical single-nucleotide mutation discovery. *Plant J* 45:684–694

Comai L, Young K, Till BJ, Reynolds SH, Greene EA, Codomo CA (2004) Efficient discovery of DNA polymorphisms in natural populations by Ecotilling. *Plant J* 37:778–786

The Daily-Production of principal field crops (2022) Statistics Canada

Dakouri A, Lamara M, Karim Md M, Wang J, Chen Q, Gossen BD, Strelkov SE, Hwang SF, Peng G, Yu F (2021) Identification of resistance loci against new pathotypes of *Plasmodiophora brassicae* in *Brassica napus* based on genome-wide association mapping. *Sci Rep* 11(1):6599

Delourme R, Falentin C, Huteau V, Clouet V, Horvais R, Gandon B, Specel S, Hanneton L, Dheu JE, Deschamps M, Margale E, Vincourt P, Renard M (2006) Genetic control of oil content in oilseed rape (*Brassica napus L.). Theor Appl Genet* 113(7):1331–1345

Del Río LE, Bradley CA, Henson RA, Endres GJ, Mckay K, Halvorson M, Porter PM, Gare DGL, Lamey HA (2007) Impact of sclerotinia stem rot on yield of canola. *Plant Dis* 91(2):191–194

deVincente MC, Tanksley SD (1993) QTL analysis of transgressive segregation in an interspecific tomato cross. *Genetics* 134(2):585–596

Diederichsen E, Frauen M, Linders EGA, Hatakeyama K, Hirai M (2009) Status and perspectives of clubroot resistance breeding in crucifer crops. *J Plant Growth Regul* 28(3):265–281

Elshire RJ, Glaubitz JC, Sun Q, Poland JA, Kawamoto K, Buckler ES, Mitchell SE (2011) A robust, simple genotyping-by-sequencing (GBS) approach for high diversity species. *PLoS One* 6(5):e19379

FAO, (2020) FAOSTAT. Food and Agriculture Organization of the United Nations, Rome

Fikere M, Barbulescu DM, Malmberg MM, Spangenberg GC, Cogan NOI, Daetwyler HD (2020) Meta-analysis of GWAS in canola blackleg (*Leptosphaeria maculans*) disease traits demonstrates increased power from imputed whole-genome sequence. *Sci Rep* 10(1):14300

FopaFomeju, B, Falentin C, Lassalle G, Manzanares-Dauleux MJ, Delourme R (2014) Homoeologous duplicated regions are involved in quantitative resistance of *Brassica napus* to stem canker. *BMC Genomics* 15(1):498

Fredua-Agyeman R, Yu Z, Hwang SF, Strelkov SE (2020) Genome-Wide Mapping of Loci Associated with Resistance to Clubroot in *Brassica napus* ssp. *napobrassica* (Rutabaga) *Accessions from Nordic Countries. Front Plant Sci* 11:742

Gao L, Gonda I, Sun H, Ma Q, Bao K, Tieman DM, Burzynski-Chang EA, Fish TL, Stromberg KA, Sacks GL, Thannhauser TW, Foolad MR, Diez MJ, Blanca J, Canizares J, Xu Y, Knapp EV, Huang S, Klee HJ, Giovannoni JJ, Fei Z (2019) The tomato pan-genome uncovers new genes and a rare allele regulating fruit flavor. *Nat Gent* 51:1044–1051

Gatrell S, Lum K, Kim J, Lei XG (2014) Non-ruminant nutrition symposium: potential of defatted microalgae from the biofuel industry as an ingredient to replace corn and soybean meal in swine and poultry diets. *J Anim Sci* 92:1306–1314

Gilchrist EJ, Sidebottom CHD, Koh CS, MacInnes T, Sharpe AG, Vaughn GW (2013) A Mutant *Brassica napus* (Canola) Population for the Identification of New Genetic Diversity via TILLING and Next Generation Sequencing. *PLoS* 8(12):e84303

Gracia AA, Mollinari M, Marconi TG, Serang OR, Silva RR Vieira MLC, Vicentini R, Costa EA, Mancini MC, Garcia MOS, Pastina MM, Gazaffi R, Martins ERF, Dahmer N, Sforca DA, Silva CBC, Bundock P, Henry RJ, Souza GM, Sluys MA, Landell MGA, Carneiro MS, Vincentz MAG, Pinto LR, Vencovsky R, Souza AP (2013) SNP genotyping allows an in-depth characterization of the genome of sugarcane and other complex autopolyploids. *Sci Rep* 3:3399

Guerra D, Morica C, Badeck F, Rizza F, Delbono S, Francia E, Milc JA, Monostori I, Galiba G, Cattivelli L, Tondelli A (2022) Extensive allele mining discovers novel genetic diversity in the loci controlling frost tolerance in barley. *Theor Appl Genet* 135:553–569

Gyawali S, Harrington M, Durkin J, Horner K, Parkin IAP, Hegedus DD, Bekkaoui D, Buchwaldt L (2016) Microsatellite markers used for genome-wide association mapping of partial resistance to *Sclerotinina sclerotiorum* in a world collection of *Brassica napus*. *Mol Breed* 36:72

Hansen JC, Schillinger WF, Sullivan TS, Paulitz TC (2018) Rhizosphere microbial communities of canola and wheat at six paired field sites. *Appl Soil Ecol* 130:185–193

Hasan J, Megha S, Rahman H (2021) Clubroot in Brassica: recent advances in genomics, breeding, and disease management. *Genome* 64(8):735–760

Hasan MJ, Rahman H (2016) Genetics and molecular mapping of resistance to *Plasmodiophora brassicae pathotypes* 2, 3, 5, 6, and 8 in rutabaga (*Brassica napus* var. *napobrassica*). *Genome* 59(10):805–815

Hess GT, Frésard L, Han K, Lee CH, Li A, Cimprich KA, Montgomery SB, Bassik MC (2016) Directed evolution using DCas9-targeted somatic hypermutation in mammalian cells. *Nat Methods* 13(12):1036–1042

Hirani AH, Gao F, Liu J, Fu G, Wu C, McVetty PBE, Duncan RW, Li G (2018) Combinations of independent dominant loci conferring clubroot resistance in all four turnip accessions (*Brassica rapa*) from the European clubroot differential set. *Front Plant Sci* 9:Article 1628

Hu D, Jing J, Snowdon RJ, Mason AS, Shen J, Meng J, Zou J (2021) Exploring the gene pool of *Brassica napus* by genomics-based approaches. *Plant Biotechnol J* 19(9):1693–1712

Hu J, Chen B, Zhao J, Zhang F, Xie T, Xu K, Gao G et al. (2022) Genomic selection and genetic architecture of agronomic traits during modern rapeseed breeding. *Nat Genet* 54(5):694–704

Hu J, Guo C, Wang B, Ye J, Liu M, Wu Z, Xiao Y, Zhang Q, Li H, King GJ, Liu K (2018) genetic properties of a nested association mapping population constructed with semi-winter and spring oilseed rapes. *Front Plant Sci* 9:1740

Hua K, Zhang J, Botella JR, Ma C, Kong F, Liu B, Zhu JK (2019) Perspectives on the application of Genome-Editing Technologies in Crop Breeding. *Mol Plant* 12:1047–1059

Jan HU, Guan M, Yao M, Liu W, Wei D, Abbadi A, Zheng M, He X, Chen H, Guan C, Nicholas RA, Snowdon RJ, Hua W, Qian L (2019) Genome-wide haplotype analysis improves trait predictions in *Brassica napus hybrids. Plant Sci* 283:157–164

Jestin C, Bardol N, Lodé M, Duffé P, Domin C, Vallée P, Mangin B, Manzanares-Dauleux M, Delourme R (2015) Connected populations for detecting quantitative resistance factors to phoma stem canker in oilseed rape (*Brassica napus L.). Mol Breed* 35(8):167

Jestin C, Lodé M, Vallée P, Domin C, Falentin C, Horvais R, Coedel S, Manzanares-Dauleux M, Delourme R (2011) Association mapping of quantitative resistance for *Leptosphaeria maculans* in oilseed rape (*Brassica napus* L.). *Mol Breed* 27(3):271–287

Jiang C, Shi J, Li R, Long Y, Wang H, Li D, Zhao J, Meng J (2014) Quantitative trait loci that control the oil content variation of rapeseed (*Brassica napus L.). Theor Appl Genet* 127(4):957–968

Karim Md M, Dakouri A, Zhang Y, Chen Q, Peng G, Strelkov SE, Gossen BD, Yu F (2020) Two clubroot-resistance genes, Rcr3 and Rcr9wa, mapped in *Brassica rapa* using bulk segregant RNA sequencing. *Int J Mol Sci* 21(14):5033

Kirkegaard J, Lilley J, Berry P, Rondanini D (2020) Canola. In: Sadras, V., Calderini, D. (Eds.), Crop Physiology Case Histories for Major Crops. Academic Press.

Komor AC, Kim YB, Packer MS, Zuris JA, Liu DR (2016) Programmable Editing of a target base genomic DNA without double-stranded DNA Cleavage. *Nature* 533(7603):420–424

Krzymanski J (1978) Double low winter rape for Poland [*Brassica napus*]. In *International Rapeseed Conference. June 12–16, 1978.* Malmoe (Sweden).

Kumar V, Paillard S, Fopa-Fomeju B, Falentin C, Deniot G, Baron C, Vallée P, Manzanares- Dauleux M, Delourme R (2018) Multi-year linkage and association mapping confirm the high number of genomic regions involved in oilseed rape quantitative resistance to blackleg. *Theor Appl Genet* 131(8):1627–1643

Kurowska M, Daszkowska-Golec A, Gruszka D, Marzec M, Szurman M, Szarejko I, Maluszynski (2011) TILLING- a shortcut in functional genomics. *J Appl Genet* 52(4):371–390

Larkan NJ, Raman H, Lydiate DJ, Robinson SJ, Yu F, Barbulescu DM, Raman R, Luckett DJ, Burton W, Wratten N, Salisbury PA, Rimmer SR, Borhan MH (2016) Multi-environment QTL studies suggest a role for cysteine-rich protein kinase genes in quantitative resistance to blackleg disease in *Brassica napus. BMC Plant Biol* 16(1):183

Lee GP, Baek NK, Park KW (2002) Genetic Mapping of Resistant Genes in *Brassica pekinensis* against *Plasmodiophora brassicae Race6. Plant Pathol J* 18(5):266–270

Lee H, Chawla HS, Obermeier C, Dreyer F, Abbadi A, Snowdon R (2020) Chromosome-scale assembly of winter oilseed rape *Brassica napus*. *Front Plant Sci* 11:496.

Li F, Chen B, Xu K, Wu J, Song W, Bancroft I, Harper AL, Trick M, Liu S, Gao G, Wang N, Yan G, Qiao, J, Li J, Li H, Xiao, X, Zhang T, Wu X (2014) Genome-wide association study dissects the genetic architecture of seed weight and seed quality in rapeseed (*Brassica napus* L). *DNA Res* 21(4):355–367

Li J, Zhao Z, Hayward A, Cheng H, Fu D. (2015) Integration analysis of quantitative trait loci for resistance to *Sclerotinia sclerotiorum* in *Brassica napus*. *Euphytica* 205:483–489

Liu S, Fan C, Li J, Cai G, Yang Q, Wu J, Yi X, Zhnag C, Zhou Y (2016) A genome-wide association study reveals novel elite allelic variations in seed oil content of *Brassica napus. Theor Appl Genet* 129(6):1203–1215

Liu Y, Yang G, Huang S, Li X, Wang X, Li G, Chi T, Chen Y, Huang X, Wang X (2021) Enhancing prime editing by Csy4-mediated processing of PegRNA. *Cell Res* 31:1134–1136

Lu K, Wei L, Li X, Wang Y, Wu J, Liu M, Zhang C, Chen Z, Xiao Z, Jian H, Cheng F, Zhang K, Du H, Cheng X, Qu C, Qian W, Liu L, Wang R, Zou Q, Ying J, Xu X, Mei J, Liang Y, Chai Y, Wan H, Ni Y, He Y, Lin N, Fan Y, Sun W, Li N, Zhou G, Zheng H, Wang X(2019). Whole-genome resequencing reveals *Brassica napus* origin and genetic loci involved in its improvement. *Nat Commun* 10(1):1–12

Ma Y, Zhang J, Yin W, Zhang Z, Song Y, Chang X (2016) Targeted AID-Mediated Mutagenesis (TAM) enables efficient genomic diversification in mammalian cells. *Nat Methods* 13(12):1029–1035

Manzanares-Dauleux MJ, Delourme R, Baron F, Thomas G (2000) Mapping of one major gene and of QTLs involved in resistance to clubroot in *Brassica napus. Theor Appl Genet* 101(5):885–891

McCouch SR, Sweeney M, Li J, Jiang H, Thomson M, Septiningish E, Edwards J, Moncada P, Xiao J, Garris A, Tai T, Martinez C, Tohme J, Sugiono M, McClung A, Yuan LP, Ahn SN (2007) Through the genetic bottleneck: *O.*rufipogonas a source of trait-enhancing alleles for *O.sativa*. *Euphytica* 154:317–339

Nagaharu U (1935) Genome analysis in Brassica with special reference to the experimental formation of B. napus and peculiar mode of fertilization. *Jpn J Bot* 7(7):389–452

NASS: National Agricultural Statistics Service (2022). https://www.nass.usda.gov/Publications/AgCensus/2022/index.php

Nelson JW, Randolph PB, Shen SP, Evertte KA, Chen PJ, Anzalone AV, An M, Newby GA, Chen JC, Hsu A, Liu DR (2021) Engineered PegRNAs improve prime editing efficiency. *Nat Biotechnol* 40(3):402–410

Nelson MN, Nesi N, Barrero JM, Fletcher AL, Greaves IK, Hughes T, Laperche A, Snowdon R, Rebetzke GJ, Kirkegaard JA (2022) Chapter Two - Strategies to improve field establishment of canola: A review, Advances in Agronomy. *Academic Press* 175:133–177

Nguyen ML, Monakhos GF, Komakhin RA, Monakhos SG (2018) The new clubroot resistance locus is located on chromosome A05 in Chinese cabbage (*Brassica rapa L.). Russ J Genet* 54(3):296–304

Nishida K, Arazoe T, Yachie N, Banno S, Kakimoto M, Tabata M, Mochizuki M, MIyabe A, Araki M, Hara KY, Shimatani Z, Kondo A (2016) Targeted nucleotide editing using hybrid prokaryotic and vertebrate adaptive immune systems. *Science* 353(6305):aaf8729

Pageau D, Lajeunesse J, Lafond (2006) Impact de l'hernie des crucifères [*Plasmodiophora brassicae*] sur la productivité et la qualité du canola. *Canadian J Plant Pathol* 28(1):137–143

Pang W, Fu P, Li X, Zhan Z, Yu S, Piao Z (2018) Identification and mapping of the clubroot resistance gene CRd in Chinese cabbage (*Brassica rapa* ssp. *pekinensis). Front Plant Sci* 9:653

Pilet ML, Delourme R, Foisset N, Renard M (1998) Identification of loci contributing to quantitative field resistance to blackleg disease, causal agent *Leptosphaeria maculans* (Desm.) Ces. et de Not., in Winter rapeseed (*Brassica napus* L.). *Theor Appl Genet* 96(1):23–30

Qasim MU, Zhao Q, Shahid M, Samad RA, Ahmar S, Wu J, Fan C, Zhou Y (2020) Identification of QTLs containing resistance genes for sclerotinia stem rot in *Brassica napus* using comparative transcriptomic studies. *Front Plant Sci* 11:776

Qian L, Hickey LT, Stahl A, Werner CR, Hayes B, Snowdon RJ, Voss-Fels KP (2017) Exploring and harnessing haplotype diversity to improve yield stability in crops. *Front Plant Sci* 8:1534

Raghavan C, Elizabeth MCB, Wang H, Atienza G, Liu B, Qiu F, McNally KL, Leung H (2006) Rapid method for detecting SNPs on agarose gels and its application in candidate gene mapping. *Mol Breed* 19:87–101

Rahman H (2013) Review: Breeding spring canola (*Brassica napus* L.) by the use of exotic germplasm. *Canadian J Plant Sci* 93(3):363–373

Rahman H, Peng G, Yu F, Falk KC, Kulkarni M, Selvaraj G (2014) Genetics and breeding for clubroot resistance in Canadian spring canola (*Brassica napus* L.). *Canadian J Plant Pathol* 36(sup1):122–134

Rahman M, Hoque A, Roy J (2022) Linkage disequilibrium and population structure in a core collection of *Brassica napus (L.). PloS ONE* 17(3):e0250310

Rahman M, Mamidi S, del Rio L, Ross A, Kadir MM, Rahaman MM, Arifuzzaman M (2016) Association mapping in *Brassica napus* (L.) accessions identifies a major QTL for blackleg disease resistance on chromosome A01. *Mol Breed* 36(7):90

Rahman M, McClean P (2013) Genetic analysis on flowering time and root system in *Brassica napus L. Crop Science* 53(1):141–147

Raman H, McVittie B, Pirathiban R, Raman R, Zhang Y, Barbulescu Dm, Qiu Y, Liu S, Cullis B (2020) Genome-wide association mapping identifies novel loci for quantitative resistance to blackleg disease in canola. *Front Plant Sci* 11:1184

Raman R, Taylor B, Marcroft S, Stiller J, Eckermann P Coombes N, Rehman A, Lindbeck K, Luckett D, Wratten N, Batley J, Edwards D, Wang X, Raman H (2012) Molecular mapping of qualitative and quantitative loci for resistance to *Leptosphaeria maculans* causing blackleg disease in canola (*Brassica napus* L.). *Theor Appl Genet* 125(2):405- 418

Rousseau-Gueutin M, Belser C, Da Silva C, Richard G, Istace B, Cruaud C, Falentin C, Boideau F, Boutte J, Delourme R, Deniot G (2020) Long-read assembly of the *Brassica napus* reference genome Darmor-bzh. *GigaScience* 9(12):giaa137.

Roy, J (2021) Genome-wide association mapping and genomic prediction for resistance to sclerotinia stem rot in rapeseed/canola (*Brassica napus* L.) Germplasm Collections. ProQuest Diss.Theses

Roy J, del Río LE, Mendoza N, Bandillo PE, McClean, Rahman M (2022) Genetic mapping and genomic prediction of sclerotinia stem rot resistance to rapeseed/canola (*Brassica napus* L.) at seedling stage. *Theor Appl Genet* 135(6):2167–2184

Roy J, Shaikh TM, del Río Mendoza L, Hosain S, Chapara V, Rahman M (2021) Genome-wide association mapping and genomic prediction for adult stage sclerotinia stem rot resistance in *Brassica napus* (L) under field environments. *Sci Rep* 11(1):21773

Sakamoto K, Saito A, Hayashida N, Taguchi G, Matsumoto E (2008) Mapping of isolate- specific QTLs for clubroot resistance in Chinese cabbage (*Brassica rapa L. ssp. pekinensis). Theor Appl Genet* 117(5):759–767

Shahoveisi F, Oladzad A, del Rio Mendoza LE, Hosseinirad S, Ruud S, Rissato B (2021) Assessing the effect of phenotyping scoring systems and SNP calling and filtering methods on detection of QTL associated with reaction of *Brassica napus* to *Sclerotinia sclerotiorum*. PhytoFrontiers (3):135–148

Song JM, Guan Z, Hu J, Guo C, Yang Z, Wang S, Liu D, Wang B, Lu S, Zhou R, Xie WZ, Cheng Y, Zhang Y, Liu K, Yang QY, Chen LL, Guo L (2020) Eight high-quality genomes reveal pan-genome architecture and ecotype differentiation of *Brassica napus. Nat Plants* 6:34–35

Song JM, Liu DX, Xie WZ, Yang Z, Guo L, Liu K et al (2021) BnPIR: *Brassica napus* pan genome information resource for 1689 accessions. *Plant Biotechnol. J* 19(3):412

Spencer CCA, Su Z, Donnelly P, Marchini J (2009) Designing genome-wide association studies:sample size, power, imputation, and the choice of genotyping chip. *PLoS Genet* 5(5): e1000477

Strelkov SE, Manolii VP, Cao T, Xue S, Hwang SF (2007) Pathotype classification of *Plasmodiophora brassicae* and its occurrence in *Brassica napus* in Alberta, *Canada. J Phytopathol* 155(11-12):706–712

Su X, Zhu X, Song X, Xu H, Li W, Ning X, Chen Q, Guo W (2020) Genome-wide association analysis reveals loci and candidate genes involved in fiber quality traits in sea island cotton (*Gossypium barbadense*). *BMC Plant Biol* 20:289

Sun F, Fan G, Hu Q, Zhou Y, Guan M, Tong C, Li J, Du D, Qi C, Jiang L, Liu W (2017) The high-quality genome of *Brassica napus* cultivar 'ZS 11'reveals the introgression history in semi-winter morphotype. *The Plant J* 92(3):452–68

Sun F, Liu J, Hua W, Sun X, Wang X, Wang H (2016) Identification of stable QTLs for seed oil content by combined linkage and association mapping in *Brassica napus. Plant Sci* 252:388–399

Suwabe K, Tsukazaki H, Iketani H, Hatakeyama K, Kondo M, Fujimura M, Nunome T, Fukuoka H, Hirai M, Matsumoto S (2006) Simple Sequence Repeat-Based Comparative Genomics Between *Brassica rapa* and *Arabidopsis thaliana*: The genetic origin of clubroot resistance. *Genetics* 173(1):309–319

Tang S, Zhao H, Lu S, Yu L, Zhang G, Zhang Y, Yang QY, Zhou Y, Wang X, Ma W, Xie W, Guo L (2021) Genome- and transcriptome-wide association studies provide insights into the genetic basis of natural variation of seed oil content in *Brassica napus. Mol Plant* 14(3):470–487

USDA (2020) Foreign agricultural service. Oilseeds: World markets and trade reports (https://apps.fas.usda.gov/psdonline/circulars/oilseeds.pdf.)

Van de Wouw AP, Idnurm A, Davidson JA, Sprague SJ, Khangura RK, Ware RH, Lindbeck KD, Marcroft SJ (2016) Fungal diseases of canola in Australia: identification of trends, threats and potential therapies. *Australasian Plant Pathol* 45(4):415–423

Voorrips RE, Visser DL (1993) Examination of resistance to clubroot in accessions of *Brassica oleracea* using a glasshouse seedling test. *Netherlands J Plant Pathol* 99(5):269–276

Wang B, Wu Z, Li Z, Zhang Q, Hu J, Xiao Y, Cai D, Wu J, King GJ, Li H, Liu K (2018) Dissection of the genetic architecture of three seed-quality traits and consequences for breeding in *Brassica napus Plant Biotechnol J* 16(7):1336- 1348

Wang N, Chen B, Xu K, Gao G, Li F, Qioa J, Can G, Li J, Li H, Wu X (2016) Association mapping of flowering time QTLs and insight into their contributions to rapeseed growth habits. *Front Plant Sci* 7:338–348

Wang N, Qian W, Suppanz I, Wei L, Mao B, Long Y, et al. (2011) Flowering time variation in oilseed rape (*Brassica napus* L.) is associated with allelic variation in the FRIGIDA homologue BnaA.FRI.a. *J Exp Bot* 62:5641–5658

Wang X, Wang H, Long Y, Li D, Yin Y, Tian J, Li C, Liu L, Zhao W, Zhao Y, Yu L, Li M (2013) *Identification of QTLs associated with oil content in a high-oil Brassica napus cultivar and construction of a high-density consensus map for QTLs comparison in B. napus. PLoS One* 8(12):e80569

Wei D, Mei J, Fu Y, Disi JO, Li J, Qian W (2014) Quantitative trait loci analyses for resistance to *Sclerotinia sclerotiorum* and flowering time in *Brassica napus*. *Mol Breed* 34:1797–1804

Wei DY, Cui Y, He Y, Xiong Q, Qian L, Tong C, Lu G, Ding Y, Li J, Jung C, Qian W (2017) A genome-wide survey with different rapeseed ecotypes uncovers footprints of domestication and breeding. *J Exp Bot* 68(17):4791–4801

Wei L, Jian H, Lu K, Filardo F, Yin N, Liu L, Qu C, Li W, Du H, Li J (2016) Genome-wide association analysis and differential expression analysis of resistance to sclerotinia stem rot in *Brassica napus. Plant Biotechnol J* 14(6):1368–1380

Wu J, Cai G, Tu J, Li L, Liu S, Luo X, Zhou L, Fan C, Zhou Y (2013) Identification of QTLs for resistance to sclerotinia stem rot and *BnaC.IGMT5.a* as a candidate gene of the major resistant QTL SRC6 in *Brassica napus. PLoS ONE* 8:1–12

Wu J, Zhao Q, Liu S, Shahid M, Lan L, Cai G, Zhang C, Fan C, Wang Y, Zhou Y (2016) Genome-wide association study identifies new loci for resistance to sclerotinia stem rot in *Brassica napus*. Front Plant Sci 7:1418

Xiao J, Grandillo S, Ahn SN, McCouch SR, Tanksley SD, Li J, Yuan L (1996) Genes from wild rice improve yield. *Nature* 384:223–224

Xiao Z, Zhang C, Tang F, Yang B, Zhang L, Liu J, Huo Q, Wang S, Li S, Wei L, Du H, Qu C, Lu K, Li J, Li N (2019) Identification of candidate genes controlling oil content by combination of genome-wide association and transcriptome analysis in the oilseed crop *Brassica napus. Biotechnol Biofuels* 12(1):216

Xu L, Hu K, Zhang Z, Guan C, Chen S, Hua W, Li J, Wn J, Yi b, Shen J, Ma c, Tu J, Fu T (2016) Genome-wide association study reveals the genetic architecture of flowering time in rapeseed (*Brassica napus* L), *DNA Res*, 23(1):43–52

Xu R, Li J, Liu X, Shan T, Qin R, Wei P (2020) Development of plant prime-editing systems for precise genome editing. *Plant Commun* 1(3):100043

Yan T, Wang Q, Maodzeka A, Wu D, Jiang L (2020) BnaSNPDB: an interactive web portal for the efficient retrieval and analysis of SNPs among 1,007 rapeseed accessions. *Comput. Struct. Biotechnol. J* 18:2766–2773

Yin X, Yi B, Chen W, Zhang W, Tu J, Dilantha Fernando WG, Fu T (2010) Mapping of QTLs detected in a *Brassica napus* DH population for resistance to *Sclerotinia sclerotiorum* in multiple environments. *Euphytica* 173(1):25–35

Yu F, Zhang X, Peng G, Falk KC, Strelkov SE, Gossen BD (2017) Genotyping-by-sequencing reveals three QTL for clubroot resistance to six pathotypes of *Plasmodiophora brassicae* in *Brassica rapa. Sci Rep* 7(1):4516

Zhao J, Udall JA, Quijada PA, Grau CR, Meng J, Osborn TC (2006) Quantitative trait loci for resistance to *Sclerotinia sclerotiorum* and its association with a homoeologous non- reciprocal transposition in *Brassica napus L. Theor Appl Genet* 112(3):509–516

Zhou X, Dai L, Wang P, Liu Y, Xie Z, Zhang H, Xin Q, Wan L, Yang L, Yang G, Hong D (2021) Mining favorable alleles for five agronomic traits from the elite rapeseed cultivar Zhongshuang11 by QTL mapping and integration. *The Crop Journal* 9(6):1449–1459

Zhu H, Li X, Xi D, Zhai W, Zhang Z, Zhu Y (2019) Integrating long noncoding RNAs and mRNAs expression profiles of response to *Plasmodiophora brassicae* infection in Pakchoi (*Brassica campestris* ssp. *chinensisMakino*). *PLoS ONE* 14(12):e0224927

Zou J, Mao L, Qiu J, Wang M, Jia L, Wu D, Fan L (2019) Genome-wide selection footprints and deleterious variations in young Asian allotetraploid rapeseed. *Plant Biotechnol J* 17(10):1998–2010

4 The QTL Identification and Allele Mining for Molecular Design Breeding of Rapeseed

Jinjin Jiang and Youping Wang

4.1 INTRODUCTION

Rapeseed (*Brassica napus* L., AACC, 2n=38) is a polyploid formed ~7,500 years ago, which was derived from natural hybridization between two diploid species, *B. rapa* (AA, 2n=20) and *B. oleracea* (CC, 2n=18; Chalhoub et al. 2014). It is an allopolyploid with a short history and narrow genetic background, and the 'true' ancestor, as well as the wild type of rapeseed, has not been confirmed yet. As the second largest oilseed crop in the world, *B. napus* has prominent values in edible and industrial oil production and animal food production. Recently, it has been developed as an important vegetable (i.e. rapeseed stalk) and ornamental (i.e. colorful genotypes) crop. As reported, rapeseed might have originated and first been cultivated in Europe, then migrated to and was widely cultivated in Australia, Western Canada, and China. But artificial selection and breeding for desired traits (i.e. low erucic acid and glucosinolate content) caused less genetic variation in the current germplasms of rapeseed. Thus, it is necessary to broaden the genetic diversity of rapeseed through interspecific and intergeneric crosses, physically-/chemically-induced and natural mutations, which would also be helpful for mining alleles that control valuable traits of rapeseed. In the past century, genetic analysis and identification of alleles controlling important agronomic traits (i.e. oil content, seed size, silique length, and plant height) and adaptation of rapeseed were only successful in putative quantitative trait locus (QTL) localization through marker-assisted selection (MAS). Hitherto, the genome, pangenome, and transcriptome data of rapeseed are available, which will greatly facilitate the genetic and gene functional analysis of rapeseed (Chalhoub et al. 2014; Liu et al. 2021; Song et al. 2021b). Lately, whole-genome resequencing of 991 rapeseed accessions collected from 39 countries in the world (including 658 winter types, 145 semi-winter types, and 188 spring types), resequencing of 418 core collections representing the phenotypic diversity and geographic distribution of *B. napus* in the world, and genome-wide association studies (GWAS) of 56 agronomic traits revealed a large number of associated loci and candidate genes, as well as putative markers including single nucleotide polymorphisms (SNPs) and insertion/deletions (InDels) for MAS during rapeseed breeding (Wu et al. 2019; Hu et al. 2022). These sequencing and GWAS data will also facilitate the haplotype analysis and allele mining of interested traits in rapeseed. Although more and more alleles related to the agronomic traits of rapeseed have been identified, there are still lots of alleles to be identified, which will be vital for rapeseed improvement in the future.

4.2 DECIPHERING THE QTLS AND ALLELES CONTROLLING SEED QUALITY OF *B. NAPUS*

4.2.1 GENES REGULATING SEED OIL CONTENT OF *B. NAPUS*

Rapeseed produces ~13% of edible oil in the world, and it is speculated that an increase of 1% in oil content equals a 2.2%–2.3% increase in rapeseed yield. Thus, breeding rapeseeds with increased

DOI: 10.1201/9781003385035-4

oil content is economically important. Many QTLs controlling the seed oil content (SOC) of rapeseed have been reported, mainly using bi-parental linkage populations. The oil content of rapeseed is mainly controlled by maternal genotype, and the related loci were reported to have an additive effect of the genes (Guo et al. 2017). The epistatic effect and QTL-environment interactions have also been identified for SOC (Zhao et al. 2008a; Wurschum et al. 2012). The QTLs for improved oil content dispersed in different rapeseed lines or cultivars could result in transgressive segregation in the progenies, even if the parents with similar SOC were used for crossing (Delourme et al. 2006b; Chao et al. 2017). Besides, the SOC was negatively correlated with protein content; thus, it is challenging to simultaneously improve both traits in rapeseed (Zhao et al. 2006b; Chao et al. 2017). Therefore, identification of QTLs and alleles controlling seed oil and protein content was preferred for rapeseed breeding (Table 4.1).

Previously, 14 and 10 QTLs of SOC in two different doubled haploid (DH) populations were found, and some of these QTLs were more consistent across different genetic backgrounds, while some QTLs were specific to one genetic background with putative original alleles. This indicated that it is necessary to combine favorable alleles at different QTLs to increase the SOC of rapeseed (Delourme et al. 2006b). Cao et al. (2010) identified a novel QTL for the erucic acid content of rapeseed based on a QTL map of the Tapidor and Ningyou7 (TN) DH populations with improved resolution. *BnA8.FAE1*, an allele for the decreased oil and erucic acid content in seeds of modern cultivars, was located with this QTL. Besides, seven independent QTLs for SOC were identified near two QTLs of erucic acid content on linkage group A8. The alleles from *B. napus* cv. Tapidor in all the QTLs were identified with negative effects on either erucic acid or oil content. And five of the Tapidor alleles might originate from the forage rape 'Liho', which was taken as the origin of low seed erucic acid content. Smooker et al. (2011) compared different methods for QTL mapping, and identified 34 QTLs for fatty acid (FA) content, including 14 and 20 QTLs in the A- and C-genomes, respectively. They found 13 loci with novel alleles inherited from the progenitors of resynthesized *B. napus*, which were useful for modulating the desaturation of polyunsaturated fatty acids. An ortholog of *FAD2* coincided with the anticipated position and was proved to be the main contributor to the oleic acid content of rapeseed. Sun et al. (2012) constructed a DH genetic map containing 19 linkage groups with 4.30 cM in average between markers and covering 2,265.54 cM. They identified 12 QTLs associated with SOC and four candidate genes, encoding the PsbR subunit of photosystem II (*PS*II), fatty acyl-CoA reductases (FARs)–FAR1, subunit A of the trimeric protein ATP-citrate lyase (ACL)–ACLA, and a transcript regulated by the circadian clock. A QTL for SOC on linkage group A1 (OilA1) was narrowed to 6.9 and 4.3 cM (1.4 Mb) in two generations (BC5F3:4 and BC5F4) of near-isogenic lines (NILs). This QTL was absent in 80% of the rapeseed cultivars in China, and would be interesting for breeding rapeseed with improved SOC (Chen et al. 2013b). Using a TN DH population and its F2 population, Jiang et al. (2014) identified 46 distinct QTL regions controlling SOC on 16 of the 19 linkage groups. A consensus QTL *cqOCA10b* has been detected and delimited into a ~614 kb region on chromosome A10, which was consistently identified in all three environments of a recombinant inbred line (RIL) population and explained 9.8%–12.8% of the SOC variation (Chen et al. 2017). Chao et al. (2017) reported 67 QTLs for SOC, 38 QTLs for protein content, and 38 genomic regions from bulked segregant analysis (BSA) were overlapped with the SOC-QTLs. Finally, candidate genes related to acyl-lipid, seed storage, and the primary carbohydrate pathway were differentially expressed between the two parents of the DH lines used for mapping. Gu et al. (2017) found the oil body (OB) size was positively correlated with the SOC but negatively related to the glucosinolate and protein content. Many genes related to triacylglycerol (TAG) synthesis, OB-membrane proteins, and the cell progress regulatory pathway were located in the confidence interval of QTLs for SOC, FA compositions, and protein content. The advantageous allele of a novel yellow seed QTL, *cqSC-A09*, could increase the SOC, reduce the fiber content and pigmentation in the seed of rapeseed (Chao et al. 2022a).

Besides traditional QTL mapping, GWAS analysis coupling genome re-sequencing, the 60K SNP array, and transcriptome data have greatly promoted the identification of loci and alleles referring to SOC. Sun et al. (2016b) identified 40 QTLs and 29 associations for SOC by linkage analysis and GWAS in different environments. And 32 consensus QTLs were obtained, including seven overlapping QTLs that explained 10.23%–24.45% of the phenotypic variation, and allele-specific markers were developed for SOC breeding. Through GWAS analysis, Tang et al. (2021) deciphered the genetic variation of SOC in 505 rapeseed inbred lines grown in eight environments, including 27 QTLs that have been reported and novel QTLs such as *qOC.A05.1*, *qOC.A09.1*, *qOC.C05.4*, *qOC.A09.4*, and *qOC.A09.5*. Transcriptome-wide association studies (TWAS) revealed 690 genes associated with the SOC QTLs, including candidates for transcription factors (TFs), flavonoid biosynthesis, lipid metabolism, and transporters. This paper provided oil-related loci and candidate genes for rapeseed breeding with improved SOC by pyramiding favorable alleles. The relatives in other genus of Brassicaceae, such as *Sinapis alba*, possess mainly traits that are valuable for rapeseed breeding. Through GWAS and QTL analysis of 204 rapeseed accessions and RIL lines of *B. napus-S. alba* somatic hybrids, Zhao et al. (2022) constructed a high-density linkage map with 2,799 markers covering 1,835.21 cM. A novel and major QTL *qA07.SOC* was identified for SOC, and different haplotypes of *BnaA07.NRT1*, *BnaA07.STR18*, and *BnaA07g12880D* were identified as candidates in this QTL. Cui et al. (2021) confirmed the function of genes screened from GWAS and co-expression network analysis of acyl-lipid metabolism in developing siliques of 71 rapeseed lines, and mutation of four genes reduced SOC, while overexpression of 10 genes significantly increased SOC in Arabidopsis. They found genes in the photosynthesis/carbon metabolism pathways were co-expressed with genes involved in the FA and TAG biosynthesis pathways.

4.2.2 The QTLs and Alleles Controlling Fatty Acid Composition in Rapeseed

Besides SOC, genetic dissection in FA biosynthesis is also important to improve the seed oil quality and accumulation in rapeseed. FA in rapeseed includes C16:0, C18:0, C18:1, C18:2, C18:3, C20:1, and C22:1. The FA composition of rapeseed affects the seed's nutritional and industrial values, as well as germination. Many QTLs, alleles, and candidate genes controlling FA composition have been reported (Table 4.1).

Presently, most rapeseed cultivars produce oil with 55%–65% C18:1 and 8%–12% C18:2. Rapeseeds with increased C18:1 and reduced C18:3 are preferred by breeders. Hu et al. (2006) reported a major and a minor QTL for high C18:1 and two major QTLs for low C18:3 in a DH population of rapeseed. They found *fad2* in the N5 linkage group and *fad3c* in the N14 linkage group that were the candidates for high C18:1 and low C18:3, respectively. Based on the SNPs identified in *fad2* and *fad3c*, molecular markers were developed to help breed rapeseed with improved FA composition. Wang et al. (2015b) identified 72 QTLs and many QTL pairs of epistatic interactions correlated with the 10 FAs. In total, 234 genes in FA metabolism were located within the 47 QTLs, including genes encoding 50 enzymes and 5 regulatory factors (WRI1, FUS3, LEC1, LEC2, and ABI3). Using a DH population and parents cultivated in six environments, Teh & Mollers (2016) identified one to six QTLs for nine phytosterol traits, two to six QTLs for four FAs, five QTLs for SOC, four QTLs for protein content of defatted meal, and three QTLs for seed weight. The putative candidate genes were also identified by referring to the *B. rapa* genome. To comprehensively acknowledge the genes regulating oil content (OC) and FA content, Raboanatahiry et al. (2017) combined the 335 OC-QTLs and FA-QTLs reported from nine rapeseed populations and mapped 439 putative genes of FA and TAG biosynthesis to the QTL loci. Finally, they found 218 genes located in 162 QTLs, including 76 genes inside 57 overlapped QTLs; 61 out of the 76 genes were candidates for oil content. Besides, 10 genes were predicted to be the main regulators of FA and oil biosynthesis. Bao et al. (2018) detected 406 QTLs for eight FA components, explaining 1.49%–45.05% of the phenotypic variation. Among them, 91 unique and 204 consensus QTLs were detected in multiple environments and taken as pleiotropic QTLs, respectively. Besides, 22 environmental-specific and 74 stable QTLs were confirmed. Comparative genomic analysis and homolog annotation revealed 61 genes involved in acyl lipid metabolism; a deeper dissection of the molecular regulatory mechanism of these genes involved in FA metabolism would be valuable to rapeseed breeding. Chen et al. (2018) mapped 16 QTLs for C18:1, 22 QTLs for C18:3, 21 QTLs for C16:0, C18:0, and C18:2, respectively. Using meta-analysis, these loci were integrated into 71 pleiotropically unique QTLs. *uuqA5–6* and *uuqA5–7* were two major QTLs that simultaneously affected the FAs (besides C18:0) in most environments. The homologs of *fatty acid desaturase 2* (*FAD2*) and *glycerol-3-phosphate sn-2-acyltransferase 5* (*GPAT5*) in *uuqA5–6*, and *phosphatidic acid phosphohydrolase 1* (*PAH1*) in *uuqA5–7* were identified as candidates. Based on conditional mapping analysis, they identified 49, 30, 48, 60, and 45 consensus QTLs for C16:0, C18:0, C18:1, C18:2, and C18:3, respectively. A total of 128 unique QTLs were then integrated from the 232 conditional consensus QTLs. Finally, 63 unique loci were identified by both mapping methodologies, and 65 additional unique loci were specifically identified by conditional mapping. Zhao et al. (2019b) reported a novel QTL (*OLEA9*) of oleic acid content on the A9 chromosome, combing GWAS analysis of 375 low erucic acid germplasms and SNP chip analysis of a 150 DH population. Three candidate genes were identified, of which *BnaA09g39570D* encoded a metal ion-binding protein with major effects on C18:1 content. This new QTL can increase C18:1 to 80% together with the *FAD2* mutant allele, the only stable QTL of oleic acid content reported before. Fu et al. (2021) also identified two *BnFAD2* alleles on A- and C-chromosomes that were located in the QTL intervals, which explained 45%–52% and 15%–44% of the C18:1 variation, respectively. The *BnFAD2* alleles (*BnFAD2.A05* and *BnFAD2.C05*) contained two novel SNPs for the C421T and G1073E mutations in the parents of the F2:3 lines, respectively. Furthermore, the combination of the two alleles increased 88% of the C18:1 content in *B. napus*.

4.2.3 Major Genes Controlling Yellow Seed

Yellow seed is a quantitative trait with improved oil and protein content and less lignin, fiber, pigments, and polyphenols than the black-seeded genotype (Yu 2013). Presently, the yellow seed trait (YST) has been studied but not fully elucidated for the complex genome and genetic background, as well as the sensitivity to environmental variation (i.e. temperature, fertilizers; Jiang et al. 2019; Niu et al. 2020). Besides, the molecular mechanisms of YST related to other seed traits have not been elucidated (Table 4.1). Badani et al. (2006) reported a major QTL for YST on C08, which was co-located with a major QTL of acid detergent fiber (ADF) and then confirmed by Zhang et al. (2011b). Another major QTL on A09 controlling seed color and fiber content was identified (Liu et al. 2012; Stein et al. 2013). This locus was closely linked to a major QTL for seed acid detergent lignin (ADL) content, which contained a candidate gene *BnaA.CCR1.A9* with a sense mutation in the first exon (Liu et al. 2012). Through GWAS analysis, two homologs on C08 and A09 regulating seed coat color were identified (Wang et al. 2017). Several minor QTLs of YST were also identified on C02, C05, C06, C07, A01, A04, A07, and A08 of rapeseed (Wang et al. 2017; Gacek et al. 2021). Recently, Chao et al. (2022a) reported a major QTL *cqSC-A09* for YST, and the advantageous allele could increase the SOC and reduce the pigment and fiber content of the seed in *B. napus*. Among the 648 genes in *cqSC-A09*, *BnaA09.GH3.3*, *BnaA09.JAZ1*, and *BnaA09.LOX3* were major candidates for YST (Chao et al. 2022a).

4.2.4 Genetic Control of Lignin and Fiber Content in Seed of *B. napus*

The ADL content in rapeseed is negatively correlated with oil and protein content. Based on the BSA-seq of a RIL population from the cross of GH06 (yellow seed, low ADL) and P174 (black seed, high ADL), Liu et al. (2012) identified a major QTL for seed ADL content and found a functional mutation in *Bna.CCR1* homologs from GH06 that might be an effective allele controlling ADL content in rapeseed. Afterwards, the same research group identified 11 QTLs for YST and fiber content across four environments and constructed a QTL map with high density and improved resolution. The ADL loci mentioned above were narrowed on the interval of chromosome A09, and a minor QTL for ADL content variation was found on C05. They also detected three and five loci for the variation of hemicellulose and cellulose content, respectively (Liu et al. 2013). Behnke et al. (2018) found a major QTL on C05 that explained 81% of the ADL content variation. This locus was co-segregated with QTL for the oil and protein content of defatted rapeseed meal, with opposite additive effects. Seed lignin content (LC), cellulose content (CC), and hemicellulose content (HC) were negatively correlated with oil content (OC). On A09, QTLs of LC, CC, HC, and OC were co-localized and explained 9.87%–48.50% of the phenotypic variations. Miao et al. (2019) found 7 co-localized QTLs for fiber component and OC, including a unique and novel QTL, *uqA9–12*, that could affect LC, CC, and OC simultaneously but with opposite additive effects (Miao et al. 2019). Gacek et al. (2021) characterized 187,835 SNP markers through high-resolution skim genotyping by sequencing (SkimGBS) and identified 11 QTLs for seed quality traits. Key candidates in cell wall, lignin, and flavonoid biosynthesis and carbohydrate metabolic processes were screened, including *TT10/LAC15*, *TT4*, cellulase *GH5*, and *SUC2*.

4.2.5 Alleles Controlling Glucosinolate (GSL) Content in Rapeseed

Hitherto, only a few QTLs of seed GSL content in rapeseed have been mapped by researchers (Table 4.1). And the cultivars with low GSL content (12–15 μmol/g) were selected from a homozygous recessive state at all known loci. Mining novel QTL and alleles for low GSL content will be helpful to further reduce GSL content in rapeseed. Toroser et al. (1995) reported two major loci (*GSL-1* and *GSL-2*) and three minor loci (*GSL-3*, *GSL-4*, and *GSL-5*) for total aliphatic glucosinolates in seeds. These QTLs played with additive effects and explained 71% of the variation in total seed glucosinolates. Feng et al. (2012) reported 105 metabolite QTLs (mQTLs), including 58 mcQTLs (metabolite QTL for complex products) and 47 msQTLs (metabolite QTL for specific products), of GLS in the leaf and seed of *B. napus* and comprehensively revealed the important genetic variations and unknown potential regulatory relationships involved in glucosinolate synthesis. Schatzki et al. (2014) mapped three QTLs for napin and two loci for cruciferin in winter rapeseed, which explained 47% and 35% of the napin and cruciferin variance, respectively. Besides, a major locus on N19 for GSL content was overlapped with the QTL for napin and cruciferin content. Through GWAS and candidate quantitative trait loci (cQTL) analysis, 17 and 5 loci for seed GSL and hemicellulose content were identified, respectively. Four and six significant cQTLs for seed glucosinolate and hemicellulose content were identified, respectively. The SNP markers for these seed quality traits were stable across different environments and also overlapped with the QTLs reported previously (Gajardo et al. 2015). Fu et al. (2015a) identified 43 QTLs for seed GSL content in a low-GSL genetic background and seven environments in China and Germany. In these QTL loci, 23 orthologue genes (i.e. *BnaIPMDH1* and *BnaAPR3*) of Arabidopsis that participated in GSL metabolism were identified. Four QTLs explaining 4.70%–9.93% phenotypic variation for GSL content between eco-environments (GSLE) were identified on A04, A06, and A09. Three QTLs on A02, A04, and A09 exhibited different effects between the German and Chinese eco-environments, which were sensitive to the GSLE (He et al. 2018). Liu et al. (2020) identified a major QTL, *qGSL-C2*, that explained 30.88%–72.87% of the different seed GSL contents in five environments. *BnaC2.MYB28*, a candidate gene that encoded the R2R3-MYB protein and functioned in aliphatic methionine-derived GSL synthesis, was located in *qGSL-C2*. Rahman and Kebede (2021) mapped a QTL on C5 for SOC and a QTL on C9 for GSL content, which were derived from

TABLE 4.1
Genes and QTLs Controlling Seed Quality Traits of *Brassica napus*.

Genes/QTLs	Detail of the Traits Studied	Mapping Population, Size	Method (Linkage or Association Mapping)	References
fad2, fad3c	Fatty acid composition	Quantum×DMS100, 604 DH	LA	Hu et al. (2006)
LEC1, LEC2, FUS3, WRI1, ABI3	Fatty acid composition	Tapidor×Ningyou-7, 202 DH	LA	Wang et al. (2015b)
uuqA5–6(FAD2, GPAT5), uuqA5–7(PAH1)	Fatty acid composition	APL01×Holly, 189 RILs	LA	Chen et al. (2018)
OLEA9(BnaA09g39570D)	Fatty acid composition	375 low erucic acid accessions; ZP1×D126, 150 DH	LA, GWAS	Zhao et al. (2019b)
BnFAD2.A05, BnFAD2.C05	Fatty acid composition	FC81×E183, 190 F2:3; FM5×FM6, 114 F2	LA	Fu et al. (2021)
GSL-1, GSL-2, GSL-3, GSL-4, GSL-5	GSL content	Major×Stellar, 105 DH	LA	Toroser et al. (1995)
BnaIPMDH1, BnaAPR3	GSL content	Express×SWU07, 261 DH	LA	Fu et al. (2015a)
qGSL-C2(BnaC2.MYB28)	GSL content	G120×9172, 151 DH	LA	Liu et al. (2020)
BnaC09.MYB28, BnaA09.APK1, BnaC09.SUR1, BnaC02.GTR2a	GSL content	N53-2×KenC-8, 300 DH	LA	Chao et al. (2022b)
BnA8.FAE1	Oil content	Tapidor×Ningyou, 202 DH	LA	Cao et al. (2010)
FAD2	Oil content	Resynthesized *B. napus*×European *B. napus* cultivars, 92 DH	LA	Smooker et al. (2011)
BnG30, BnG36, BnG40, BnG44	Oil content	zy036×51070, 112 DH	LA	Sun et al. (2012)
qA07.SOC(BnaA07.STR18, BnaA07.NRT1, BnaA07g12880D)	Oil content	GRG2462×GRD328, 158 RILs, F2:11	LA, GWAS	Zhao et al. (2022)
OilA1	Oil content	SGDH267×Gaoyou, BC5F3:4, BC5F4	LA	Chen et al. (2013b)
cqOCA10b	Oil content	M201×352, 192 RILs	LA	Chen et al. (2017)
qOC.A05.1, qOC.A09.1, qOC.A09.4, qOC.A09.5, qOC.C05.4	Oil content	505 accessions	GWAS	Tang et al. (2021)
cqSC-A09	Oil content and pigment content	N53-2×KenC-8, 300 DH	LA	Chao et al. (2022a)
uqA9–12	Lignin and fiber content	N53-2×KenC-8, 300 DH	LA	Miao et al. (2019)
Bn-A01-p26914512	Seed erucic acid and GSL content	215 accessions	GWAS	Hatzig et al. (2018)
GH5, TT10/LAC15, TT4, SUC2	Seed quality traits	M305×Z114, 78 DH	LA	Gacek et al. (2021)
BnaA.CCR1.A9	Yellow seed and ADL content	GH06×P174, 232 RILs	LA	Liu et al. (2012)

Chinese kale and not affected by growth conditions. Besides, the low-GSL QTL on C9 exhibited a positive effect on the protein content of rapeseed. Recently, Chao et al. (2022b) reported four QTL hotspot regions (QTL-HRs) for seed glucosinolate content (SGC) variation on A09, C02, C07, and C09, covering seven known major QTLs and four novel major QTLs. They found the SGC was primary controlled by the superimposition of the advantage allele in the above-mentioned QTL-HRs and identified several candidate genes (i.e. *BnaC09.MYB28*, *BnaC09.SUR1*, *BnaA09.APK1*, and *BnaC02.GTR2a*) in GSL pathways. Based on the GWAS analysis of 215 winter varieties in 50 years of rapeseed breeding progress, a QTL at locus Bn-A01-p26914512 was identified for mean germination time that co-localized with a minor QTL of SGC and seed erucic acid content (Hatzig et al. 2018).

4.3 QTLS AND ALLELES CONTROLLING DISEASE RESISTANCE IN RAPESEED

Rapeseed is susceptible to many diseases that cause significant yield loss, such as blackleg, clubroot, and Sclerotinia stem rot, which were caused by *Leptosphaeria maculans*, *Plasmodiophora brassicae*, and *Sclerotinia sclerotiorum*, respectively. The major loci, alleles, and molecular mechanisms for resistance to these biotic stresses are still not well clarified in rapeseed due to the quick adaptation of fungal pathogens, as well as the lack of resistant germplasm (Table 4.2).

TABLE 4.2
Genes and QTLs Controlling Biotic Stress Resistances of *Brassica napus*.

Genes/QTLs	Detail of the Traits Studied	Mapping Population, Size, Number of Environment(s) Studied	Method (Linkage or Association Mapping)	References
CRK	Blackleg resistance	Topas AG-Castle, 242 DH; Topas AV-Sapphir, 104 DH	LA	Larken et al. (2016)
Rlm9	Blackleg resistance	Lorenz × Adriana, 60 DH; Lorenz × Alpaga, 60 DH; Lorenz × DK Cabernet, 60 DH; Lorenz × AGalileo 54 DH; Lorenz × King 10, 60 DH; Lorenz × JN, 60 DH	GWAS	Vollrath et al. (2021)
Rlm13	Blackleg resistance	CB-Telfer × ATR-Cobbler, 464 F2	LA	Raman et al. (2021)
Rlm3, Rlm4	Blackleg resistance	RP04 × Ag-Outback, 177 DH	LA	Raman et al. (2020)
BLMR2(BnaA10g11280D/ BnaA10g11290D), Rlm2	Blackleg resistance	Surpass 400 × Westar, BC4F4; Glacier × Westar, BC4F2	LA	Zhang et al. (2021)
Crr2, Crr4	Clubroot resistance	A9709 × G004, 94 F2	LA	Suwabe et al. (2003)
PbBa1.1	Clubroot resistance	C59-1 × ECD04, 115 BC1F1	LA	Chen et al. (2013a)
CRc	Clubroot resistance	K10 × Q5; C9 × 6R	LA	Sakamoto et al. (2008)
Rcr8	Clubroot resistance	T19 × ACDC, 92 BC1S1	LA	Yu et al. (2017)
qBrCR38-1	Clubroot resistance	CR38 × CS22, 292 F2	LA	Zhu et al. (2019)

(Continued)

TABLE 4.2 (*Continued*)
Genes and QTLs Controlling Biotic Stress Resistances of *Brassica napus*.

Genes/QTLs	Detail of the Traits Studied	Mapping Population, Size, Number of Environment(s) Studied	Method (Linkage or Association Mapping)	References
PbBn_di_A02, PbBn_di_A04, PbBn_di_C03, PbBn_rsp_C03	Clubroot resistance	Aviso × Montego, 114 DH	LA	Botero-Ramirez et al. (2020)
CRk, Crr3, CRd	Clubroot resistance	11SR0099 × 12DH0001, 116 DH	LA	Fredua-Agyeman et al. (2021)
Rcr10ECD01, Rcr9ECD01	Clubroot resistance	ECD01 × DH16516, 82 DH	LA	Yu et al. (2022)
BnaA03g29290D, BnaA03g29310D, BnaA08g10100D, BnaA08g09220D, BnaA08g10540D	Clubroot resistance	1CA1446.476-A1296 × A04–73NA, 90 DH	LA	Wang et al. (2022)
BnaA02T0335400WE, BnaA03T0374600WE, BnaA03T0262200WE, BnaA03T0464700WE	Clubroot resistance	A07–29NI × Rutabaga Brookfield-9005 P1	LA	Adhikary et al. (2022)
LRA9, SRC6	Sclerotinia stem rot resistance	Huashuang 5 × J7005, 190 DH	LA	Wu et al. (2013)
DSRC4, DSRC6, DSRC8	Sclerotinia stem rot resistance	448 accessions (420 accessions from Asia, 18 accessions from Europe, 5 accessions from North America, and 5 accessions from Australia)	GWAS	Wu et al. (2016)
GSTU	Sclerotinia stem rot resistance	448 accessions	GWAS	Wei et al. (2016)
qSSE2-1, qFTE2-1, qSS2–8, qSS2–9	Sclerotinia stem rot resistance	888-5 × M083, 244 F10 RILs	LA	Zhang et al. (2019)
SRA9a, SRC2a, SRC3a	Sclerotinia stem rot resistance	J964 × J902,181 DH	LA	Qasim et al. (2020)

4.3.1 Blackleg Disease

Many loci controlling resistance to *L. maculans* have been mapped in rapeseed, such as *Rlm1–10*, *BLMR1*, *BLMR2*, *LepR1* to *LepR4* (Yu et al. 2005, 2008; Delourme et al. 2006a; Long et al. 2011; Yu et al. 2013; Larkan et al. 2013; Dandena et al. 2019). A few resistance genes, *LepR2*, *LepR3*, *Rlm2*, and *BLMR2*, have been mapped, and the highly resistant *LepR3* and *Rlm2* were allelic and functioned as receptors (Larkan et al. 2015). Larken et al. (2016) identified three QTL loci through single- and multi-environment analysis and found eight candidate genes with homology to the *cysteine-rich receptor-like protein kinase* (*CRK*) of *A. thaliana*. But the effect of these genes was impaired after application in cultivation (Kutcher et al. 2007; Rouxel et al. 2003). For instance, *LepR3* from Surpass 400 provided race-specific resistance but suffered a major defeat in France and Australia (Sprague et al. 2006). Raman et al. (2018) also found *Rlm9* was not effective in conferring resistance in Australian field conditions (Raman et al. 2018). Thus, it is urgent to screen and integrate quantitative and qualitative genes for durable resistance to blackleg, since quantitative genes have been reported to have better effective resistance over time than qualitative resistance in other crops (i.e. rice, barley; van Berloo et al. 2001; Maeda et al. 2006; Zhang et al. 2016).

Huang et al. (2016) detected 17 QTLs involved in 'field' resistance against *L. maculans* using a mapping population grown in five winter rapeseed field experiments. These QTLs explained 51% of the phenotypic variation, while individual QTLs explained 2%–25% of the resistance variation, and only two to nine QTLs were detected in each experiment. They found six QTLs were stable and less sensitive to environmental factors, which would be important for breeding rapeseed with durable resistance. Vollrath et al. (2021) revealed high levels of single-nucleotide and structural variants within the *R* gene *Rlm9*, which may play novel roles in quantitative disease resistance in rapeseed. Raman et al. (2018) reported 27 significant QTLs that explained between 2.14% and 10.13% of the blackleg resistance variance; seven of them were repeatedly detected across different experiments, and eight QTLs were detected across Australia, France, and the United Kingdom growing environments. They found several *R* genes encoding nucleotide-binding leucine-rich repeat (LRR) receptors located in these QTL regions. A new allele on chromosome C03, *Rlm13*, and 24 candidate genes have been identified, conferring resistance to blackleg. This was the first *R* gene for blackleg resistance located on the C-genome of rapeseed (Raman et al. 2021). This group also reported two race-specific resistance genes, *Rlm3* and *Rlm4*, with 21 markers associated with the quantitative resistance loci. And three candidate genes were annotated as protein kinase and jumonji proteins that are involved in defense resistance. But the *Rlm3* and *Rlm4* reported in this study were not effective in Australian field conditions (Raman et al. 2020). Through GWAS and linkage-based mapping of rapeseeds grown in multiple years, Kumar et al. (2018) identified eight loci significantly associated with resistance to blackleg, including five loci located within QTL support intervals on A06, A08, A09, C01, and C04. And 657 pairs of homologs were screened in the QTL regions. Lately, Zhang et al. (2021a) predicted that *BnaA10g11280D/BnaA10g11290D* encoding cytochrome P450 proteins, located in the *BLMR2* locus that was resistant to all isolates tested, might be candidates of blackleg resistance in rapeseed. On the other hand, genome-wide comparison and co-localization analysis confirmed that the nucleotide-binding sites (NBS)-encoding genes, one of the largest plant resistance gene families, envolved rapidly and included significant variation after polyploidization of rapeseed (Fu et al. 2019). Pangenome analysis also revealed that the core *R*-genes harbored more SNPs than variable genes; 1,749 resistance gene analogs (RGAs) have been identified, and 106 of them were linked to blackleg resistance QTLs (Dolatabadian et al. 2020). These variations in resistance genes would be helpful for genomics-based allele screening and improved disease resistance.

4.3.2 Clubroot Resistance

Clubroot is an important disease in *Brassica* crops that is caused by a soil-borne pathogen. The pathogen survives in soil, and the resting spores sustain themselves for a long period, making it challenging to manage the disease through cultural techniques or chemical drugs. The affected rapeseeds appeared to have abnormal root development and impaired water and nutrient uptake, thus leading to growth inhibition, plant wilting, premature ripening, or death. Unlike the diploid parents, the rapeseed germplasm sources with clubroot resistance are limited. Hitherto, a few QTLs or genes of clubroot resistance have been reported on six chromosomes of *B. rapa*, such as *PbBa1.1* and *Crr2* on A01 (Suwabe et al. 2003; Chen et al. 2013a), *Rcr8* and *CRc* on A02 (Sakamoto et al. 2008; Yu et al. 2017), *Bra.CR.a*, *Bra.CR.c*, *Crr3*, *CRa*, *CRb*, *CRb^{kato}*, *CRd*, *CRk*, *PbBa3.1* to *3.3*, *Rcrl*, *Rcr2*, *Rcr4*, and *Rcr5* on A03 (Matsumoto et al. 1998; Hirai et al. 2004; Piao et al. 2004; Sakamoto et al. 2008; Chen et al. 2013a; Chu et al. 2014; Pang et al. 2014; Yu et al. 2016; Huang et al. 2017; Yu et al. 2017; Hirani et al. 2018; Huang et al. 2019), *Crr4* on A06 (Suwabe et al. 2003), *qBrCR38-1* on A07 (Zhu et al. 2019), *Crrl*, *CRs*, *PbBa8.1*, *Bra.CR.b*, *Rcr3*, *Rcr9/Rcr9^{wa}*, and *qBrCR38-2* on A08 (Chen et al. 2013a; Yu et al. 2017; Hirani et al. 2018; Laila et al. 2019; Zhu et al. 2019; Karim et al. 2020). However, only *Crr1*, *CRa*, and *CRb^kato^* encoding toll-interleukin-1 receptor, nucleotide binding site, and leucine-rich repeat (TIR-NBS-LRR, TNL) proteins, respectively, have been cloned (Ueno et al. 2012; Hatakeyama et al. 2013; Hatakeyama et al. 2017). The resistance genes have also been identified in *B. oleracea* (Dakouri et al. 2018; Peng et al. 2018). In rapeseed, Werner et al. (2008) created a genetic map using AFLP and simple sequence repeats (SSR) markers, and 19 QTLs of clubroot resistance were mapped on linkages N02, N03, N08, N13, N15, N16, and N19 with specific resistance to different isolates. The linkage analysis of a DH population from the cross between 'Aviso' × 'Montego' allowed the identification of three QTLs (*PbBn_di_A02*, *PbBn_di_A04*, and *PbBn_di_C03*) controlling disease index (DI) and *PbBn_rsp_C03* controlling resting spores per plant (RSP). *PbBn_rsp_C03* was partially overlapped with *PbBn_di_C03* in a nucleotide-binding leucine-rich repeat (NLR) gene-containing region (Botero-Ramirez et al. 2020). Fredua-Agyeman et al. (2021) found the clubroot resistance of *B. napus* cv. 'Tosca' was not effective after introgressing into a canola line '11SR0099'. Based on linkage analysis and QTL mapping, they found a resistance region (0.88–0.95 Mb) on the A03 chromosome of 'Tosca' in nine clubroot isolates, which contained the *CRk*, *Crr3*, and *CRd* genes. According to the SNP chip, maker-assisted selection, and RNA-seq analysis, Kopec et al. (2021) found the duplicated *TNL* gene is important for the resistance in the *Crr3* region of *B. napus* cv. 'Tosca'. Through the resistance analysis of four major clubroot isolates (3A, 3D, 3H, and 5X) on a DH population from the BC2 plants of the interspecific hybrids between *B. rapa* turnip cv. ECD01 (resistant, R) and canola line DH16516 (susceptible, S), as well as the parental lines, Yu et al. (2022) identified two major QTLs, $Rcr10^{ECD01}$ on A03 and $Rcr9^{ECD01}$ on A08, and 37 candidate genes encoding disease resistance proteins or involving in defense responses. These alleles in *B. napus* were introgressed from *B. rapa*, which possessed valuable resistance sources for clubroot. Wang et al. (2022) also mapped two major QTLs located on the A03 and A08 chromosomes, conferring additive resistance to pathotypes 3H, 3A, and 3D. And the decreased expression of *sugar transporter* genes (*BnaA03g29290D* and *BnaA03g29310D*), increased expression of the *TNL* genes (*BnaA08g10100D*, *BnaA08g09220D*, and *BnaA08g10540D*) in these loci may play important roles in clubroot resistance of rapeseed. Adhikary et al. (2022) integrated the differentially abundant proteins (DAPs) between clubroot resistant (CR) and susceptible (CS) canola lines with the reported QTLs associated with eight CR loci and found *BnaA02T0335400WE*, *BnaA03T0374600WE*, *BnaA03T0262200WE*, and *BnaA03T0464700WE* were possible candidates mediating clubroot responses.

4.3.3 Sclerotinia Stem Rot Resistance

Stem rot caused by *S. sclerotiorum* is one of the most devastating fungal diseases in rapeseed and imposes huge yield losses each year worldwide. Zhao and Meng (2003) identified 13 loci of *Sclerotinia* resistance by one-way ANOVA and six QTLs with MapMaker-QTL. They found three of the six QTLs were associated with leaf resistance at the seedling stage, and three QTLs corresponded to resistance at the mature plant stage. They also concluded that both single-locus QTLs and epistatic interactions played important roles in *Sclerotinia* resistance in rapeseed. Eight genomic regions affecting *Sclerotinia* resistance were detected in the HUA population, derived from a cross between a partially resistant Chinese winter line (Hua dbl2) and a susceptible European spring line (P1804). One region was detected in the MS population, derived from a partially resistant French winter cultivar (Major) and a susceptible Canadian spring cultivar (Stellar). Each QTL explained 6%–22% of the variance in *Sclerotinia* resistance (Zhao et al. 2006a). Wu et al. (2013) reported 10 QTLs for stem resistance (SR) at the mature stage and three QTLs for leaf resistance (LR) at the seedling stage of rapeseed grown in multiple environments. These QTLs were located on nine linkage groups (LGs), and two major QTLs, *LRA9* on LG A9 and *SRC6* on LG C6, were repeatedly detected across all environments. The homologous gene of *AtIGMT5* (*At1g76790*) related to *SRC6* might be the candidate for *Sclerotinia* resistance, and *BnaC.IGMT5.a* showed a polymorphism between parental lines of the DH population. Then, they identified 26 SNPs corresponding to three *Sclerotinia* resistance loci, *DSRC4*, *DSRC6*, and *DSRC8*, and 39 candidate genes were predicted based on the GWAS combining with the differentially expressed genes. Among the three loci, *DSRC6* was located within a confidence interval reported previously, while *DSRC4* and *DSRC8* were novel loci for *Sclerotinia* resistance (Wu et al. 2016). Based on the GWAS analysis of 347 rapeseed accessions, Wei et al. (2016) identified 17 significant associated loci for *Sclerotinia* resistance on chromosomes A8 and C6, including 5 loci on A8 and 12 on C6. The SNPs identified on A8 were located in a 409-kb haplotype block, and those on C6 were consistent with previous QTL mapping results. Twenty-four candidate genes of *Sclerotinia* resistance were identified in both the SNP-trait association and transcriptome sequencing analyses, including a tau class *glutathione S-transferase* (*GSTU*) gene cluster. Zhang et al. (2019) detected 30 *Sclerotinia* resistance QTLs and 22 flowering time QTLs, including some QTLs associated with both traits. For example, the stable major SRR QTL *qSSE2-1* was observed to overlap with the stable major QTL *qFTE2-1*. The confidence interval of *qSSE2-1* was observed to overlap with those of *qSS2–8* and *qSS2–9*. Qasim et al. (2020) reported 17 QTLs of *Sclerotinia* resistance in three different seasons, but no common QTL was identified in all three years. There were three QTLs (*SRA9a*, *SRC2a*, and *SRC3a*) that appeared in two seasons, with phenotypic effect variances of 14.75% and 11.57%, 7.73% and 6.81%, 7.49% and 10.38%, respectively. They also proved a negative correlation between flowering time and *Sclerotinia* resistance, and identified 36 putative candidate genes upregulated in resistant lines compared to susceptible lines.

4.3.4 Other Disease Resistance

Turnip yellows virus (TuYV) is transmitted by the peach-potato aphid (*Myzus persicae*) and causes severe yield losses in oilseed rape. Hitherto, a genetic resource with resistance to TuYV has been identified in the resynthesized *B. napus* line 'R54'. Hackenberg et al. (2020) found the Korean spring oilseed rape and rapeseed variety Yudal were partially resistant to TuYV, while the dwarf French winter oilseed rape line Darmor-*bzh* was susceptible. Using the established Darmor-*bzh*×Yudal DH population, they identified a major QTL on ChrA04 explaining 36% and 18% of the phenotypic variation in two independent experiments, which was verified in the segregating BC1 population. Besides, a minor QTL on ChrC05 was identified in one of the experiments, but not observed in the BC1 population. Interestingly, the markers co-segregated with the 'R54' resistance were not conserved in Yudal, suggesting an independent origin of the TuYV resistances. Light leaf spot disease is caused by *Pyrenopeziza brassicae*, one of the most economically damaging pathogens of rapeseed and has a high ability to evolve for the reproduction system and airborne ascospores. Breeding of rapeseed with host resistance is inadequate to minimize the yield losses caused by *P. brassicae*. Karandeni Dewageet al. (2022) used a DH population from a secondary gene pool (ancestral genomes) of rapeseed and identified four QTLs with different effects on *P. brassicae* resistance, including a QTL on C1 with a major effect on inhibiting asexual sporulation.

4.4 QTLS AND ALLELES CONTROLLING YIELD AND YIELD-RELATED TRAITS OF RAPESEED

The rapeseed yield is mainly determined by silique number per plant, seed number per silique (SNPS), and seed weight (Chen et al. 2007; Lu et al. 2017) and is affected by plant height (PH), length of main inflorescence, branch number, initiation height of the first effective branch, seed density, and silique density (Cai et al. 2014; Wang et al. 2016b). Presently, researches on yield and yield-related traits of rapeseed are mainly focused on the identification of loci and candidate genes, while the functional analysis of genes regulating yield traits is less reported. Besides, these traits were significantly correlated with each other, and the reported loci could manipulate multiple yield-related traits to coordinately regulate rapeseed yield (Table 4.3).

4.4.1 Plant Architecture

Plant architecture refers to the length and number of stems, branches, and inflorescences. The rapeseed main stem controls the plant architecture, water and nutrient transport, determines the inflorescence number, and affects yield-related traits (i.e. silique number). Rapeseed inflorescence is a typical primary raceme, and usually contributes 15%–25% of seed yield in transplanting cultivation, which is even more prominent in the mechanized cultivation system. Yi et al. (2006) reported the 17 QTLs controlling yield-related traits of rapeseed, including three loci for plant height, four loci for branch initiation height, three loci for length of main inflorescence, two loci for the number of the first effective branch, a locus for silique number, and one locus for seed weight, as well as three loci for yield per plant. Using a DH population of rapeseed, Chen et al. (2007) constructed a genetic map with SSR and SRAP markers and identified the QTLs for plant height, branch initiation height, length of main inflorescence, silique length and density, and the number of the first effective branch. Increasing the number of main stems is helpful to improve the rapeseed yield. On the A7 linkage group of the TN DH population, two genes (*BnA7.SUT1.a* and *BnA7.SUT1.b*) encoding sucrose transporters were identified with a correlation to the number of effective first branches, silique number per plant, and seed weight. Comparative mapping also revealed 37 rice and rapeseed homologues located in similar yield-related QTL regions (Li et al. 2011; Li et al. 2012). Zhao et al. (2019c) also found a mutant with a multi-main stem (MMS) in the KN DH population. They identified 43 QTLs for MMS that explained 2.95%–14.9% of the stem variation and two stable QTLs (*cqMMS.A3-2* and *cqMMS.C3–5*) in the winter and semi-winter growth conditions. Epistatic interaction analysis confirmed that *cqMMS.C3–5* was an important locus for MMS. A total of 159 genes located in the QTL intervals were identified, including 148 Arabidopsis homologs, such as *shoot meristemless* (*STM*), *cytokinin response factors* (*CRF5*), *cytokinin oxidase* (*CKX4*), *gibberellin-regulated* (*RDK1*), and auxin-regulated genes (*ARL*, *STV1*, and *IAR4*). Moreover, six common candidates (*RPT2A*, *HLR*, *CRK*, *LRR-RLK*, *AGL79*, and *TCTP*) for the MMS phenotype were identified with QTL and DEG analysis. Cai et al. (2014) identified loci for six yield-related traits in rapeseed and found 5, 14, 6, 9, 2, and 7 markers linked with branch initiation height, plant height, silique length, seed weight, length of main inflorescence, and seed number per silique, respectively. Among them, 11 markers were consistent with previous reports (Chen et al. 2007; Radoev et al. 2008; Shi et al. 2009; Basunanda et al. 2010; Fan et al. 2010; Yang et al. 2012). Based on the SNP chip analysis of 155 rapeseed germplasms, Luo et al. (2015) identified the SNPs explaining the complex trait of harvest index, including 2, 6, 11, and 9 SNPs correlated with plant height, branch number, yield per plant, and harvest index, respectively. Cai et al. (2016) dissected the genetic loci controlling plant architecture and yield-related traits in *B. napus*; a total of 163 unique QTLs for plant architecture and plant yield-related traits were mapped. And 190 plant architecture related candidate genes were identified; three candidate genes, *Bna.A02.CLV2*, *Bna.A09.SLY2*, and *Bna.C07.AHK4*, were cloned for effectiveness analysis. They also found the QTLs for plant architecture were more stable than the QTLs for yield-related traits, and a few QTLs of plant architecture showed multiple effects on yield-related traits. Shen et al. (2018) obtained a DH population from a cross between Y689 (one *Capsella bursa-pastoris*-derived rapeseed intertribal introgression) and Westar (*B. napus* cultivar) and found three stem-related traits, plant height (PH), branch initiation height (BIH), and stem diameter (SD), were positively correlated with one another and with flowering time. They identified 72 consensus QTLs and 49 unique QTLs through a two-round strategy of QTL meta-analysis. And 19 major QTLs, including 11 novel QTLs on A02 and A07, were detected for the stem-related traits. In the genomic regions of the major QTLs, genes involved in phytohormone biosynthesis and signaling, flower development, and cell differentiation were identified as candidate genes, including *BnaA02g02560* (an ortholog of *AtGASA4*, which plays as a candidate for PH, SD and flowering time), and *BnaA02g08490* (an ortholog of *AtGNL*, associated with PH, BIH, and flowering time). Multi-inflorescence is a desirable trait for the genetic improvement of rapeseed. Zhang et al. (2018a) obtained a stable multi-inflorescence mutant 'M01' by selfing a multi-inflorescent F2 individual for four generations, which was derived from the cross between spring accession '506' and semi-winter accession 'SWU01' with single-inflorescence. They mapped a major QTL for multi-inflorescence on A05 by integrating QTL mapping, microarray analysis, and whole-genome sequencing. This will be valuable for gene cloning and molecular breeding of multi-inflorescence in rapeseed.

4.4.2 Plant Height

Plant height is a crucial element in plant architecture that influences rapeseed yield. Butruille et al. (1999) identified four QTLs on the N2, N3, N12, and N13 linkage groups that controlled the plant height of *B. napus*. Wang et al. (2015a) studied the developmental behavior factors affecting plant height of rapeseed DH lines grown in six experiments and identified 20 QTLs at maturity, 50 QTLs by the conditional mapping method, and 50 QTLs by unconditional mapping strategies. Li et al. (2016) used the GWAS data of 472 rapeseed accessions in three consecutive years and identified eight QTLs for plant height and five QTLs for primary branch number. But only two QTLs of plant height on A05 and a QTL of branch number on C07 were novel loci. In the genomic regions of these loci, candidate genes involved in flower development, phytohormone biosynthesis, metabolism, and signaling were identified. Zhao et al. (2019a) reported a new locus (*DS-4*) controlling the dwarf trait of *B. napus* and a candidate gene, *BnaC05.IAA7*, encoding Aux/IAA7 was identified with a P87L substitution in the *ds-4* mutant. Dong et al. (2021) found a semi-dwarf mutant (*sdw-e*) of rapeseed and identified two QTLs of PH on A10 using a F2 population of *sdw-e* and ZS11-HP. In the major QTL *qPHA10*, genes involved in phytohormone metabolism and cell wall division (*BnaA10g08290D*, *BnaA10g09290D*, and *BnaA10g08230D*) were predicted as candidates by integrating whole genome sequencing and transcriptome difference analysis.

4.4.3 Branch Angle and Number

Branching is important to the adaptation and yield potential of crops, but the genetic regulation underlining branching in rapeseed is poorly understood. Based on the GWAS data, Sun et al. (2016a) identified 56 SNPs correlated with the branch angle of rapeseed, and putative candidates were screened, such as *SRG4*, *LAZY1*, and *TIR1*. Through BSA and next-generation sequencing, Wang et al. (2016a) identified a major QTL on A06 (*ba1*) for branch angle, including 82 genes located in the *ba1* region. A homolog of *AtYUCCA6* (*BnaA0639380D*) in small- and big-branch angle rapeseed lines (Purler and Huyou19) was identified with sequence variations. The *BnaA0639380D* expression was significantly higher in Purler than in Huyou19, which might be due to the genomic mutations in Huyou19. This agreed with the increased IAA content in Purler. He et al. (2017) successfully identified a locus (*BnaC03g63480D*) on C03 through GWAS, QTL mapping, and gene expression analysis, which was correlated with the branch number of rapeseed. Li et al. (2017) reported 46 and 38 significantly associated loci of rapeseed branch angle using a mixed linear model (MLM) and a multi-locus random-SNP-effect mixed linear model (MRMLM), respectively. Numerous candidate genes were identified by annotating them against *A. thaliana* homologs. Among them, some candidate genes were first correlated with branch angle in rapeseed, such as *TAC1*, *SGR1*, *SGR3*, and *SGR5*. Li et al. (2020) found a natural mutant, *db1*, which exhibited an ultrahigh branching density. BSA and the SNP chip analysis revealed two major QTLs, *qDBA09* and *qDBC06*. The *qDBA09* locus was proven to be a single Mendelian factor, and the dense branching was controlled by a single recessive gene. Using a BC1F3 population, they mapped the locus into a 270 kb interval on A09 that contained 27 predicted genes, including three genes (i.e. *BnaA09g22690*, *BnaA09g22670D*, and *BnaA09g22680D*) potentially involved in regulating the dense branching trait of *db1*. A candidate branch-controlling gene *BnaA09.ELP6* for *qDBA09* was reported based on the predicted gene function, sequence comparisons, and co-segregation analysis. Based on the analysis of a BC1F1 population derived from dense branching 2 (dense branching line) and L72 (normal branching line), Li et al. (2022) mapped a major QTL, *qDB.A03*, for branching-related traits through BSA analysis, which was fine-mapped to a 256 kb region. Finally, 56 genes in this QTL, including 8 genes controlling the branching trait of rapeseed, were identified.

4.4.4 Silique-Related Traits

Silique (also named pod) is an important physiological and reproductive organ that determines seed yield and has been considered as a breeding target trait in rapeseed. Silique length is positively correlated with SNPS and seed weight. SNPS is the consequence of ovule initiation and ovule/seed development. Genetic analysis of silique-related traits is valuable for rapeseed breeding. Zhang et al. (2012) successfully mapped a major locus, *qSS.C9*, controlling the seed number per silique of *B. napus*. Through GWAS analysis of 520 *B. napus* accessions, 294 SNPs significantly associated with harvest index-related traits and 33 related candidate genes were identified, such as *Bna.SGR5*, that might be involved in silique development (Lu et al. 2016). Wang et al. (2016b) and Zhao et al. (2016) mapped the QTLs for six silique-related traits (seed number per silique, silique length, silique breadth, silique thickness, seed density, and silique volume) and seed yield- or yield-related traits, and then identified 25 genes (i.e. *TTG2* and *AP2*) that putatively regulate silique traits via comparative genomic analysis. Afterwards, Lu et al. (2017) reported the QTLs for 7 yield-determining traits, including main inflorescence pod number, branch pod number, pod number per plant, seed number per pod, thousand seed weight, main inflorescence yield, and branch yield. And 14 candidates (i.e. *Bna.MYB83*, *Bna.SPL5*, and *Bna.ROP3*) were identified in assistance with RNA-seq data of samples pooled from four high-yielding or low-yielding accessions from the 520 rapeseed materials. Shi et al. (2015) found that two major QTLs (*qPN.A06-1* and *qSN.A06-1*) of pod number and seed number per pod were co-localized with opposite effects on rapeseed yield. Only *qPN.A06-1* was confirmed and narrowed to 180 kb by regional association analysis, which included 33 candidate genes. Yang et al. (2017) identified 45 QTLs related to the seed number per silique, seed weight, and silique length, and screened 38 candidate genes controlling silique traits, including *TTG2*, *AP2*, *AGL*, *SMZ*, and *AFB2*. Ye et al. (2017) reported 15 QTLs correlated with the silique number of rapeseed and identified 647 DEGs located in these QTL loci through RNA-seq analysis. Using two DH populations (ZR and RR), Wang et al. (2019) constructed a high-dense genetic map for each population, identified *qSL_ZR_A09* and *qSL_RR_A09b* from two populations, respectively. These QTLs were stable in four environments, and 26 genes in these loci were identified with SNPs/Indels in parents. Two candidate genes (*BnaA09g41180D* and *BnaA09g41380D*) were selected, of which *BnaA09g41180D*, encoding homologs of Arabidopsis fasciclin-like arabinogalactan proteins (FLA3), was identified with expression variations and taken as the main candidate for *qSL_ZR_A09*. Zhu et al. (2020) found the natural variation of SNPS was mainly controlled by maternal and embryonic genotypes, which are co-determined by ovule number per ovary, fertile ovule ratio, ovule fertilization rate, and fertilized ovule development rate. Using the F2:3 population derived from Zhongshuang11 and No. 73290, they mapped two QTLs for SNPS, of which a major QTL, *qSN.A6*, was fine-mapped. They also identified *qSN.A7* that was fine-mapped using a BC4F2 segregating population. They found the difference in SNPS between two types of homologous NILs was determined by the embryonic genotypic effect. And the regulatory mechanism of *qSN.A7* was completely different from the cloned *qSS.C9* and *qSN.*

A6 reported before. A major locus of ovule abortion and SNPS was precisely mapped to an 80.1 kb region on A8 of rapeseed, and two genes, *BnaA08g07940D* and *BnaA08g07950D*, encoding photosystem I subunit F and zinc transporter 10 precursor, respectively, were identified as the most likely candidate genes (Jiao et al. 2021). Wang et al. (2021) detected four major QTLs (*qSLA7-2*, *qSLA9-1*, *qSLA9-3*, and *qSLC8-2*) of silique length on A07, A09, and C08. Candidate genes, such as *BnaA9.CP12-2*, *BnaA9.NST2*, *BnaA7.MYB63*, and *BnaA7.ARF17*, were predicted to have functions in starch and sucrose metabolism, plant hormone signal transmission, and phenylpropanoid biosynthesis. Based on a recombinant inbred population, Zhao et al. (2021) identified 120 consistent QTLs of 5 silique-related traits, including 23, 25, 29, 22, and 21 QTLs for silique length, breadth, thickness, SNPS, and silique volume, respectively. Among them, 13 QTLs were detected in different environments. Homologous gene analysis and re-sequencing of parental lines revealed 12 candidate genes underlying the 13 QTLs.

4.4.5 Seed-Related Traits

Seed size and weight are important factors affecting the seed yield of rapeseed. Two major QTLs of seed weight, *TSWA7a* and *TSWA7b*, were identified by Fan et al. (2010), and the specific markers (*BnMINI3a* and *BnTTG2a*) were developed for maker-assisted breeding of rapeseed. Wang & Guan (2010) reported 12 QTLs linked to silique, seed number, and seed weight of rapeseed. Zhang et al. (2011a) identified five, two, and three QTLs for silique length, seed number per silique, and seed weight, respectively. Based on the analysis of recombinant inbred lines from an inbred line with regular traits and a mutant with long silique and large seed, Yang et al. (2012) identified two major QTLs, *cqSLA9* and *cqSWA9*, for silique length and seed weight, respectively. The two loci were both located in the *uqA9* region. Li et al. (2014) found the QTLs for seed weight and silique length were co-located in rapeseed, and two integrated QTLs, *uq.A09-1* and *uq.A09-3*, were observed with opposite additive effects. Through comparative analysis, 20 QTLs for silique length and 21 QTLs for seed weight were identified, among which 17 QTLs on A09 were homologous with two QTLs on C08 (Fu et al. 2015b). Based on the SLAF-seq (specific locus amplified fragment sequencing) and BSA analysis of DH lines derived from large- and small-seed germplasms of *B. napus*, Geng et al. (2016) identified four SLAF markers correlated with thousand-seed weight, and four candidate genes were confirmed to be located in these regions, including genes involved in cell division, seed development, and IAA biosynthesis. Afterwards, they screened 272 differentially expressed genes between large and small seeds, which also included genes that participated in cell differentiation and seed development (Geng et al. 2018). Using a double haploid population derived from the cross between an elite line ZY50 and a pol cytoplasmic male sterility restorer line 7–5, Wang et al. (2020a) performed QTL mapping of the genetic mechanism of thousand-seed weight (TSW) and identified six major QTLs. A major QTL *cqSW.A03-2* was detected across different environments. Based on the analysis of a set of near-isogenic lines, *cqSW.A03-2* was identified as a single Mendelian factor and the ZY50 allele at *cqSW.A03-2* showed a positive effect on TSW. Finally, *cqSW.A03-2* was mapped into a 61.6-kb region through fine mapping, and 18 candidate genes were predicted. Candidate gene association analysis and expression analysis indicated a histidine kinase gene (*BnaA03G37960D*) as a candidate gene for this locus. Zhang et al. (2021c) identified a QTL *qSW.C9* for the thousand-seed weight of rapeseed using a DH population constructed from the parental accessions HZ396 and Y106. They narrowed *qSW.C9* into a candidate causal locus with a 266 kb interval and revealed a 24,482 bp deletion in HZ396, which contained six candidate genes that affected cell proliferation and expansion through multiple signaling pathways. Haplotype analysis in a rapeseed diversity panel confirmed the efficiency of two specific markers of *qSW.C9* in improving the seed weight of rapeseed.

4.4.6 Other QTLs and Genes Regulating Rapeseed Yield Trait, Heterosis, and Harvest Index

Quijada et al. (2006) and Udall et al. (2006) identified six and five yield-related QTLs originated from winter genotype and undomesticated germplasms (i.e. resynthesized rapeseed), respectively. These QTLs were significantly correlated with the yield traits of hybrid spring rapeseed. Based on the phenotypic data of rapeseed DH and F2 populations grown in 10 environments, Shi et al. (2009) identified 85 QTLs related to yield and 785 QTLs of yield-related traits. This research group re-mapped the QTLs correlated with the heterosis of yield-correlated traits (i.e. branch number and seed yield), using the same data from three environments (Shi et al. 2011). Radoev et al. (2008) also reported a few QTLs related to yield traits when mapping the loci controlling heterosis of rapeseed. Using DH and backcrossing populations, Basunanda et al. (2010) reported the QTLs of rapeseed heterosis, including QTLs correlated with yield-related traits. Similarly, 69 SNPs correlated with plant height and 36 SNPs related to seed yield were identified from the SNP phenotyping of 158 rapeseed accessions (Schiessl et al. 2015). Luo et al. (2017) analyzed the QTLs controlling 22 traits in 19 independent experiments of a segregating rapeseed population, and a total of 1904 QTLs, including 80 seed yield-related QTLs, were identified. Finally, they integrated 525 QTLs of seed yield through association analysis and QTL meta-analysis and found 81.5% of these QTLs were pleiotropic. Using the introgression lines from *B. napus-Orychophragmus violaceus* hybrids, Raboanatahiry et al. (2018) assembled the QTLs for seed-yield and yield-related traits of rapeseed into one unique map. A total of 972 QTLs were integrated into the physical map of rapeseed, 'Darmor-*bzh*', where 198 loci were overlapped. They also identified 147 candidate genes in 65 regions where 131 QTLs were overlapped, which might affect 9 yield traits (Zhao et al. 2016). Chao et al. (2019) analyzed 11 harvest index (HI)-related traits in rapeseed and found 160 QTL, 163 epistatic loci pairs for HI, and three closely related traits: seed yield (SY), biomass yield (BY), and plant height (PH). Among them, two, five, and three major QTLs were correlated with HI, SY, and PH, respectively. The candidate genes with CDS variation underlying these loci were identified, including a negative regulator of lateral branching (*BnaA02g14010D*, homologous to *OsTB1*) and a regulator of seed width and weight (*BnaA02g18890D*, homologous to *OsGW2*). Plant density defines vegetative architecture and the competition for light between individuals. Using a panel of 152 spring-type accessions and a double-haploid population of 99 lines from a cross between the cultivars Lynx and Monty, Menendezet al. (2021) analyzed 12 growth and yield traits at two contrasting plant densities, using GWAS and QTL mapping. They found time to flower, biomass at harvest, plant height, silique and seed numbers, and seed yield were significantly associated but were independent of plant density. According to the function of candidate genes in *A. thaliana*, they confirmed that TCP1 promoted plant growth independently of plant density, and HY5 increased biomass and seed yield specifically at high plant density. PIN genes accelerated flowering in a density-dependent manner, while FT, HY5, and TCP1 accelerated flowering in a density-independent manner. Liang et al. (2022) analyzed the correlation, network, and principal component for 21 yield-related traits in the BnaZN-RIL population of rapeseed and confirmed that rapeseed yield was determined by a complex trait network with key contributors. They established a high-density SNP linkage map and identified 134 consensus QTLs for 21 traits, and all the QTLs were incorporated into an interconnecting network with many hub QTLs. Four representative hub-QTLs were further dissected to identify the target or candidate genes that governed the causal relationships between the relevant traits. Although many QTLs and candidates regulating yield and yield-related traits have been reported in rapeseed through QTL mapping, GWAS analysis, RNA-seq, and comparative analysis with genomic data from rapeseed, only a few genes of alleles have been proven to have functions in regulating yield and yield-related traits of *B. napus*. Based on QTL mapping and map-based cloning, Liu et al. (2010) found the *BnRGA* mutation was related to the semi-dwarf phenotype in *B. napus*, which encoded a DELLA protein that functioned as a GA signaling repressor. Both overexpression of mutated *BnRGA* and specific mutagenesis of Arabidopsis homolog caused dwarf phenotypes in Arabidopsis. Li et al. (2015) cloned *BnaC9.SMG7b* in the *qSS.C9*QTL region and proved its function in regulating the formation of functional female gametophytes, thus proving that *BnaC9.SMG7b* positively regulates seed number per silique in rapeseed (Zhang et al. 2012). Liu et al. (2015) identified a major QTL on A09 that controlled seed weight and silique length. They found a deletion of the seventh exon of *ARF18* in zy72360, a rapeseed line with a large seed and long silique. The ARF18 mutation lost its binding activity because of failure to form a homodimer and finally affected silique development and seed size. Through QTL mapping, Fan et al. (2014) identified a *BrCLV* mutation related to the multilocular silique trait in *B. rapa*. Then, they constructed *BnCLV3* mutants through CRISPR/Cas9 and found the mutants had increased leaves, carpel, seed per silique, and seed weight compared to the control (Yang et al. 2018).

4.5 ALLELES CONTROLLING FLOWERING TIME OF RAPESEED

Growth period-related traits, such as flowering time, are important to rapeseed yield, oil quality, and plant adaptation to different environments. During rapeseed domestication and breeding, flowering time has been selected for its substantial impacts on determining the life-cycle period, yield, and seed quality of rapeseed. Developing early flowering and maturity rapeseed varieties is an important breeding objective. As reported, flowering time is highly influenced by the environment. Wang et al. (2009) identified six *BnFT* paralogues in rapeseed that were mapped to six distinct regions. Four of the six regions were located in the inverted duplicated regions on A7 and C6. The coding sequences (CDSs) of *BnFTs* were identified with high similarity to *AtFT*, but two *BnFTs* (*BnA2.FT* and *BnC2.FT*) lacked the distinctive CArG box in the first intron, which was necessary for binding FLC protein in Arabidopsis. Besides, they found three *BnFTs* (*BnA2.FT*, *BnC6.FT.a*, and *BnC6.FT.b*) were corelated with two major QTLs for flowering time. And the specific alleles of *BnC6.FT*.a and *BnC6.FT.b* in near isogenic lines were positive regulators of rapeseed flowering. Hou et al. (2012) reported that *BnFLC.A10* was the main candidate gene in flowering time QTL *qFT10–4* and the allelic diversity caused by MITE insertion/deletion in the upstream of *BnFLC.A10* was responsible for the differentiation of winter and spring rapeseeds. Raman et al. (2013) indicated that the flowering time of rapeseed was a complex trait that was controlled by at least 20 QTL loci and might be regulated by candidate genes *Flowering Locus C* (*FLC*), *FLT*, *Vernalisation Insensitive 3*, *Apetala1*, *Cauliflower*, *Curly Leaf*, *Short Vegetative Phase*, *Ga3Oxidase*, and *Leafy*. Li et al. (2018) identified 55 consensus QTLs of flowering time by combining the phenotype and genomic data of DH lines grown in spring, semi-winter, and winter environments, including 12 environment-stable and 43 environment-specific QTLs. Among them, six major QTLs were found, including two specific QTLs in the spring environment and four stable QTLs in winter and semi-winter conditions. Among them, 18 QTLs were co-localized with other reported QTLs. A total of 151 genes in flowering regulatory pathways were screened, including *BnaC03g32910D* (*CO*), *BnaA02g12130D* (*FT*), and *BnaA03g13630D* (*FLC*). InDels and mutations in some candidate genes were also identified, which might lead to gene functional variation. Jian et al. (2019) reported 27 QTLs among 6 environments of FT in a RIL population, including a stable major QTL on C02 that was found in all environments. In conjunction with RNA-seq, 105 flowering time-related DEGs with putative functions in the circadian clock/photoperiod, autonomous pathway, hormone, and vernalization pathways were screened, including eight FT-related genes in these loci, such as *Pseudo Response Regulator 7* (*PRR7*) and *FY*. Schebenet al. (2020) performed double digest restriction-site associated DNA sequencing (ddRAD) on an F2 population from a spring type (early-flowering) and a winter type (late-flowering). They constructed a map with 14,630 SNPs that covered 3981.31 cM, and a QTL on C2 was identified, including eight genes such as *FLC*. Xu et al. (2020) used SSR and SLAF-seq to map the QTLs for flowering time of rapeseed grown in four environments. Three target intervals, *FTA09*, *FTA10*, and *FTC05*, were identified. Based on the SNPs and InDels between two parents, 186 genes with sequence variations in these intervals were identified; 14 of them were related to

TABLE 4.3
Genes and QTLs Controlling Yield and Yield-Related Traits of *Brassica napus*.

Genes/QTLs	Detail of the Traits Studied	Mapping Population, Size, Number of Environment(s) Studied	Method (Linkage or Association Mapping)	References
BnaYUCC6	Branch angle	Purler × Huyou19, 277 F2	LA	Wang et al. (2016a)
TAC1, SGR1, SGR3, SGR5	Branch angle	472 rapeseed accessions	GWAS, Associative transcriptomics	Li et al. (2017)
SRG4, LAZY1, TIR1	Branch angle and number	530 accessions	GWAS, Associative transcriptomics	Sun et al. (2016)
BnaC03g63480D	Branch number	327 accessions, SWU07 × Express, 261 DH	LA, GWAS	He et al. (2017)
qDBA09(BnaA09g22690, BnaA09g22670, BnaA09g22680D, BnaA09.ELP6), qDBC06	Branch number	*db1* × T109, 7,785 BC1F3	LA	Li et al. (2020)
qDB.A03	Branch number	*db2* × L72,256 BC1F1	LA	Li et al. (2022)
q18DFF.A03-2, q18MT.A03-2, q17DFF.A05-1, q18FP.C04, q17DIF.C05, q17GP.C09, BnaSOC1.A05-Haplb, BnaLNK2.C06-Hapla	Days to flowering	588 accessions	GWAS	Wang et al. (2020)
FT.A07.1, FT.A08, FT.C06, FT.C07, FRI, FLC, SPL15, NF-YC10	Days to flowering	373 accessions	GWAS, Associative transcriptomics	Helal et al. (2021)
qFTA06, qFTC08	Days to flowering	NTS57 × CY12, 174 F2	LA	Liu et al. (2022)
BnFTs (BnA2.FT, BnC6.FT.a and BnC6.FT.b)	Flowering time	Tapidor × Ningyou, 202 DH	LA	Wang et al. (2009)
qFT10–4 (BnFLC.A10)	Flowering time	TN DH043 × Ningyou7, 9,000 BC5F1	LA	Hou et al. (2012)
Vernalisation Insensitive 3, Apetala1, Cauliflower, Flowering Locus C, Flowering Locus T, Curly Leaf, Short Vegetative Phase, Ga3Oxidase, Leafy	Flowering time	Skipton × Ag-Spectrum, 186 DH	LA	Raman et al. (2013)
BnaC03g32910D (CO), BnaA02g12130D (FT), BnaA03g13630D (FLC)	Flowering time	N53-2 × KenC-8, 348 DH	LA	Li et al. (2018)
Pseudo Response Regulator 7 (PRR7), FY	Flowering time	GH06 × P174, 172 RILs	LA	Jian et al. (2019)
FTA09, FTA10, FTC05	Flowering time	GW × DZ, 170 F2a; YL17 × SY17, 80 F2b	LA	Xu et al. (2020)
BnaFLC.A02, BnaFT.A02	Flowering time	Cabriolet × Darmor, 720 F2	LA	Tudor et al. (2020)

(Continued)

TABLE 4.3 (*Continued*)
Genes and QTLs Controlling Yield and Yield-Related Traits of *Brassica napus*.

Genes/QTLs	Detail of the Traits Studied	Mapping Population, Size, Number of Environment(s) Studied	Method (Linkage or Association Mapping)	References
PIF4, ELF4, CIB4, COL9	Flowering time	Bing409 (B409) × Zhongshuang8 (ZS8), 148 DH	LA	Song et al. (2021)
BnaC09g41990D(CO)	Flowering time	523 accessions	GWAS	Xu et al. (2016)
BnaA02g14010D, BnaA02g18890D	Lateral branching, seed width and weight	N53-2 × KenC-8, 348 DH	LA	Chao et al. (2019)
BnLL1(BnLMI1)	Leaf development and shape	Yuye 87 × Zheyou 50	BSA	Ni et al. (2015)
BnLLA10(BnA10.LMI1, BnA10.ROC)	Leaf development and shape	Z9 × HY	LA	Hu et al. (2018 and 2020)
GRFs, AS2s, KNATs	Leaf development and shape	GH06 × P174, 172 RILs	LA	Jian et al. (2017)
cqMMS.A3-2, cqMMS.C3–5	Multi-main stem	N53-2 × KenC-8, 384 DH	LA	Zhao et al. (2019)
BnA7.SUT1.a, BnA7.SUT1.b	Plant architecture	S-1300 × Eagle, F2:3; Teqing × Zhenxian97, 186 RILs	LA	Li et al. (2011 and 2012)
Bna.A02.CLV2, Bna.C07.AHK4, Bna.A09.SLY2	Plant architecture	Huashuang 5 × J7005, 254 H	LA	Cai et al. (2016)
BnaA02g02560, BnaA02g08490	Plant architecture	Y689 × Westar, 208 DH	LA	Shen et al. (2018)
DS-4(BnaC05.IAA7)	Plant height	*ds-4* × ZY821, F2 BC1F2	LA, BSA	Zhao et al. (2019a)
BnaA10g08290D, BnaA10g09290D, BnaA10g08230D	Plant height	sdw-e and ZS11-HP, 200 F2	LA	Dong et al. (2021)
BnDWF/DCL1	Plant height and leaf shape	ZS11 × *Bndwf/dcl1*, 109 BC1	LA	Wang et al. (2016c)
BnaA3.IAA2, BnaA3.IAA7	Plant height and leaf shape	ZS11 × *sca*	LA	Li et al. (2019)
qPN.A06-1, qSN.A06-1	Pod number and seed number per pod	Zhongshuang11 × No.73290, 184 F2	LA	Shi et al. (2015)
qSS.C9(BnaC9.SMG7b)	Seed number per silique	HZ396 Y106, 17,554 BC4F2	GWAS	Zhang et al. (2012), Li et al. (2015)
qSN.A6, qSN.A7	Seed number per silique	Zhongshuang11 × No. 73290, 150 BC3F2	LA	Zhu et al. (2020)
BnaA08g07940D, BnaA08g07950D	Seed number per silique	C4–146 × C4–58B, 290 DH	LA	Jiao et al. (2021)
TSWA7a, TSWA7b	Seed weight	SW Hickory × JA177, 238 DH; J7046 × J7005, 190 F2	LA	Fan et al. (2010)

(Continued)

TABLE 4.3 (*Continued*)
Genes and QTLs Controlling Yield and Yield-Related Traits of *Brassica napus*.

Genes/QTLs	Detail of the Traits Studied	Mapping Population, Size, Number of Environment(s) Studied	Method (Linkage or Association Mapping)	References
cqSWA9	Seed weight	S1 S2,86 SS RILs; No2127 × S1, 192 NS RILs	LA	Yang et al. (2012).
uq.A09-1, uq.A09-3	Seed weight and silique length	Zhongshuang11 × No.73290, 184 F2	LA	Li et al. (2014)
ARF18	Seed weight and silique length	zy72360 × R1, 380 RILs	LA	Liu et al. (2015)
BnRGA	Semi-dwarf	*ds-1* × ZS4, 486 F2	LA	Liu et al. (2010)
Bna.SGR5	Silique development	520 accessions	GWAS, Associative transcriptomics	Lu et al. (2016)
TTG2, AP2	Silique development	N53-2 × KenC-8, 348 DH	LA	Zhao et al. (2016)
TTG2, AP2, AGL, SMZ, AFB2	Silique development, pod number and seed number per pod	No. 1167 × HZ396, 167 DH	LA	Yang et al. (2017)
qSL_ZR_A09 (*BnaA09g41180D*), *qSL_RR_A09b*	Silique development, pod number and seed number per pod	ZS11 × R11, 280 DH; R1 × R2, 95 DH	LA	Wang et al. (2019)
qSLA7-2, qSLA9-1, qSLA9-3, qSLC8-2	Silique length	172 RILs, 1,090 accessions	LA, GWAS, etaGWAS	Wang et al. (2021)
cqSW.A03-2(BnaA03G37960D)	Thousand-seed weight	ZY50 × 7-5, 189 DH	LA	Wang et al. (2020)
qSW.C9	Thousand-seed weight	HZ396 × Y106, 373 BC3F2, 355 BC3F2	LA	Zhang et al. (2021b)
TCP1, PIN, FT, HY5	Time to flowering, biomass at harvest, plant height, silique and seed number, seed yield	152 accessions, Lynx-037DH × Monty-028DH, 99DH	LA, GWAS	Menendez et al. (2021)
BnUC1(CFLAP2,RL14)	Up-curved leaf shape	NJAU5734 × ZS11, BC6F2	LA	Yang et al. (2019)
IAA2/7	Up-curved leaf shape	NJAU5737 × ZS11, BC5, BC6, BC5F2	LA	Huang et al. (2020)
Bna.MYB83, Bna.SPL5, Bna.ROP3	Yield-determining traits	520 accessions	GWAS, Associative transcriptomics	Lu et al. (2017)

the photoperiod, circadian clock, vernalization, and gibberellin pathways. They also developed six InDel markers in the three intervals, which could be helpful to further genetic selection in rapeseed breeding. Tudor et al. (2020) proved the vernalization of European winter rapeseed was determined by the allelic variation at a 10 Mbp region on chromosome A02, which included orthologues of the key floral regulators *BnaFLC.A02* and *Flowering Locus T* (*BnaFT.A02*). They found the sequence polymorphism of *BnaFLC.A02* and *BnaFT.A02*, especially in the *cis*-regulatory regions, caused different gene expression patterns in response to vernalization treatment. The allelic variation of *BnaFT.A02* was correlated with flowering time in the absence of vernalization, while the *BnaFLC.A02* variation was related to flowering time under vernalizing conditions. This indicated that the selection of *BnaFLC.A02* and *BnaFT.A02* variations would facilitate the selection of rapeseed varieties that could adapt to different climates. Using the SSR markers and 60 K SNP array of Brassica, Song et al. (2021a) identified five consensus QTLs (including two major loci) of flowering time from a DH population grown in three environments. One major locus on A03 was specifically identified in the semi-winter region, and another major QTL on C08 was found in all environments. Four flowering time-related genes were found in association with RNA-seq analysis, such as *PIF4*, *ELF4*, *CIB4*, and *COL9*.

Through GWAS analysis of flowering time in 523 rapeseed cultivars and inbred lines grown in eight environments, Xu et al. (2016) reported 41 SNPs associated with flowering time, and 12 SNPs were located in the QTLs reported before. Twenty-five candidate genes were identified for controlling the flowering time of rapeseed. The most significant SNPs next to Bn-scaff_16362_1-p380982, 13 kb from a candidate gene *BnaC09g41990D* encoding CONSTANS (CO), were identified and might be important in the photoperiod flowering pathway. Zhou et al. (2018) analyzed the SNP loci and candidate genes associated with the timing of initial flowering, maturity, and final flowering, as well as the flowering period, in 300 rapeseed-inbred lines grown in two environments. GWAS analysis with the 201,817 SNP markers developed from SLAF-seq revealed 131 SNPs strongly linked with the investigated traits. Among them, 40 SNPs were located in the confidence intervals or physically adjacent to the reported flowering time QTL or SNPs. They detected 35 and 90 elite allelic variants for early maturing and long flowering periods, respectively. Besides, 57 candidate genes in or around the linkage groups were screened for putative functions in regulating flowering time. Through GWAS analysis of 300–368 canola accessions, Raman et al. (2019) identified many SNPs correlated with variation in flowering time and response to photoperiod across multiple locations, including variations in the vicinity of *Flowering Locus T* (*FT*) paralogs and related regulators. Complementary QTL and eQTL mapping of the DH population also detected consistent genomic regions close to the *FT* paralogs that were associated with flowering time and yield-related traits. They also found the markers linked to the *FT* paralogs were associated with variation in multiple traits, including flowering time, plant emergence, shoot biomass, and grain yield.

Hitherto, QTLs and regulators of days to flowering (DTF) have been reported, but few reports on budding, bolting, and the interval between bolting and DTF were available. To dissect the genetic regulation of growth period traits, Wang et al. (2020b) analyzed the QTLs of days to initial flowering (DIF), days to final flowering (DFF), flowering period (FP), maturity time (MT), and whole growth period (GP) through GWAS and linkage mapping. Finally, 146 SNPs and 83 QTLs were identified using the RIL population of GH06 and P174, two parental lines with different flowering time and other growth period traits. Among these loci, 19 were pleiotropic SNPs correlated with multiple traits, and 6 QTLs (*q18DFF.A03-2*, *q18MT.A03-2*, *q17DFF.A05-1*, *q18FP.C04*, *q17DIF.C05*, and *q17GP.C09*) were consistently identified using both mapping strategies. By integrating the DEGs into the QTLs, 12 genes related to growth period traits were screened, including seven candidates with polymorphic sites in the CDS and the upstream 2 kb sequence. Furthermore, the haplotypes *BnaSOC1.A05-Haplb* and *BnaLNK2.C06-Hapla* may be used for further earliness molecular breeding of rapeseed. Helal et al. (2021) identified 15 and 37 QTLs regulating flowering time based on the SNP and haplotype-based GWAS analyses, respectively. Three and eight environmentally stable QTLs were detected by SNP and haplotype-based GWAS, respectively. Ten genomic regions were selected by integrating the two approaches and co-localizing the four traits, among which the regions *FT.A07.1*, *FT.A08*, *FT.C06*, and *FT.C07* were novel. A total of 197 genes controlling flowering time were detected, including 14 highly expressed DEGs between early- and late-flowering samples, such as *FRI*, *FLC*, *SPL15*, and *NF-YC10*. Liu et al. (2022b) performed RNA-seq of two parents after vernalization at low temperatures and combined the DEGs with QTL mapping of an F2 population. They identified two QTLs, a major QTL *qFTA06* on A06 and a minor QTL *qFTC08* on C06, associated with flowering time. Six candidate genes participating in regulating flowering time through the circadian clock/photoperiod, auxin and ABA signaling, cold signal transduction, and vernalization pathways were identified.

4.6 GENES CONTROLLING LEAF DEVELOPMENT AND SHAPE OF RAPESEED

Rapeseed has many germplasms with variations in leaf shape, and the leaf shapes of most rapeseed cultivars are entire and serrate leaves only a few germplasms with lobed leaves have been reported recently (Ni et al. 2017; Hu et al. 2020). Besides the novel germplasms with leaf shape variation that derived from crosses between *B. napus-O. violaceus* and *B. napus-Rorippa indica* (Zhao et al. 2008b; Tang et al. 2008), rolled leaf and winkled leaf mutants have been created through chemical mutagenesis (Wang et al. 2016c; Li et al. 2019; Huang et al. 2020). As reported, leaf shape variation, such as lobed leaves and curling leaves, could improve light transmittance, gas exchange, and photosynthesis efficiency, thus allowing high-density planting and hybrid production (Xu et al. 2018). Appropriate leaf shape variation is helpful to improve rapeseed yield through increasing planting density (Huang et al. 2020). Presently, genetic analysis and gene cloning of leaf shape characters have been reported in rapeseed, but there are still many regulators to be identified and studied.

Ni et al. (2015 and 2017) mapped a *BnLL1* locus correlated with the leaf lobes of rapeseed and found the candidate gene *BnLMI1* affected leaf margins by regulating the characteristics of meristems. Overexpression of *BnLMI1* in *A. thaliana* showed deeper and more serration in the leaf margin. Hu et al. (2018 and 2020) identified a major QTL, *BnLLA10*, controlling the lobed leaves in rapeseed, and confirmed the biological functions of two candidate genes, *BnA10.LMI1* and *BnA10.ROC*, in regulating leaf shape, through overexpression and mutation of these genes in rapeseed. Using the recombinant inbred lines from GH06 (less lobes) and P174 (more lobes) lines, Jian et al. (2017) mapped 31 QTLs correlated with leaf traits of *B. napus*, such as petiole length (PL), lobe number (PN), lamina width (LW), lamina length (LL), lamina size ratio (LL/LTL), and leaf total length (LTL). Then, they screened 33 candidate genes in tandem with RNA-seq analysis, including *GRFs*, *AS2s*, and *KNATs*.

Presently, research on curling leaves is mainly in monocotyledons, while genetic and gene analysis of rapeseed curling leaves is barely reported. Wang et al. (2016c) obtained a rapeseed mutant from EMS mutants with dwarf height and a down-curved leaf. Based on the backcross and selfcross populations from the mutant and ZS11, they identified a QTL *BnDWF/DCL1* with multiple effects on plant height and leaf shape through SNP chip and SSR analysis. Yang et al. (2019) constructed near-isogenic lines of up-curved leaf mutant *NJAU5734* and ZS11 and mapped a locus *BnUC1* linked with the up-curved leaf shape. In this QTL, two candidate genes might be important in regulating leaf shape variation in rapeseed. *CFLAP2* negatively regulated cuticle development, and *RL14* affected the formation of the secondary walls of thick-walled cells and mesophyll cells. From the same research group, Huang et al. (2020) used the backcross population of ZS11 and another up-curved leaf mutant, *NJAU5737*, and identified a locus *BnUC2* controlling the curling leaf and dwarf trait of rapeseed. Furthermore, *IAA2/7*-encoding Aux/IAA proteins were identified as candidate genes in this QTL region, which were revealed by comparative sequencing to be mutated in *NJAU5737* and might be involved in auxin signaling. Li et al. (2019) identified a rapeseed mutant *sca* with dwarf height, curled leaf, low branch initiation height, short internodes, small branch and silique angle. Based on QTL mapping analysis, they obtained two candidate genes *BnaA3.IAA2* and *BnaA3.IAA7*, among which *BnaA3.IAA7* was identified with a G-to-A mutation in the second exon that might be related to the phenotype variation of the *sca* mutant. Overexpression of *BnaA3.iaa7* increased plant height and decreased branch angle of rapeseed, and the transgenic plants also exhibited wrinkled cotyledon and true leaves. Zhao et al. (2019a) reported a new QTL *DS-4* controlling plant height and the down-curved leaf of rapeseed, and a candidate gene *BnaC05.IAA7* with a C-to-T mutation on the second exon was related to the phenotype variation. The gene mutation blocked the interaction of BnaC05.iaa7 with Transport Inhibitor Response 1 in the presence of auxin, which might affect the plant architecture of rapeseed through regulating auxin signaling.

4.7 QTLS AND ALLELES CONTROLLING RAPESEED TOLERANCE TO ABIOTIC STRESSES

Rapeseed growth and production are severely affected by abiotic stresses (i.e. salt, drought, osmotic, temperature, and nutrient stresses). Rapeseed oil production is greatly affected by the temperature during seed maturation. Presently, a few genes and QTLs related to the abiotic stress responses of rapeseed have been reported (Table 4.4). Zhu et al. (2012) constructed two near-isogenic lines (NIL-9 and NIL-1) that mainly differed in a QTL *qOC.C2.2* influencing oil content under high temperatures during seed maturation. In this locus, six DEGs among the NILs might be responsible

TABLE 4.4
Genes and QTLs Related to Abiotic Stress Responses of *Brassica napus*.

Genes/QTLs	Detail of the Traits Studied	Mapping Population, Size, Number of Environment(s) Studied	Method (Linkage or Association Mapping)	References
qOC.C2.2	Oil content under high temperature	Tapidor × Ningyou, 927 BC4	LA	Zhu et al. (2012)
BnaC03g32780D, BnaC03g37030D, BnaC09g27300D	Drought stress	N53-2 × KenC-8, 300 DH	LA	Gad et al. (2020)
qLPRA-A06c	Root angle at LP	Tapidor × Ningyou, 182 DH	LA	Duan et al. (2021)
BnaA3.NIP5;1	Low-B tolerance	QY10 × W10, BC5F3	Genotype analysis	He et al. (2021)
BnaA03g30320D, BnaA10g11500D, BnaC03g38360D, BnaC06g30030D	Al tolerance	254 accessions	GWAS	Zhou et al. (2022)
BnaA03g40290D, BnaA06g07530D, BnaA09g06240D, BnaA09g06250D, BnaC02g10720D	Low-temperature stress	442 accessions	GWAS	Luo et al. (2021)
qLTGA9-1, qLTGC1-1	Low-temperature stress	Huyou21 × 3429, 574 F2	LA, BSA	Zhu et al. (2021)
qOWRTA07, qOWRLA07, BnaA07G0198300ZS	Low-temperature stress	NTS57 × CY12, 348 F2, 174 F2:3	LA	Liu et al. (2022)

for the genetic divergence that was unaffected by temperature. Based on the F2:3 population from 2,205 (a salt tolerance line with normal leaf) and 1,423 (a salt sensitive line with lobed leaf), Zhang et al. (2018b) identified a major QTL on linkage group 10 and developed two intron polymorphic (IP) markers and one sequence-characterized amplified region (SCAR) marker for this QTL region. They also mapped the lobed-leaf gene to a region of 15.701–15.817 M on A10, containing a leaf morphological development-related gene, *Bra009510*, which was identified with eight amino acid variations between the two parents, and with significantly expressional difference between the two parents under salt stress. This study indicated that the lobed-leaf genes were associated with salt tolerance in rapeseed. Using 170 recombinant inbred lines of *B. napus*, Di et al. (2020) analyzed the QTLs of seed in response to ABA and screened the DGEs in *B. napus* ZS11 embryos under different ABA treatments (10, 20, and 30 mg/L). Finally, 10 QTLs explaining 2.70%–6.73% of the phenotypic variation under ABA treatment were identified, and 1,495, 3,332, and 3,868 DEGs were identified under different ABA treatments. These DEGs mainly participated in the responses to plant hormone signal transduction, starch and sucrose metabolism, cutin, suberine, and wax biosynthesis, and phenylpropanoid biosynthesis processes. Gad et al. (2020) indicated that the root-to-shoot length ratio was effective in phenotyping drought tolerance at the rapeseed seedling stage and identified 39 QTLs, which were integrated into 36 consensus QTLs. Finally, 18 QTLs were taken as loci that affected the drought stress index (DSI) of four traits (germination percentage, root length, shoot length, and root-to-shoot length ratio). Besides, 256 candidate genes were detected, including 128 genes with SNPs/InDels between the parents of the DH lines used for mapping, among which, some genes were involved in the drought stress responses (i.e. *BnaC03g32780D*, *BnaC03g37030D*, and *BnaC09g27300D*).

Plant root angle determines the root architecture in different soil layers and influences the uptake of nutrients such as phosphorus (P). Genetic variation in root angle has been reported in a linkage mapping population (BnaTNDH population) and an association panel of rapeseed grown with low P (LP) or optimal P (OP) conditions. Based on the linkage mapping analysis of the BnaTNDH population, nine QTLs at LP, and three loci at OP for root angle, and five loci for relative root angle (RRA) have been reported by Duan et al. (2021). GWAS analysis revealed 11 SNPs that correlated with the root angle at LP (LPRA). A QTL for LPRA, *qLPRA-A06c*, was co-located with the confidence region of Bn-A06-p14439400. Furthermore, a QTL cluster on C01 was identified for the root angle and primary root length (PRL), and 87 genes on A06 and 192 genes on C01 were located in the region; 14 genes involved in auxin asymmetric redistribution and root developmental were predicted as candidates (Duan et al. 2021). As the second leading crop for vegetable oil in the world, rapeseed production is critically dependent on boron (B) supplies, which are essential for vascular plants. Previously, *BnaA3.NIP5;1*, encoding a boric acid channel, was identified as a B-efficient candidate gene in rapeseed through QTL fine mapping. He et al. (2021) found genetic variation in *BnaA3.NIP5;1* is a key determinant of low-B tolerance in rapeseed. Increased *BnaA3.NIP5;1* expression in rapeseed exhibited improved low-B tolerance in both the seedling and maturity stages. *BnaA3.NIP5;1* is preferentially polar-localized in the distal plasma membrane of lateral root cap (LRC) cells and transports B into the root tips to promote root growth under B-deficiency conditions. A CTTTC tandem repeat in the 5′UTR of *BnaA3.NIP5;1* altered the expression level of the gene, which is tightly associated with plant growth and seed yield. Rapeseed varieties carrying *BnaA3.NIP5;1Q* allele significantly improved seed yield, indicating the elite allele of *BnaA3.NIP5;1* is valuable in breeding low-B-tolerant rapeseed cultivars. In acid soils, exchangeable aluminum (Al) is an environmental stress for the development and yield of rapeseed. The plant response to Al toxicity is complicated, and only a few QTLs have been studied in rapeseed. Zhou et al. (2022) used SLAF-seq and GWAS on inbred lines of rapeseed with significant genetic variation and different Al-tolerances. They found 43 SNPs closely associated with 8 tolerance traits at the seedling stage, and 777 candidates were identified. In combination with the DEGs from RNA-seq of the Al-tolerant line (ATL) and the AI sensitive line (ASL), they found 64 genes differentially expressed between rapeseed lines with different AI tolerances, such as *BnaA03g30320D*, *BnaA10g11500D*, *BnaC03g38360D*, and *BnaC06g30030D*.

Luo et al. (2021) investigated 13 seed germination- and emergence-related traits of 442 rapeseed accessions grown under normal and low-temperature conditions, using the stress tolerance index for the phenotype of each trait. GWAS analysis with SNP markers from genome resequencing revealed 22 QTLs significantly associated with seed vigor under low-temperature stress. In these QTL loci, 62 genes were identified with functions in stress tolerance, and most of them were involved in DNA repair, RNA translation, ubiquitination and degradation of protein reserves, the antioxidant system, plant hormones, and signal transduction. The effective variation and haplotype analysis revealed that *BnaA03g40290D*, *BnaA06g07530D*, *BnaA09g06240D*, *BnaA09g06250D*, and *BnaC02g10720D* might be valuable for further study in regulating seed vigor under cold stress. Rapeseed is also easily affected by low-temperature stress, which could delay seed germination, increase seedling mortality, and adversely affect rapeseed growth and production. Breeding of winter rapeseed cultivars with cold tolerance is critically important for winter survival and yield formation in some regions (i.e. northern China). Zhu et al. (2021) used a tolerant rapeseed cultivar (Huyou21), which was crossed with a susceptible rapeseed line (3,429). Two QTLs (*qLTGA9-1* on A09 and *qLTGC1-1* on C01) for low-temperature germination (LTG) were identified using the QTL-seq and SSR analysis on the F2 mapping population. A total of 69 genes on *qLTGA9-1* were identified, including six specific genes with moderate- or high-effect function variants. Furthermore, 133 genes on *qLTGC1*-1, including five specific genes with moderate-effect function variants, were identified. Liu et al. (2022a) constructed an F2 population and an F2:3 population of two parents, NST57 and CY12, for QTL mapping of cold tolerance in rapeseed. They obtained a genetic linkage map including 1,017 markers covering 793.53 cM and identified 16 QTLs for two cold-tolerance indicators related to overwintering success. Two major QTLs, *qOWRTA07* and *qOWRLA07*, were identified in overlapping regions, and *BnaA07G0198300ZS* was found as an important candidate gene for overwintering rate (OWR), which was more highly expressed in NST57 than in CY12.

4.8 GENETIC CONTROL OF OTHER TRAITS OF *B. NAPUS*

Chlorophyll is an important factor in plant photosynthesis and yield formation. In rapeseed, studies on the initial location of chlorophyll in seed embryos and pericarps have been reported, but only a few reports on the mapping of chlorophyll QTLs were available. A QTL locus of chlorophyll content on chromosome A02, *cqSPDA2*, was mapped and narrowed to a 152 kb region between Indel15 and SSR2, which included *BnaA02g30260D*, *BnaA02g30290D*, and *BnaA02g30310D* that might be involved in chlorophyll synthesis (Ye et al. 2020). Uniform and rapid seed germination is helpful to improve mechanized rapeseed production in modern agriculture. But the genetic mechanism of seed germination was still unclear in rapeseed. Yu et al. (2021) used RILs from the varieties 'APL01' (lower germination rate) and 'Holly' (higher germination rate) and mapped five loci for germination energy (GE) and six loci for germination percentage (GP). Among them, *qGE.C3* for GE and *qGP.C3* for GP were co-colocalied in a 28.5–30.5 cM interval on C3, which was a novel and major QTL for seed germination.

Improving oil and protein quality for food and feed purposes is an important goal in rapeseed breeding. Phytosterols in rapeseed have been used to enrich food products, while sinapate esters are not favored in rapeseed protein utilization. Increasing the phytosterol content of oil and lowering the sinapate ester content of meal could increase the value of the oilseed rape crop. Amar et al. (2008) identified three QTLs for total phytosterol content and four QTLs for sinapate ester content. These QTLs were confirmed to have a negative correlation with erucic acid, and the QTLs in linkage groups N8 and N13 were identified as having the strongest additive effects, indicating that the effective alleles for low erucic acid content increase phytosterol and sinapate ester content. Tocopherols are important antioxidants in vegetable oils; when present as vitamin E, tocopherols are an essential nutrient for humans and livestock. Wang et al. (2012) identified 33 unique QTLs for tocopherol content and composition in rapeseed. Seven of these QTLs were co-localized with candidate sequences associated with tocopherol biosynthesis. GWAS analysis of

142 rapeseed accessions revealed 61 loci that were significantly associated with tocopherol content and composition, 11 of which were localized within the confidence intervals of tocopherol QTL.

Petal size is important to the ornamental value of rapeseed, but the molecular mechanisms controlling petal size in *B. napus* remain unclear. Qian et al. (2021) identified 16 significant SNPs associated with petal size, including the most significant SNPs on A05 and C06, through GWAS analysis of 588 rapeseed accessions over three consecutive years. In combination with the RNA-seq data of two accessions with significant differences in petal size, they identified 52 DEGs controlling petal size variation in rapeseed, such as *BnaA05.RAP2.2*, that are involved in the ethylene signaling pathway.

4.9 CONCLUSION AND PERSPECTIVES

In conclusion, many QTLs and genes related to seed quality, agronomic traits, and stress resistances have been identified in rapeseed, including candidate loci controlling oil content, FA composition, yellow seed coat, lignin, fiber and glucosinolate content, blackleg, clubroot and Sclerotinia resistance, plant shape, silique and seed traits, as well as abiotic stress responses. Although the SNP chip and genome references have greatly facilitated QTL identification and allele mining in *B. napus*, the complexity caused by genome duplication and variations (such as InDels, presence/absence variations (PAVs), and copy number variations (CNVs) still hindered the genetic analysis of important traits in *B. napus*. Thus, pangenomes covering more genetic diversity in rapeseed germplasm would be valuable for allele identification for important traits (Edwards and Batley 2022). Furthermore, many of the reported QTLs and candidate genes were identified using linkage analysis and GWAS under simple environmental conditions, which may not be suitable for complex growth conditions. Besides, rapeseed is sensitive to environmental conditions that may affect the accuracy of phenotyping data, and many candidate loci were correlated with different traits. Genetic analysis of multi-traits under multi-environments would be appropriate for *B. napus*. On the other hand, epigenetic regulation has been barely reported in *B. napus*, which is also associated with the genome dynamics and homolog expression pattern and is important to the combination of valuable alleles in elite rapeseed varieties (Zhang et al. 2021b; Boideau et al. 2022).

ABBREVIATION

Abbreviation	Full name
ACL	ATP-citrate lyase
ADF	Acid detergent fiber
ADL	Acid detergent lignin
AFLP	Amplified fragment length polymorphism
Al	Aluminum
ASL	Al-sensitive line
ATL	Al-tolerant line
B	Boron
BIH	Branch initiation height
BSA	Bulked segregant analysis
BY	Biomass yield
C16:0	Palmitic acid
C18:0	Stearic acid
C18:1	Oleic acid
C18:2	Linoleic acid
C18:3	Linolenic acid

C20:1	Eicosenoic acid
C22:1	Erucic acid
CC	Cellulose content
CDS	Coding sequence
CKX	Cytokinin oxidase
CNV	Copy number variation
CO	CONSTANS
cQTL	Candidate quantitative trait loci
CR	Clubroot resistant
CRF	Cytokinin response factor
CRK	Cysteine-rich receptor-like protein kinase
CS	Clubroot susceptible
DAP	Differentially abundant protein
DEG	Differentially expressed gene
ddRAD	Double digest restriction-site associated DNA sequencing
DFF	Days to final flowering
DH	Doubled haploid
DI	Disease index
DIF	Days to initial flowering
DSI	Drought stress index
DTF	Days to flowering
FA	Fatty acid
FAD2	Fatty acid desaturase 2
FAR	Fatty acyl-CoA reductase
FLA	Fasciclin-like arabinogalactan protein
FLC	Flowering locus C
FP	Flowering period
FT	Flowering locus T
GE	Germination energy
GP	Whole growth period
GP	Germination percentage
GPAT5	Glycerol-3-phosphate sn-2-acyltransferase 5
GSL	Glucosinolate
GSLE	GSL content between eco-environment
GSTU	Glutathione S-transferase
GWAS	Genome-wide association study
HC	Hemicellulose content
HI	Harvest index
HPT	High-throughput phenotyping
InDels	Insertion/deletions
IP	Intron polymorphic
LC	Lignin content
LGs	Linkage groups
LL	Lamina length
LP	Low P
LPRA	Root angle at LP
LR	Leaf resistance

LRC	Lateral root cap
LRR	Leucine-rich repeat
LTG	Low-temperature germination
LTL	Leaf total length
LW	Lamina width
MAS	Marker-assisted selection
mcQTL	Metabolite QTL for complex product
MLM	Mixed linear model
MMS	Multi-main stem
mQTL	Metabolite QTL
MRMLM	Multi-locus random-SNP-effect mixed linear model
msQTL	Metabolite QTL for specific product
MT	Maturity time
NBS	Nucleotide-binding site
NIL	Near-isogenic line
NLR	Nucleotide-binding leucine-rich repeat
OB	Oil body
OC	Oil content
OilA1	Oil content on linkage group A1
OP	Optimal P
OWR	Overwintering rate
P	Phosphorus
PAH1	Phosphatidic acid phosphohydrolase 1
PAV	Presence/absence variation
PH	Plant height
PL	Petiole length
PN	Lobe number
PRL	Primary root length
PRR	Pseudo response regulator
PSII	Photosystem II
QTL	Quantitative trait locu
QTL-HR	QTL hotspot region
RFLP	Restriction fragment length polymorphism
RGA	Resistance gene analog
RIL	Recombinant inbred line
RRA	Relative root angle
RSP	Resting spores per plant
SCAR	Sequence characterized amplified region
SD	Stem diameter
SGC	Seed glucosinolate content
SkimGBS	Skim genotyping by sequencing
SLAF-seq	Specific locus amplified fragment sequencing
SNP	Single nucleotide polymorphism
SNPS	Seed number per silique
SOC	Seed oil content
SR	Stem resistance
SSR	Simple sequence repeat
STM	Shoot meristemless
SY	Seed yield

TAG	Triacylglycerol
TF	Transcription factor
TIR-NBS-LRR	Toll-interleukin-1 receptor, nucleotide binding site, and leucine-rich repeat
TN	Tapidor and Ningyou7
TNL	TIR-NBS-LRR
TSW	Thousand-seed weight
TuYV	Turnip yellows virus
TWAS	Transcriptome-wide association study
YST	Yellow seed trait

REFERENCES

Adhikary D, Mehta D, Uhrig RG, Rahman H, Kav NNV (2022) A proteome-level investigation into *Plasmodiophora brassicae* resistance in *Brassica napus* canola. *Front Plant Sci* 13:860393

Amar S, Ecke W, Becker HC, Mollers C (2008) QTL for phytosterol and sinapate ester content in *Brassica napus* L. collocate with the two erucic acid genes. *Theor Appl Genet* 116:1051–1061

Badani AG, Snowdon RJ, Wittkop B, Lipsa FD, Baetzel R et al (2006) Colocalization of a partially dominant gene for yellow seed colour with a major QTL influencing acid detergent fibre (ADF) content in different crosses of oilseed rape (*Brassica napus*). *Genome* 49:1499–1509

Bao BH, Chao HB, Wang H, Zhao WG, Zhang LN et al (2018) Stable, environmental specific and novel QTL identification as well as genetic dissection of fatty acid metabolism in *Brassica napus*. *Front Plant Sci* 9:1018

Basunanda P, Radoev M, Ecke W, Friedt W, Becker HC et al (2010) Comparative mapping of quantitative trait loci involved in heterosis for seedling and yield traits in oilseed rape (*Brassica napus* L.). *Theor Appl Genet* 120:271–281

Behnke N, Suprianto E, Mollers C (2018) A major QTL on chromosome C05 significantly reduces acid detergent lignin (ADL) content and increases seed oil and protein content in oilseed rape (*Brassica napus* L.). *Theor Appl Genet* 131:2477–2492

Boideau F, Richard G, Coriton O, Huteau V, Belser C et al (2022) Epigenomic and structural events preclude recombination in *Brassica napus*. *New Phytol* 234:545–559

Botero-Ramirez A, Laperche A, Guichard S, Jubault M, Gravot A et al (2020) Clubroot symptoms and resting spore production in a doubled haploid population of oilseed rape (*Brassica napus*) are controlled by four main QTLs. *Front Plant Sci* 11:604527

Butruille DV, Guries RP, Osborn TC (1999) Linkage analysis of molecular markers and quantitative trait loci in populations of inbred backcross lines of *Brassica napus L. Genetics* 153:949–964

Cai D, Xiao Y, Yang W, Ye W, Wang B et al (2014) Association mapping of six yield related traits in rapeseed (*Brassica napus* L.). *Theor Appl Genet* 127:85–96

Cai G, Yang Q, Chen H, Yang Q, Zhang C et al (2016) Genetic dissection of plant architecture and yield-related traits in *Brassica napus*. *Sci Rep* 6:21625

Cao ZY, Tian F, Wang NA, Jiang CC, Lin B et al (2010) Analysis of QTLs for erucic acid and oil content in seeds on A8 chromosome and the linkage drag between the alleles for the two traits in *Brassica napus*. *J Genet Genomics* 37:231–240

Chalhoub B, Denoeud F, Liu S, Parkin IA, Tang H et al (2014) Plant genetics. Early allopolyploid evolution in the post-neolithic *Brassica napus* oilseed genome. *Science* 345:950–953

Chao H, Guo L, Zhao W, Li H, Li M (2022a) A major yellow-seed QTL on chromosome A09 significantly increases the oil content and reduces the fiber content of seed in *Brassica napus*. *Theor Appl Genet* 135:1293–1305.

Chao H, Raboanatahiry N, Wang X, Zhao W, Chen L et al (2019) Genetic dissection of harvest index and related traits through genome-wide quantitative trait locus mapping in *Brassica napus* L. *Breed Sci* 69:104–116

Chao HB, Li HX, Yan SX, Zhao WG, Chen K et al (2022b) Further insight into decreases in seed glucosinolate content based on QTL mapping and RNA-seq in *Brassica napus* L. *Theor Appl Genet* 135:2969–2991

Chao HB, Wang H, Wang XD, Guo LX, Gu JW et al (2017) Genetic dissection of seed oil and protein content and identification of networks associated with oil content in *Brassica napus*. *Sci Rep* 7:46295

Chen F, Zhang W, Yu K, Sun L, Gao J et al (2018) Unconditional and conditional QTL analyses of seed fatty acid composition in *Brassica napus* L. *BMC Plant Biol* 18:49

Chen J, Wang B, Zhang YL, Yue XP, Li ZH et al (2017) High-density ddRAD linkage and yield-related QTL mapping delimits a chromosomal region responsible for oil content in rapeseed (*Brassica napus* L.). *Breed Sci* 67:296–306

Chen JJ, Jing J, Zhan ZX, Zhang T, Zhang CY et al (2013a) Identification of novel QTLs for isolate-specific partial resistance to *Plasmodiophora brassicae* in *Brassica rapa*. *PLoS One* 8:e85307

Chen W, Zhang Y, Liu X, Chen B, Tu J et al (2007) Detection of QTL for six yield-related traits in oilseed rape (*Brassica napus*) using DH and immortalized F2 populations. *Theor Appl Genet* 115:849–858

Chen Y, Qi L, Zhang X, Huang J, Wang J et al (2013b) Characterization of the quantitative trait locus *OilA1* for oil content in *Brassica napus*. *Theor Appl Genet* 126:2499–2509

Chu MG, Song T, Falk KC, Zhang XG, Liu XJ et al (2014) Fine mapping of *Rcr1* and analyses of its effect on transcriptome patterns during infection by *Plasmodiophora brassicae*. *BMC Genomics* 15:1166

Cui Y, Zeng X, Xiong Q, Wei D, Liao J et al (2021) Combining quantitative trait locus and co-expression analysis allowed identification of new candidates for oil accumulation in rapeseed. *J Exp Bot* 72:1649–1660

Dakouri A, Zhang XG, Peng G, Falk KC, Gossen BD et al (2018) Analysis of genome-wide variants through bulked segregant RNA sequencing reveals a major gene for resistance to *Plasmodiophora brassicae* in *Brassica oleracea*. *Sci Rep* 8:17657

Dandena HB, Zhang Q, Zhou TS, Hirani AH, Liu Z et al (2019) Analysis of quantitative adult plant resistance to blackleg in *Brassica napus*. *Mol Breed* 39:124

Delourme R, Chevre AM, Brun H, Rouxel T, Balesdent MH et al (2006a) Major gene and polygenic resistance to *Leptosphaeria maculans* in oilseed rape (*Brassica napus*). *Eur J Plant Pathol* 114:41–52

Delourme R, Falentin C, Huteau V, Clouet V, Horvais R et al (2006b) Genetic control of oil content in oilseed rape (*Brassica napus* L.). *Theor Appl Genet* 113:1331–1345

Di F, Wang T, Ding Y, Chen X, Wang H et al (2020) Genetic mapping combined with a transcriptome analysis to screen for candidate genes responsive to abscisic acid treatment in *Brassica napus* embryos during seed germination. *DNA Cell Biol* 39:533–547

Dolatabadian A, Bayer PE, Tirnaz S, Hurgobin B, Edwards D et al (2020) Characterization of disease resistance genes in the *Brassica napus* pangenome reveals significant structural variation. *Plant Biotechnol J* 18:969–982

Dong Z, Alam M. K, Xie M, Yang L, Liu J et al (2021) Mapping of a major QTL controlling plant height using a high-density genetic map and QTL-seq methods based on whole-genome resequencing in *Brassica napus*. *G3* 11:jkab118

Duan X, Wang X, Jin K, Wang W, Liu H et al (2021) Genetic dissection of root angle of *Brassica napus* in response to low phosphorus. *Front Plant Sci* 12:697872

Edwards D, Batley J (2022) Graph pangenomes find missing heritability. *Nat Genet 54*:919–920

Fan C, Cai G, Qin J, Li Q, Yang M et al (2010) Mapping of quantitative trait loci and development of allele-specific markers for seed weight in *Brassica napus*. *Theor Appl Genet* 121:1289–1301

Fan C, Wu Y, Yang Q, Yang Y, Meng Q et al (2014) A novel single-nucleotide mutation in a *CLAVATA3* gene homolog controls a multilocular silique trait in *Brassica rapa* L. *Mol Plant* 7:1788–1792

Feng J, Long Y, Shi L, Shi JQ, Barker G et al (2012) Characterization of metabolite quantitative trait loci and metabolic networks that control glucosinolate concentration in the seeds and leaves of *Brassica napus*. *New Phytol* 193:96–108

Fredua-Agyeman R, Hwang SF, Zhang H, Falak I, Huang XQ et al (2021) Clubroot resistance derived from the European *Brassica napus* cv. 'Tosca' is not effective against virulent *Plasmodiophora brassicae* isolates from Alberta, Canada. *Sci Rep* 11:14472

Fu Y, Lu K, Qian L, Mei J, Wei D et al (2015a) Development of genic cleavage markers in association with seed glucosinolate content in canola. *Theor Appl Genet* 128:1029–1037

Fu Y, Mason AS, Zhang Y, Yu H (2021) Identification and development of KASP markers for novel mutant *BnFAD2* alleles associated with elevated oleic acid in *Brassica napus*. *Front Plant Sci* 12:715633.

Fu Y, Wei D, Dong H, He Y, Cui Y et al (2015b) Comparative quantitative trait loci for silique length and seed weight in *Brassica napus*. *Sci Rep* 5:14407

Fu Y, Zhang YF, Mason AS, Lin BG, Zhang DQ et al (2019) NBS-encoding genes in *Brassica napus* evolved rapidly after allopolyploidization and co-localize with known disease resistance loci. *Front Plant Sci* 10:26

Gacek K, Bayer PE, Anderson R, Severn-Ellis AA, Wolko J et al (2021) QTL genetic mapping study for traits affecting meal quality in winter oilseed rape (*Brassica napus* L.). *Genes* 12:1235

Gad M, Chao H, Li H, Zhao W, Lu G et al (2020) QTL mapping for seed germination response to drought stress in *Brassica napus*. *Front Plant Sci* 11:629970

Gajardo HA, Wittkop B, Soto-Cerda B, Higgins EE, Parkin IAP et al (2015) Association mapping of seed quality traits in *Brassica napus* L. using GWAS and candidate QTL approaches. *Mol Breed* 35:143

Geng X, Dong N, Wang Y, Li G, Wang L et al (2018) RNA-seq transcriptome analysis of the immature seeds of two *Brassica napus* lines with extremely different thousand-seed weight to identify the candidate genes related to seed weight. *PLoS One* 13:e0191297

Geng X, Jiang C, Yang J, Wang L, Wu X et al (2016) Rapid identification of candidate genes for seed weight using the SLAF-seq method in *Brassica napus*. *PLoS One* 11:e0147580

Gu JW, Chao HB, Wang H, Li YH, Li DR et al (2017) Identification of the relationship between oil body morphology and oil content by microstructure comparison combining with QTL analysis in *Brassica napus*. *Front Plant Sci* 7:1989

Guo L, Si P, Wang N, Wen J, Yi B et al (2017) Genetic effects and genotype × environment interactions govern seed oil content in *Brassica napus* L. *BMC Genet* 18:1

Hackenberg D, Asare-Bediako E, Baker A, Walley P, Jenner C et al (2020) Identification and QTL mapping of resistance to Turnip yellows virus (TuYV) in oilseed rape, *Brassica napus*. *Theor Appl Genet* 133:383–393

Hatakeyama K, Niwa T, Kato T, Ohara T, Kakizaki T et al (2017) The tandem repeated organization of *NB-LRR* genes in the clubroot-resistant *CRb* locus in *Brassica rapa* L. *Mol Genet Genomics* 292:397–405

Hatakeyama K, Suwabe K, Tomita RN, Kato T, Nunome T (2013) Identification and characterization of *Crr1a*, a gene for resistance to clubroot disease (*Plasmodiophora brassicae Woronin*) in *Brassica rapa* L. *PLoS One* 8:e54745

Hatzig S, Breuer F, Nesi N, Ducournau S, Wagner MH et al (2018) Hidden effects of seed quality breeding on germination in oilseed rape (*Brassica napus* L.). *Front Plant Sci* 9:419

He M, Wang S, Zhang C, Liu L, Zhang J et al (2021) Genetic variation of *BnaA3.NIP5;1* expressing in the lateral root cap contributes to boron deficiency tolerance in *Brassica napus*. *PLoS Genet* 17:e1009661

He Y, Wu D, Wei D, Fu Y, Cui Y et al (2017) GWAS, QTL mapping and gene expression analyses in *Brassica napus* reveal genetic control of branching morphogenesis. *Sci Rep* 7:15971

He YJ, Fu Y, Hu D. X, Wei DY, Qian W (2018) QTL mapping of seed glucosinolate content responsible for environment in *Brassica napus*. *Front Plant Sci* 9:891

Helal M, Gill RA, Tang M, Yang L, Hu M et al (2021) SNP- and haplotype-based GWAS of flowering-related traits in *Brassica napus*. *Plants* 10:2475

Hirai M, Harada T, Kubo N, Tsukada M, Suwabe K et al (2004) A novel locus for clubroot resistance in *Brassica rapa* and its linkage markers. *Theor Appl Genet* 108:639–643

Hirani AH, Gao F, Liu J, Fu G. H, Wu CR et al (2018) Combinations of independent dominant loci conferring clubroot resistance in all four turnip accessions (*Brassica rapa*) from the European clubroot differential set. *Front Plant Sci* 9:1628

Hou J, Long Y, Raman H, Zou X, Wang J et al (2012) A Tourist-like MITE insertion in the upstream region of the *BnFLC.A10* gene is associated with vernalization requirement in rapeseed (*Brassica napus* L.). *BMC Plant Biol* 12:238

Hu J, Chen B, Zhao J, Zhang F, Xie T et al (2022) Genomic selection and genetic architecture of agronomic traits during modern rapeseed breeding. *Nat Genet* 54:694–704

Hu L, Zhang H, Sun Y, Shen X, Amoo O et al (2020) *BnA10.RCO*, a homeobox gene, positively regulates leaf lobe formation in *Brassica napus* L. *Theor Appl Genet* 133:3333–3343

Hu L, Zhang H, Yang Q, Meng Q, Han S et al (2018) Promoter variations in a homeobox gene, *BnA10.LMI1*, determine lobed leaves in rapeseed (*Brassica napus* L.). *Theor Appl Genet* 131:2699–2708

Hu X, Sullivan-Gilbert M, Gupta M, Thompson SA (2006) Mapping of the loci controlling oleic and linolenic acid contents and development of *fad2* and *fad3* allele-specific markers in canola (*Brassica napus* L.). *Theor Appl Genet* 113:497–507

Huang C, Yang M, Shao D, Wang Y, Wan S et al (2020) Fine mapping of the *BnUC2* locus related to leaf up-curling and plant semi-dwarfing in *Brassica napus*. *BMC Genomics* 21:530

Huang YJ, Jestin C, Welham SJ, King GJ, Manzanares-Dauleux MJ et al (2016) Identification of environmentally stable QTL for resistance against *Leptosphaeria maculans* in oilseed rape (*Brassica napus*). *Theor Appl Genet* 129:169–180

Huang Z, Peng G, Gossen BD, Yu F (2019) Fine mapping of a clubroot resistance gene from turnip using SNP markers identified from bulked segregant RNA-Seq. *Mol Breed* 39:131

Huang Z, Peng G, Liu X, Deora A, Falk KC et al (2017) Fine mapping of a clubroot resistance gene in Chinese cabbage using SNP markers identified from bulked segregant RNA sequencing. *Front Plant Sci* 8:1448

Jian H, Yang B, Zhang A, Zhang L, Xu X et al (2017) Screening of candidate leaf morphology genes by integration of QTL mapping and RNA sequencing technologies in oilseed rape (*Brassica napus* L.). *PLoS One* 12:e0169641

Jian H, Zhang A, Ma J, Wang T, Yang B et al (2019) Joint QTL mapping and transcriptome sequencing analysis reveal candidate flowering time genes in *Brassica napus* L. *BMC Genomics* 20:21

Jiang CC, Shi JQ, Li RY, Long Y, Wang H et al (2014) Quantitative trait loci that control the oil content variation of rapeseed (*Brassica napus* L.). *Theor Appl Genet* *127*:957–968

Jiao Y, Zhang K, Cai G, Yu K, Amoo O et al (2021) Fine mapping and candidate gene analysis of a major locus controlling ovule abortion and seed number per silique in *Brassica napus* L. *Theor Appl Genet* 134:2517–2530

Karandeni Dewage CS, Cools K, Stotz HU, Qi A, Huang YJ et al (2022) Quantitative trait locus mapping for resistance against *Pyrenopeziza brassicae* derived from a *Brassica napus* secondary gene pool. *Front Plant Sci* 13:786189

Karim MM, Dakouri A, Zhang Y, Chen Q, Peng G et al (2020) Two clubroot-resistance genes, *Rcr3* and *Rcr9*(wa), mapped in *Brassica rapa* using bulk segregant RNA sequencing. *Int J Mol Sci* 21:5033

Kopec PM, Mikolajczyk K, Jajor E, Perek A, Nowakowska J et al (2021) Local duplication of *TIR-NBS-LRR* gene marks clubroot resistance in *Brassica napus* cv. Tosca. *Front Plant Sci* 12:639631

Kumar V, Paillard S, Fopa-Fomeju B, Falentin C, Deniot G et al (2018) Multi-year linkage and association mapping confirm the high number of genomic regions involved in oilseed rape quantitative resistance to blackleg. *Theor Appl Genet* 131:1627–1643

Kutcher HR, Keri M, McLaren DL, Rimmer SR (2007) Pathogenic variability of *Leptosphaeria maculans* in western Canada. *Can J Plant Pathol* 29:388–393

Laila R, Park JI, Robin AHK, Natarajan S, Vijayakumar H et al (2019) Mapping of a novel clubroot resistance QTL using ddRAD-seq in Chinese cabbage (*Brassica rapa* L.). *BMC Plant Biol* 19:13

Larkan NJ, Lydiate DJ, Parkin IAP, Nelson MN, Epp DJ et al (2013) The *Brassica napus* blackleg resistance gene *LepR3* encodes a receptor-like protein triggered by the *Leptosphaeria maculans* effector AVRLM1. *New Phytol* 197:595–605

Larkan NJ, Ma L, Borhan MH (2015) The *Brassica napus* receptor-like protein RLM2 is encoded by a second allele of the *LepR3/Rlm2* blackleg resistance locus. *Plant Biotechnol J* 13:983–992

Larkan NJ, Raman H, Lydiate DJ, Robinson SJ, Yu F et al (2016) Multi-environment QTL studies suggest a role for cysteine-rich protein kinase genes in quantitative resistance to blackleg disease in *Brassica napus*. *BMC Plant Biol* 16:183

Li B, Gao J, Chen J, Wang Z, Shen W et al (2020) Identification and fine mapping of a major locus controlling branching in *Brassica napus*. *Theor Appl Genet* 133:771–783

Li B, Wang T, Guo Y, Liu X, Deng L et al (2022) Fine mapping of *qDB.A03*, a QTL for rapeseed branching, and identification of the candidate gene. *Mol Genet Genomics* 297:699–710

Li B, Zhao W, Li D, Chao H, Zhao X et al (2018) Genetic dissection of the mechanism of flowering time based on an environmentally stable and specific QTL in *Brassica napus*. *Plant Sci* 277:296–310

Li F, Chen B, Xu K, Gao G, Yan G et al (2016) A genome-wide association study of plant height and primary branch number in rapeseed (*Brassica napus*). *Plant Sci* 242:169–177

Li F, Ma C, Chen Q, Liu T, Shen J et al (2012) Comparative mapping reveals similar linkage of functional genes to QTL of yield-related traits between *Brassica napus* and *Oryza sativa*. *J Genetics* 91:163–170

Li F, Ma C, Wang X, Gao C, Zhang J et al (2011) Characterization of sucrose transporter alleles and their association with seed yield-related traits in *Brassica napus* L. *BMC Plant Biol* 11:168

Li H, Li J, Song J, Zhao B, Guo C et al (2019) An auxin signaling gene *BnaA3.IAA7* contributes to improved plant architecture and yield heterosis in rapeseed. *New Phytol* 222:837–851

Li H, Zhang L, Hu J, Zhang F, Chen B et al (2017) Genome-wide association mapping reveals the genetic control underlying branch angle in rapeseed (*Brassica napus* L.). *Front Plant Sci* 8:1054

Li N, Shi J, Wang X, Liu G, Wang H (2014) A combined linkage and regional association mapping validation and fine mapping of two major pleiotropic QTLs for seed weight and silique length in rapeseed (*Brassica napus* L.). *BMC Plant Biol* 14:114

Li S, Chen L, Zhang L, Li X, Liu Y et al (2015) BnaC9.SMG7b functions as a positive regulator of the number of seeds per silique in *Brassica napus* by regulating the formation of functional female gametophytes. *Plant Physiol* 169:2744–2760

Liang H, Ye J, Wang Y, Wang X, Zhou XR et al (2022) Systematic trait dissection in oilseed rape provides a comprehensive view, further insight, and exact roadmap for yield determination. *Biotechnol Biofuels Bioprod* 15:38

Liu C, Wang J, Huang T, Wang F, Yuan F et al (2010) A missense mutation in the VHYNP motif of a DELLA protein causes a semi-dwarf mutant phenotype in *Brassica napus*. *Theor Appl Genet* 121:249–258

Liu D, Yu L, Wei L, Yu P, Wang J et al (2021) BnTIR: an online transcriptome platform for exploring RNA-seq libraries for oil crop *Brassica napus*. *Plant Biotechnol J* 19:1895–1897

Liu J, Hua W, Hu Z, Yang H, Zhang L et al (2015) Natural variation in *ARF18* gene simultaneously affects seed weight and silique length in polyploid rapeseed. *Proc Natl Acad Sci USA* 112:E5123–E5132

Liu L, Qu C, Wittkop B, Yi B, Xiao Y et al (2013) A high-density SNP map for accurate mapping of seed fibre QTL in *Brassica napus* L. *PLoS One* 8:e83052

Liu L, Stein A, Wittkop B, Sarvari P, Li J et al (2012) A knockout mutation in the lignin biosynthesis gene *CCR1* explains a major QTL for acid detergent lignin content in *Brassica napus* seeds. *Theor Appl Genet* 124:1573–1586

Liu Y, Zhou X, Yan M, Wang P, Wang H et al (2020) Fine mapping and candidate gene analysis of a seed glucosinolate content QTL, *qGSL-C2*, in rapeseed (*Brassica napus* L.). *Theor Appl Genet* 133:479–490

Liu Z, Dong X, Cao X, Xu C, Wei J et al (2022a) QTL mapping for cold tolerance and higher overwintering survival rate in winter rapeseed (*Brassica napus*). *J Plant Physiol* 275:153735

Liu Z, Dong X, Zheng G, Xu C, Wei J et al (2022b) Integrate QTL mapping and transcription profiles reveal candidate genes regulating flowering time in *Brassica napus*. *Front Plant Sci* 13:904198

Long YM, Wang ZN, Sun ZD, Fernando DWG, McVetty PBE et al (2011) Identification of two blackleg resistance genes and fine mapping of one of these two genes in a *Brassica napus* canola cultivar 'Surpass 400'. *Theor Appl Genet* 122:1223–1231

Lu K, Peng L, Zhang C, Lu J, Yang B et al (2017) Genome-wide association and transcriptome analyses reveal candidate genes underlying yield-determining traits in *Brassica napus*. *Front Plant Sci* 8:206

Lu K, Xiao Z, Jian H, Peng L, Qu C et al (2016) A combination of genome-wide association and transcriptome analysis reveals candidate genes controlling harvest index-related traits in *Brassica napus*. *Sci Rep* 6:36452

Luo T, Zhang Y, Zhang C, Nelson MN, Yuan J et al (2021) Genome-wide association mapping unravels the genetic control of seed vigor under low-temperature conditions in rapeseed (*Brassica napus* L.). *Plants* 10:426

Luo X, Ma C, Yue Y, Hu K, Li Y et al (2015) Unravelling the complex trait of harvest index in rapeseed (*Brassica napus* L.) with association mapping. *BMC Genomics* 16:379

Luo Z, Wang M, Long Y, Huang Y, Shi L et al (2017) Incorporating pleiotropic quantitative trait loci in dissection of complex traits: seed yield in rapeseed as an example. *Theor Appl Genet* 130:1569–1585

Maeda H, Matsushita K, Iida S, Sunohara Y (2006) Characterization of two QTLs controlling resistance to rice stripe virus detected in a Japanese upland rice line, Kanto 72. *Breed Sci* 56:359–364

Matsumoto E, Yasui C, Ohi M, Tsukada M (1998) Linkage analysis of RFLP markers for clubroot resistance and pigmentation in Chinese cabbage (*Brassica rapa* ssp. pekinensis). *Euphytica* 104:79–86

Menendez YC, Sanchez DH, Snowdon RJ, Rondanini DP, Botto JF (2021) Unraveling the impact on agronomic traits of the genetic architecture underlying plant-density responses in canola. *J Exp Bot* 72:5426–5441

Miao LY, Chao HB, Chen L, Wang H, Zhao WG et al (2019) Stable and novel QTL identification and new insights into the genetic networks affecting seed fiber traits in *Brassica napus*. *Theor Appl Genet* 132:1761–1775

Ni X, Liu H, Huang J, Zhao J (2017) *LMI1-like* genes involved in leaf margin development of *Brassica napus*. *Genetica* 145:269–274

Ni XY, Huang JX, Ali B, Zhou WJ, Zhao JY (2015) Genetic analysis and fine mapping of the *LOBED-LEAF 1* (*BnLL1*) gene in rapeseed (*Brassica napus* L.). *Euphytica* 204:29–38

Pang WX, Liang S, Li XN, Li PP, Yu S et al (2014) Genetic detection of clubroot resistance loci in a new population of *Brassica rapa*. *Hortic Environ Biote* 55:540–547

Peng L, Zhou L, Li Q, Wei D, Ren X et al (2018) Identification of quantitative trait loci for clubroot resistance in *Brassica oleracea* with the use of *Brassica* SNP microarray. *Front Plant Sci* 9:822

Piao ZY, Deng YQ, Choi SR, Park YJ, Lim YP (2004) SCAR and CAPS mapping of *CRb*, a gene conferring resistance to *Plasmodiophora brassicae* in Chinese cabbage (*Brassica rapa* ssp *pekinensis*). *Theor Appl Genet* 108:1458–1465

Qasim MU, Zhao Q, Shahid M, Samad RA, Ahmar S et al (2020) Identification of QTLs containing resistance genes for Sclerotinia stem rot in *Brassica napus* using comparative transcriptomic studies. *Front Plant Sci* 11:776

Qian M, Fan Y, Li Y, Liu M, Sun W et al (2021) Genome-wide association study and transcriptome comparison reveal novel QTL and candidate genes that control petal size in rapeseed. *J Exp Bot* 72:3597–3610

Quijada PA, Udall JA, Lambert B, Osborn TC (2006) Quantitative trait analysis of seed yield and other complex traits in hybrid spring rapeseed (*Brassica napus* L.): 1. Identification of genomic regions from winter germplasm. *Theor Appl Genet* 113:549–561

Raboanatahiry N, Chao H, Dalin H, Pu S, Yan W et al (2018) QTL alignment for seed yield and yield related traits in *Brassica napus*. *Front Plant Sci* 9:1127

Raboanatahiry N, Chao HB, Guo LX, Gan JP, Xiang J et al (2017) Synteny analysis of genes and distribution of loci controlling oil content and fatty acid profile based on QTL alignment map in *Brassica napus*. *BMC Genomics* 18:776

Radoev M, Becker HC, Ecke W (2008) Genetic analysis of heterosis for yield and yield components in rapeseed (*Brassica napus* L.) by quantitative trait locus mapping. *Genetics* 179:1547–1558

Rahman H, Kebede B (2021) Mapping of seed quality traits in the C genome of *Brassica napus* by using a population carrying genome content of *B. oleracea* and their effect on other traits. *Plant Genome* 14:e20078

Raman H, Raman R, Diffey S, Qiu Y, McVittie B et al (2018) Stable quantitative resistance loci to blackleg disease in canola (*Brassica napus* L.) over continents. *Front Plant Sci* 9:1622

Raman H, Raman R, Eckermann P, Coombes N, Manoli S et al (2013) Genetic and physical mapping of flowering time loci in canola (*Brassica napus* L.). *Theor Appl Genet* 126:119–132

Raman H, Raman R, Qiu Y, Yadav AS, Sureshkumar S et al (2019) GWAS hints at pleiotropic roles for *FLOWERING LOCUS T* in flowering time and yield-related traits in canola. *BMC Genomics* 20:636

Raman H, Raman R, Qiu Y, Zhang YY, Batley J et al (2021) The *Rlm13* gene, a new player of *Brassica napus-Leptosphaeria maculans* interaction maps on chromosome C03 in canola. *Front Plant Sci* 12:654604

Raman R, Diffey S, Barbulescu DM, Coombes N, Luckett D et al (2020) Genetic and physical mapping of loci for resistance to blackleg disease in canola (*Brassica napus* L.). *Sci Rep* 10:4416

Rouxel T, Penaud A, Pinochet X, Brun H, Gout L et al (2003) A 10-year survey of populations of *Leptosphaeria maculans* in France indicates a rapid adaptation towards the *Rlm1* resistance gene of oilseed rape. *Eur J Plant Pathol* 109:871–881

Sakamoto K, Saito A, Hayashida N, Taguchi G, Matsumoto E (2008) Mapping of isolate-specific QTLs for clubroot resistance in Chinese cabbage (*Brassica rapa* L. ssp *pekinensis*). *Theor Appl Genet* 117:759–767

Schatzki J, Ecke W, Becker HC, Mollers C (2014) Mapping of QTL for the seed storage proteins cruciferin and napin in a winter oilseed rape doubled haploid population and their inheritance in relation to other seed traits. *Theor Appl Genet* 127:1213–1222

Scheben A, Severn-Ellis AA, Patel D, Pradhan A, Rae SJ et al (2020) Linkage mapping and QTL analysis of flowering time using ddRAD sequencing with genotype error correction in *Brassica napus*. *BMC Plant Biol* 20:546

Schiessl S, Iniguez-Luy F, Qian W, Snowdon RJ (2015) Diverse regulatory factors associate with flowering time and yield responses in winter-type *Brassica napus*. *BMC Genomics* 16:737

Shen Y, Xiang Y, Xu E, Ge X, Li Z (2018) Major co-localized QTL for plant height, branch initiation height, stem diameter, and flowering time in an alien introgression derived *Brassica napus* DH population. *Front Plant Sci* 9:390

Shi J, Li R, Qiu D, Jiang C, Long Y et al (2009) Unraveling the complex trait of crop yield with quantitative trait loci mapping in *Brassica napus*. *Genetics* 182:851–861

Shi J, Li R, Zou J, Long Y, Meng J (2011) A dynamic and complex network regulates the heterosis of yield-correlated traits in rapeseed (*Brassica napus* L.). *PLoS One* 6:e21645

Shi J, Zhan J, Yang Y, Ye J, Huang S et al (2015) Linkage and regional association analysis reveal two new tightly-linked major-QTLs for pod number and seed number per pod in rapeseed (*Brassica napus* L.). *Sci Rep* 5:14481

Smooker AM, Wells R, Morgan C, Beaudoin F, Cho K et al (2011) The identification and mapping of candidate genes and QTL involved in the fatty acid desaturation pathway in *Brassica napus*. *Theor Appl Genet* 122:1075–1090

Song J, Li B, Cui Y, Zhuo C, Gu Y et al (2021a) QTL mapping and diurnal transcriptome analysis identify candidate genes regulating *Brassica napus* flowering time. *Int J Mol Sci* 22:7559

Song JM, Liu DX, Xie WZ, Yang Z, Guo L et al (2021b) BnPIR: *Brassica napus* pan-genome information resource for 1689 accessions. *Plant Biotechnol J* 19:412–414

Sprague SJ, Balesdent MH, Brun H, Hayden HL, Marcroft SJ et al (2006) Major gene resistance in *Brassica napus* (oilseed rape) is overcome by changes in virulence of populations of *Leptosphaeria maculans* in France and Australia. *Eur J Plant Pathol* 114:33–40

Stein A, Wittkop B, Liu L. Z, Obermeier C, Friedt W et al (2013) Dissection of a major QTL for seed colour and fibre content in *Brassica napus* reveals colocalization with candidate genes for phenylpropanoid biosynthesis and flavonoid deposition. *Plant Breed* 132:382–389

Sun C, Wang B, Wang X, Hu K, Li K et al (2016a) Genome-wide association study dissecting the genetic architecture underlying the branch angle trait in rapeseed (*Brassica napus* L.). *Sci Rep* 6:33673

Sun F, Liu J, Hua W, Sun X, Wang X et al (2016b) Identification of stable QTLs for seed oil content by combined linkage and association mapping in *Brassica napus*. *Plant Sci* 252:388–399

Sun MY, Hua W, Liu J, Huang SM, Wang XF et al (2012) Design of new genome- and gene-sourced primers and identification of QTL for seed oil content in a specially high-oil *Brassica napus* cultivar. *PLoS One* 7:e47037

Suwabe K, Tsukazaki H, Iketani H, Hatakeyama K, Fujimura M et al (2003) Identification of two loci for resistance to clubroot (*Plasmodiophora brassicae Woronin*) in *Brassica rapa* L. *Theor Appl Genet* 107:997–1002

Tang J, Dai XL, Zhang Z, Song LQ, Zou XF et al (2008) Study on major agronomy and yield characters of new rapeseed lines from distant hybridization between *B. napus* L. and *Rorippa motana* (Wall.) Small. *Acta Agric Jiangxi* 20:35–37

Tang S, Zhao H, Lu SP, Yu LQ, Zhang GF et al (2021) Genome- and transcriptome-wide association studies provide insights into the genetic basis of natural variation of seed oil content in *Brassica napus*. *Mol Plant* 14:470–487

Teh L, Mollers C (2016) Genetic variation and inheritance of phytosterol and oil content in a doubled haploid population derived from the winter oilseed rape Sansibar ×× Oase cross. *Theor Appl Genet* 129:181–199

Toroser D, Thormann CE, Osborn TC, Mithen R (1995) RFLP mapping of quantitative trait loci controlling seed aliphatic-glucosinolate content in oilseed rape (*Brassica napus* L). *Theor Appl Genet* 91:802–808

Tudor EH, Jones DM, He Z, Bancroft I, Trick M et al (2020) QTL-seq identifies *BnaFT.A02* and *BnaFLC.A02* as candidates for variation in vernalization requirement and response in winter oilseed rape (*Brassica napus*). *Plant Biotechnol J* 18:2466–2481

Udall JA, Quijada PA, Lambert B, Osborn TC (2006) Quantitative trait analysis of seed yield and other complex traits in hybrid spring rapeseed (*Brassica napus* L.): 2. Identification of alleles from unadapted germplasm. *Theor Appl Genet* 113:597–609

Ueno H, Matsumoto E, Aruga D, Kitagawa S, Matsumura H et al (2012) Molecular characterization of the *CRa* gene conferring clubroot resistance in *Brassica rapa*. *Plant Mol Biol* 80:621–629

vanBerloo R, Aalbers H, Werkman A, Niks RE (2001) Resistance QTL confirmed through development of QTL-NILs for barley leaf rust resistance. *Mol Breed* 8:187–195

Vollrath P, Chawla HS, Alnajar D, Gabur I, Lee H et al (2021) Dissection of quantitative blackleg resistance reveals novel variants of resistance gene *Rlm9* in elite *Brassica napus*. *Front Plant Sci* 12:749491

Wang F, Guan CY (2010) Molecular mapping and identification of quantitative trait loci for yield components in rapeseed (*Brasscia napus* L.). *Hereditas* 32:271–277

Wang H, Cheng H, Wang W, Liu J, Hao M et al (2016a) Identification of *BnaYUCCA6* as a candidate gene for branch angle in *Brassica napus* by QTL-seq. *Sci Rep* 6:38493

Wang H, Yan M, Xiong M, Wang P, Liu Y et al (2020a) Genetic dissection of thousand-seed weight and fine mapping of *cqSW.A03-2* via linkage and association analysis in rapeseed (*Brassica napus* L.). *Theor Appl Genet* 133:1321–1335

Wang H, Zaman Q. U, Huang W, Mei D, Liu J et al (2019) QTL and candidate gene identification for silique length based on high-dense genetic map in *Brassica napus* L. *Front Plant Sci* 10:1579

Wang J, Fan Y, Mao L, Qu C, Lu K et al (2021) Genome-wide association study and transcriptome analysis dissect the genetic control of silique length in *Brassica napus* L. *Biotechnol Biofuels* 14:214

Wang J, Long Y, Wu B, Liu J, Jiang C et al (2009) The evolution of *Brassica napus FLOWERING LOCUS T* paralogues in the context of inverted chromosomal duplication blocks. *BMC Evol Biol* 9:271

Wang J, Xian X, Xu X, Qu C, Lu K et al (2017) Genome-wide association mapping of seed coat color in *Brassica napus*. *J Agric Food Chem* 65:5229–5237

Wang T, Wei L, Wang J, Xie L, Li YY et al (2020b) Integrating GWAS, linkage mapping and gene expression analyses reveals the genetic control of growth period traits in rapeseed (*Brassica napus* L.). *Biotechnol Biofuels* 13:134

Wang X, Chen L, Wang A, Wang H, Tian J et al (2016b) Quantitative trait loci analysis and genome-wide comparison for silique related traits in *Brassica napus*. *BMC Plant Biol* 16:71

Wang X, Wang H, Long Y, Liu L, Zhao Y et al (2015a) Dynamic and comparative QTL analysis for plant height in different developmental stages of *Brassica napus* L. *Theor Appl Genet* 128:1175–1192

Wang XD, Long Y, Yin YT, Zhang CY, Gan L et al (2015b) New insights into the genetic networks affecting seed fatty acid concentrations in *Brassica napus*. *BMC Plant Biol* 15:91

Wang XX, Zhang CY, Li LJ, Fritsche S, Endrigkeit J et al (2012) Unraveling the genetic basis of seed tocopherol content and composition in rapeseed (*Brassica napus* L.). *PLoS One* 7:e50038

Wang Y, Chen W, Chu P, Wan S, Yang M et al (2016c) Mapping a major QTL responsible for dwarf architecture in *Brassica napus* using a single-nucleotide polymorphism marker approach. *BMC Plant Biol* 16:178

Wang Z, Megha S, Kebede B, Kav NNV, Rahman H (2022) Genetic and molecular analysis reveals that two major loci and their interaction confer clubroot resistance in canola introgressed from rutabaga. *Plant Genome* 15:e20241

Wei L, Jian H, Lu K, Filardo F, Yin N et al (2016) Genome-wide association analysis and differential expression analysis of resistance to Sclerotinia stem rot in *Brassica napus*. *Plant Biotechnol J* 14:1368–1380

Werner S, Diederichsen E, Frauen M, Schondelmaier J, Jung C (2008) Genetic mapping of clubroot resistance genes in oilseed rape. *Theor Appl Genet* 116:363–372

Wu D, Liang Z, Yan T, Xu Y, Xuan L et al (2019) Whole-genome resequencing of a worldwide collection of rapeseed accessions reveals the genetic basis of ecotype divergence. *Mol Plant* 12:30–43

Wu J, Cai G, Tu J, Li L, Liu S et al (2013) Identification of QTLs for resistance to Sclerotinia stem rot and *BnaC.IGMT5.a* as a candidate gene of the major resistant QTL *SRC6* in *Brassica napus*. *PLoS One* 8:e67740

Wu J, Zhao Q, Liu S, Shahid M, Lan L et al (2016) Genome-wide association study identifies new loci for resistance to Sclerotinia stem rot in *Brassica napus*. *Front Plant Sci* 7:1418

Wurschum T, Liu WX, Maurer HP, Abel S, Reif JC (2012) Dissecting the genetic architecture of agronomic traits in multiple segregating populations in rapeseed (*Brassica napus* L.). *Theor Appl Genet* 124:153–161

Xu L, Hu K, Zhang Z, Guan C, Chen S et al (2016) Genome-wide association study reveals the genetic architecture of flowering time in rapeseed (*Brassica napus* L.). *DNA Res* 23:43–52

Xu P, Ali A, Han B, Wu X (2018) Current advances in molecular basis and mechanisms regulating leaf morphology in rice. *Front Plant Sci* 9:1528

Xu Y, Zhang B, Ma N, Liu X, Qin M et al (2020) Quantitative trait locus mapping and identification of candidate genes controlling flowering time in *Brassica napus* L. *Front Plant Sci* 11:626205

Yang M, Huang C, Wang M, Fan H, Wan S et al (2019) Fine mapping of an up-curling leaf locus (*BnUC1*) in *Brassica napus*. *BMC Plant Biol* 19:324

Yang P, Shu C, Chen L, Xu J, Wu J et al (2012) Identification of a major QTL for silique length and seed weight in oilseed rape (*Brassica napus* L.). *Theor Appl Genet* 125:285–296

Yang Y, Shen Y, Li S, Ge X, Li Z (2017) High density linkage map construction and QTL detection for three silique-related traits in *Orychophragmus violaceus* derived *Brassica napus* population. *Front Plant Sci* 8:1512

Yang Y, Zhu K, Li H, Han S, Meng Q et al (2018) Precise editing of *CLAVATA* genes in *Brassica napus* L. regulates multilocular silique development. *Plant Biotechnol J* 16:1322–1335

Ye J, Liu H, Zhao Z, Xu L, Li K et al (2020) Fine mapping of the QTL *cqSPDA2* for chlorophyll content in *Brassica napus* L. *BMC Plant Biol* 20:511

Ye J, Yang Y, Chen B, Shi J, Luo M et al (2017) An integrated analysis of QTL mapping and RNA sequencing provides further insights and promising candidates for pod number variation in rapeseed (*Brassica napus* L.). *BMC Genomics* 18:71

Yi B, Chen W, Ma C, Fu T, Tu J (2006) Mapping of quantitative trait loci for yield and yield components in *Brassica napus* L. *Acta Agronomca Sinica* 32:676–682

Yu CY (2013) Molecular mechanism of manipulating seed coat coloration in oilseed *Brassica* species. *J Appl Genet* 54 :135–145

Yu F, Gugel RK, Kutcher HR, Peng G, Rimmer SR (2013) Identification and mapping of a novel blackleg resistance locus *LepR4* in the progenies from *Brassica napus* × *B. rapa* subsp. *sylvestris*. *Theor Appl Genet* 126:307–315

Yu F, Lydiate DJ, Rimmer SR (2005) Identification of two novel genes for blackleg resistance in *Brassica napus*. *Theor Appl Genet* 110:969–979

Yu F, Lydiate DJ, Rimmer SR (2008) Identification and mapping of a third blackleg resistance locus in *Brassica napus* derived from *B. rapa* subsp. *sylvestris*. *Genome* 51:64–72

Yu F, Zhang X, Huang Z, Chu M, Song T et al (2016) Identification of genome-wide variants and discovery of variants associated with *Brassica rapa* clubroot resistance gene *Rcr1* through bulked segregant RNA sequencing. *PLoS One* 11:e0153218

Yu F, Zhang Y, Wang J, Chen Q, Karim MM et al (2022) Identification of two major QTLs in *Brassica napus* lines with introgressed clubroot resistance from turnip cultivar ECD01. *Front Plant Sci* 12:785989

Yu FQ, Zhang XG, Peng G, Falk KC, Strelkov SE et al (2017) Genotyping-by-sequencing reveals three QTL for clubroot resistance to six pathotypes of *Plasmodiophora brassicae* in *Brassica rapa*. *Sci Rep* 7:4516

Yu K, He Y, Li Y, Li Z, Zhang J et al (2021) Quantitative trait locus mapping combined with RNA sequencing reveals the molecular basis of seed germination in oilseed rape. *Biomolecules* 11:1780

Zhang FQ, Huang JY, Tang MQ, Cheng XH, Liu YY et al (2019) Syntenic quantitative trait loci and genomic divergence for Sclerotinia resistance and flowering time in *Brassica napus*. *J Integr Plant Biol* 61:75–88

Zhang L, Li S, Chen L, Yang G (2012) Identification and mapping of a major dominant quantitative trait locus controlling seeds per silique as a single Mendelian factor in *Brassica napus* L. *Theor Appl Genet* 125:695–705

Zhang L, Yang G, Liu P, Hong D, Li S et al (2011a) Genetic and correlation analysis of silique-traits in *Brassica napus* L. by quantitative trait locus mapping. *Theor Appl Genet* 122:21–31

Zhang Q, Dandena H, McCausland M, Liu H, Liu Z et al (2021a) Genetic analysis of a horizontal resistance locus *BLMR2* in *Brassica napus*. *Front Plant Sci* 12:663868

Zhang Q, Guan P, Zhao L, Ma M, Xie L et al (2021b) Asymmetric epigenome maps of subgenomes reveal imbalanced transcription and distinct evolutionary trends in *Brassica napus*. *Mol Plant* 14:604–619

Zhang X, Huang Q, Wang P, Liu F, Luo M et al (2021c) A 24,482-bp deletion is associated with increased seed weight in *Brassica napus* L. *Theor Appl Genet* 134:2653–2669

Zhang XH, Peng G, Kutcher HR, Balesdent MH, Delourme R et al (2016) Breakdown of *Rlm3* resistance in the *Brassica napus-Leptosphaeria maculans* pathosystem in western Canada. *Eur J Plant Pathol* 145:659–674

Zhang Y, Li Q, Cui Y, Liu Z, Chen Z et al (2018a) Genetic characterization and fine mapping for multi-inflorescence in *Brassica napus* L. *Theor Appl Genet* 131:2311–2319

Zhang Y, Li X, Chen W, Yi B, Wen J et al (2011b) Identification of two major QTL for yellow seed color in two crosses of resynthesized *Brassica napus* line No. 2127-17. *Mol Breed* 28:335–342

Zhang Y, Xu A, Lang L, Wang Y, Liu X et al (2018b) Genetic mapping of a lobed-leaf gene associated with salt tolerance in *Brassica napus* L. *Plant Sci* 269:75–84

Zhao B, Wang B, Li Z, Guo T, Zhao J et al (2019a) Identification and characterization of a new dwarf locus *DS-4* encoding an Aux/IAA7 protein in *Brassica napus*. *Theor Appl Genet* 132:1435–1449

Zhao CJ, Xie ML, Liang LB, Yang L, Han HS et al (2022) Genome-wide association analysis combined with quantitative trait loci mapping and dynamic transcriptome unveil the genetic control of seed oil content in *Brassica napus* L. *Front Plant Sci* 13:929197

Zhao J, Meng J (2003) Genetic analysis of loci associated with partial resistance to *Sclerotinia sclerotiorum* in rapeseed (*Brassica napus* L.). *Theor Appl Genet* 106:759–764

Zhao J, Udall JA, Quijada PA, Grau CR, Meng J et al (2006a) Quantitative trait loci for resistance to *Sclerotinia sclerotiorum* and its association with a homeologous non-reciprocal transposition in *Brassica napus* L. *Theor Appl Genet* 112:509–516

Zhao JY, Becker HC, Zhang DQ, Zhang YF, Ecke W (2006b) Conditional QTL mapping of oil content in rapeseed with respect to protein content and traits related to plant development and grain yield. *Theor Appl Genet* 113:33–38

Zhao JY, Dimov Z, Becker HC, Ecke WG, Mollers C (2008a) Mapping QTL controlling fatty acid composition in a doubled haploid rapeseed population segregating for oil content. *Mol Breed* 21:115–125

Zhao Q, Wu J, Cai G, Yang Q, Shahid M et al (2019b) A novel quantitative trait locus on chromosome A9 controlling oleic acid content in *Brassica napus*. *Plant Biotechnol J* 17:2313–2324

Zhao W, Chao H, Zhang L, Ta N, Zhao Y et al (2019c) Integration of QTL mapping and gene fishing techniques to dissect the multi-main stem trait in rapeseed (*Brassica napus* L.). *Front Plant Sci* 10:1152

Zhao W, Wang X, Wang H, Tian J, Li B et al (2016) Genome-wide identification of QTL for seed yield and yield-related traits and construction of a high-density consensus map for QTL comparison in *Brassica napus*. *Front Plant Sci* 7:17

Zhao X, Yu K, Pang C, Wu X, Shi R et al (2021) QTL analysis of five silique-related traits in *Brassica napus* L. across multiple environments. *Front Plant Sci* 12:766271

Zhao ZG, Hu TT, Ge XH, Du XZ, Ding L et al (2008b) Production and characterization of intergeneric somatic hybrids between *Brassica napus* and *Orychophragmus violaceus* and their backcrossing progenies. *Plant Cell Rep* 27:1611–1621

Zhou H, Xiao X, Asjad A, Han D, Zheng W et al (2022) Integration of GWAS and transcriptome analyses to identify SNPs and candidate genes for aluminum tolerance in rapeseed (*Brassica napus* L.). *BMC Plant Biol* 22:130

Zhou Q, Han D, Mason AS, Zhou C, Zheng W et al (2018) Earliness traits in rapeseed (*Brassica napus*): SNP loci and candidate genes identified by genome-wide association analysis. *DNA Res* 25:229–244

Zhu HF, Zhai W, Li XF, Zhu YY (2019) Two QTLs controlling clubroot resistance identified from bulked segregant sequencing in pakchoi (*Brassica campestris* ssp. *chinensis* Makino). *Sci Rep* 9:9228

Zhu J, Wang W, Jiang M, Yang L, Zhou X (2021) QTL mapping for low temperature germination in rapeseed. *Sci Rep* 11:23382

Zhu Y, Cao Z, Xu F, Huang Y, Chen M et al (2012) Analysis of gene expression profiles of two near-isogenic lines differing at a QTL region affecting oil content at high temperatures during seed maturation in oilseed rape (*Brassica napus* L.). *Theor Appl Genet* 124:515–531

Zhu Y, Ye J, Zhan J, Zheng X, Zhang J et al (2020) Validation and characterization of a seed number per silique quantitative trait locus *qSN.A7* in rapeseed (*Brassica napus* L.). *Front Plant Sci* 11:68

5 Mining Alleles for Creating Environment Resilient Sunflower

Aleksandra Radanović and Renate Horn

5.1 INTRODUCTION

The genetic diversity present in sunflower (*Helianthus annuus* L.) that can be explored for allele mining can be best realized in the background of the botany in the genus *Helianthus*, the domestication of sunflowers, and the history of sunflower breeding, which explains the occurrence of bottlenecks (Burke et al. 2002, 2005).

Common sunflower (*H. annuus* L.) is a widespread cultivated species of sunflower that belongs to the broad and genetically divergent genus *Helianthus* within the Asteraceae family. The *Helianthus* genus comprises about 70 species, mostly perennial with about a dozen annual forms, having an emblematic composite flower head resembling the sun (Heiser et al. 1969; Berglund 2007; Britannica 2021). The average plant height is from 1.8 to 3 m, ranging from 0.3 to 4.5 m or even more. The common sunflower is a robust annual self-incompatible insect-pollinated plant having a hairy stem and broad, coarsely toothed leaves. However, annual sunflowers have shallow, string-like roots, in contrast to perennials that have deeper roots with rhizomes and tubers (USDA-NRCS 2006; Britannica 2021). The composite head has two types of flowers: sterile ray flowers and hermaphrodite disc flowers. Petal-like ray flowers surround the disc flowers in the center of the head. Ray florets give a beautiful, characteristic sunny yellow corolla that sunflowers are famous for (Fernández-Luqueño et al. 2014). Sunflower, because of its beauty and resemblance to the sun and sunshine is considered as the plant of positivity, joy, warmth, happiness, friendship, and adoration. Common sunflower is an open land species, a species of moist grassland. A relatively short growth season, 90–120 days from sowing to harvesting for varieties of early maturation and 120–160 days for late varieties, makes it convenient for agricultural production (Fernández-Luqueño et al. 2014). The haploid chromosome number is 17 (diploid 2n = 34). Tetra- and hexaploid forms have been found as well (Atlagić 2004). Nikolai Ivanovich Vavilov, a Russian geneticist and the founder of the Theory of the Centers of Origin from 1924 to 1933, identified eight Centers of Plant Origin (CPO), i.e. Centers of Plant Diversity (CPD), since CPOs are characterized by the greatest genetic variability within the species, and wild forms exist together with cultivated forms. Though sunflower is the only cultivar among the most economically important crops that comes from North America, it was overlooked by Vavilov (Atlagić 2004; Hummer 2015; Hummer and Hanckok 2015). The reason might be that the broader importance of sunflower as a valuable commodity did not start to rise on the global market before the Second World War. Plant breeders have achieved remarkable improvements in *H. annuus* L., particularly in the second half of the 20th century. By means of intra- and interspecific hybridization, favorable gene recombination has been achieved within the genus, and the resulting improved genetic variability has led to higher oil content, increased resistance to biotic and abiotic stresses, and the identification of cytoplasmic male sterility utilized in the production of commercial hybrids. The sunflower crop exhibited a rapid evolution on a global scale, from 9.6 million hectares producing 10 million tons of seeds in 1975 to 27 million hectares producing 52 million tons in 2018 (Pilorgé 2020). The top average seed yield could reach up to 3–4 t/ha. The first three ranking states according to average

DOI: 10.1201/9781003385035-5

seed yield are Israel with an average seed yield of 4.4 t/ha on a total production area of 1,100 ha, Uzbekistan with 4.1 t/ha on 10,186 ha, and Serbia with 3.3 t/ha on 219,404 ha. However, the largest producers are located in the Russian Federation with 15.4 million tons per year and Ukraine with 15.3 million tons, which represents about 55% of the global annual sunflower production, reaching up to 56.1 million tons (https://www.atlasbig.com/en-us). Today, 14% of the world's seed oil comes from sunflowers. Sunflower oil is considered a high-quality oil, having a light color, bland flavor, and high smoke point, as well as a lack of linoleic acid and a high content of unsaturated fatty acids. Beside its use for oil, sunflowers are grown for edible seeds. At the moment, sunflower is considered the most profitable and economic oilseed crop, globally ranked as the third most important oilseed, the fourth vegetable oil after soybeans, rapeseed, and safflower, as well as the third oilseed meal among the sources of protein feed (Adeleke and Babalola 2020; Pilorgé 2020).

Fossil remains from the Eocene in Patagonia, as well as in Africa and Australia, indicate that ancestral *Asteraceae* populations had been a part of the general geoflora of Southern Godwana even before the separation of continents. That longevity, survival through durability, and adaptability could be the reason that the Asteraceae family today is the largest among plants, harboring 25,000 species in 1,600 genera all over the world (Barreda et al. 2010; Rolnik and Olas 2021). Hence, there is still perplexity in scientific circles about the CPO of the common sunflower. According to some sources, *H. annuus* is a North American native. At first, the CPO of the *Helianthus* genus was placed in the Western United States of America (USA). There are three main subspecies of *H. annuus* L., namely, the "wild" *H. annuus* ssp. *lenticularis*, a branched, small-headed phenotype with small achenes and very few rays. This "wild" form is widespread in the Western USA. The second form, *H. annuus* ssp. *annuus*, "weed" sunflower, is branched as well; however, heads and achenes are larger with more numerous rays. The "weed" form was mostly found in the Midwest of the US. The third form, *H. annuus* ssp. *macrocarpus*, the "giant" unbranched form with a big single head, large achenes, and numerous rays, has been cultivated because of edible seeds. American indigenous tribes, Hopi and Navajo in particular, had grown and cultivated sunflowers ever since prehistoric times in today's states of Arizona and New Mexico, as far back as 3,000 B.C. They consumed raw seeds or pounded them to make flat cakes. Sunflower had been used as a medicinal plant for curing chest pain. Native Americans spread cultivated sunflower further to the east of the US, spreading this crop to the Dakotas, Kentucky, Iowa, Tennessee, Ohio, Pennsylvania, and Ontario (Canada). Sunflower was grown in the Mississippi-Missouri Basin 2,800 years ago. This cultivated form, today, does not exist outside of cultivation and is grown worldwide. According to some points of view, sunflowers had been domesticated even before corn (Heiser 1955; Putt 1997). However, there are some newer research results brought out between 1997 and 2000 that indicate the hypothesis of Mexican sunflower domestication. The San Andre's archeological site, about 10 km south of the Gulf of Mexico, in Tabasco, revealed evidence that sunflower was grown in Southern Mexico about 2800 B.C., as well as in Western El Salvador by 400 B.C., but further investigations are required to firmly establish Mexico's CPO of *Helianthus* (Lentz et al. 2001; Heiser 2007; Lentz et al. 2008). Earlier molecular investigations, conducted by isozyme and chloroplast DNA (cpDNA) analysis, suggested "a single origin of the domesticated sunflower from a very limited gene pool" (Rieseberg and Seiler 1990). Hinting that the place of the initial cultivation could be the American Midwest, the authors stated that molecular evidence was not conclusive enough to mark the exact area of the first sunflower domestication event. Recent results based on DNA sequence diversity analysis using single nucleotide polymorphisms (SNPs), presented by Park and Burke (2020), advocate a "single domestication event" that took place in the Great Plains region, i.e. in central North America. A greater reduction of genetic variability followed by narrower novel genetic variability in the cultivated sunflower population enclosed by wild populations was observed, suggesting a "domestication bottleneck" approximately 5,000 years ago. In the pre-Columbian era, sunflowers in their wild forms could be found in today's Central Mexico, passing through the USA and reaching up to Southern Canada. The spots of domestication are still the subject of debate, but their cultivation has undoubtedly been conducted by indigenous tribes. They found sunflowers very economical, as all parts of the plant had some utility.

Primarily, seeds were used for direct consumption, as a component of the diet, as oil for cooking, or as ointment for skin or hair. Stems were utilized for constructing hoods above the fire places for making traditional Hopi food, piki. Sunflower was used for medical and ceremonial purposes as a dye for textiles and basketry (Heiser 1951; Putt 1997). In the post-Columbian era, sunflowers reached Europe in the 16th century. The first contacts between Europe and America were made by Spaniards. Their conquistadors brought the sunflower to Spain at the beginning of the 16th century. However, different types of sunflowers that have been traced in Europe throughout the 16th century suggest that introductions from the New World happened at different times and in various places. At first, sunflower was used as an ornamental garden plant; later, its use in diets began. Up to the end of the 16th century, sunflowers spread in gardens all over Western Europe, first in Italy and France, and later in Belgium, England, Germany, the Netherlands, and Switzerland. By that time, some botanists and physicians had brought attention to the sunflower as a medical plant. In the 17th century, sunflowers moved further to the east, reaching Russia in the early 18th century. Credits for this introduction went to Peter the Great, who obtained the original seeds from the Netherlands. The first patent for getting oil from sunflower seeds was granted in England in 1716. However, sunflower oil production on a commercial scale only started in Russia between 1830 and 1840. The sunflower production in Russia was boosted because of high seed oil demands. The Orthodox Church has quite strict rules of Lenten fasting, and all meals of animal origin are forbidden for certain periods of time during a year (Charlet et al. 1987). The sunflower seed oil very conveniently filled up this gap in the diet. Hence, from 1860 on, a selection for high oil content has been successfully conducted in Russia, increasing seed oil content from 28% to nearly 50%. Russian breeders achieved the first high-yielding sunflower varieties that had been spread all over the world. The introduction of male sterility by interspecific hybridization from *H. petiolaris* (prairie sunflower) and *H. annuus*, as well as the identification of corresponding fertility restorer genes at the beginning of the 1970s, opened the way for an enormous increase in the production of sunflower as hybrids in the mid-seventies. The heterotic effects that hybrids exhibited provided additional yield and oil content enhancement, as well as improvements in disease resistance. Nowadays, sunflower is the second largest hybrid crop after corn (Rieseberg et al. 1994; Putt 1997; Seiler et al. 2017).

However, there were certain opinions that sunflower breeding has reached its plateau for a number of important agronomic traits (Rieseberg and Seiler 1990). Intensive breeding and selection of desirable genotypes using the same selection criteria over and over again could, after a while, deplete the available genetic variability. The requirements are growing higher and higher due to market competition, climate change, and more frequent abiotic and novel biotic stresses, not to mention the growth of the global human population, which is demanding more and more food. The first routine step is the detection of desirable genes within the same species, also called the primary gene pool (older bred varieties, landraces, etc.). Then, if the requirements are not met, plant breeders should mine deeper, reaching taxonomically more distant relatives commonly found in spontaneous relatives, wild, i.e. native populations, relying on Vavilov's law of homologous series in variation. The concept was brought out by N. I. Vavilov in the early 20s based on his research on plant variability and the CPO. The common sunflower is a good example of successful gene introgression from wild relatives representing the secondary gene pool to the cultivated gene pool. The introgression of cytoplasmic male-sterility (*cms*) genes from *H. petiolaris* to *H. annuus*, by classical interspecific hybridization at the end of the 1960s (Leclercq 1969) and fertility restorer (*Rf*) genes from *H. petiolaris* and some tamed populations of *H. annuus* from Arizona, the Dakotas, New Mexico, etc., at the beginning of the 1970s has led to the introduction of commercial sunflower hybrids that permitted an increase in agricultural production to a new and more intensive, industrial level (Kinman 1970). A number of desirable gene recombination, abiotic and biotic stress resistance, herbicide tolerance, and modified fatty acid profiles have been obtained from wild relatives, representing secondary gene pool mining (Enns et al. 1970; Warburton et al. 2017). Among thirteen major crops globally grown for food, sunflower is fifth in the utilization of useful genes from wild relatives for genetic variation improvement of the cultivated form (Seiler et al. 2017). Climate change is forcing sunflower breeders to intensify efforts to create novel variability resilient enough to withstand increasing production irregularities caused by biotic and abiotic stresses (Miladinović et al. 2019). Molecular genetic approaches in modern sunflower breeding significantly improve monitoring, gene mining in genetic resources, use of the gene pool, and gene introgression in order to meet increased challenges in food production placed by human population increment and climate change (Dimitrijević and Horn 2018).

5.2 STRATEGY FOR ALLELE MINING IN SUNFLOWER

Especially, for the new demands in breeding climate-resilient sunflower hybrids, allele mining will help to make the breeding process more efficient and less time-consuming. In allele mining, plant genetic resources are screened for novel or superior alleles that can be used for crop improvement. Allele mining nowadays not only addresses differences in the coding region but also in the 5′- and 3′-regions in order to detect causes of expression differences (Kumar et al. 2010). Allele mining requires several steps, and considerable progress has been made in sunflowers in this direction in the last few decades. In other crop species, like rice or barley, allele mining has already been successfully applied to identify new alleles and integrate them into breeding programs (Kumar et al. 2010; Ashkani et al. 2015).

First, germplasm must be assembled and maintained to represent the broad genetic diversity of the crop (Figure 5.1). These accessions need to be molecularly characterized and screened for agronomically important traits. Next, the traits must be mapped on genetic maps to obtain their location in the genome. Finally, a subset of accessions representing most of the genetic variability must be selected for further investigations to keep costs at an acceptable limit. An annotated reference genome is essential to identifying candidate genes in relevant genomic regions. Genes can also be cloned by map-based cloning approaches and sequenced.

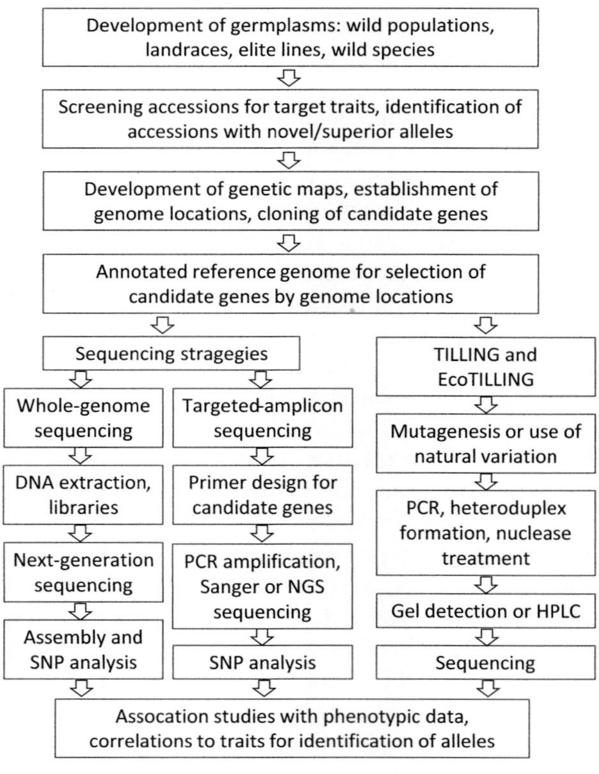

FIGURE 5.1 General concept for allele mining in sunflower.

The other options are whole-genome sequencing (WGS) strategies, which can be applied to selected accessions representing the most genetic diversity. Next-generation sequencing (NGS) technologies have reduced the cost of sequencing considerably, so WGS today represents the most interesting strategy. In addition, targeted-induced local lesions in genomes (TILLING) and Ecotype-TILLING (EcoTILLING) can be applied to detect new alleles (Comai et al. 2004; Till et al. 2006). Special software programs are then available to perform SNP analyses (Kumar et al. 2010; Ashkani et al. 2015). These SNPs can then be used to identify haplotypes and alleles. Correlation and association studies using the different trait phenotypes finally allow the identification of novel and superior alleles, which can then be introduced by developing markers for marker-assisted selection (MAS) or by genomic selection. In the following, the state of the art for sunflowers and the different steps in this general concept will be outlaid.

5.3 PLANT GENETIC RESOURCES AVAILABLE FOR ALLELE MINING IN SUNFLOWER

Decades of sunflower breeding have led to the narrowing down of the genetic diversity in the genetic collections used as gene well for creating novel and improved genetic variability through the breeding process. Thus, for the mining of desirable alleles in order to broaden and renew the available gene pool in breeding programs, breeders have to reach out to all available plant genetic resources. These would first cover the sunflower cultivated pool, as it is the easiest and quickest source, not only for discovering but also for introducing desirable alleles into the existing breeding material. Older genetic breeding material could still offer some specific useful genes for creating desirable ideotypes. If this older breeding gene pool is not good enough to fulfill the requirements, the next gene pool to search for could be local and domesticated sunflower populations. This material covers wild forms that have been tamed and purposely grown in past times, but with no human intervention in genetic design, as well as populations subjected to natural selection and breeding, resulting in landraces and sunflower populations commonly well adapted to their local environment. However, most frequently, breeders needed to expand the mining to the secondary gene pool. For instance, tolerance or resistance to most diseases and pests has been discovered among natural populations of wild species within the genus *Helianthus* (Seiler et al. 2017; Terzić et al. 2020a). In addition, this is also true for other desirable traits associated with agronomic and seed-oil characteristics, abiotic stress tolerance, and genes essential for hybrid breeding such as *cms* and fertility restoration (Seiler and Rieseberg 1997; Seiler et al. 2017).

However, broadening the genetic variability of cultivated breeding material by introducing desirable traits from taxonomically more or less distant wild species is not only time-consuming but also quite challenging if conventional breeding is used. It involves interspecific, sometimes even intergeneric hybridization (Reyes-Valdes et al. 2005), using parents frequently with different numbers of chromosomes, even with different genomes. Still, even if some offspring were obtained, overcoming sterility, incompatibility, and other eventual obstacles, the contribution of the cultivated parent to the offspring could be seriously compromised by the genetic contribution of the wild parent. Hybridization involves haploid genomes of both parents and, along with one or two desirable traits from the wild parent, goes with all sorts of undesirable and unwanted traits that require years of backcrossing to be cleaned up, sometimes with very uncertain results due to linkage drags. Cytogenetic methods in hybridization involving aneuploidy, chromosomal substitutions, and additions could diminish the problem of unwanted genetic burden coming from wild relatives, but commonly entail uncertainties due to complications from chromosomal aberrations and time-consuming breeding pathways. Moreover, most species belonging to the secondary gene pool remain unused because abortion of embryos occurs in the early stages of development (Seiler and Rieseberg 1997). In these cases, even the use of *in vitro* regeneration usually fails to overcome this problem. Despite these drawbacks, the secondary gene pool, with its large number of wild species in the genus *Helianthus*, is still extensively used in sunflower pre-breeding programs.

5.3.1 Cultivated Sunflower Pools

Collections of cultivated sunflowers comprised of different genotypes (open-pollinated varieties - OPV, local populations, landraces, and inbred lines) are mostly maintained in breeding institutions, either in the public or private sector, as well as in gene banks. The United States Department of Agriculture (USDA) sunflower gene bank in Ames, Iowa, USA, which is a part of the U.S. National Plant Germplasm System (NPGS), is holding a vast collection of cultivated sunflowers, representing most of the sunflower genetic variability, donated or collected over time from all over the world. It was established in the late 1940s, with the first accessions mainly originating from Turkey (Seiler and Marek 2011). Today, this collection consists of open-pollinated varieties of different origins; landraces collected in the Southwest US and Spain in the late 1900s; as well as inbred lines developed as early as the 1960s at Texas A&M University, USA. Inbred lines, partial inbred lines, and populations were additionally donated by the USDA Agricultural Research Service (ARS) sunflower breeding program established in Fargo, North Dacota, in the 1980s (Terzić et al. 2020a). Moreover, this collection possessed 360 pre-bred sunflower lines, partially inbred accessions that were developed by the University of British Columbia, Canada. Information on accessible genotypes can be found on the following website: https://www.ag.ndsu.edu/fss/ndsu-varieties/usda-sunflower-inbred-lines. The accessions are publicly available for research and educational purposes; thus, they represent an invaluable source of genetic variability that can be freely accessed, examined, exploited, and finally used for allele mining and the introduction of favorable alleles into desirable genetic backgrounds, leading to the development of superior sunflower genotypes. Within this collection, a sunflower association panel, the UGA-SAM1 core collection, was established, which represents 90% of the genetic diversity present in the gene pool of cultivated sunflower (Chapman et al. 2008; Mandel et al. 2011). Subsequently, UGA-SAM1 was utilized for association mapping, gene discovery, and phenotyping (Mandel et al. 2013; Nambeesan et al. 2015; Horn et al. 2019; Terzic et al. 2020b).

On January 1st, 2020, France's new National Research Institute for Agriculture, Food and Environment (INRAE) was formed, merging the National Institute for Agricultural Research (INRA) and the National Research Institute of Science and Technology for the Environment and Agriculture (IRSTEA) (Arpin et al. 2020; INRAE 2020). INRAE is a complex public organization formed to carry out various scientific research within the framework of national policy, among which are the examination, utilization, restoration, and preservation of genetic resources and biodiversity. Within the INRAE system, there is a Biological Resource Center for Sunflower (Centre de Ressources Biologiques - CRB) in Toulouse, France. All the research related to sunflowers in INRAE is conducted at the CRB Occitanie-Toulouse Center. The Laboratory of Plant-Microbe Interactions (LIPM, INRA Occitanie-Toulouse Research Center) is responsible for accessions' preservation, propagation, and distribution of *H. annuus*, its wild relatives, and relative species (https://www.lipme.fr/). According to the INRAE research unit in genomics and bioinformatics, the Unité de Recherche Génomique Info (URGI), LIPM holds 2,170 entries of *H. annuus* L., and the list is given with accession number, name, taxon, biological status, and country at https://urgi.versailles.inra.fr/siregal/siregal/accessionAction.do?grcId=33&taxonIds=11457. The sunflower collection at INRA consists of 3,933 accessions (Meena and Sujatha 2022). Within this collection, there are 403 cultivated populations (OPV, landraces, pools, or synthetics developed as a base for breeding programs) and 2,300 inbred lines. Accessions are not only of French origin but also gathered through cooperation with different institutions around the world. Among the available accessions, there are 1,690 maintainers (B lines) of CMS PET1 cytoplasm, with male sterile A forms being available for 560 of them, and 640 fertility restorers (R lines). Most of the accessions are available for breeding and research purposes under a Material Transfer Agreement (MTA) (Filippi et al. 2020; Terzić et al. 2020a).

The collection located at the Institute of Field and Vegetable Crops (IFVC) in Novi Sad, National Institute of the Republic of Serbia, has perhaps the greatest collection of cultivated sunflowers. Within this collection, there are OPV, synthetic populations of divergent types, and over 7,000 inbred lines that have been developed for breeding purposes (Jocić et al. 2012; Atlagić and Terzić 2014; Hladni et al. 2017; Dimitrijević and Horn 2018; Terzić et al. 2020a). Moreover, decades of breeding at IFVC have led to the creation of numerous hybrids, of which 700 have been registered either by IFVC alone or in cooperation with other institutions from other countries and companies (Anđelković et al. 2020).

The oldest sunflower gene bank is located at the Federal Research Center N. I. Vavilov All-Russian Institute of Plant Genetic Resources (VIR) under the Ministry of Science and Higher Education of Russian Federation [https://www.vir.nw.ru/en/], holding 2,230 accessions of cultivated sunflowers, including landraces, local varieties, cultivars, inbred lines, and recombinant inbred lines (RILs) that were obtained and/or developed from national and international breeding programs, as well as collecting expeditions (Gavrilova et al. 2014). These sunflower lines include several groups: (i) 189 lines with a different manifestation of morphological characters; (ii) 120 R lines; 20 CMS lines as well as their fertile analogs; (iii) 46 lines with resistance to different races of downy mildew; (iv) 90 lines out of 362 of the genetic collection are entries exhibiting genetic variability in seed storage protein content. Moreover, this gene bank saves the first CMS-based hybrids preserved as hybrid populations and lines.

Dobrudzha Agricultural Institute General Toshevo (DAI), as the sole sunflower breeding institution in Bulgaria, possesses a collection of over 6,000 inbred lines and numerous hybrids, some of which are registered in Bulgaria and abroad (Georgiev et al. 2016, 2018).

The Active Germplasm Bank of the Manfredi Experimental Station (AGB-IM) Instituto Nacional de Tecnología Agropecuaria (INTA) in Argentina is a part of the National Network of Genetic Resources and possesses approximately 1,200 accessions of cultivated sunflower. This collection contains OPV and composite populations, as well as inbred lines. INTA germplasm comes from a wide range of geographical origins and is well adapted to a wide range of environmental variation (Filippi et al. 2017, 2020). At INTA, an association mapping population (AMP-IL) was created, comprising 137 inbred lines developed to represent an equilibrium of genetic variability and good adaptation to immediate agro-ecological conditions (Filippi et al. 2015, 2017).

The collection of 196 accessions of confectionary-type landraces collected across Spain is kept at the Centre of Plant Genetic Resources of the National Institute for Agricultural and Food Research and Technology (CRF-INIA) (Dimitrijević and Horn 2018; Terzić et al. 2020a).

The National Seed Gene Bank of Turkey keeps a collection of over 932 sunflower accessions of both oilseed and confectionary sunflowers, including Turkish landraces and sunflowers collected from fields, farmer storage, and markets (Tan and Kaya 2019).

5.3.2 Crop Wild Relatives

The use of crop wild relatives (CWR) for sunflower improvement is a time-consuming process that requires a lot of resources and manual labor – starting with collecting wild relatives, their maintenance, testing, and mining for desirable alleles to the valorization in breeding programs (Radanović et al. 2019). Existing gene banks provide quite a bit of genetic variability for breeders to mine for desirable traits. The greatest gene bank of wild sunflowers was established in the mid-1970s in Bushland, Texas, and later moved to Ames, Iowa. This is the USDA-ARS NPGS sunflower gene bank that possesses samples of all annual and perennial species except for *H. niveus ssp. niveus*, endemic to Baja California, Mexico (Terzić et al. 2020a). This collection is comprised of 1,057 wild *H. annuus* samples, 636 samples of other wild annual species, and 904 wild perennial samples. Accessions were collected mainly in the U.S.; however, some entries originate from Mexico, Canada, Argentina, Australia, Mozambique, Moldova, and the former Soviet Union. This collection and the access to the seeds it provides to the global research community were the cornerstones for the development of other gene banks and breeding programs around the world (Seiler and Jan 2010). The majority of other sunflower gene banks possess some wild species samples either collected in the U.S. or received from the USDA-ARS. The gene flow from USDA-ARS as a source to end users for the first decade of this Millennium had reached a level of 10,603 packets of sunflower CWR seeds in 520 requests from the U.S. or abroad that were processed between 2001 and 2011. The next period from 2011 to 2015 brought an increase in sunflower CWR seed requests to up to 775, and consequently, 8,600 seed packets were sent. That makes 1,295 requests, resulting in 19,203 sunflower CWR seed packets that were sent from USDA-ARS to users within the U.S. and abroad from 2001 to 2015 (Seiler and Marek 2011; Marek 2016). This illustrates not only the growing importance of sunflowers but also the need for genetic variability to handle the new challenges in sunflower breeding caused by climate-dependent changes in biotic and abiotic stresses, as well as growing market demands. This requires intensifying gene mining to broaden genetic variability and propel sunflower breeding programs. Moreover, especially in the past and during the establishment of the collections, which took years and decades, significant exchange of material between different gene banks was evident. Thus, the collection of wild sunflowers at INRAE contains accessions obtained from the USDA, IFVC, Eastern Europe (Bulgaria, Russia), Australia, South America, etc. Today, this collection has 369 accessions of wild *H. annuus* as well as 66 and 230 accessions of other annual and perennial *Helianthus* species, respectively. Fifty-nine accessions from INRA, Montpellier, France, represented the starting point for the establishment of the collection of wild sunflowers at IFVC, Novi Sad, Serbia, in the 1980s, after which several collecting expeditions to the U.S. and Montenegro led to the expansion of the collection as well as to the exchange with other gene banks (Atlagić et al. 2006; Atlagić and Terzić 2015). Today, the IFVC collection contains 185 accessions of eight annual *Helianthus* species and 332 accessions of 21 perennial *Helianthus* species (Terzić et al. 2020a).

Despite the origin of *H. annuus* being North America, the oldest gene bank at VIR, established in the 1920s, possesses 442 wild accessions of five annual and 19 perennial *Helianthus* species (Terzić et al. 2020a). Another gene bank located in Eastern Europe, DAI in Bulgaria, was established thanks to the accessions obtained from USDA-ARS in the 1970s and later enriched with accessions gathered through collection trips to the U.S. in cooperation with the USDA. Today, this collection possesses 175 and 215 accessions from seven annual and 30 perennial *Helianthus* species, respectively (Terzić et al. 2020a).

An advantage of keeping sunflower accessions in different gene banks is that all gene banks possess sunflower genotypes adapted to a certain degree to local climatic conditions. As sunflowers can grow in a large variety of climatic conditions, breeders can mine for specific alleles within a large spectrum of genotypes; thus, a lot of variability can be found in different gene banks. The majority of accessions can either be freely accessed or can be obtained under the conditions of an MTA. Within gene banks and breeding institutions, extensive research on the successfulness of crossings between the wild and cultivated pools was performed (Atlagić 2004; Jocić et al. 2015).

Wild relatives played a key role in switching from varietal sunflowers to hybrids, giving chance to the utilization of heterotic effects in yield increments of 1.3% per year in the last 30 years of the 20th century (Jocić et al. 2015). Maximum heterozygosity in sunflower hybrids has led to better plasticity, e.g. adaptation ability, compared to varietal sunflowers with traits fixed in maximum homozygosity. These achievements would not be possible without the utilization of the wild sunflower gene pool. The introduction of the cytoplasmatic male sterility gene (PET1) in *H. annuus* from the wild relative *H. petiolaris* by means of classic interspecific hybridization was a turning point for boosting commercial sunflower production (Leclercq 1969).

5.4 SUNFLOWER REFERENCE GENOME – A "GAME" CHANGER

The scientific sunflower community had been anticipating the appearance of the sunflower reference genome for a long time, and after quite some waiting, the first sunflower genome assembly was published in 2017 (Badouin et al. 2017). The major obstacles to producing a high-quality reference genome for sunflowers lay in the size of the genome (3.6 Gb), and the numerous repetitive sequences it contained, which made the first assembly rather difficult. The advancements in biotechnology and bioinformatics tools enabled the production of sequences longer than the known repeats and allowed the assembly of the inbred line HanXRQ, a maintainer oil line. Soon after, the sequence of the sunflower line HA412-HO was reported (https://www.heliagene.org/HA412.v1.1.bronze.20141015/). The availability of the annotated sunflower reference genome (https://www.heliagene.org/HanXRQr2.0-SUNRISE/) was a "game" changer because now the molecular markers flanking the traits of interest in the genetic maps can be used to define the exact locations in the sunflower genome. Depending on the size of the region defined by the markers, a reasonable number of candidate genes for allele mining can now be specifically addressed, which had been impossible so far. Five years after the first reference sequence was presented, sequences of the genomes of seven additional cultivated genotypes (maintainers and restorers) and one wild *H. annuus* PI659440 have been published (Huang et al. 2022). Assemblies and annotations of the genomes are available at https://www.heliagene.org/ and https://sunflowergenome.org/ for the PacBio and Illumina genomes, respectively.

Whole genome sequencing of well-characterized accessions allows efficient allele mining for traits of interest. These resources have now been provided by a huge re-sequencing effort (Hübner et al. 2019). Genomes of 287 cultivated lines representing the UGA-SAM1-population, 17 Native American landraces, and 11 wild species compatible with *H. annuus*, represented by 189 wild accessions, have been sequenced with 5×–25× coverage (Hübner et al. 2019). The sequences are available via the Sequence Read Archive (SRA) in NCBI for the cultivated lines under Bioproject PRJNA353001 and for the wild and landraces under PRJNA397453 (Hübner et al. 2019). On the web site (https://www.sunflowergenome.org), the data set of the SNP (vcf format), the pan-genome sequences (fasta format), and the corresponding statics (bed file format) can be accessed. Using the genome locations delimited by the genetic markers for specific traits, an intensive work of allele mining can now be started, which will reveal valuable alleles for improved breeding regarding abiotic and biotic stress.

Using this enormous data set, a sunflower 'pan-genome' was created, comprising 61,205 genes (Hübner et al. 2019). The pan-genome shows great variability in gene presence/absence, with 27% of them varying between the genomes. It was estimated that approximately 10% of the cultivated genome was derived from wild germplasm introgression, with 1.5% of genes coming only through these introgressions (Baute et al. 2015; Hübner et al. 2019; Owens et al. 2019). Introgressions had a great impact on the phenotypic variability found within the association panel, the UGA-SAM1 population, which was created with the goal of capturing as much genetic variability in the cultivated sunflower pool as possible (Mandel et al. 2013). Introgressions had a positive impact on increasing abiotic stress resistance while at the same time reducing yield and quality traits. The majority of introgressions came from the primary gene pool (approximately 3%) compared to 0.1% from the secondary pool. Restorer lines showed more introgressions than female maintainer lines (Hübner et al. 2019).

5.5 SEARCHING FOR PHENOTYPIC VARIATION

Allele mining requires the restriction of a small number of genotypes (accessions and genomes) from a larger collection based on either phenotypic information or sequence data. The availability of the annotated sunflower reference genome and the genome information of the re-sequenced accessions will greatly facilitate allele mining in sunflower. However, mining for traits demands precise phenotyping for the association or correlation of traits with novel/superior alleles. There are two approaches to searching for genes/traits: top-bottom and bottom-up genetic approaches. While in the first approach, the starting point is the phenotype, which is exploited to reveal the genetic mechanism(s) and gene(s) responsible for the phenotype, the latter approach would involve analyzing datasets and searching known sequences in the genome database to identify unique sequences and haplotypes in outstanding accessions (Leung et al. 2015).

As was previously shown, there are numerous accessions of wild and cultivated sunflowers available for allele mining. Thus, creating core collections can be of great help in analyzing sunflower genetic diversity and easing the discovery of desirable traits. Several association panels have been created to capture as much genetic variability as possible. The first one was created from USDA-ARS-NCRPIS gene bank cultivated sunflower accessions (Brothers and Miller 1999). Later, Chapman et al. (2008) and Mandel et al. (2011) generated a pool consisting of 433 cultivated accessions and 24 wild populations to analyze candidate genes involved in sunflower evolution, genetic diversity, and population structure, respectively. This first international sunflower association panel, the UGA-SAM1 population, consists of 288 accessions (259 and 29 accessions, which are available from the GRIN of USDA-NPGS and INRAE, respectively), which captured 90% of the genetic diversity and was created and exploited for association studies (Mandel et al. 2013; Nambeesan et al. 2015). It was phenotyped by morphological descriptors (Terzić et al. 2020b). However, breeders are frequently looking for rare alleles that might not be part of existing association panels but need to be searched in a larger pool of genotypes. Accessions of different gene banks have been extensively screened for the presence of abiotic and biotic stress resistance genes (Škorić and Rajčan 1992; Terzić et al. 2010; Tančić et al. 2012; Tančić-Živanov et al. 2021) as well as quality and yield properties (Ćuk 1982; Velasco et al. 2014; Pérez-Vich et al. 2018). However, most of the applied methods are labor-intensive and time-consuming, especially if a large panel of genotypes has to be screened.

New tools have emerged for high-throughput screening due to recent technological developments. Nowadays, with the availability of whole-genome sequences of crops, including sunflower, the following goal would be to describe the whole phenotypes, decipher key functional loci, and get more insight into the genetic structure of crops (Yang et al. 2020). Different phenomics tools enable quicker, non-destructive, and more precise detection and measurement of desirable traits as an important step in allele mining. Nowadays, there are quite some high-throughput and intelligent platforms available for screening small plants but also some for larger plants. However, they are still unreachable to many crop researchers. Thus, phenotyping networks have been established on different national and international levels in order to attempt to bridge this gap. Such networks on the national level are Deutsches Pflanzen Phänotypisierungs-Netzwerk (DPPN - https://dppn.plant-phenotyping-network.de/), Austrian Plant Phenotyping Network (APPN), while on the international level, the European Plant Phenotyping Network (EPPN - http://www.plant-phenotyping-network.eu/ and EPPN2020 - https://eppn2020.plant-phenotyping.eu/), International Plant Phenotyping Network (IPPN - https://www.plant-phenotyping.org/), North American Plant Phenotyping Network (NaPPN-http://nappn.plantphenotyping.org/), and others have been established. Among the most investigated crops using such more advanced systems are wheat (30%), maize (13%), rice (13%), Arabidopsis (13%), and barley (7%) (Kondić-Špika et al. 2022).

In sunflower, certain advancements have been made in using more automated systems in phenotyping; however, much more effort is needed. Recently, an automated seed phenotyping device for analysis of seed color and shape was developed (Aliiev 2020), while Hasson et al. (2021) combined an image processing phenotyping platform and Genome-wide Association Studies (GWAS) to discover genes involved in the regulation of different seed characteristics. Different imaging techniques have been applied to assess both abiotic and biotic stress in sunflower shoots as quick and efficient, non-destructive methods for plant phenotyping. Ochogavía et al. (2014) exploited spectrophotometric quantification of photosynthetic leaf pigments to analyze the level of tolerance of sunflowers to the herbicide imazapyr. The authors used bioinformatics-based color Tomato Analyzer® test software and exploited the observed relationship of herbicide-induced chlorosis with leaf color and variation of chlorophyll concentration. Thermal imaging and blue-green fluorescence (BGF) were used for capturing sunflower leaves during an underground broomrape attack (Ortiz-Bustos et al. 2017). Recently, Flower Color Image Analysis (FloCIA), a classification of ray floret color, has been developed by exploiting digital imaging and automatic classification (Zorić et al. 2020). Healthy plants exhibited higher BGF emissions compared to the infected plants, whose leaves were warmer due to a reduction in transpiration and the closure of stomata. Heliaphen, an outdoor phenotyping facility located in INRAE station, Auzeville, France, has been developed and used for high-throughput drought tolerance analysis in sunflower (Blanchet et al. 2018; Gosseau et al. 2019). The main focus of phenotyping was the shoot, leaving an important organ, the root, neglected. Radanović et al. (2018, 2022a) exploited an automated phenotyping platform (GROWSCREEN-Rhizo) and image processing software (GrowScreen-Root) for analysis of the root architecture under optimal water and drought conditions in cultivated and wild sunflower accessions with the goal of setting up a basis for breeding of modified root characteristics. Recently, Le Ru et al. (2021) developed a RhizOSun system for automated counting of broomrape tubercles by exploiting rhizotrons and ImageJ software, thus enabling reduced time for evaluation of the broomrape resistance level in the analyzed sunflower genotypes.

5.5.1 Mining for Alleles for Biotic Stress Resistance

The most extensive studies in sunflower, so far, have been conducted in the field of discovering biotic stress resistance alleles and introducing them into the cultivated pool, as most of the resistance genes have been originally detected in the wild pool. This is an everlasting circle, as the exploitation of a newly introduced resistance gene puts a certain evolutionary pressure on the plant pathogen, leading to the appearance of a new, more virulent race and making breeders look for new resistance sources. Discovering the alleles and their introduction can be a long process, as it requires screening a large sample of sunflower genotypes, both cultivated and wild populations, and searching for desirable alleles. Then, the allele is introduced into the breeding material either by conventional means or with the help of biotechnological tools, mainly molecular markers. When it comes to breeding for downy mildew, rust, and even broomrape resistance, this job has been eased a bit by the fact that the majority of resistance genes, which have been discovered so far, are dominantly inherited, with a few exceptions. However, resistance to *Sclerotinia*, *Phoma*, and *Phomopsis* is quantitatively inherited and thus requires more time and work for the introduction of new alleles into the breeding material. Designing modern sunflower hybrids aims at pyramiding several pest resistance alleles, thus providing sustainable sunflower production. An overview of the efforts put into discovering new resistance alleles is given in this section.

Phomopsis, stem canker caused by Diaporthe/*P. helianthi*, a very damaging widespread disease of sunflower, has been fought with resistance genes from *H. tuberosus* and *H. argophyllus*, as well as *H. debilis*, *H. petiolaris*, and *H. exilis*. Wild annual sunflower species such as *H. annuus, H. argophyllus, H. debilis*, and *H. petiolaris*, as well as perennial species such as *H. decapetalus, H. divaricatus, H. eggertii, H. giganteus, H. xlaetiflorus, H. mollis, H. nuttallii, H. scaberrimus, H. pauciflorus, H. salicifolius*, and *H. tuberosus*, offered resistance genes to the downy mildew (Miladinović et al. 2019). *H. mollis* has been described and used as a donor for *Sclerotinia sclerotiorum* resistance (Rönicke et al. 2004) and *Orobanche cumana* resistance (Labrousse et al. 2001).

5.5.1.1 Downy Mildew Resistance

Downy mildew is a severe disease caused by the oomycete *Plasmopara halstedii* Farl. Berl. & deToni. Originally, the downy mildew pathogen *Peronospora halstedii* was described by Farlow (1883) as attacking numerous species such as *Eupatorium* and perennial sunflower species *H. tuberosus*, *H. doronicoides* (*H. mollis x giganteus*), and *H. strumosus*. This pathogen taxon was later transferred to a new genus under the name *Plasmopara halstedii* (Farlow) Berlese & de Toni (1888), still being described as affecting perennial sunflowers. At the end of the 19th century, the first report on downy mildew occurrence on *H. annuus* appeared, and in the 1920s, downy mildew quickly spread, presenting a severe threat to US sunflower production (Henry and Gilbert 1924; Young and Morris 1927). Until today, downy mildew has been a major constraint on sunflower production worldwide.

Initial efforts to find resistant genotypes included screening of perennial *Helianthus* species and introduction of resistance genes into the cultivated sunflower by interspecific hybridization (Pustovoit 1966; Leclercq et al. 1970). The introduction of resistance into open-pollinated variety (OPV) breeding material was quite labor-intensive and time-consuming, while the introduction of sunflower hybrid breeding facilitated this process to a certain extent (Molinero-Ruiz 2022). With a strong interaction and co-evolution of *P. halstedii* and sunflower, numerous races and resistance genes were found. So far, more than 40 different resistance genes, named *Pl* genes, have been described (Pl_v–Pl_z, Pl_1–Pl_{36}, and Pl_{Arg}). The development of molecular tools enabled localization of the desirable resistance genes, identification or design of molecular markers linked to the alleles, and use of some of the markers in MAS. The recent re-sequencing of the association panel UGA-SAM1 population and the simultaneous screening for *Plasmopara halstedii* race 734 have now identified candidate genes on several chromosomes, allowing targeted allele mining (Hübner et al. 2019). These results support former molecular studies, which have also shown that most of the resistance genes are clustered on several sunflower chromosomes (Dimitrijević and Horn 2018). Due to the existence of such a variety of resistance genes, breeders now aim at pyramiding resistance genes in hope of achieving sustainable resistance to downy mildew. For this purpose, allele mining and the development of allele-specific molecular markers can be of great assistance.

An important source of downy mildew resistance alleles that were greatly exploited is located on *Chr8,* which harbors Pl_2, Pl_6, and Pl_7, with the Pl_6 gene being the most exploited in sunflower breeding as it conferred resistance to all races of downy mildew at the time, except for race 304. Pl_6 originates from wild *H. annuus* and was first described by Miller and Gulya (1991). Thus, a lot of effort was put into creating different types of markers for its detection, such as markers belonging to the TIR-NBS-LRR (Toll-interleukin-like receptor–nucleotide-binding site–leucine-rich repeat) class of RGA (resistance-gene analog) and sequence-tagged site (STS) markers (Bouzidi et al. 2002) (Table 5.1). Pankovic et al. (2007) utilized some of the previously mentioned markers and identified restriction sites in a fragment common for resistant and susceptible parents to develop codominant cleaved amplified polymorphic sequence (CAPS) markers. Restriction enzymes *Hha*I and *Rsa*I revealed unique bands, R1–1215 bp, R2–890 bp and R1–630 bp, R2–600 bp, and R3–430 bp, respectively, in the resistant genotypes, enabling an even quicker detection of homozygous lines

TABLE 5.1
Overview of Resistances Genes for Downy Mildew, Rust and *Orobanche*, the Resistance Sources, Locations, and Markers Used for Their detection in Sunflower

Chromosome	Resistance Gene	Source	Markers Linked to the Resistance Gene	Distance (cM)	Reference
Chr1	Pl_{Arg}	Arg1575-2[1,2]	NSA_007595 [4]	0.01	Dußle et al. (2004) [1]
		RHA 419 [3]	NSA_001835 [4]	0.01	Wieckhorst et al. (2010) [2]
		RHA 464 [4]	ORS 716 [2,3]	0.3 [2]; 0.0[3]	Imerovski et al. (2014) [3] Qi et al. (2017) [4]
			ORS 662 [1,2,3]	1.9 [1]; 0.3[2]; 0.0[3]	
			ORS 675 [3]	0.0	
			RGC52a, RGC52b, RGC151 [2]	0.0	
	Pl_{13}	HA-R5	ORS1008 [1,2]	0.9[1]; 1.8[2]	Mulpuri et al. (2009) [1] Liu et al. (2012a) [2]
			HT636 [2]	2.2	
	Pl_{14}	29004	RGC203	1.6	Bachlava et al. (2011)
			RGC188	2.6	
	Pl_{16}	HA-R4	ORS1008	0.0	Liu et al. (2012a)
			HT636	0.3	
Chr2	R_5	HA-R2	SFW03654 [2]	0.6	Qi et al. (2012a) [1]
			ORS653a [1]	1.8	Qi et al. (2015b) [2]
			NSA_000267 [2]	2.2	
			ORS1197-2 [1]	3.3	
	Pl_{18}	PI 494573	ORS203	0.4	Qi et al. (2016)
			SFW03060	0.9	
			SFW03883	0.9	
			CRT214	1.1	
Chr3	Or_5	RPG01[1]	RTS05 [1]	5.6	Lu et al. (2000) [1]
		PHD[2]	CRT 392 [2]	6.2	Tang et al. (2003) [2]
			ORS1036 [2]	7.5	
	or3.1	P-96	ZVG406	4.6	Pérez-Vichet al. (2004)
	orab-vl-8	AB-VL-8	ORS683	1.5	Imerovski et al. (2016)

(Continued)

TABLE 5.1 (*Continued*)
Overview of Resistances Genes for Downy Mildew, Rust and *Orobanche*, the Resistance Sources, Locations, and Markers Used for Their detection in Sunflower

Chromosome	Resistance Gene	Source	Markers Linked to the Resistance Gene	Distance (cM)	Reference
Chr4	Pl_{17}	HA 458	C4_5711524[2]	0.03	Qi et al. (2015a) [1]
			SPB0001 [2]	0.03	Ma et al. (2019)[2]
			ORS963[1]	0.8	
			SFW04052[1]	2.1	
	Pl_{19}	PI435414	C4_6676629 [2]	0.04	Zhang et al. (2017) [1]
			C4_6711381 [2]	0.18	Ma et al. (2019) [2]
			NSA_003564 [1]	0.6	
			NSA_006089[1]	0.6	
	Pl_{33}	TX16R wild *H. annuus*	SFW04901/ SFW04052/ ORS644/ORS963[1]	0.0	Liu et al. (2019) [1], Ma et al. (2022) [2]
				0.0	
			C4_5641353/ C4_5671004/ SPB006[2]	0.0	
	Or_{SII}	PHSC1102	M-4_30.40	0.7	Martin-Sanz et al. 2020
	Or_{Deb2}	DEB2	AX-105399507	0.2	Fernández-Aparicio et al. (2022)
			AX-105525205/AX-105938724	0.7	
Chr7	*HaOr7*	LSR002 proprietary line	SF3178	1.7	Duriez et al. (2019)
			SF3131	1.7	
Chr8	Pl_1	RHA 266	S017H3-3	5.6	Mouzeyar et al. (1995)
			S124EI-2	7.1	
	Pl_6	HA335 [1]	Hap2R	0.0	Pankovic et al. (2007)
		HA 336 [2]	Hap3	0.0	
			CAPS *Hha*I	0.0	
			CAPS *Rsa*I	0.0	
	Pl_{20}	PI 494578	SFW02745/SFW09076/S8_11272025/ S8_11272046	0.0	Ma et al. (2017)
	R_1	RHA 279	$SCT06_{950}$	0.0	Lawson et al. (1998)
	R_{15}	HA-R8	SFW01920/SFW00128/SFW05824	0.3	Ma et al. (2018)
			NSA_008457	0.7	
Chr11	R_{12}	RHA 464	NSA_003426/ NSA_004155/ NSA_000064[2]	0.83	Gong et al. (2013b) [1] Talukder et al. (2014) [2]
			CRT275 [2]	1.0	
			ORS1227 [1]	3.3	
			ZVG53 [1]	9.6	
	R_{14}	PH3	NSA_000064	0.7	Zhang et al. (2016)
			ORS1227	1.6	
			ORS542	3.5	
			ZVG53	6.9	
Chr13	Pl_5	XRQ	Ha-NT5R1	4.8	Radwan et al. (2004)
			Ha-NT5S3	6.4	
			Ha-NT5S1/Ha-NT5S2/ Ha-NT5R2	13.6	
	Pl_8	QIR8 (Pl_8 derived from RHA 340) [1]	RGC251 [2]	0.3	Radwan et al. (2004) [1]
			RGC15 [2]	0.4	Bachlava et al. (2011) [2]
			NSA_000423/SFW01497/SFW08875[3]	0.4	Qi et al. (2017) [3]
		RHA 340 [2,3]	Ha-NT8R3/Ha-NT8R4 [1]	3.5	
			Ha-NT8R5/Ha-NT8R6 [1]	4.8	
			Ha-NTIR11gEI [1]	6.3	

(*Continued*)

TABLE 5.1 (*Continued*)
Overview of Resistances Genes for Downy Mildew, Rust and *Orobanche*, the Resistance Sources, Locations, and Markers Used for Their detection in Sunflower

Chromosome	Resistance Gene	Source	Markers Linked to the Resistance Gene	Distance (cM)	Reference
	Pl_{21}	RHA 274	ORS630	10.3	Vincourt et al. (2012)
	Pl_{34}	RHA 428 from *H. annuus*	SFW08875/SFW09223 ORS191/ORS316	0.3 0.9	Talukder et al. (2019)
	Pl_{36}	803–1 from *H. tuberosus*	SFW01497/SFW05453/ SFW05832/ SFW08875	0.0	Qi and Cai (2022)
	R_4	HA-R3 (1,2)	SFW05240/SFW05630/SFW06095/ SFW08283(2) SFW01497/SFW05630/ SFW08875 (2) ORS581 (1) ZVG61 (1)	0.6 0.7 0.8 2.1	Qi et al. (2011b) (1) Qi et al. (2015b) (2)
	R_{HAR6}	HA-R6	ZVG61 ORS581	0.7 1.5	Bulos et al. (2013)
	R_{13a}	HA-R6	C13_194757055(3) C13_195501970/ C13_195522913/ C13_195526945/ C13_195556768(3) SFW05743 (2) RGC15/16 (1) ORS316/ZVG61 (1) ZVG61 (2) SUN14 (1)	0.13 0.35 0.2 0.3 0.4 0.4 0.8	Gong et al. (2013a) (1) Qi et al. (2015b) (2), Ma et al. (2022) (3)
	R_{13b}	RHA 397	SUN14 (1) SFW00757 (2) RGC15/16 (1) ZVG61(1) C13_194268343/ C13_194735854/ C13_194757055 (3) ORS316 (1) SFW04275/SFW04317/ SFW05743 (2) ZVG61, ORS316 (2)	0.0 0.0 0.1 0.3 0.43 2.4 5.9	Gong et al. (2013a) (1) Qi et al. (2015b) (2) Ma et al. (2022) (3)
	R_{adv}	'Advance' (1) RHA 340(2)	$SCX20_{600}$ (1) RGC260(2) ORS316(2)	0.0 0.2 3.0	Lawson et al. (1998) (1) Bachlava et al. (2011) (2)
	P_{u6}	P386	ORS316 ORS224	2.5 4.8	Bulos et al. (2014)
	R_{11}	Rf ANN-1742	ORS728 ORS45	0.3 1.0	Qi et al. (2012b)
	R_{16}	TX16R	C13_194722668(2) C13_195512786/ C13_195552917/ C13_195605372(2) SFW08875 (1) SFW04317 (1)	0.18 0.13 1.8 1.1	Lui et al. (2019) (1), Ma et al. (2022) (2)

(*Continued*)

TABLE 5.1 (*Continued*)
Overview of Resistances Genes for Downy Mildew, Rust and *Orobanche*, the Resistance Sources, Locations, and Markers Used for Their detection in Sunflower

Chromosome	Resistance Gene	Source	Markers Linked to the Resistance Gene	Distance (cM)	Reference
	R_{17}	KP193 from South Africa	SFW01497/SFW08875	1.9	Qi et al. (2021)
			C13_196165992/C13_196166091/ C13_196166100/C13_196166143/ C13_196166162	0.7	
	R_{18}	KP199 from South Africa	SFW04317	0.0	Qi et al. (2021)
			C13_196165992/C13_196166091/ C13_196166100/C13_196166143/ C13_196545298/C13_196545312	0.8	
			SFW01497/SFW05453	1.9	
	$Or_{ab\text{-}vl\text{-}8}$	AB-VL-8	ORS683	1.5	Imerovski et al. (2016)
			ORS657	4.7	
Chr14	R_2	MC 29 (USDA)	SFW01272	1.8	Qi et al. (2015c)
			NSA_002316	2.8	
			SFW00211	2.9	

carrying the Pl_6 gene and thus enabling the faster creation of numerous sunflower lines and hybrids harboring this gene (Dimitrijević et al. 2010; Jocić et al. 2010). As, Pl_6 lost its efficiency, breeders began exploiting Pl_8 and Pl_{Arg}, both originating from *H. argophyllus*.

Pl_8 is located in a region on the lower end of *Chr13* (193–196 Mb) and is a part of a cluster Pl_8/Pl_{34}/Pl_{36} (Qi and Cai 2022). Numerous markers were developed and exploited for the detection of Pl_8. Among the first ones were STS and closely linked RGC markers, RGC251 and RGC15/16, belonging to the group of single-strand conformational polymorphism (SSCP) type of markers (Radwan et al. 2004; Bachlava et al. 2011; Dimitrijević et al. 2011). By exploiting a biparental cross HA 434 × RHA 340 and, in addition, an association panel of 548 diverse sunflower accessions, Qi et al. (2017) identified two SNP markers, NSA_000423 [G/A] and NSA_002220 [G/A], with great diagnostic potential for Pl_8 when the restorer line RHA 340 is used as a donor of this gene. Qi and Cai (2022) identified four common SNP markers for Pl_8, Pl_{34}, and Pl_{36} in three maps: SFW06095, SFW01497, SFW08875, and NSA_002251, while identifying four SNPs (SFW01497, SFW05453, SFW05832, and SFW08875) co-segregating with Pl_{36}. Of the four markers, SFW08875, together with SFW09223, were mapped 0.3 cm distal from Pl_{34} (Talukder et al. 2019).

On the lower end of *Chr13*, several additional downy mildew resistance genes/clusters were mapped, distal from the Pl_8/Pl_{34}/Pl_{36} cluster: Pl_{22}, Pl_5/Pl_{21}, and Pl_{31}/Pl_{32} originating from wild sunflower species such as *H. tuberosus*, *H. annuus*, and *H. argophyllus*. By analyzing the linkage of molecular markers associated with different *Pl* genes and positioning them on the sunflower reference genome HanXRQr1.0 assembly, Pl_{22} was located in the 181–183 Mb region, while clusters Pl_5/Pl_{21} and Pl_{31}/Pl_{32} were positioned in the 186 Mb and 188–196 Mb regions, respectively (Qi and Cai 2022). Regarding the large number of *Pl* genes in these regions as well as the recombination events that vary between different crosses, some differences in distances between the markers and mapped *Pl* genes may occur. By using common and publicly available simple sequence repeat (SSR) markers, Pl_5 and Pl_{21} were mapped 8 cM apart from each other, between ORS316 and ORS630, while Pl_{22} and Pl_{31} were mapped above and Pl_{32} below both markers. Pl_{22} was mapped distal from ORS316 and ORS630 (Vincourt et al. 2012; Pecrix et al. 2018).

Two major *Pl* gene clusters are located on *Chr1*: Pl_{13}, Pl_{14}, Pl_{16}, and Pl_{25} are clustered on the lower end, while a cluster containing Pl_{Arg}, Pl_{23}, Pl_{24}, and Pl_{35} is positioned distant from it (Qi et al. 2017, 2019; Pecrix et al. 2018). Out of the *Pl* genes located on *Chr1*, Pl_{Arg} is the most frequently used in sunflower breeding as it confers resistance to all downy mildew races in North America (Qi et al. 2019, 2022). Several types of molecular markers have been developed for its detection, including SSRs, resistance gene candidates (RGCs), and SNPs (Wieckhorst et al. 2010; Imerovski et al. 2014; Qi et al. 2017). The SSR marker ORS716 proved to be a very successful marker for the identification of the Pl_{Arg} gene in diverse material, amplifying a 303-bp-band in resistant lines and 325 bp-band in susceptible lines. Another interesting marker was ORS675, which amplified a 245-bp-band in a susceptible line, while no homologous site at this locus was amplified in the resistant line (Imerovski et al. 2014). Moreover, two publicly available SNPs, NSA_007595 [G/A] and NSA_001835 [A/G], enabled narrowing down of the Pl_{Arg} locus to an area of 2.83 Mb (Qi et al. 2017). The authors exploited a segregating F_2 population obtained by crossing RHA340 (donor of Pl_8) and RHA464 (donor of Pl_{Arg} and R_{12}) for developing homozygous lines in which the three resistance genes were pyramided. SNP markers significantly speeded up the process of exploitation of the desirable alleles and their introduction into breeding material.

Besides the previously described Pl_{Arg}, Pl_{15}, Pl_{17}-Pl_{20}, and Pl_{35} confer resistance to all downy mildew races in North America, while Pl_{23}-Pl_{32} confer resistance to all downy mildew races in Europe (Zhang et al. 2017; Ma et al. 2017; Pecrix et al. 2018; Gilley et al. 2020; Qi et al. 2022). The boost of newly discovered *Pl* genes enabled the discovery of another cluster of *Pl* genes located on the upper end of *Chr4*. In this region, Pl_{19} was mapped downstream of Pl_{17} and Pl_{33} when the positions of the genes were compared to the common SSR marker ORS963 and by exploiting a sequence-based chromosome walking strategy (Ma et al. 2019; Liu et al. 2019). Liu et al. (2019) reported four co-segregating SSR and SNP markers for Pl_{33} (SFW04901, SFW04052, ORS644, and ORS963) and Ma et al. (2022) added another three co-segregating SNP markers (C4_5641353, C4_5671004, and SPB006). Ma et al. (2019) also identified six and eight diagnostic markers for Pl_{17} and Pl_{19}, respectively. The diagnostic markers were further tested in the 96-line panel and could differentiate between the mapped genes and other previously mapped *Pl* genes, including Pl_{Arg}, Pl_1–Pl_3, Pl_6–Pl_{13}, Pl_{15}–Pl_{21}, Pl_{33}, and Pl_{34}. Pecrix et al. (2018) mapped Pl_{27} in the Pl_{17} region and Pl_{28} and Pl_{29} in the vicinity of Pl_{19} by using an AXIOM® genotyping array of 49,449 SNPs. Pl_{28} and Pl_{29} were flanked by *AX-105633491* and *AX-105852138* and *AX-105890384* and *AX-105933470*, respectively. Resistance genes from this region are mostly derived from wild *H. annuus*, except for Pl_{27} and Pl_{28}, which were introduced from *H. tomentosus* into cultivated sunflower.

Not that many *Pl* genes have been annotated. Sequences of Pl_8 and Pl_{14} belong to the coiled-coil nucleotide-binding and leucine-rich repeat (CC-NBS-LRR) class of resistance genes, while sequences of Pl_6 and Pl_{Arg} belong to the TIR-NBS-LRR class (Radwan et al. 2004; 2011: Bouzidi et al. 2002). The use of genotype-by-sequencing and whole genome re-sequencing approaches, as well as the available sunflower reference genome, enabled the identification of some more candidate genes. Thus, Ma et al. (2020) proposed a candidate gene *HanXRQChr02g0048181* for Pl_{18} with a predicted function to encode a protein with a typical leucine-rich repeat (NLR) domain for disease resistance, and a putative gene with an NLR domain for Pl_{20}. The proposed function of the gene encoding Pl_{19} is an RNA methyltransferase family protein, as this is the only gene found in the region of Pl_{19} in the sunflower reference genome (Ma et al. 2019), while 25 different genes related to disease resistance with NBS-LRR domains were found in the region between SNPs flanking Pl_{36} (Qi and Cai 2022). Pecrix et al. (2019) proposed *PhRXLRC01* and Pl_{22} as an avirulence/resistance gene. A whole genome sequencing approach allowed the identification of *HanXRQChr04g0095641* as a potential candidate gene for Pl_{33} (Ma et al. 2022). As mentioned before, the re-sequencing of the UGA-SAM1 population also allowed the identification of candidate genes for down mildew resistance against race 734 (Hübner et al. 2019).

For all the newly discovered genes, SNP markers have been developed, thus enabling the fast introduction of multiple *Pl* genes into sunflower breeding material. However, no comprehensive allele sequences have been defined, probably because the identified SNPs could not be aligned to different alleles because of the short reads used in sequencing. The value of molecular markers becomes even greater considering that some genes such as Pl_{Arg}, Pl_{17}-Pl_{19}, and Pl_{35} cannot be differentiated using downy mildew assays as differentiating virulent *P. halstedii* races have not yet been detected for these genes, making the application of molecular markers the only option for gene pyramiding (Qi et al. 2019). It is easier to pyramid genes located on different chromosomes than ones on the same chromosome. Nowadays, breeders have a great variety of resistance genes to choose from for breeding sustainable downy mildew resistance. For a proper selection, breeders need to take into account the *P. halstedii* races present in a particular production area – i.e., depending on race variability in a certain growing region, breeders must mine genes from available gene pools in order to create corresponding resistant genotypes, primarily for the prevalent pathogens and their races, and if possible, even broader than that.

5.5.1.2 Rust Resistance

As sunflower, *Puccinia helianthi* (a fungus causing rust) is native to North America. It was first discovered in 1822 in specimens collected in the southeastern U.S. (Bailey 1923; Putt and Sackston 1957). Since then, it has been reported on all continents where sunflowers are grown, and today, jointly with downy mildew, it is considered one of the two most damaging diseases of sunflowers globally (Lui et al. 2019). Similar to downy mildew, the main sources of rust resistance have been found in different genetic resources. All known rust resistance genes have been identified in cultivated sunflower and wild annual species, mainly in *H. annuus*, *H. argophyllus*, *H. praecox*, and *H. petiolaris,* with the exception of the perennial *H. tuberosus* (Seiler et al. 2017). Interspecific hybridizations have been extensively used for the introduction of rust resistance genes into the cultivated pool. In 1916, the Russian scientist Sazyperow made the first efforts to produce an interspecific hybrid between *H. annuus* and *H. argophyllus* with the aim of developing rust-resistant sunflowers (Cockerell, 1929). Due to the crop-pathogen co-evolution, breeders have always been in search of novel rust resistance genes to overcome the existing pathogen race variability and strength.

Dominantly inherited rust resistance genes are designated *R* genes. Almost 20 genes have been described so far, while 15 have been mapped on chromosomes 2, 8, 11, 13, and 14. Most of these *R* genes are clustered together. In addition, in some of the major clusters, *R* genes are combined with *Pl* or *Rf* (fertility restoration) genes (Table 5.1). Two of the largest and most complex clusters are located, as previously mentioned, on chromosomes 8 and 13. Two *R* genes, R_1 and R_{15}, are clustered with six *Pl* genes on the upper end of *Chr8*, corresponding to the largest NLR cluster described in sunflower (Ma et al. 2017, 2018; Radwan et al. 2008). A co-segregating sequence characterized amplified region (SCAR) marker, $SCT06_{950}$, amplified a 950-bp-band in resistant plants containing R_1 gene and was proven to be very efficient in detecting this gene in different genetic backgrounds, except for line MC29 possessing R_2 and R_{10} (Lawson et al. 1998). However, this gene lost its effectiveness of providing resistance to all rust races, and currently, on this chromosome, only R_{15} is still working against all rust races present in North America (Ma et al. 2018). Resistance was derived from a landrace collected from Havasupai Indian Reservation, Arizona, US, and was ultimately introduced into inbred line HA-R8, which was one of the first released to confer resistance to rust race 777 (Miller and Gulya 2001). For the detection of R_{15}, four co-segregating SNP markers, SFW01920 [A/C], SFW00128 [A/C], SFW05824 [T/G], and NSA_008457 [A/G], are available, thus making this gene very desirable for *R* gene pyramiding and obtaining durable resistance to rust in sunflower.

Two sub-clusters of *R* genes are found on the lower end of *Chr13*, corresponding to the second largest cluster of NLR loci (Radwan et al. 2008). Three *R* genes, R_{adv}, Pu_6, and R_{11}, are clustered with three *Rf* genes on sub-cluster I, while six *R* genes R_4, R_{13a}, R_{13b}, R_{16}, R_{17}, and R_{18} and seven *Pl* genes are members of the sub-cluster II (Gong et al. 2013a; Liu et al. 2019; Talukder et al. 2019; Ma et al. 2021; Qi et al. 2021). Of the *R* genes in these clusters, R_{11}, R_{13a}, R_{13b}, and R_{16} are resistant to all races reported in North America, while the recently mapped R_{17} and R_{18} are resistant to the most virulent and prevalent races (Moreno et al. 2012; Gong et al. 2013b; Qi et al. 2021).

Concerning sub-cluster I, different types of markers were created for the detection of the genes. Tightly linked RGC and SSR markers, as well as co-segregating SCAR markers, were developed for the detection of R_{adv} (Lawson et al. 1998; Bachlava et al. 2011). SCAR marker $SCX20_{600}$ amplified a 600-bp-band in proprietary resistant sunflower line P2 and hybrid 'Advance', while failing to amplify a band in susceptible genotypes. Additionally, the marker was verified on a panel containing rust differentials, public and proprietary inbred lines, and hybrids. While SSR markers for detection of Pu_6 were too distant to be used in MAS (Bulos et al. 2014), R_{11} can be detected by the use of closely linked ORS72 and ORS45 (Qi et al. 2012b).

Differentiation of R_{13a}, R_{13b}, and R_{16} is facilitated using molecular markers available for their detection and introgression into sunflower breeding material. Initially, R_{13a} (in inbred line HA-R6) and R_{13b} (in inbred line RHA 397) were characterized as two alleles of a single gene as they shared PCR profiles of the common markers exploited for their detection (Gong et al. 2013a; Qi et al. 2015a). The two common markers, ORS316 and ZVG16, however, showed different amplification patterns between R_{13} and R_4 genes. Recent exploitation of the WGS approach showed that R_{13a} and R_{13b} are independent genes (Ma et al. 2022). Comparative mapping identified SNP markers differentiating between R_{13a} and R_{13b}. Recently mapped, R_{16} can be detected by using flanking markers SFW08875 [T/C] and SFW04317 [A/G] (Lui et al. 2019) or four other SNP markers flanking R_{16} with 0.18 cM (C13_194722668) and 0.13 cM (C13_195512786, C13_195552917, and C13_195605372), respectively (Ma et al. 2022). Four NLR motifs containing disease-related genes are annotated in the region of R_{13a} and R_{16}, for which allele mining can now be performed (Ma et al. 2022). For the rust resistance genes, R_{17} and R_{18}, seven and nine co-segregating SNPs were detected, respectively (Qi et al. 2021). The authors identified 25 plant disease defense-related genes in the 3.4 Mb (R_{17} and R_{18}) region in the HanXRQ genome annotation.

Chromosome 11 carries the remaining two *R* genes still effective against all rust races, R_{12} and R_{14}. These genes are located in the middle of the chromosome, and the resistance was derived from two different wild *H. annuus* accessions (Gong et al. 2013b; Talukder et al. 2014; Zhang et al. 2016). Currently, SNS markers are available for the detection and discrimination of the two genes and their introduction into sunflower breeding material. Zhang et al. (2016) identified six common markers on *Chr11* in a cross HA 434 × PH3, while mapping R_{14} on this chromosome with the reference map of Gong et al. (2013b), who mapped R_{12} in a cross HA89 × RHA464.

The remaining *R* genes are not clustered. R_2 and R_5 are located on *Chr14* and 2, respectively. R_2 is flanked by SNP markers, NSA_002316 and SFW01272, with a distance of 2.8 cM and 1.8 cM, respectively. However, NSA_002316 [T/C] cannot universally be used in MAS as a lot of susceptible genotypes amplify the same allele as the resistant line MC 29. Thus, the use of adjacent SNP SFW00211 [T/C] is recommended to be utilized in the detection of R_2 jointly with SFW01272 [A/G] to avoid false positive results (Qi et al. 2015c). Different types of molecular markers are available for the detection of R_5, with the closest being the SNP marker SFW03654 [A/C], 0.6 cM distant from the gene (Qi et al. 2015b).

Pyramiding of the *R* genes is underway, and breeders have several different genes to choose from for the introduction of the desirable *R* alleles into the breeding material. One of the examples would be the pyramiding of the R_2 gene located on *Chr14* and R_{13a} located on *Chr13* to obtain a double-resistant confectionary sunflower line (Qi et al. 2015c). Qi et al. (2015b) identified five and three SNP markers surrounding the R_5 and R_{13a} genes, respectively, for higher throughput detection of the genes in creating double-resistant lines by gene pyramiding.

5.5.1.3 Broomrape Resistance

Sunflower broomrape (*Orobanche cumana* Wallr.) is a host-specific holoparasitic plant that parasitizes on sunflower roots to obtain its nutritional requirements and can ultimately cause plant/host death. In infested fields, it can lead to up to a complete yield loss; thus, it is regarded as one of the most important constraints in sunflower production. Sunflower broomrape was first observed in the 1890s in the Saratov district of Russia, while the first resistant cultivar was found in 1918 (Škorić 2012). Soon after, in 1926, this resistance was prevailed by broomrape, and since then, breeders are in search of new resistance alleles that will overcome the ever-emerging new broomrape races. While the initial broomrape resistance found in the USSR came from *H. tuberosus* (Pustovoit et al. 1976), nowadays broomrape resistances have been found in both annual and perennial species, including *H. deserticola*, *H. maximiliani*, *H. grosseserratus*, *H. divaricatus*, *H. debilis*, and others (Seiler et al. 2017).

Today, broomrape can be found in Southeast Europe and regions around the Black Sea in which sunflower is grown, as well as in Israel, North Africa, and China (Molinero-Ruiz et al. 2015). Broomrape races are named by using letters, with eight races (A–H) being reported so far (Kaya 2014; Cvejić et al. 2020). Resistance to broomrape in sunflowers is mainly vertical, controlled by one major dominant gene, with several exceptions of horizontal (qualitative) resistance cases reported. The nature of resistance is thus very suitable for sunflower hybrid breeding. Five broomrape races A to E were reported to be controlled by the dominant genes Or_1-Or_5, respectively (Vranceanu et al. 1980).

Imerovski et al. (2013) exploited the contingency coefficient test and identified significantly associated alleles amplified by SSR markers and broomrape resistance genes. Thus, Or_2 was associated with the ORS1114_260 allele, while Or_4 and Or_6 genes with ORS1036_240 and ORS1114_265, and ORS665_281 and ORS1114_264, respectively. The dominantly inherited resistance gene Or_5, conferring resistance to races A–E, was mapped on the telomeric region of *Chr3* (Lu et al. 2000; Tang et al. 2003). Pérez-Vich et al. (2004) mapped several quantitative trait loci (QTL) conferring resistance to races E and F, including a major QTL *or3.1* conferring resistance to race E on *Chr3*. This chromosome was also reported to harbor a recessive *orab-vl*-8 gene conferring resistance to races higher than F and a major recessive QTL conferring resistance to race G, *or3.2* (Imerovski et al. 2016, 2019). SSR marker ORS683 was the closest marker mapped with 1.5 cM to *orab-vl*-8 on *Chr3*, while CAPS markers (C3_9713 and C3_9702) were developed for facilitating the introgression of the *or3.2* in different crosses. *HanXRQChr03g0076321* has been identified as a candidate gene for *or3.2*; however, further confirmation is needed to verify this hypothesis.

In addition to two recessive genes controlling resistance to race F (*or6* and *or7*) (Rodríguez-Ojeda et al. 2001), dominant resistance was also found for the Romanian race F (Or_6) and the Spanish race F and higher virulence ($HaOr_7$) (Pacureanu-Joita et al. 2004; Duriez et al. 2019; Martin-Sanz et al. 2020). $HaOr_7$ is the only cloned broomrape resistance gene so far. It was mapped on *Chr7* between two SNP markers, SF3178 and SF3131, encoding a LRR receptor-like kinase protein (Duriez et al. 2019). This protein is truncated in broomrape susceptible genotypes, while it is fully functional in the resistant genotype, conferring resistance in the early stages of the sunflower-broomrape interaction. It is homologous to the immune receptor Xa21, which is well described in rice as conferring resistance to bacterial blight.

Dominant resistance to race G, as well as races F_{GV} and E, was found in *H. debilis* subsp. *tardiflorus*. The gene originating from this introgression was designated as *OrDeb2* and was located on *Chr4*, flanked by the two SNP markers, AX-105399507 and AX-105525205/AX-105938724 (0.3 cM and 0.7 cM, respectively) (Velasco et al. 2012; Fernández-Aparicio et al. 2022). This chromosome also harbors another broomrape resistance gene conferring resistance to races F and G, Or_{SII} (Hassan et al. 2008; Martin-Sanz et al. 2020). Or_{SII} is a partly dominant gene, also known as "System II resistance", found in proprietary germplasm. Or_{SII} is highly associated with four SNP markers: DHAI000240, DHAG000732, DHAI007796, and DHAI007334 (Gao et al. 2019). Later on, the SNP marker, M-4_30.40, was mapped 0.7 cM distal from the gene (Martin-Sanz et al. 2020). Molecular studies suggested that *OrDeb2* and Or_{SII} are non-allelic but are located at a tightly linked position on the upper segment of *Chr4* (Fernández-Aparicio et al. 2022). This region is also rich in *Pl* genes, as discussed previously. Besides, the origin of the two genes and the mode of inheritance, *OrDeb2* and Or_{SII} also differ in their modes of action. *OrDeb2* is involved in post-attachment resistance by blocking broomrape development at the cortex, while Or_{SII} has a more delayed activation, disabling the emergence of the parasite from the soil and finishing its life cycle.

When mining for broomrape alleles and introducing them into breeding material, breeders have fewer options compared to downy mildew and rust; however, combining major and minor broomrape resistance genes, as well as genes with a different mode of action, may lead to sustainable broomrape resistance in sunflower. Recently, a large-scale screening for broomrape resistance of 71 accessions, including wild sunflowers and relatives from 16 *Helianthus* species, identified 18 accessions from nine species with broomrape resistance (Chabaud et al. 2022).

5.5.2 Mining for Herbicide Tolerance

Weeds present a significant limitation for sunflower production, with yield losses reaching up to 70% in weed-infected fields (Blamey et al. 1997). Thus, the use of herbicides inhibiting the activity of the large, catalytic subunit of the acetohydroxyacid synthase (AHASL) has been greatly exploited in controlling some of the most significant broad-leaved weeds in sunflower, such as *Ambrosia artemisiifolia* and *Xanthium strumarium*, as well as broomrape. Imidazolinones (IMIs) and sulfonylureas (Sus) are the two most commonly used AHAS-inhibiting classes of herbicides. The emergence and discovery of herbicide-resistant populations of weedy sunflowers in soybean fields in the USA, ANN-PUR and ANN-KAN treated with imazethapyr (IMI) or chlorsulfuron (SU), respectively, enabled the creation of herbicide-tolerant sunflower lines and hybrids by applying conventional breeding (Al-Khatib et al. 1998; Baumgartner et al. 1999; Miller and Al-Khatib 2002, 2004; Jocić et al. 2011). Out of the three *AHASL* genes mapped on chromosomes 2, 6, and 9, only mutations discovered in the *AHASL1* locus on *Chr9* conferred herbicide tolerance (Kolkman et al. 2004). This multiple allelic locus has been heavily exploited for sunflower hybrid breeding. Alleles, *AHASL1-1* and *AHASL1–2*, are characterized by a C>T mutation in codons 205 and 197 (adopting the *Arabidopsis* nomenclature) conferring moderate IMI tolerance and high Sus tolerance, respectively. Both alleles were identified in naturally grown common sunflowers and exploited for commercial use as Imisun/Clearfield® and Sures technologies, respectively. Discrimination between the wild-type allele (*ahasl1*) and the two alleles mentioned was enabled by applying SSR markers, which detect differences in the ACC repeats present in the locus (Kolkman et al. 2004). Additional SSR primer combinations were also used for the identification of allele variations in the *AHASL1* locus. Bulos et al. (2013) exploited the primer combination p-*AHAS18*/p-*AHAS19* for detecting different *AHASL1* alleles in the cultivated sunflower pool, while Jacob et al. (2017) used the *AHAS16*/*AHAS17* primer combination in the wild pool. Moreover, Kolkman et al. (2004) developed a co-segregating SNP marker for the detection of the C>T mutation in codon 205, speeding up the process of identification of partially IMI-tolerant sunflower genotypes, where an additional modifier gene was involved in IMI tolerance for which no marker was available (Bruniard and Miller 2001). Further on, CAPS markers for the detection of the point mutation in codon 197 (*AHASL1–2* allele) were developed for the detection of tribenuron-methyl-tolerant sunflowers by exploiting either of the three restriction enzymes, *Kpn2I*, *BseDI*, or *BcnI*, to digest the amplified PCR product (Dimitrijević 2013; Dimitrijević et al. 2014). Cross-tolerance to different AHAS-inhibiting herbicides was detected in wild Argentinian sunflower population and introduced into cultivated sunflower. The conventionally developed line RW-B, carrying the *AHASL1–4* allele, harbors a G->T mutation in codon 574 (Sala and Bulos 2012).

Chemical mutagenesis of cultivated sunflower gave rise to the development of new technologies in sunflower hybrid breeding: ClearfieldPlus® and ExpressSun®. ClearfieldPlus® was developed by ethyl methanesulfonate (EMS) treatment of the maintainer line BTK47, leading to an induced artificial mutation in codon 122 [G/A], resulting in high IMI tolerance (Sala et al. 2008a, b). Detection of this new allele, *AHASL1–3*, was enabled using SNP and CAPS markers (Bulos et al. 2013). One of the SNP primer combinations, p-*AHASNidF*/p-*AHAS122TWT*, amplified a 195-bp-band for the wild allele, while amplifying a PCR product of 186 and 204 bp in genotypes carrying *AHASL1-1* and *AHASL1–2*, respectively. No amplification occurred in genotypes carrying *AHASL1–3* allele, while this allele was amplified by primer combination p-*AHASNidF*/p-*AHAS122TMU* yielding a 195-bp-band and no band was amplified in other genotypes carrying different *AHASL1* alleles.

5.5.3 Induced Mutation – Driving Force for Allele Mining for Quality Traits

The increase in oil content and an improvement in oil quality are among the most important goals in sunflower breeding to ensure the raising of its market value. Worldwide, twelve percent of edible oil comes from sunflower, which is mainly used for cooking and as a salad dressing, in addition to its seeds being consumed as a tasty snack (Rauf et al. 2017). The exploitation of conventional breeding to improve oil content and quality gave these results. However, further progress was hampered due to the low genetic variation observed in elite germplasm and the polygenic nature of the traits. However, mutation breeding gave significant results in this field. The most well-known and important trait being altered by mutation breeding is, without doubt, the increase in oleic acid content (OAC). The first and most used source of elevated OAC comes from the cultivar Pervenets; therefore, the mutation underlying the trait, which later on was discovered, is at times called the "Pervenets" mutation (Soldatov 1976). By mutagenesis treatment of seeds of the variety VNIMK 8931 with 0.5% dimethyl sulfate (DMS) solution, Soldatov was able to identify sunflower plants in the M3 progeny with 70% OAC (range between 60% and 80%), and ultimately, this Pervenets mutated population (high oleic acid content - HOAC) has been widely used to develop varieties containing more than 83% OAC. The introduction of the increased OAC trait was, to a certain extent, eased by the mode of inheritance of the trait. While the majority of oil properties are polygenic traits, OAC can be regarded as a semi-qualitative trait, even though it is still affected by environmental conditions and the genetic background of the germplasm into which the trait is introduced (Izquierdo and Aguirrezábal 2008; Ferfuia et al. 2015; Hlisnikovský et al. 2015; Regitano Neto et al. 2016; Dimitrijević and Horn 2018). Ambiguous reports on the mode of inheritance of OAC can be found, mainly concerning the number of genes involved, spanning from one to several. The genes involved in OAC inheritance are named *Ol* genes (Urie 1984; Fick 1984; Fernández et al. 1999; Lacombe et al. 2004; Ferfuia and Vannozzi 2015; Premnath et al. 2016; Dimitrijević et al. 2017). On the molecular level, the chemical mutagenesis caused a partial duplication of the *FAD2-1* allele, resulting in the silencing of the gene (Lacombe et al. 2002; Schuppert et al. 2006). The *FAD2-1* gene encodes FAD2 (oleoyl phosphatidylcholine desaturase), an enzyme catalyzing the synthesis of linoleic acid from oleic acid. By silencing this gene, oleic acid is accumulated. A variety of molecular markers have been designed to detect the *Ol* mutation, which had been mapped on *Chr14* in a region co-segregating with the SSR markers ORS391, ORS832, and ORS1180 and flanked by the SSR and INDEL (Insertion/Deletion) markers, CRT20, CRT25, and ZVG64 (Pérez-Vich et al. 2002; Schuppert et al. 2006) (Table 5.2). Furthermore, Schuppert et al. (2006) developed a dominant

TABLE 5.2
Overview of Oil Quality Trait Sources, Locations, and Markers for Their Detection in Sunflower

Chromosome	Oil Quality Gene	Source	Markers Linked	Distance (cM)	Reference
Chr1	Tph_1	SRA16(1)	ORS605/ORS610 (1)	0.0	Tang et al. (2006) (1)
		T589(2)	ORS222/ORS598/ ORS1093(2)	0.0	Vera-Ruiz et al. (2006) (2)
Chr8	Tph_2	LG24 (1)	ORS780 (1)	0.5	Hass et al. (2006) (1)
		IAST-540(2)	ORS599 (2)	1.9	García-Moreno et al. (2006) (2)
			ORS312 (3)	3.6	
Chr14	*Ol (FAD2-1)*	RHA280 × RHA801(1)	ORS391/ORS832/ORS1180(1)	0.0	Schuppert et al., (2006) (1)
			F15–690 (2)	7.0	Dehmer and Friedt (1998) (2)
		HA-OL 9 (2)	AC10–765 (2)	7.2	

INDEL marker for tracking the *Ol* mutation as well as identifying variation downstream of the gene (five INDELs and 49 SNPs) and provided co-dominant INDEL and SSR markers. Following research by Lacombe et al. (2009), the number of available co-dominant and dominant markers for *Ol* mutation detection has increased. Some of the markers were validated by different research groups in a variety of sunflower genotypes (Nagarathna et al. 2011; Bilgen 2016; Dimitrijević et al. 2016). Out of the chosen markers that were previously developed for the detection of *Ol* mutation, Dimitrijević et al. (2017) reported that the F4/R1 INDEL marker for testing the presence or absence of the tandem FAD2-1 repeat was the most efficient in MAS for OAC (Schuppert et al. 2006). This marker amplified a 650-bp-band in high-oleic sunflower genotypes while failing to amplify a band in low-oleic genotypes. In addition, TILLING platforms, especially with regard to fatty acid biosynthesis, were established for sunflower (Sabetta et al. 2011; Kumar et al. 2013).

Another important trait in oil that may be regarded as somewhat neglected in comparison to fatty acids is the tocopherol content (vitamin E) responsible for antioxidant activity. There are four tocopherol forms: α, β, γ, and δ. Antioxidant activity increases from α to δ, while vitamin E activity decreases (Demurin et al. 1996; Fernández-Martínez et al. 2007). Common sunflower oil contains predominantly α-tocopherol. Designing sunflowers with specific fatty acid and tocopherol profiles provides sunflower oil for different purposes. Thus, oil convenient for frying is produced by combining the high oleic trait with increased γ- and/or δ-tocopherols (Márquez-Ruiz et al. 2014). Natural variation in tocopherol content in sunflower germplasm is low; thus, exploitation of induced mutations can lead to an increase in available variation (Velasco et al. 2004a). Inheritance of tocopherol composition is controlled by two independent major loci (Tph_1 and Tph_2) (Vera-Ruiz et al. 2005; García-Moreno et al. 2006; Gubaev et al. 2022). However, there are reports about the existence of gene modifiers with minor effects and about the influence of the genetic background (Demurin et al. 1996; Vera-Ruiz et al. 2005; García-Moreno et al. 2006; Hass et al. 2006; García-Moreno et al. 2012). The Tph_1 gene, also known as *m* locus, is located on the upper end of *Chr1* and encodes 2-methyl-6-phytyl-1,4-benzoquinone methyltransferase (MPBQ-MT) (Tang et al. 2006; Vera-Ruiz et al. 2006; Gubaev et al. 2022). Two independent sources of naturally increased β-tocopherol content were reported in two sunflower lines, LG15 and T589, developed at VNIIMK and the Institute of Sustainable Agriculture (CSIC) (Demurin 1993; Velasco et al. 2004b). Mutated recessive *tph1* alleles cause an increase in β-tocopherol content by partial disruption of the α-tocopherol synthesis in seeds (Demurin 1993). Tang et al. (2006) mapped two co-segregating SSR markers, ORS605 and ORS610, to the locus in the cross NMS373 × SRA16, while Vera-Ruiz et al. (2006) mapped three co-segregating SSR markers, ORS222, ORS598, and ORS1093, in CAS-12 × T589. Common SSR markers of the two maps, ORS965 and ORS716, were mapped with 12.1 and 19 cM, and with 9.9 and 11.9 cM distant from the locus by Tang et al. (2006) and by Vera-Ruiz et al. (2006), respectively. Recently, Gubaev et al. (2022) mapped two SNPs, S1_9228105 and S1_4480460, at a physical distance of 0.5 and 5.3 Mb to the Thp_1 locus (NCBI Gene ID: 110937001) in the crosses VK195 × VK303 and VK876 × VK101, respectively.

Recessive alleles at the Tph_2 locus, also known as *g* locus, are responsible for increased γ-tocopherol content. This locus corresponds to the γ-tocopherol methyltransferase (γ-TMT) gene located on *Chr8* (Hass et al. 2006; García-Moreno et al. 2006, 2012). Mutation in this locus causes disruption in α-tocopherol synthesis, thus leading to the accumulation of γ-tocopherol. Sources of increased γ-tocopherol content can be found in the VNIIMK and CSIC collections: LG17, T2100, IAST-1, and IAST-540 (Demurin et al. 1996; Velasco et al. 2004a, b). García-Moreno et al. (2006) reported two flanking SSR markers to the Tph_2 locus, ORS312 and ORS599, which were mapped 3.6 and 1.9 cM away from the locus in the cross CAS-12 × IAST-540, respectively, while Hass et al. (2006) mapped ORS780 0.5 cM from the *g* locus in the cross R112 × LG24. The locus was shown to carry two genes (NCBI GeneIDs: 110872346 and 110872347) that encode γ-TMT1 and γ-TMT2 involved in the accumulation of γ- and δ-tocopherol (Hass et al. 2006; Gubaev et al. 2022). Hass et al. (2006) found 173 SNPs and 33 INDELs in γ-TMT sequences and identified five haplotypes. To detect some of the variation, SSCP and INDEL markers were developed. However, the INDEL markers were not polymorphic in the study of García-Moreno et al. (2012), who instead developed the γ-TMT_F1/F2/R24 INDEL marker that enabled discrimination between alleles of the Tph_2 locus in two high γ-tocopherol lines, IAST-1 and nmsT2100. The newly developed INDEL marker amplified a band of 1,460 and 1,420 bp, respectively. While the tph_2 allele in nmsT2100 is stable, the *tph2a* allele in IAST-1 is affected by modifying genes, which correspond to a duplication of the γ-TMT loci in the sunflower genome (García-Moreno et al. 2012). The physically closest SNPs, S8_21613153 and S8_21536716, were mapped 0.34 and 0.42 Mb from the Tph_2 locus in the crosses VK195 × VK303 and VK876 × VK101, respectively (Gubaev et al. 2022).

Gubaev et al. (2022) developed markers for the detection of SNPs associated to *Tph1* and *Tph2* loci based on the LOD confidence interval and the ability to target unique DNA sequences in the genome. Thus, two markers were developed for the detection of the loci in the VK195 × VK303 population: S1_55196434 and S8_30578572, and two for VK876 × VK101 population: S1_71748138 and S8_23941299. The exploitation of these markers will facilitate the introduction of the chosen alleles and significantly accelerate the process of designing sunflower genotypes with changed oil properties regarding the tocopherol content, especially bearing in mind that recessive alleles are responsible for developing these traits. However, as demonstrated, all markers need to be tested and validated in donor and recipient sunflower lines before their successful exploitation in MAS.

5.5.4 Mining for Fertility Restoration

For five decades, sunflower production has been based on hybrid production; thus, the initial identification of efficient cytoplasmic male sterility (CMS) and corresponding fertility restoration (*Rf*) genes in the 1960s and 1970s was a turning point in sunflower breeding. Since its discovery in an interspecific hybrid of *H. petiolaris* and *H. annuus* by Leclercq (1969), CMS PET1 has become the most common source of CMS cytoplasm used worldwide, despite the existence of more than 70 CMS sources (Rauf et al. 2020). On the molecular level, CMS PET1 is characterized by the expression of a 16-kDa-protein as the result of the co-transcription of the *atpA* gene and a new CMS-specific *orfH522* (Horn et al. 1991; Köhler et al. 1991). Fertility restoration of PET1 cytoplasm was first discovered by Kinman (1970) in the line T66006-2-1-B (Rf_1 gene). In addition, the Rf_2 gene was found in the crosses of T66006-2-1-B and MZ01398. This Rf_2 gene seems to be present in most sunflower lines; thus, *Rf1* gene is the main one restoring male fertility (Vranceanu and Stoenescu 1978; Leclercq 1984; Serieys 2005). As the most widely exploited gene for hybrid production, Rf_1 has ever since been the subject of numerous studies (Table 5.3). Different types of markers have been developed for its detection, as markers greatly facilitate and speed up the process of hybrid breeding and production.

Initial mapping of Rf_1 included the use of random amplified polymorphic DNA (RAPD) and amplified fragment length polymorphism (AFLP) markers, which were then enriched by newly developed AFLP and STS markers (Horn et al. 2003; Sajer et al. 2020, respectively) (Table 5.3). Two SCAR markers, developed from two closely linked RAPD markers associated with the Rf_1 gene, were designed for the detection of the Rf_1 gene in MAS (Horn et al. 2003). These markers, HRG01 and HRG02, were exploited in a diverse panel of wild and cultivated sunflowers. HRG01 was more efficient in detecting the Rf_1 gene in annual species, in which a band of approximately 450 bp was amplified, while HRG02 was recommended for use in perennial species in which it amplified a band of approximately 750 bp (Markin et al. 2017). Moreover, a multiplex TaqMan assay was designed for the detection of Rf_1 and *orfH522* genes in the same reaction. Further marker types developed for the detection of the Rf_1 gene included a codominant CAPS marker, H13, which mapped with 7.7 cM distal from Rf_1 (Kusterer et al. 2005) as well as a target region amplification polymorphism (TRAP) marker, K11F05Sa12–160, amplifying a 160-bp-band mapped to the gene with proximal 0.4 cM (Yue et al. 2010). By testing this marker in an USDA sunflower panel (177 accessions), the authors reported a limited potential for this marker in MAS. The authors also mapped SSR marker ORS511 with 3.7 cM distal to Rf_1. Recent efforts in identifying an Rf_1-candidate gene and developing Rf_1-associated markers exploited the advantage of

TABLE 5.3
Overview of Fertility Restoration Genes Sources, Locations, and Markers for Their Detection in Sunflower

Chromosome	Fertility Restoration Gene	Source	Markers Linked	Distance (cM)	Reference
Chr3	Rf_4	HA89	ORS13	0.9	Feng and Jan (2008)
	Rf_6	HA 89	ORS13	1.6	Liu et al. (2013)
		HA 821	ORS13	1.6	
Chr7	Rf_3	RHA340	ORS004	9.3	Abratti et al. (2008)
	Rf_3	RHA280	ORS328	0.7	Liu et al. (2012b)
Chr13	Rf_1	RHA325 (1)(2)(4)	K11F05Sa12–160 (3)	0.4	Horn et al. (2003) (1)
		RHA439(3)	E32M36–155R (2)	0.7	Kusterer et al. (2005) (2)
			HRG01 (1)	0.8	Yue et al. (2010) (3)
			OP-K13 454 (1)(2)	0.8 (1)0.9(2)	Sajer et al. (2020) (4)
			E62M52_249A(4)	1.0	
			HRG02 (1)	2.0	
			OPY10_740(1)	2.0	
			ORS511(3)	3.7	
			H13 (2)	7.7	
	Rf_5	Rf ANN-1742	ORS728,	1.3	Qi et al. (2012b)
			ORS995	4.5	
	Rf_7	RHA 428	SFW04577/ SFW04100/ NSA_008361/ SFW01741 SFW01515	0.0	Talukder et al. (2019)
	Rf-PET2	IH-51	E40M61_166A	4.4	Sajer et al. (2020)
			E41M59_410A/ E44M61_122A/ E45M52_321A	4.9	
undesignated	*Rf_PEF1*	LC 1064	E39M51_300R	3.9	Schnabel et al. (2008)
			E44M56_112A	6.0	

the published sunflower genome sequence (Badouin et al. 2017). A genome-wide association study (GWAS) revealed 21 candidate genes in a 7.72 Mb region, with most genes belonging to the pentatricopeptide repeat (PPR) gene family and one identified probable aldehyde dehydrogenase gene (Goryunov et al. 2019), which was also the proposed function of the Rf_1-candidate gene *HanXRQChr13g0419821* by Owens et al. (2019). The number of candidate genes was narrowed down to three PPR genes (*HanXRQChr13g0419621*, *HanXRQChr13g0418841*, and *HanXRQChr13g0418861*) by Horn et al. (2019). Within these three genes, 10 significantly associated SNPs to the Rf_1 gene were identified in a panel of 59 sunflower lines by amplicon-targeted sequencing. Different types of molecular markers were developed for the detection of some of the SNPs, such as PCR amplification of multiple specific alleles (PAMSA) and two dominant SCAR markers (Horn et al. 2019). These three gel-based markers were validated in a panel of 557 cultivated sunflower accessions. The PAMSA marker amplified a 170-bp-band in the restorer and a 155-bp-band in the maintainer lines. The two SCAR markers, PPR621.5R and PPR621.5M, were developed for the amplification of one specific SNP. Thus, the Rf_1-associated PPR621.5R marker amplified a 164-bp-band in restorer lines and failed to amplify a band in the maintainer lines, while PPR621.5R amplified a 164-bp-band in the maintainer lines but did not amplify a band in tested restorer lines (Horn et al. 2019). Recently, Radanović et al. (2022b) developed four co-dominant kompetitive allele-specific PCR (KASP) markers for the detection of Rf_1-associated SNPs found in the three PPR candidate genes identified by Horn et al. (2019). The designed markers cover a 3-Mb-region of the Rf_1 locus and were proven useful in detecting a recombination event that occurred in one plant of the developed EMS population obtained by chemical mutagenesis of F_1-hybrids of the cross cmsHA342 × RHA325. Recombination in this Rf_1 locus might be responsible for the phenotype change from fertile to male-sterile. Thus, this will be further exploited for potential identification of the Rf_1 gene, which so far has proven to be very difficult due to the complexity of the region. Polivanova et al. (2021) identified six haplotypes among tested restorer lines, examining 25 Rf_1 potential candidate genes using PCR gene-specific primers. On the other hand, Talukder et al. (2019) identified 24 SNPs associated with the Rf_1 gene that were transferred jointly as a haplotype into numerous Rf_1-bearing fertility restorer lines. However, these SNPs were also detected in three USDA-released maintainer lines. Besides Rf_1, these authors examined the remaining two other *Rf* genes located on the lower end of *Chr13*: Rf_5 and Rf_7. The restorer gene Rf_5 was initially mapped between the two SSR markers, ORS728 and ORS995, which are also linked to the rust resistance gene R_{11} (Qi et al. 2012b). One SSR marker, ORS316, common for Rf_5 and Rf_7, was mapped with 1.2 and 5.3 cM proximal to the genes, respectively, while ORS511 represents a common marker for Rf_1 and Rf_7, mapping with 3.7 and 0.9 cM distal to the genes, respectively (Yue et al. 2010; Qi et al. 2012b; Talukder et al. 2019). Consequently, these *Rf* genes are most likely clustered together as well as clustered with some *R* and *Pl* genes, as previously outlined. Out of the 24 marker alleles significantly associated with the Rf_1 gene, five and seven were preserved in the Rf_5 and Rf_7 genes, respectively (Talukder et al. 2019). Further molecular studies should reveal the true relationship between the genes. So far, it seems that Rf_5 and Rf_7 differ from Rf_1 in genomic composition, but there is still a possibility that Rf_7 might be an alternate *Rf* source (Talukder et al. 2019). Ma et al. (2021) proposed two Rf_5 candidate genes (Ha412v1r1_13g048240 and Ha412v1r1_13g048260), both belonging to the PPR gene family. A recent study revealed another *Rf* gene that mapped close to *Rf1*, *Rf-PET2* (Sajer et al. 2020). This gene was mapped between ORS630 and ORS1030, previously linked to the Rf_1 gene (Kusterer et al. 2005). *Rf-PET2* is a fertility restorer gene of both PET2 and PET1 cytoplasm; thus, it can be explored for expanding the sunflower hybrid breeding pool and potentially reducing the possible risk of pathogen attacks. Detailed allele mining is required to identify the different genes behind fertility restoration.

Restorer of fertility gene Rf_3, differing from Rf_1 and Rf_2 genes, was mapped on *Chr7*. It was discovered in the confectionary sunflower line RHA 280 (Jan and Vick 2007) and associated with five previously developed SSR markers (ORS328, ORS331, ORS928, ORS966, and ORS1092) and three newly developed expressed sequence tag (EST)-SSRs (HT-619-1, HT619-2, and HT1013), with the closest SSR being the co-dominant ORS328 at a distance of 0.7 cM (Liu et al. 2012b). This gene was clearly differentiated from a previously reported Rf_3 gene discovered in the downy mildew restorer line RHA340, also named Rf_3. Its origin can be tracked down to *H. argophyllus* (Abratti et al. 2008). This gene was mapped between SSR ORS004 and the proprietary INDEL marker IN0182. The allelic relationship between the two genes should be examined; however, initial molecular studies showed that ORS004 and ORS460 used for mapping the Rf_3 gene in a cross between RHA340 and ZENB8 by Abratti et al. (2008) were monomorphic in the mapping population CMS HA 89–3149 9 × RHA 280 (Liu et al. 2012b). Similarly, a major ORS328 band associated with Rf_3 from RHA280 was not amplified in *Rf3* originating from RHA340.

Despite the existence of 72 CMS sources discovered so far, corresponding *Rf* genes have only been discovered for about half of them (Serieys 2005; Liu et al. 2012b). The CMS *GIG2-*Rf_4, CMS *514A-*Rf_6, CMS *PEF1-Rf_PEF1*, and CMS *PET2-Rf-PET2* systems present four alternative sources for hybrid breeding possessing available molecular markers for tracing *Rf* genes. The first alternative system was discovered by Feng and Jan (2008), who mapped the Rf_4 gene near three SSR markers, ORS1114, ORS13 and ORS822–3, and one RFLP-derived STS-marker STS10B1 on the distal region of *Chr3*, with the closest marker being ORS1114, which mapped at a distance of 0.9 cM from the gene. SSR marker ORS1114, together with ORS13, were the closest markers mapped with 1.6 cM from Rf_6 on *Chr3*. The two genes have a different origin; hence, Liu et al. (2013) proposed that these are two different genes. According to genomic in situ hybridization (GISH) analysis, the fertility restoration trait is associated with a small translocation from *H. angustifolius* that occurred in the terminal part of the *Chr3*. Schnabel et al. (2008) created an AFLP linkage group with E39M51_300R and E44M56_112A being the closest markers linked to the *Rf_PEF1* gene. This gene originates from LC 1064, a restorer line of PEF1 and PET1 cytoplasm (Horn and Friedt 1997). The two CMS sources can also be discriminated against using newly developed CMS-specific molecular markers (Schnabel et al. 2008). The fourth CMS/Rf-system is the well-characterized CMS PET2, where recombination events at the *atp9* gene lead to duplication and the creation of two new open reading frames, *orf288* and *orf231*, involved in male sterility (Reddemann and Horn 2018). The *Rf-PET2* restorer gene is located next to *Rf1* on *Chr13*, between the SSR markers ORS1030 and ORS630 (Sajer et al. 2020).

Further efforts to analyze *Rf* genes in sunflowers will develop in two directions. One would be to finally decipher the Rf_1 gene, to gain more knowledge about its function, and to design ideally functional markers for its detection. The second direction will be the exploitation of mainly CWR to obtain more *Rf* sources and to diversify the existing pool of CMS-Rf systems to avoid linkage drags due to the use of only Rf_1.

5.5.5 Mining for Abiotic Stress

Responses and breeding aspects regarding abiotic stress (drought, salinity, heat, cold, and mineral toxicity) have been intensively reviewed by Škorić (2009, 2016), with an emphasis on drought stress also by Rauf (2008) and Hussain et al. (2018). Sunflower is regarded as a moderately drought-tolerant crop because of its deep tap root (Angadi and Entz 2002; Dwyer 2015). Drought represents the most challenging abiotic stress for sunflower crop production. More than 50% of yield losses have been observed in sunflowers and changes in the climate will increase the losses in the future (Bošnjak 2004). Effects of drought depend on the developmental stage (germination, seedling, and flowering) and on the severity (low, moderate, and severe), as well as the frequency and duration of the water deficit periods. Drought in sunflowers leads to severe losses in yield and oil quality. Screening for drought tolerance has been performed on the field (Gholinezhad et al. 2014; Safavi et al. 2015; Sarvari et al. 2017; Gul et al. 2022) as well as by *in vitro* screening using polyethylene glycol 6000 (PEG) (Fulda et al. 2011; Vassilevska-Ivanova et al. 2014; Hussain et al. 2017; Liang et al. 2017; Razzaq et al. 2017). Traits like deeper rooting depth, an increase in water uptake by the root, higher tolerance to osmotic pressure, reduced transpiration rates, and good recovery seem to be correlated with drought tolerance in sunflowers (Škorić 2016). Wild species in the genus *Helianthus* have been a rich source for abiotic stress tolerances in sunflowers; e.g., *H. argophyllus* has been explored for drought tolerance (Belhassen et al. 1996; Hussain et al. 2017) and *H. paradoxus* for salt tolerance (Lexer et al. 2004).

H. petiolaris reacts to water deficiency by deepening roots, which represents a desirable trait in breeding programs of sunflower hybrids to be more tolerant to water deficiency (Sobrado and Turner 1986). *H. petiolaris* appeared to be a perspective source of broadening genetic variability for heat stress as well (Miladinović et al. 2019). There are more sources of desirable genes in wild populations that could enlarge the genetic diversity in sunflower breeding programs, particularly in respect of abiotic stress, which has come into the scope of contemporary research due to climate changes. Hence, *H. argophyllus, H. anomalus, H. deserticola*, and *H. nuttallii* could be perspective germplasm for mining genes for heat stress as well as drought tolerance. Interspecific hybridization between *H. paradoxus* and *H. annuus* gave lines with enhanced tolerance to high sodium content in soil, i.e., salinity stress (Lexer et al. 2004). The study of eleven genes revealed a Ca-dependent protein kinase as a putative candidate gene for salt tolerance as it mapped to a corresponding QTL. Two more genes, an endoplasmic reticulum-type calcium adenosine triphosphatase and a transcriptional regulator, resulted in increased stress-induced fitness originating from *H. paradoxus* (Lexer et al. 2004).

Numerous candidate genes for drought tolerance involving phytohormones, especially abscisic acid (ABA) and ethylene, biosynthesis and signal transduction, as well as osmoprotectants, antioxidative protection, water transport, staying green, and recovery, have been identified in plants (reviewed in Krannich et al. 2015; Wang et al. 2021), in addition to a network of transcription factors responsible for regulating drought response (Shinozaki and Yamaguchi-Shinozaki 2007, Joshi et al. 2016). In sunflower, research studies on drought response and drought tolerance have also revealed a number of potential candidate genes whose natural variation might be interesting for breeding improved sunflower hybrids.

Comparing the proteome patterns of two sunflower lines differing in drought sensitivity, two proteins (glyceraldehyde-3-phosphate dehydrogenase and phosphoglycerate kinase) distinguishing the genotypes, 23 mostly down-regulated proteins in the susceptible genotype and five proteins in the tolerant genotypes were spotted (Castillejo et al. 2008). The susceptible line was characterized by the reduction of proteins from the photosynthesis apparatus and the carbohydrate metabolism. Fulda et al. (2011) identified six proteins induced more than 2-fold in leaves as a drought response to treatment with 20% PEG: a Cu/Zn-superoxide dismutase (SOD) (2.1-fold), a glycine-rich RNA binding protein (2.2-fold), an unknown protein (2.3-fold), a vegetative storage protein (2.5-fold), a fructokinase 3 (3.3-fold), and a putative caffeoyl-CoA 3-O-methyltransferase (4.5-fold). Soluble proteins were separated on two-dimensional gel electrophoresis, and protein spots were identified by matrix-assisted laser desorption/ionization time-of-flight mass spectrometry (MALDI-TOF) mass spectrometry. The increased expression makes the identified proteins interesting candidates for allele mining with regard to drought tolerance. Whereas both Castillejo et al. (2008) and Fulda et al. (2011) addressed seedling drought stress, the work of Ghaffari et al. (2017) looked at protein changes under drought stress during the flowering stage. Oxygen-evolving enhancer (OEE) and ferredoxin-NADP reductase (FNR), which were suppressed in both the tolerant (BGK 329) and the sensitive line (RGK 21) under drought conditions, were identified as primary sensors of water deficit. The reduction of these two proteins plays a role in the protection of the electron-transfer chain in the chloroplast by avoiding the production of reactive oxygen species (ROS) (Ghaffari et al. 2017). Higher abundance of the Cu/Zn-SOD and sucrose synthase might be relevant for the drought tolerance of line BGK 329 by allowing a more efficient scavenging of ROS and a better conservation of energy and maintenance of the cell structure integrity, respectively. In addition, the differential expression of the cytosolic NADP malic enzyme and ATP synthase involved in water conservation and energy might be important. The putative primary sensors as well as the differentially expressed proteins between RGK 21 and BGK 329 would be interesting candidates for an allele mining approach.

Furthermore, dehydrin genes (*HaDhn1* and *HaDhn2*) were shown to be upregulated under drought in the tolerant line R1 (Cellier et al. 1998) and a gene coding a dehydrin-like protein was one of the nine genes associated with drought tolerance in the RNA sequencing approach (Liang et al. 2017). Transcriptome analysis of two sunflower lines revealed the drought response mechanism and identified candidate genes (Wu et al. 2022b). In addition, transcription factors encoded by the homeobox-leucine zipper gene *Hahb-4* (Gago et al. 2002; Dezar et al. 2005) and *HaNAC* genes (Li et al. 2021) have been connected to drought responses in sunflowers, the latter also to salt stress.

Water use efficiency (WUE) and leaf carbon isotope discrimination (CID) were analyzed in sunflowers under different drought regimes (Adiredjo et al. 2014a, b). WUE is calculated as a relation between biomass (BM) and water consumption (cumulative water transpired - CWT) and is measured at the end of the trials. Using a recombinant inbred population of 150 lines from the cross XRQ and PSCS8, QTLs for WUE and CID were localized on a genetic map based on SNP markers derived from an Infinium HD iSAelctBeadChip (Infinium). Nine QTLs controlling WUE were detected depending on the drought scenario, whereas the eight QTLs for CID occurred independent of the drought treatments. Co-localizations were observed for three QTLs of CID and WUE. Stable genomic locations for CID were located in linkage groups 6 and 13 (Adiredjo et al. 2014a). The SNP markers flanking the QTL now allow the exact positioning of the QTL in the annotated sunflower reference genome. The regions under the identified QTL for WUE and CID represent a rich source for candidate genes to be screened for natural allelic variation, which can then be tested for association with drought tolerance.

The richness of variation within wild sunflower species and the vast variability of environments they inhabit offer the possibility of mining genes in this naturally occurring gene pool for resistance to different stresses, from heat to cold, frost to drought, salinity, flooding, and other abiotic stresses (Miladinović et al. 2019). Whereas the introduction of major resistance genes from wild species to the cultivated germplasm by interspecific hybridization is relatively easy, tolerances/resistances caused mainly by minor gene effects (QTL) require more efforts in gene identification and introgression into the cultivated sunflower. Using modern molecular genetic approaches, these obstacles should be overcome in the future.

5.6 FUTURE PROSPECTS OF ALLELE MINING IN SUNFLOWER

Selection is an important step in the breeding process. Selecting genotypes with desirable phenotypic outcomes (positive selection) or eliminating undesirable genotypes (negative selection) results in the same – the gradual narrowing of genetic variability in breeding programs. The economy, i.e., market demands, is driving breeders to conduct selection under the same or similar criteria over and over again. Breeders are in demand to deliver high-yielding varieties with high oil content, resistance to prevalent diseases, and high stability of desirable traits in given environmental conditions. After a while, the gene stock becomes exhausted, and mining for the renewal of operative genetic variability is highly required in order to boost a breeding program and to continue to produce competitive varieties for the seed market. The first gene pool to dig into is a genetic variability that has been previously used for breeding. This gene pool includes old varieties that still may hold some interesting genes that could be used for modern variety improvement; varieties and lines from older or/and contemporary breeding programs that have not been registered but could possess usable genes or superior alleles for particular traits that have recently come into focus of breeders' interest. If desirable genetic variability cannot be found in breeding material, the next gene pool to mine consists of local populations and landraces that are commonly instinctively bred or just selected out of a broader population. The described primary gene pool can still contribute to modern breeding programs, with some usable genes coming from good adaptation abilities to particular regional environments. Old varieties, lines, and landraces represent a gene source that is the easiest to handle for broadening or refreshing genetic variability in contemporary breeding programs because of their high cross-compatibility with modern lines, which leads to vital and fertile progenies. The other convenience is that old varieties, lines, and even landraces have, to some extent, refined genomes due to previously conducted breeding processes or just instinctive selection. This means that gene donors that come out of this gene pool in conventional hybridization with contemporary lines would not greatly spoil the genetic background of the progeny, and eventually, after several backcrosses, breeders might isolate the desirable trait from the donors, preserving all the other favorable traits of the modern acceptor line. Unfortunately, challenges posed by the dramatic environmental changes resulting from continuous climate developments and by market demands require deeper mining in the secondary gene pool of wild relatives. Here, breeders could face interspecific or even intergeneric hybridizations, i.e., crosses of different species or even genera, respectively. Crosses of taxonomically distant parents in classical hybridizations bring up sterility problems in the progenies, which, as a rule, grow with the increasing taxonomical distance of the parents. Differences in chromosome numbers as well as chromosomal and genome incompatibilities make the introduction of the desirable gene from the wild into a modern genetic background more difficult, time-consuming, and consequently costlier. Even if the sterility problem can be resolved, the obtained progeny could be affected by the deleterious effect of the genetic background of the wild parent mixed in hybridization with the modern parental line. Cleaning the genetic background through selection over generations of backcrosses to isolate the desirable trait from the wild donor in the preserved advanced genome of the modern line could be an additional burden for the breeding program. However, despite all the problems and complications, sunflower is one of the leading cultivars that has been improved through time, with genetic variation introduced from wild relatives.

Allele mining, which can now be started on a large scale after the re-sequencing of the UGA-SAM1 population and the wild accessions, will open a new era of breeding possibilities in sunflowers (Figure 5.2). So far, for most traits in sunflower, no specific candidate genes have been addressed, and no superior alleles have been identified. Knowledge about the different alleles behind a trait might help to better understand natural or artificial variations occurring in different crosses and to better understand the function of the underlying genes.

In other crops, e.g., rice, wheat or barley, allele mining has already been widely explored, not only with regard to disease resistance genes (Kumar et al. 2010; Ashkani et al. 2015; Yu et al. 2022) but also for candidate genes for drought tolerance. For drought stress assessment, proline accumulation has always been an indicator. Sequencing 49 barley genotypes comprising cultivars, landraces, and wild barley accessions, the promoter region of the pyrroline-5-carboxylate synthase (P5CS1), a central enzyme of proline biosynthesis, was analyzed (Shrestha et al. 2022). Allele mining revealed novel allelic series in the P5CS1 promoter region, which were correlated with expression analyses of the gene and proline accumulation. A newly developed near-isogenic line (NIL-143) showed increased proline accumulation, better recovery from drought, and better yield stability (Shrestha et al. 2022). In rice, the *Deeper Rooting 1* (*DRO1*) gene, present under a major QTL on chromosome 9, plays an important role in drought survival by enlarging the root angle (Singh et al. 2021). Allele mining identified 16 variants displaying a whole set of alleles, from functional to truncated non-functional alleles. All "yield under drought" (DTY) genotypes were characterized by specific SNPs leading to changes in the amino acid sequence encoded by this gene. This knowledge can now be used to accelerate breeding for drought tolerance (Singh et al. 2021). In wheat, allele mining of the genes encoding the ABA receptor TaPYL4 revealed specific haplotypes for each of the three genomes A, B, and D (Wu et al. 2022a). The haplotype *TaPYL4-2A-Hap2* represents a superior allele for breeding in Chinese wheat, as it is significantly associated with reduced plant height and a higher drought tolerance coefficient (DTC). Haplotype *TaPYL4-2B-Hap1* was identified as the favorable allele for increasing grain numbers under drought conditions, whereas the *TaPYL4-2D* haplotypes showed no correlations with the investigated agronomic traits (Wu et al. 2022a). CAPS and Derived Cleaved Amplified Polymorphic Sequence (dCAPS) markers were developed for MAS. Another study investigated the drought tolerance transcription activator gene *OsDREB1F* in rice (Singh et al. 2015). Natural allelic diversity was confirmed, and one variant characterized by three SNPs in the coding region of the gene seems to be associated with drought tolerance. In chickpeas, allele mining was performed for the well-known drought-responsive gene *dehydrin* (*DHN*) (Kumar et al. 2020). Indels and SNPs specifically present only in drought-tolerant genotypes could be identified. Also, in sunflower, a drought-tolerant line R1

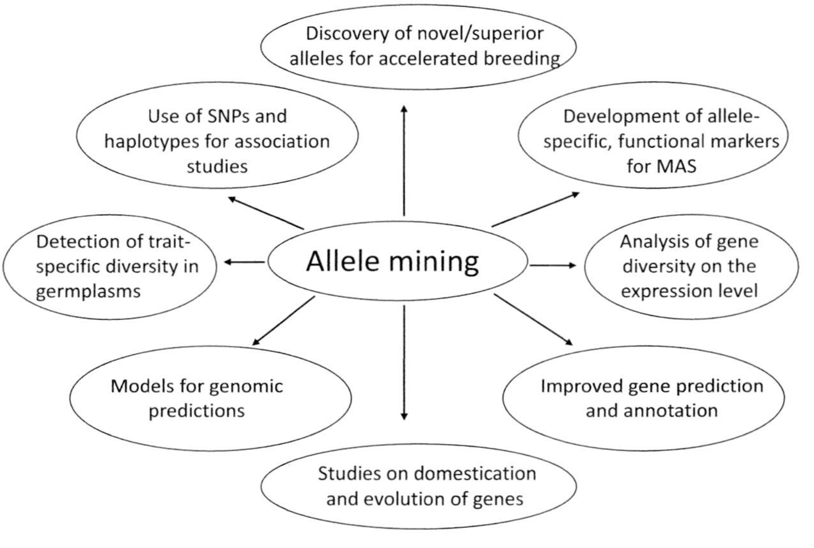

FIGURE 5.2 Application and use of allele mining for breeding.

showed an increase in the transcription of *HaDhn1* and *HaDnh2*, which was correlated with a positive drought-adaptive reaction (Cellier et al. 1998), making these two genes interesting candidates for mining natural variants with regard to drought tolerance in sunflower.

Breeding for climate-resilient sunflowers will require efficient allele mining, either based on whole genome sequence analyses or targeted amplicon sequencing addressing interesting candidate genes for abiotic and biotic stress. In the last decades, sunflowers have made large progress in the direction of allele mining by gaining all the basic requirements, such as characterized selected germplasm accessions, genetic maps, annotated reference genome and the availability of a large number of re-sequenced and publicly available accessions for the phenotyping of relevant traits in sunflowers. With the resources in place, sunflower breeding can now make considerable progress in the direction of environmentally resilient sunflower hybrids in a more efficient and less time-consuming manner.

Acknowledgements: This study was supported by CROPINNO Grant No. 101059784, by Ministry of Education, Science and Technological Development of Republic of Serbia, grant number 451–03–9/2021-14/200032, COST Action CA 19125, the Science Fund of the Republic of Serbia, through IDEAS project "SMARTSUN" grant number 7732457. This work was done as a part of activities of Center of Excellence for Innovations in Breeding of Climate Resilient Crops - Climate Crops, Institute of Field and Vegetable Crops, Novi Sad, Serbia and as part of activities at the Department of Plant Genetics at the University of Rostock.

REFERENCES

Abratti G, Bazzalo ME, Leon A (2008) Mapping a novel fertility restoration gene in sunflower. In: Proceedings of the 17th International Sunflower Conference, Cordoba, Spain, June 8-12, 2008, ISA, Paris, France, pp 617–621

Adeleke BS, Babalola OO (2020) Oilseed crop sunflower (*Helianthus annuus*) as a source of food: Nutritional and health benefits. *Food Sci Nutr* 8(9):4666–4684

Adiredjo AL, Navaud O, Lamaze T, Grieu P (2014a) Leaf carbon isotope discrimination as an accurate indicator of water-use efficiency in sunflower genotypes subjected to five stable soil water contents. *J Agron Crop Sci* 200:416–424

Adiredjo AL, Navaud O, Munos S, Langdale N, Lamaze T, Grieu P (2014b) Genetic control of water use efficiency and leaf carbon isotope discrimination ins sunflower (*Helianthus annuus* L.) subjected to two drought scenarios. *PLoS One* 9(7):e101218

Aliiev E (2020) Automatic phenotyping test of sunflower seeds. *Helia* 43(72):51–66

Al-Khatib K, Baumgartner JR, Peterson, DE, Currie RS (1998) Imazethapyr resistance in common sunflower (*Helianthus annuus*). *Weed Sci* 46(4):403–407

Anđelković V, Cvejić S, Jocić S, Kondić-Špika A, Marjanović Jeromela A, Mikić S, Prodanović S, Radanović A, Savić Ivanov M, Trkulja D, Miladinović D (2020) Use of plant genetic resources in crop improvement-example of Serbia. *Genet Resour Crop E* 67(8):1935–1948

Angadi SV, Entz MH (2002) Root system and water use patterns of different height sunflower cultivars. *Agron J* 94(1):136–145

Arpin I, Barreteau O, Caranta C, Ducrot C, Garin P, Hannachi M, Maillet I (2020) Séminaire d'échanges sur les pratiques de recherches interdisciplinaires à INRAE, 7-8 Janvier 2020, Synthèse, 22 p

Ashkani S, Yusop MR, Shabanimofrad M, Azadi A, Ghasemzadeh A, Azizi P, Latif MA (2015) Allele mining strategies: principles and utilization for blast resistance genes in rice (*Oryza sativa* L.). *Curr Issues Mol Biol* 17:57–74

Atlagić J (2004) Roles of interspecific hybridization and cytogenetic studies in sunflower breeding. *Helia* 27(41):1–24

Atlagic J, Terzić S (2014) Sunflower genetic resources-interspecific hybridization and cytogenetics in prebreeding. In: Arribas JI (ed) *Sunflowers: Growth and Development, Environmental Influences and Pests/Diseases*. Nova Science Publishers, New York, USA, pp 95–130

Atlagić J, Terzić S (2015) The challenges of maintaining a collection of wild sunflower (*Helianthus*) species. *Genet Resour Crop E* 63:1219–1236

Atlagić J, Terzić S, Škorić D, Marinković R, VasiljevićLj, Panković D (2006) The wild sunflowers collection in Novi Sad. *Helia* 29 (44):55–64

Bachlava E, Radwan OE, Abratti G, Tang S, Gao W, Heesacker AF, Bazzalo ME, Zambelli A, Leon A, Knapp SJ (2011) Downy mildew (*Pl8* and *Pl14*) and rust (*Radv*) resistance genes reside in close proximity to tandemly duplicated clusters of non-TIR-like NBS-LRR encoding genes on sunflower chromosomes 1 and 13. *Theor Appl Genet* 122:1211–1221

Badouin H, Gouzy J, Grassa JC, Murat F, Evan Staton S, Cottret L, Christine Lelandais-Brière C, Owens GL, Carrère S, Mayjonade B, Legrand L, Gill N, Kane NC, Bowers JE, Hubner S, Bellec A, Bérard A, Bergès H, Blanchet N, Boniface M-C, Brunel D, Catrice O, Chaidir N, Claudel C, Donnadieu C, Faraut T, Fievet G, Helmstetter N, King M, Knapp SJ, Lai Z, Le Paslier M-C, Lippi Y, Lorenzon L, Mandel JR, Marage M, Marchand G, Marquand E, Bret-Mestries E, Morien E, Nambeesan S, Nguyen T, Pegot-Espagnet P, Pouilly N, Raftis F, Sallet E, Schiex T, Thomas J, Vandecasteele C, Varès D, Vear F, Vautrin S, Crespi M, Mangin B, Burke JM, Salse J, Muños S, Vincourt P, Rieseberg LH &Langlade NB (2017) The sunflower genome provides insights into oil metabolism, flowering and Asterid evolution. *Nature* 546(7656):148–152

Bailey DL (1923) Sunflower rust. Minn Agr Exp Sta Tech Bull 16

Barreda V, Palazzesi L, Tellería M, Katinas L, Crisci J, Bremer K, Passalia M, Corsolini R, Brizuela R, Bechis F (2010) Eocene patagonia fossils of the daisy family. *Science* 329(5999):1621

Baumgartner JR, Al-Khatib K, Currie RS (1999) Survey of common sunflower (*Helianthus annuus*) resistance to imazethapyr and chlorimuronin northeast Kansas. *Weed Technol* 13(3):510–514

Baute GJ, Kane NC, Grassa CJ, Lai Z, Rieseberg LH (2015) Genome scans reveal candidate domestication and improvement genes in cultivated sunflower, as well as post-domestication introgression with wild relatives. *New Phytol* 206(2):830–838

Belhassen E, Castiglioni VPR, Chimenti C, Griveau Y, Jamaux I, Steinmetz A (1996) Looking for physiological and molecular markers of leaf cuticular transpiration using interspecific crosses between *Helianthus argophyllus* and *Helianthus annuus*. ISA-Symposium II: Drought tolerance in sunflower. Beijing PR China, June 14, pp 39–45

Berglund DR (2007) *Sunflower Production, North Dakota Agricultural Experiment Station and North Dakota State University Extension Service Extension Publication A-1331* (EB-25 Revised). North Dakota State University, Fargo.

Berlese AN, De Toni JB (1888) Phycomycetae. In: Saccardo PA (ed) Sylloge Fungorum Omnium Hucusque Cognitorum. Vol 7:181–322

Bilgen BB (2016) Characterization of sunflower inbred lines with high oleic acid content by DNA markers. In: Kaya Y, Hasancebi S (eds) *Proceedings of the 19th International Sunflower Conference*, Edirne, Turkey, 29-05.03.06.2016. ISA, Paris, France, pp 662–668

Blamey FPC, Zollinger Rand Schneiter A (1997) Sunflower production and culture. In: Schneiter AA (ed) *Sunflower Technology and Production, Agronomy Monographs* Vol. 35. ASA, CSSA, SSSA, pp 595–670

Blanchet N, Casadebaig P, Debaeke P, Duruflé H, Gody L, Gosseau F, Langlade NB, Maury P (2018) Data describing the eco-physiological responses of twenty-four sunflower genotypes to water deficit. *Data Brief* 21:1296–1301

Bošnjak DJ (2004) Drought and its relation to field crop production in Vojvodina Province. In: *Proceedings of the Institute of Field and Vegetable Crops*, Novi Sad 40, pp 45–55 [orig. in Serbian]

Bouzidi MF, Badaoui S, Cambon F, Vear F, De Labrouhe D, Nicolas P, Mouzeyar S (2002) Molecular analysis of a major locus for resistance to downy mildew in sunflower with specific PCR-based markers. *Theor Appl Genet* 104:592–600

Britannica T. Editors of Encyclopaedia (2021, September 10). Sunflower. Encyclopedia Britannica. https://www.britannica.com/plant/sunflower-plant

Brothers ME, Miller JF (1999) Core subset for the cultivated sunflower collection. In: Kroh L (ed) *Proceedings of the 21st Sunflower Research Workshop*, Fargo, ND, 14–15 June. 1999. Natl Sunflower Assoc, Bismarck, ND, USA, pp 124–127

Bruniard JM, Miller JF (2001) Inheritance of imidazolinone herbicide resistance in sunflower. *Helia* 24:11–16

Bulos M, Sala CA, Altieri E, Ramos ML (2013) Marker assisted selection for herbicide resistance in sunflower. *Helia* 59:1–16

Bulos M, Vergani PN, Altieri E (2014) Genetic mapping, marker assisted selection and allelic relationships for the *Pu6* gene conferring rust resistance in sunflower. *Breed Sci* 64:206–212

Burke JM, Knapp SJ, Rieseberg LH (2005) Genetic consequences of selection during the evolution of cultivated sunflower. *Genetics* 171:1933–1940

Burke JM, Tang S, Knapp SJ, Rieseberg LH (2002) Genetic analysis of sunflower domestication. *Genetics* 161:1257–1267

Castillejo MA, Maldonado AM, Ogucta S, Jorrin JV (2008) Proteomic analysis of responses to drought stress in sunflower (*Helianthus annuus* L.) leaves by 2DE gel electrophoresis and mass spectrometry. *Open Proteom J* 1:59–71

Cellier F, Conejero G, Breitler J-C, Casse F (1998) Molecular and physiological responses to water deficit in drought-tolerant and drought sensitive lines of sunflower. *Plant Physiol* 116:319–328

Chabaud M, Auriac M-C, Boniface M-C, Delgrange S, Folletti T, Jardinaud M-F, Legendre A, Perez-Vich B, Pouvreau J-B, Velasco L, Devavault P, Munos S (2022) Wild *Helianthus* species: a reservoir of resistance genes for sustainable pyramidal resistance to broomrape in sunflower. *Front Plant Sci* 13:1038684

Chapman MA, Pashley CH, Wenzler J, Hvala J, Tang S, Knapp SJ, Burke JM (2008) A genomic scan for selection reveals candidates for genes involved in the evolution of cultivated sunflower (*Helianthus annuus*). *Plant Cell* 20(11):2931–2945

Charlet LD, Kopp DD, Oseto CY (1987) Sunflowers: Their history and associated insect community in the northern Great Plains. *Bull Entomol Soc Amer* 33(2):69–75

Cockerell TDA (1929) Hybrid sunflowers. *Am Nat* 63:470–475

Comai L, Young K, Till BJ, Reynolds SH, Greene EA, Codomo CA, Enns LC, Johnson JE, Burtner C, Odden AR (2004) Efficient discovery of DNA polymorphisms in natural populations by Ecotilling. *Plant J* 37:778–786

Ćuk L (1982) Variability in oil content in seed of *Helianthus* spp. In: *Proceedings of the 10th International Sunflower Conference*, Surfers Paradise, Australia, March 14–18, 1982, ISA, Paris, France, p. 211

Cvejić S, Radanović A, Dedić B, Jocković M, Jocić S, Miladinović D (2020) Genetic and genomic tools in sunflower breeding for broomrape resistance. *Genes* 11(2):152

Dehmer KJ, Friedt W (1998) Development of molecular markers for high oleic acid content in sunflower (*Helianthus annuus L.*). *Ind Crops Prod* 7 (2–3):311–315

Demurin Y (1993) Genetic variability of tocopherol composition in sunflower seeds. *Helia* 16(18):59–62

Demurin Y, Skoric D, Karlovic D (1996) Genetic variability of tocopherol composition in sunflower seeds as a basis of breeding for improved oil quality. *Plant Breed* 115(1):33–36

Dezar CA, Gago GM, Gonzalez DH, Chan RL (2005) *Hahb-4*, a sunflower homeobox-leucine zipper gene, is a developmental regulator and confers drought tolerance to *Arabidopsis thaliana* plants. *Transgen Res* 14:429–440

Dimitrijević A (2013) Use of molecular markers and *in vitro* methods for detection of tribeuronmethyl resistance in sunflower, PhD these, Faculty of Agriculture, University of Belgrade, Serbia, 92 p

Dimitrijević A, Horn R (2018) Sunflower hybrid breeding: from markers to genomic selection. *Front Plant Sci* 8:2238

Dimitrijević A, Imerovski I, Miladinović D, Cvejić S, Jocić S, Zeremski T, Sakač Z (2017) Oleic acid variation and marker-assisted detection of Pervenets mutation in high- and low-oleic sunflower cross. *Crop Breed Appl Biotechnol* 17:229–235

Dimitrijević A, Imerovski I, Miladinović D, Jocić S (2014) Development of CAPS marker for marker assisted selection of tribenuron-methyl tolerant sunflower genotypes. *Book of abstracts of V Congress of the Serbian Genetic Society*, Belgrade, Serbia, 28.09.-02.10, p 351

Dimitrijević A, Imerovski I, Miladinović D, Jocić S, Miklič V (2011) Use of molecular markers in identification of non-TIR-NBS-LRR RGA linked to downy mildew resistance locus *Pl5/Pl8* in sunflower *(Helianthus annuus* L.). In: *Proceedings of the 19th Symposium of the Serbian Plant Physiology Society*, BanjaVrujci, Serbia, 13–15 July 2011. p 80

Dimitrijević A, Imerovski I, Miladinović D, Jocković M, Cvejić S, Jocić S, Zeremski T, Sakač Z (2016) Screening of the presence of *ol* gene in NS sunflower collection. In Kaya Y, Hasancebi S (eds) *Proceedings of the 19th International Sunflower Conference*, Edirne, Turkey, 29 May -3 June 2016, ISA, Paris, France, pp 661–667

Dimitrijević A, Imerovski I, Miladinović D, Tančić S, Dušanić N, Jocić S, Miklič V (2010) Use of SSR markers in identification of sunflower isogenic lines in late generations of backcrossing. *Helia* 33:191–198

Duriez P, Vautrin S, Auriac M-C, Bazerque J, Boniface M-C, Callot C, Carrère S, Cauet S, Chabaud M, Gentou F, Lopez-Sendon M, Paris C, Pegot-Espagnet P, Rousseaux JC, Pérez-Vich B, Velasco L, Bergès H, Piquemal J, Muños S (2019) A receptor-like kinase enhances sunflower resistance to *Orobanche cumana*. *Nat Plants* 5(12):1211–1215

Dußle CM, Hahn V, Knapp SJ, Bauer E (2004) *Plarg* from *Helianthus argophyllus* is unlinked to other known downy mildew resistance genes in sunflower. *Theor Appl Genet* 109:1083–1086

Dwyer BP (2015) Evaluation of agricultural sunflower plants as drought tolerant option for New Mexico growers. *United States*. https://doi.org/10.2172/1504206

Enns H, Dorrell DG, Hoes JA, Chubb WO (1970) Sunflower research, a progress report. In: *Proceedings of the 4th International Sunflower Conference,* Memphis, TN, USA, 23–25 June, 1970, ISA, Paris, France, pp 162–167

Farlow WG (1883) Note on some species in the third and eleventh centuries of Ellis's North American Fungi. *Proc Am Acad Arts Sci* 18(10):71–73

Feng J, Jan C-C (2008) Introgression and molecular tagging of *Rf4*, a new male fertility restoration gene from wild sunflower *Helianthus maximiliani* L. *Theor Appl Genet* 117:241–249

Ferfuia C, Turi M Vannozzi GP (2015) Variability of seed fatty acid composition to growing degree-days in high oleic acid sunflower genotypes. *Helia* 38:61–78

Ferfuia C, Vannozzi GP (2015) Maternal effect on seed fatty acid composition in a reciprocal cross of high oleic sunflower (*Helianthus annuus* L.). *Euphytica* 205:325–336

Fernández H, Baldini M, Olivieri AM (1999) Inheritance of high oleic acid content in sunflower oil. *J Plant Breed Genet* 53:99–103

Fernández-Aparicio M, DelMoral L, Muños S, Velasco L, Pérez-Vich B (2022) Genetic and physiological characterization of sunflower resistance provided by the wild-derived Or_{Deb2} gene against highly virulent races of *Orobanche cumana* Wallr. *Theor Appl Genet* 135(2):501–25

Fernández-Luqueño F, López-Valdez F, Miranda-Arámbula M, Rosas-Morales M, Pariona N, Espinoza-Zapata R (2014) An Introduction to the Sunflower Crop. In: Arribas JJ (ed), *Sunflowers: Growth and Development, Environmental Influences and Pests/Diseases*, Nova Science, pp 1–18

Fernández-Martínez JM, Pérez-Vich B, Velasco L, Domínguez J (2007) Breeding for specialty oil types in sunflower. *Helia* 30(46):75–84

Fick GN (1984) Inheritance of high oleic acid in the seed oil of sunflower. In: *Proceedings of the 6th Sunflower Research Workshop*, Jan 1984, National Sunflower Association, Bismarck, ND, USA, p. 9

Filippi CV, Aguirre N, Rivas JG, Zubrzycki J, Puebla A, Cordes D, Moreno MV, Fusari CM, Alvarez D, Heinz RA, Hopp HE (2015) Population structure and genetic diversity characterization of a sunflower association mapping population using SSR and SNP markers. *BMC Plant Biol* 15(1):1–2

Filippi CV, Merino GA, Montecchia JF, Aguirre NC, Rivarola M, Naamati G, Fass MI, Álvarez D, Di Rienzo J, Heinz RA, Contreras Moreira B, Lia VV, Paniego NB (2020) Genetic diversity, population structure and linkage disequilibrium assessment among international sunflower breeding collections. *Genes (Basel)* 11(3):283

Filippi CV, Zubrzycki JE, Di Rienzo JA, Quiroz F, Fusari CM, Alvarez D, Maringolo CA, Cordes D, Escande A, Hopp HE, Heinz RA (2017) Phenotyping sunflower genetic resources for *Sclerotinia* head rot response: assessing variability for disease resistance breeding. *Plant Dis* 101(11):1941–1948

Fulda S, Mikkat S, Stegmann H, Horn R (2011) Physiology and proteomics of drought stress acclimation in sunflower (*Helianthus annuus* L.). *Plant Biol* 13:632–642

Gago GM, Almoguera C, Jordano J, Gonzalez DH, Chan RL (2002) *Hahb-4*, a homeobox-leucine zipper gene potentially involved in abscisic acid-dependent responses to water stress in sunflower. *Plant Cell Environ* 25(5):633–640

Gao W, Ripley VL, Aradhya CC, Meyer DH, Velasco L, Benson RM, Vich BP, Erickson AL, Martinez JMF, Ren R, Avery M (2019) Molecular markers associated with *Orobanche* resistance in sunflower; U.S. Patent Application n15/946; United States patent: Alexandria, VA, USA:105

García-Moreno MJ, Fernández-Martínez JM, Velasco L, Pérez-Vich B (2012) Genetic basis of unstable expression of high gamma-tocopherol content in sunflower seeds. *BMC Plant Biol* 12(1):1–4

García-Moreno MJ, Vera-Ruiz EM, Fernández-Martínez JM, Velasco L, Pérez-Vich B (2006) Genetic and molecular analysis of high gamma-tocopherol content in sunflower. *Crop Sci* 46(5):2015–2021

Gavrilova VA, Rozhkova VT, Anisimova IN (2014) Sunflower genetic collection at the Vavilov Institute of Plant Industry. *Helia* 37(60):1–16

Georgiev G, Encheva V, Encheva Y, Nenova N, Valkova D, Peevska P, Georgiev G (2018) Breeding of sunflower (*Helianthus annuus* L.) at Dobrudzha Agricultural Institute-General Toshevo. *J Agricult Food Environ Sci* 72(2):15–22

Georgiev G, Encheva V, Nenova N, Encheva Y, Valkova D, Peevska P, Penchev E (2016) Production potential of new sunflower hybrids developed at Dobrudzha Agricultural Institute-General Toshevo. In: *Proceedings of the 19th International Sunflower Conference*, Edirne, Turkey, 29 May - 3 June 2016, ISA, Paris, France, pp 441–453

Ghaffari M, Toorchi M, Valizadeh M, Shakiba M (2017) Proteomics prospects of tolerance of sunflower (*Helianthus annuus*) to drought stress during flowering stage. *Crop Pasture Sci* 68:457–463

Gholinezhad E, Darvishzadeh R, Bernousi I (2014) Evaluation of drought tolerance indices for selection of confectionery sunflower (*Helianthus annuus* L.) landraces under various environmental conditions. *Notulae Botanicae Horti Agrobotanici Cluj-Napoca* 42:187–201

Gilley MA, Gulya TJ, Seiler GJ, Underwood W, Hulke BS, Misar CG, Markell SG (2020) Determination of virulence phenotypes of *Plasmopara halstedii* in the United States. *Plant Dis* 104:2823–2831

Gong L, Gulya TJ, Markell SG, Hulke BS, Qi LL (2013a) Genetic mapping of rust resistance genes in confection sunflower line HA-R6 and oilseed line RHA397. *Theor Appl Genet* 126(8):2039–2049

Gong L, Hulke BS, Gulya TJ, Markell SG, Qi LL (2013b) Molecular tagging of a novel rust resistance gene *R12* in sunflower (*Helianthus annuus* L.). *Theor Appl Genet* 126(1):93–99

Goryunov DV, Anisimova IN, Gavrilova VA, Chernova AI, Sotnikova EA, Martynova EU, Boldyrev SV, Ayupova AF, Gubaev RF, Mazin PV, Gurchenko EA (2019) Association mapping of fertility restorer gene for CMS PET1 in sunflower. *Agronomy* 9:49

Gosseau F, Blanchet N, Varès D, Burger P, Campergue D, Colombet C, Gody L, Liévin JF, Mangin B, Tison G, Vincourt P, Casadebaig P, Langlade N (2019) Heliaphen, an outdoor high-throughput phenotyping platform for genetic studies and crop modeling. *Front Plant Sci* 9:1908

Gubaev R, Boldyrev S, Martynova E, Chernova A, Kovalenko T, Peretyagina T, Goryunova S, Goryunov D, Mukhina Z, Ben C, Gentzbittel L, Khaitovich P, Demurin Y (2022) Genetic mapping of loci involved in oil tocopherol composition control in Russian sunflower (*Helianthus annuus* L.) lines. *G3: Genes Genomes Genet* 12(4):jkac036

Gul A, Chandio AA, Siyal SA, Rehman A, Xiumin W (2022) How climate change is impacting the major yield crops of Pakistan? An exploration from long- and short-run estimation. *Environ Sci Pollut Res Int* 29(18):26660–26674

Hass CG, Tang S, Leonard S, Traber MG, Miller JF, Knapp SJ (2006) Three non-allelic epistatically interacting methyltransferase mutations produce novel tocopherol (vitamin E) profiles in sunflower. *Theor Appl Genet* 113(5):767–782

Hassan ESAR, Hoeft E, Li Z, Tulsieram L (2008) Genetic markers for *Orobanche* resistance in sunflower. US Patent Publication US/2008/0178325A1

Hasson H, Dudhe MY, Mandel T, Warschefsky E, Rieseberg LH, Hübner S (2021) Image processing and genome-wide association studies in sunflower identify loci associated with seed-coat characteristics. *bioRxiv*. Preprint, posted April 15, 2021, accessed 30. Sep. 2022: https://doi.org/10.1101/2021.04.15.439933

Heiser CB (1951) The sunflower among the North American Indians. *Proc Am Phil Soc* 95(4):432–448

Heiser CB (1955) The origin and development of the cultivated sunflower. *Am Biol Teach* 17(5):161–167

Heiser CB (2007) The sunflower (*Helianthus annuus*) in Mexico: further evidence for a North American domestication. *Genet Resour Crop E* 55(1):9–13

Heiser, Ch B, Smith, DM, Clevenger, SB, Martin, WC (1969) The North American Sunflowers (*Helianthus*). *Mem Torrey Bot Club*, 22:1–219

Henry AW, Gilbert HC (1924) Important fungus diseases of the common sunflower. *Minneosta Studies Plant Sci* 5:285–305

Hladni N, Terzić S, Mutavdžić B, Zorić M (2017) Classification of confectionary sunflower genotypes based on morphological characters. *J Agric Sci* 155(10):1594–1609

Hlisnikovský L, Kunzová E, Hejcman M, Škarpa P, Zukalová H, Menšík L (2015) The effect of climate, nitrogen and micronutrients application on oiliness and fatty acid composition of sunflower achenes. *Helia* 38:221–39

Horn R, Friedt W (1997) Fertility restoration of new CMS sources in sunflower (*Helianthus annuus* L.) *Plant Breed.* 166:317–322

Horn R, Köhler R, Zetsche K (1991) A mitochondrial 16 kDa protein is associated with cytoplasmic male sterility in sunflower. *Plant Mol Biol* 17:29–36.

Horn R, Kusterer B, Lazarescu E, Prüfe M, Friedt W (2003) Molecular mapping of the *Rf1* gene restoring pollen fertility in PET1-based F1-hybrids in sunflower (*Helianthus annuus* L.). *Theor Appl Genet* 106:599–606.

Horn R, Radanović A, Fuhrmann L, Sprycha Y, Hamrit S, Jocković M, Miladinović D, Jansen C (2019) Development and validation of markers for the fertility restorer gene *Rf1* in sunflower. *Int J Mol Sci* 20(6):1260

Huang K, Jahani M, Gouzy J, Legendre A, Carrere S, Lazaro-Guevara JM, Segovia EG, Todesco M, Mayjonade B, Rodde N, Cauet S, Dufau I, Staton E, Pouilly N, Boniface M-C, Tapy C, Mangin B, Duhnene A, Gautier V, Poncet C, Donnadieu C, Mandel T, Hübner S, Burke JM, Vautrin S, Bellec A, Owens GL, Langlade N, Muños S, Rieseberg HL (2022) The genomics of linkage drag in sunflower. *bioRxiv*. Preprint, posted June 7, 2022, accessed Sep. 30, 2022; https://doi.org/10.1101/2022.06.07.495047

Hübner S, Bercovich N, Todesco M, Mandel JR, Odenheimer J, Ziegler E, Lee JS, Baute GJ, Owens GL, Grassa CJ, Ebert DP (2019) Sunflower pan-genome analysis shows that hybridization altered gene content and disease resistance. *Nat Plants* 5(1):54–62

Hummer EK (2015) In the footsteps of Vavilov: Plant diversity the and now: *HortScience* 50(6):784–788

Hummer EK, Hanckok FJ (2015) Vavilovian centers of plant diversity: Implications and impacts. *HortScience* 50(6):780–783

Hussain M, Farooq S, Hasan W, Ul-Allah S, Tanveer M, Farooq M, Nawaz A (2018) Drought stress in sunflower: Physiological effects and its management through breeding and agronomic alternatives. *Agric Water Manag* 201:152–166

Hussain MM, Rauf S, Riaz MA, Al-Khayri JM, Monneveux P (2017) Determination of drought tolerance related traits in *Helianthus argophyllus*, Helianthus annuus, and their hybrids. *Breed Sci* 67:257–267

Imerovski I, Dedić B, Cvejić S, Miladinović D, Jocić S, Owens GL, Tubić NK, Rieseberg LH (2019) BSA-seq mapping reveals major QTL for broomrape resistance in four sunflower lines. *Mol Breed* 39:41

Imerovski I, Dimitrijević A, Miladinović D, Dedić B, Jocić S, Kočiš-Tubić N, Cvejić S (2016) Mapping of a new gene for resistance to broomrape races higher than F. *Euphytica* 209:281–289

Imerovski I, Dimitrijević A, Miladinović D, Dedić B, Jocić S, Kovačević B, Obreht D (2013) Identification of PCR markers linked to different *Or* genes in sunflower. *Plant Breed* 132(1):115–20

Imerovski I, Dimitrijević A, Miladinović D, Jocić S, Dedić B, Cvejić S, Šurlan-Momirović G (2014) Identification and validation of breeder-friendly DNA markers for *Plarg* gene in sunflower. *Mol Breed* 34:779–788

INRAE (2020) Annual Report (Ed. coord. Aliette Maillard/ÉdithLegouy). INRAE, Paris, France

Izquierdo NG, Aguirrezábal LA (2008) Genetic variability in the response of fatty acid composition to minimum night temperature during grain filling in sunflower. *Field Crops Res* 106:116–25

Jacob J, Sujatha M, Varaprasad SK (2017) Screening of cultivated and wild *Helianthus* species reveals herbicide tolerance in wild sunflowers and allelic variation at *Ahasl1* (acetohydroxyacid synthase 1 large subunit) locus. *Plant Genet Res* 15(5):421–429

Jan C-C, Vick BA (2007) Inheritance and allelic relationship of fertility restoration genes for seven new sources of male-sterile cytoplasm in sunflower. *Plant Breed* 126:213–217

Jocić S, Cvejić S, Hladni N, Marinković R, Miladinović D, Miklić V, Dedić D, Imerovski I, Dimitrijević A, Cirić M, Jocković M (2012) Present achievements in sunflower breeding. In: *Proceedings of 46th Meeting of Agronomists of Serbia*, Zlatibor, Serbia, pp 7–19

Jocić S, Cvejić S, Hladni N, Miladinović D, Mikliĉ V (2010) Development of sunflower genotypes resistant to downy mildew. *Helia* 33:173–180

Jocić S, Malidža G, Cvejić S, Hladni N, Mikliĉ V, Škorić D (2011) Development of sunflower hybrids tolerant to tribenuronmethyl. *Genetika-Belgrade* 43(1):175–182

Jocić S, Miladinović D and Kaya Y (2015) Breeding and genetics of sunflower. In: Martinez-Force E, Dunford NT, Salas J *(eds) Sunflower: Chemistry, Production, Processing, and Utilization*. AOCS Press: Urbana, IL, USA, pp 1–26

Joshi R, Wani SH, Singh B, Bohra A, Dar ZA, Lone AA, Pareek A, Singla-Pareek SL (2016) Transcription factors and plant respsonse to drought stress: Current understanding and future directions. *Front Plant Sci* 7:1029

Kaya Y (2014) Current situation of sunflower broomrape around the world. In: *Proceedings of the 3rd International Symposium on Broomrape (Orobanche spp.) in Sunflower*, Cordoba, Spain, 3–6 June 2014, ISA, Paris, France, pp 9–18

Kinman LM (1970) New developments in the USDA and state experiment station sunflower breeding programs. In: *Proceedings of the 4th International Sunflower Conference*, Memphis, TN, USA, pp 181–183

Kolkman JM, Slabaugh MB, Bruniard JM, Berry S, Bushman BS, Olungu C, Maes N, Abratti G, Zambelli A, Miller JF, Leon A (2004) Acetohydroxyacid synthase mutations conferring resistance to imidazolinone or sulfonylurea herbicides in sunflower. *Theor Appl Genet* 109(6):1147–1159

Köhler RH, Horn R, LösslA, Zetche K (1991) Cytoplasmic male sterility in sunflower is correlated with the co-transcription of a new open reading frame with the *atpA* gene. *Mol Gen Genet* 227:369–376Kondić-Špika A, Mikić S, Mirosavljević M, Trkulja D, Jeromela AM, Rajković D, Radanović A, Cvejić S, Glogovac S, Dodig D, Božinović S, Šatović Z, Lazarević B, Šimić D, Novoselović D, Vass I, Pauk J, Miladinović D (2022) Crop breeding for changing climate in the Pannonian region: towards integration of modern phenotyping tools. *J Exp Bot* 73(15):5089–5110

Krannich CT, Maletzki L, Kurowsky C, Horn R (2015) Network candidate genes in breeding for drought tolerant crops. *Int J Mol Sci* 16:16378–16400

Kumar APK, Boualem A, Bhattacharya A, Parikh S, Desai N, Zambelli A, Leon A, Chatterjee M, Bendahmane A (2013) SMART - Sunflower mutant population and reverse genetic tools for crop improvement. *BMC Plant Biol* 13:38

Kumar GR, Sakthivel K, Sundaram RM, Neeraja CN, Balachandran SM, Rani NS, Viraktamath BC, Madhav MS (2010) Allele mining in crops: Prospects and potentials. *Biotechnol Adv* 28:451–461

Kumar T, Tiwari N, Bharadwaj C, Sarker A, Pappula SPR, Singh S, Singh M (2020) Identification of allelic variation in drought responsive dehydrin gene based on sequence similarity in chickpea (*Cicer arietinum* L.). *Front Genet* 11:584527

Kusterer B, Horn R, Friedt W (2005) Moelcular mapping of the fertility restoration locus *Rf1* in sunflower and development of diagnostic markers for the restorer gene. *Euphytica* 143:35–42

Labrousse P, Arnaud MC, Serieys H, Berville A, Thoulam P (2001) Several mechanisms are involved in *Helianthus* resistance to *Orobanche cumana* Wallr. *Ann Bot* 88:859–868

Lacombe S, Kaan F, Griveau Y, Bervillé A (2004) The Pervenets high oleic mutation: methodological studies. *Helia* 27:41–54

Lacombe S, Leger S, Kaan F, Bervillé A, Sas M (2002) Genetic, molecular and expression features of the Pervenets mutant leading to high oleic acid content of seed oil in sunflower. *OCL* 9:17–23

Lacombe S, Souyris I, Bervillé AJ (2009) An insertion of oleate desaturase homologous sequence silences via siRNA the functional gene leading to high oleic acid content in sunflower seed oil. *Mol. Genet. Genomics* 281:43–54.Lawson WR, Goulter KC, Henry RJ, Kong GA, Kochman JK (1998) Marker-assisted selection for two rust resistance genes in sunflower. *Mol Breed* 4:227–234

Leclercq P (1969) Une sterilite cytoplasmique chezletournesol. *Ann Amelior Plant* 19:99–106.

Leclercq P, Cazderon Y, Dauge M (1970) Selection pour la resistance au mildiou du tournesol a partir d'hybrides topinambour x tournesol. *Ann Amelior Plant* 20(3):363–373

Leclercq P (1984) Identification de gènes de restauration de fertilité sur cytoplasmic stérilisants chez le tournesol de genes de restauration de fertilite sur cytoplasmic sterilisants chez de tournesol. Agronomie 4:573–576

Lentz DL, Pohl MD, Alvarado JL, Tarighat S, Bye R (2008) Sunflower (*Helianthus annuus* L.) as pre-Columbian domesticate in Mexico. *Proc Nat Acad Sci USA* 107(17):6232–6237

Lentz DL, Pohl MED, Pope KO, Wyatt AR (2001) Prehistoric sunflower (*Helianthus annuus* L.) domestication in Mexico. *Econ Bot* 55(3):370–376

Le Ru A, Ibarcq G, Boniface MC, Baussart A, Muños S, Chabaud M (2021) Image analysis for the automatic phenotyping of *Orobanche cumana* tubercles on sunflower roots. *Plant Methods* 17(1):1–4

Leung H, Raghavan C, Zhou B, Oliva R, Choi IR, Lacorte V, Jubay ML, Cruz CV, Gregorio G, Singh RK, Ulat VJ, Borja FN, Mauleon R, Alexandrov NN, McNally KL, Sackville Hamilton R (2015) Allele mining and enhanced genetic recombination for rice breeding. *Rice* 8(1):34

Lexer C, Lai Z, Rieseberg LH (2004) Candidate gene polymorphisms associated with salt tolerance in wild sunflower hybrids: Implications for the origin of *Helianthus paradoxus*, a diploid hybrid species. *New Phytol* 161(1):225–233

Li W, Zeng Li W, Zeng Y, Yin F, Wei R, Mao X (2021) Genome-wide identification and comprehensive analysis of the NAC transcription factor family in sunflower during salt and drought stress. *Sci Rep* 11:19865

Liang C, Wang W, Wang J, Ma j, Li C, Zhou F, Zhang S, Yu Y, Zhang L, Li W, Huang X (2017) Identification of differentially expressed genes in sunflower (*Helianthus annuus*) leaves and roots under drought stress by RNA sequencing. *Bot Stud* 58:42

Liu Z, Gulya TJ, Seiler GJ, Vick BA, Jan C-C (2012a) Molecular mapping of the Pl_{16} downy mildew resistance gene from HA-R4 to facilitate marker-assisted selection in sunflower. *Theor Appl Genet* 125:121–131

Liu Z, Mulpuri S, Feng J, Vick BA, Jan C-C (2012b) Molecular mapping of the *Rf3* fertility restoration gene to facilitate its utilization in breeding confection sunflower. *Mol Breed* 29:275–284

Liu Z, Wang DM, Feng J, Seiler GJ, Cai X, Jan CC (2013) Diversifying sunflower germplasm by integration and mapping a novel male fertility restoration gene. *Genetics* 193:727–737

Liu Z, Zhang L, Ma GJ, Seiler GJ, Jan CC, Qi LL (2019) Molecular mapping of the downy mildew and rust resistance genes in a sunflower germplasm line TX16R. *Mol Breed* 39(2):19

Lu YH, Melero-Vara JM, Garcia-Tejada JA, Blanchard P (2000) Development of SCAR markers linked to the gene *Or5* conferring resistance to broomrape (*Orobanche cumana* Wallr.) in sunflower. *Theor Appl Genet* 100:625–632

Ma G, Long Y, Song Q, Talukder ZI, Shamimuzzaman MD, Qi L (2021) Map and sequence-based chromosome walking towards cloning of the male-fertility restoration gene *Rf5* linked to R_{11} in sunflower. *Sci Rep* 11:777

Ma G, Song Q, Li X, Qi L (2020) High-density mapping and candidate gene analysis of Pl_{18} and Pl_{20} in sunflower by whole genome resequencing. *Int J Mol Sci* 21(24):9571

Ma G, Song Q, Li X, Qi L (2022) Genetic insight into disease resistance gene clusters by using sequence-based fine mapping in sunflower (*Helianthus annuus* L.). *Int J Mol Sci* 23:9516

Ma G, Song Q, Underwood WR, Zhang Z, Fiedler JD, Li X, Qi L (2019) Molecular dissection of resistance gene cluster and candidate gene identification Pl_{17} and Pl_{19} in sunflower by whole genome resequencing. *Sci Rep* 9:14974

Ma GJ, Markell SG, Song QJ, Qi LL (2017) Genotyping-by-sequencing targeting of a novel downy mildew resistance gene Pl_{20} from wild *Helianthus argophyllus* in sunflower *(Helianthus annuus* L.). *Theor Appl Genet* 130(7):1519–1529

Ma GJ, Song QJ, Markell SG, Qi LL (2018) High-throughput genotyping-by-sequencing facilitates molecular tagging of a novel rust resistance gene, R_{15}, in sunflower (*Helianthus annuus* L.). *Theor Appl Genet* 131:1423–1432

Mandel JR, Dechaine JM, Marek LF, Burke JM (2011) Genetic diversity and population structure in cultivated sunflower and a comparison to its wild progenitor, Helianthus annuus L. *Theor Appl Genet* 123(5):693–704

Mandel JR, Nambeesan S, Bowers JE, Marek LF, Ebert D, Rieseberg LH, Knapp SJ, Burke MJ (2013) Association mapping and the genomic consequences of selection in sunflower. *PLoS Genet* 9(3):e1003378

Marek LF (2016) Sunflower genetic resources. In: Kaya Y, Hasancebi S (eds) *Proceedings of the 19th International Sunflower Conference*, Edirne, Turkey, 29 May-3 June 2016. International Sunflower Association, Paris, pp 31–44

Markin N, Usatov A, Marenko M, Azarin K, Gorbachenko O, Kolokolova N, Usatenko T, Markina O, Gavrilova V(2017) Study of informative DNA markers of the *Rf1* gene in sunflower for breeding practice. *Czech J Plant Breed* 53:69–75

Márquez-Ruiz G, Ruiz-Méndez MV, Velasco J (2014) Antioxidants in frying: Analysis and evaluation of efficacy. *Eur J Lipid Sci Technol* 116(11):1441–1450.

Martin-Sanz A, Perez-Vich B, Rueda S, Fernandez-Martinez JM, Velasco L (2020) Characterization of post-haustorial resistance to sunflower broomrape. *Crop Sci* 60:1188–1198

Meena HP, Sujatha M (2022) Sunflower Breeding. In: Yadava DK, Dikshit HK, Mishra GP, Tripathi S (eds) *Fundamentals of Field Crop Breeding*. Springer Nature, Singapore, pp 971–1008

Miladinović D, Hladni N, Radanović A, Jocić S, Cvejić S (2019) Sunflower and climate change: Possibilities of adaptation through breeding and genomic selection. In: Kole C (*ed*) *Genomic Designing of Climate-Smart Oilseed Crops*. Springer, Cham., pp 173–238

Miller JF, Al-Khatib K (2002) Registration of imidazolinone herbicide-resistant sunflower maintainer (HA 425) and fertility restorer (RHA 426 and RHA 427) germplasms. *Crop Sci* 42(3):988–90

Miller JF, Al-Khatib K (2004) Registration of two oilseed sunflower genetic stocks, SURES-1 and SURES-2, resistant to tribenuron herbicide. *Crop Sci* 44(3):1037–1038

Miller JF, Gulya TJ (1991) Inheritance of resistance to race 4 of downy mildew derived from interspecific crosses in sunflower. *Crop Sci* 31(1):40–43

Miller JF, Gulya TJ (2001) Registration of three rust resistant sunflower germplasm populations. *Crop Sci* 41(2):601

Molinero-Ruiz L (2022) Sustainable and efficient control of sunflower downy mildew by means of genetic resistance: a review. *Theor Appl Genet* 135(11):3757–3771

Molinero-Ruiz L, Delavault P, Perez-Vich B, Pacureanu-Joita M, Bulos M, Altieri E, Domiguez J (2015) History of the race structure of *Orobanche cumana* and the breeding of sunflower for resistance to this parasitic weed: A review. *Span J Agri Res* 13(4):e10R01

Moreno PS, de Romano AB, Romano MC, Vergani P, Sposaro M, Bulos M, Altieri E, Ramos ML, Sala CA (2012) A survey of physiological races of *Puccinia helianthi* in Argentina. In: *Proceedings 18th International Sunflower Conference*, 27 Feb- 1 March, Mar del Plata, Argentina. ISA, Paris, France, pp 87–90

Mouzeyar S, Roeckel-Drevet P, Gentzbittel L, Philippon J, De Labrouhe DT, Vear F, Nicolas P (1995) RFLP and RAPD mapping of the sunflower *Pl1* locus for resistance to Plasmopara halstedii race 1. *Theor Appl Genet* 91:733–737

Mulpuri S, Liu Z, Feng J, Gulya TJ, Jan CC (2009) Inheritance and molecular mapping of a downy mildew resistance gene, *Pl13* in cultivated sunflower (*Helianthus annuus* L.). *Theor Appl Genet* 119:795–803

Nagarathna TK, Shadakshari YG, Ramanappa TM (2011) Molecular analysis of sunflower (*Helianthus annuus L.*) genotypes for high oleic acid using microsatellite markers. *Helia* 34:63–68

Nambeesan SU, Mandel JR, Bowers JE, Marek LF, Ebert D, Corbi J, Rieseberg LH, Knapp SJ, Burke JM (2015) Association mapping in sunflower (*Helianthus annuus* L.) reveals independent control of apical vs. basal branching. *BMC Plant Biol* 15(1):84

Ochogavía AC, Gil M, Picardi L, Nestares G (2014) Precision phenotyping of imidazolinone-induced chlorosis in sunflower. *Breed Sci* 64(4):416–421

Ortiz-Bustos CM, Pérez-Bueno ML, Barón M, Molinero-Ruiz L (2017) Use of blue-green fluorescence and thermal imaging in the early detection of sunflower infection by the root parasitic weed *Orobanche cumana*Wallr. *Front Plant Sci* 8:833

Owens GL, Baute GJ, Hubner S, Rieseberg LH (2019) Genomic sequence and copy number evolution during hybrid crop development in sunflowers. *Evol Appl* 12(1):54–65

Pacureanu-Joita M, Veronesi C, Raranciuc S, Stanciu D (2004) Parasite-host plant interaction of *Orobanche cumana* Wallr. (*Orobanche cernua* Loefl.) with *Helianthus annuus*. In: *Proc.16th Int. Sunflower Conf.*, Fargo, ND, Aug 29-Sept 2, International Sunflower Association, Paris, France, pp 171–177

Pankovic D, Radovanovic N, Jocic S, Satovic Z, Skoric D (2007) Development of co-dominant amplified polymorphic sequence markers for resistance of sunflower to downy mildew race 730. *Plant Breed* 126(4):440–444

Park B, Burke JM (2020) Phylogeography and the evolutionary history of sunflower (*Helianthus annuus* L.): Wild diversity and the dynamics of domestication. *Genes (Basel)* 11(3):266

Pecrix Y, Buendia L, Penouilh-Suzette C, Maréchaux M, Legrand L, Bouchez O, Rengel D, Gouzy J, Cottret L, Vear F, Godiard L (2019) Sunflower resistance to multiple downy mildew pathotypes revealed by recognition of conserved effectors of the oomycete *Plasmopara halstedii*. *Plant J* 97(4):730–748

Pecrix Y, Penouilh-Suzette C, Munos S, Vear F, Godiard L (2018) Ten broad spectrum resistances to downy mildew physically mapped on the sunflower genome. *Front Plant Sci* 9:1780

Pérez-Vich B, Aguirre MR, Guta B, Fernández-Martínez JM, Velasco L (2018) Genetic diversity of a germplasm collection of confectionery sunflower landraces from Spain. *Crop Sci* 58(5):1972–1981

Pérez-Vich B, Akhtouch B, Knapp SJ, Leon AJ, Velasco L, Fernández-Martínez JM, Berry ST (2004) Quantitative trait loci for broomrape (*Orobanche cumana* Wallr.) resistance in sunflower. *Theor Appl Genet* 109(1):92–102

Pérez-Vich B, Fernández-Martıínez JM, Grondona M, Knapp SJ, Berry ST (2002) Stearoyl-ACP and oleoyl-PC desaturase genes cosegregate with quantitative trait loci underlying high stearic and high oleic acid mutant phenotypes in sunflower. *Theor Appl Genet* 104(2–3):338–349

Pilorgé E (2020) Sunflower in the global vegetable oil system: situation, specificities and perspectives. *OCL* 27:34

Polivanova OB, Sivolapova AB, Goryunov DV, Fedorova AV, Sotnikova EA, Chebanova YV, Karabitsina YI, Benko NI, Demurin YN, Goryunova SV (2021) Structural diversity of sunflower (*Helianthus annuus* L.) candidate *Rf1* loci based on gene-specific PCR. *Research on Crops* 22(1):40–46.

Premnath A, Narayana M, Ramakrishnan C, Kuppusamy S, Chockalingam V (2016) Mapping quantitative trait loci controlling oil content, oleic acid and linoleic acid content in sunflower (*Helianthus annuus* L.). *Mol Breed* 36(7):106

Pustovoit GV (1966) Distant (interspecific) hybridization of sunflowers in the USSR. In: *Proceedings of the 2nd International Sunflower Conference*, Morden, Manitoba, Canada, August 17–18 1966. International Sunflower Association, Paris, France, pp 82–95

Pustovoit GV, Ilatovsky VP, Sliusar EL (1976) Results and prospects ofs unflower breeding for group immunity by interspecifichy bridization. In: *Proceeding of the 7th International Sunflower Conference*, Krasnodar, USSR, 27 June-3 July. International Sunflower Association, Paris, France, pp 193–204

Putt ED (1997) Early history of sunflower. In: Schneiter AA (ed) *Sunflower Technology and Production. In: Agronomy Monograph Sunflower Technology and Production* Vol. 35, pp 1–19

Putt ED, Sackston WE (1957) Studies on sunflower rust: I. Some sources of rust resistance. *Can J Plant Sci* 37(1):43–54 Qi LL, Cai XW (2022) Characterization and mapping of a downy mildew resistance gene, Pl_{36}, in sunflower (*Helianthus annuus* L.). *Mol Breed* 42(2):8

Qi LL, Foley ME, Cai XW, Gulya TJ (2016) Genetics and mapping of a novel downy mildew resistance gene, Pl_{18}, introgressed from wild *Helianthus argophyllus* into cultivated sunflower (*Helianthus annuus* L.). *Theor Appl Genet* 129(4):741–752

Qi LL, Gulya TJ, Hulke BS, Vick BA (2012a) Chromosome location, DNA markers and rust resistance of the sunflower gene *R5*. *Mol Breed* 30:745–756

Qi LL, Gulya TJ, Seiler GJ, Hulke BS, Vick BA (2011a) Identifcation of resistance to new virulent races of rust in sunflowers and validation of DNA markers in the gene pool. *Phytopathology* 101:241–249

Qi LL, Hulke BS, Vick BA, Gulya TJ (2011b) Molecular mapping of the rust resistance gene R_4 to a large NBS-LRR cluster on linkage group 13 of sunflower. *Theor Appl Genet* 123:351–358

Qi LL, Long YM, Jan CC, Ma GJ, Gulya TJ (2015a) Pl_{17} is a novel gene independent of known downy mildew genes in the cultivated sunflower (*Helianthus annuus* L.). *Theor Appl Genet* 128(4):757–767

Qi LL, Long YM, Ma GJ, Markell SG (2015b) Map saturation and SNP marker development for the rust resistance genes (R_4, R_5, R_{13a}, and R_{13b}) in sunflower (*Helianthus annuus* L.). *Mol Breed* 35:196

Qi LL, Ma GJ, Li XH, Seiler GJ (2019) Diversification of the downy mildew resistance gene pool by introgression of a new gene, Pl_{35}, from wild *Helianthus argophyllus* into oilseed and confectionary sunflowers (*Helianthus annuus* L.) *Theor Appl Genet* 132(9):2553–2565

Qi LL, Ma GJ, Long YM, Hulke BS, Gong L, Markell SG (2015c) Relocation of a rust resistance gene *R2* and its marker-assisted gene pyramiding in confection sunflower (*Helianthus annuus* L.). *Theor Appl Genet* 128(3):477–488

Qi LL, Seiler GJ, Vick BA, Gulya TJ (2012b) Genetics and mapping of the R_{11} gene conferring resistance to recently emerged rust races, tightly linked to male fertility restoration, in sunflower (*Helianthus annuus* L.). *Theor Appl Genet* 125(5):921–932

Qi LL, Talukder ZI, Hulke BS, Foley ME (2017) Development and dissection of diagnostic SNP markers for the downy mildew resistance genes Pl_{Arg} and Pl_8 and marker-assisted gene pyramiding in sunflower (*Helianthus annuus* L.). *Mol Genet Genom* 292(3):551–563

Qi LL, Talukder ZI, Ma GJ, Li XH (2021) Discovery and mapping of two new rust resistance genes, R_{17} and R_{18}, in sunflower using genotyping by sequencing. *Theor Appl Genet* 134(7):2291–2301

Radanović A, Cvejić S, Dedić B, Dušanić N, Gvozdenac S, Hladni N, Jocić S, Jocković M, Kondić-Špika A, Marjanović-Jeromela A, Miklič V (2019) Genetic resources and new breeding tools for sunflower improvement. In: *Book of Abstracts, PlantEd COST Action CA18111 Genome Editing in Plants: 1stPlantEd Conference on Plant Genome Editing-State of the Art*, November 5–7, 2019, University Novi Sad, Institute of Lowland Forestry and Environment Novi Sad, Novi Sad, Serbia, p 40

Radanović A, Galinski A, Jocković M, Cvejić S, Terzić S, Jocić S, Miladinović D, Fiorani F, Nagel K (2022a) Mining root trits for sunflower drought tolerance improvement by use of an automated phenotyping platform. In: Terzić S, Miladinović D (eds) *Proceeding of the 20th International Sunflower Conference*, Novi Sad, Serbia, June 20–23, 2022, ISA, Paris, France, p 250

Radanović A, Galinski A, Miladinović D, Cvejić S, Jocić S, Terzić S, Nagel, K, Fiorani F (2018) Root phenotyping of NS sunflower. In: *Proceedings of the 7th Balkan Botanical Congress*, Novi Sad, Serbia, 10–14 September 2018; Institute of Botany and Botanical Garden Jevremovac: Belgrade, Serbia, p 165

Radanović A, Sprycha Y, Jockovic M, Sundt M, Miladinovic D, Jansen C, Horn R (2022b) KASP markers specific for the fertility restorer locus *Rf1* and application for genetic purity testing in sunflowers (*Helianthus annuus* L.). *Genes* 13:465

Radwan O, Bouzidi M-F, Mouzeyar S (2011) Molecular characterization of two types of resistance in sunflower to *Plasmopara halstedii*, the causal agent of downy mildew. *Phytopathology* 101:970–979

Radwan O, Bouzidi MF, Nicolas P, Mouzeyar S (2004) Development of PCR markers for the Pl_5/Pl_8 locus for resistance to *Plasmopara halstedii* in sunflower, *Helianthus annuus* L. from complete CC-NBS-LRR sequences. *Theor Appl Genet* 109(1):176–185

Radwan O, Gandhi S, Heesacker A, Whitaker B, Taylor C, Plocik A, Kesseli R, Kozik A, Michelmore RW, Knapp SJ (2008) Genetic diversity and genomic distribution of homologs encoding NBS-LRR disease resistance proteins in sunflower. *Mol Genet Genom* 280(2):111–125

Rauf S (2008) Breeding sunflower (*Helianthus annuus* L.) for drought tolerance. *Comm Biometry Crop Sci* 3(1):29–44

Rauf S, Jamil N, Tariq SA, Khan M, Kausar M, Kaya Y (2017) Progress in modification of sunflower oil to expand its industrial value. *J Sci Food Agric* 97(7):1997–2006.

Rauf S, Warburton M, Naeem A, Kainat W (2020) Validated markers for sunflower (*Helianthus annuus* L.) breeding. *OCL* 27:47.

Razzaq H, Tahir HN, Sadaqat HA, Sadia B (2017) Screening sunflower (*Helianthus annuus* L.) accessions under drought stress conditions, an experimental assay. *J Soil Sci Plant Nutr* 17(3):662–667

Reddemann A, Horn R (2018) recombination events involving the *atp9* gene are associated with male sterility of CMS PET2 in sunflower. *Int J Mol Sci* 19:806

Regitano Neto A, Miguel AMRDO, Mourad AL, Henriques EA, Alves RMV (2016) Environmental effect on sunflower oil quality. *Crop Breed Appl Biotechnol* 16:197–204

Reyes-Valdes MH, Martinez O, Gomez-Martinez M (2005) Intergeneric hybrid between cultivated sunflower (*Helianthus annuus* L.) and *Tithonia rotundifolia* (Mill.) Blake. *Helia* 43:61–68.

Rieseberg LH, Seiler GJ (1990) Molecular evidence and the origin and development of the domesticated sunflower (*Helianthus annus*, Asteraceae). *Econ Bot* 44(3):79–91

Rieseberg LH, Van Fossen C, Arias D, Carter RL (1994) Cytoplasmic male sterility in sunflower: origin, inheritance, and frequency in natural populations. *J Hered* 85(3):233–8

Rodríguez-Ojeda MI, Fernández-Escobar J, Alonso LC (2001) Sunflower inbred line (KI-374) carrying two recessive genes for resistance against a highly virulent Spanish population of *Orobanche cernua* Loelf/ O. cumana Wallr. Race F. In: Muselman LJ, Parker C, Verkleij JAC (eds), *Proceedings of the 7th International Parasitic Weed Symposium*, University de Nantes, Nantes, France, pp 208–211

Rolnik A, Olas B (2021) The plants of the Asteraceae family as agents in the protection of human health. *Int J Mol Sci* 22(6):3009

Rönicke S, Hahn V, Horn R, Grone I, Brahm L., Schnabl H, Friedt W (2004) Interspecific hybrids of sunflower as a source of *Sclerotinia* resistance. *Plant Breed* 123:152–157

Sabetta W, Alba V, Blanco A, Montemurro C (2011) sunTILL: a TILLING resource for gene function analysis in sunflower. *Plant Methods* 7:20

Safavi HR, Golmohammadi MH, Sandoval-Solis S (2015) Expert knowledge based modeling for integrated water resources planning and management in the Zayandehrud River Basin. *J Hydrol* 528:773–789

Sajer O, Schirmak U, Hamrit S, Horn R (2020) Mapping of the new fertility restorer gene *Rf-PET2* close to *Rf1* on linkage group 13 in sunflower (*Helianthus annuus* L.). Genes (Basel) 11(3):269

Sala CA, Bulos M (2012) Inheritance and molecular characterization of broad range tolerance to herbicides targeting acetohydroxyacid synthase in sunflower. *Theor Appl Genet* 124(2):355–364

Sala CA, Bulos M, Echarte AM (2008a) Genetic analysis of an induced mutation conferring imidazolinone resistance in sunflower. *Crop Sci* 48:1817–1822

Sala CA, Bulos M, Echarte M, Whitt SR, Ascenzi R (2008b) Molecular and biochemical characterization of an induced mutation conferring imidazolinone resistance in sunflower. *Theor Appl Genet* 118(1):105–12

Sarvari M, Darvishzadeh R, Najafzadeh R, Maleki H (2017) Physio-biochemical and enzymatic responses of sunflower to drought stress. *J Plant Physiol Breed* 7(1):105–119

Schnabel U, Engelmann U, Horn R (2008) Development of markers for the use of the PEF1 cytoplasm in sunflower hybrid breeding. *Plant Breed* 127:587–591

Schuppert GF, Tang S, Slabaugh MB, Knapp SJ (2006) The sunflower high-oleic mutant Ol carries variable tandem repeats of FAD2-1, a seed-specific oleoyl-phosphatidyl choline desaturase. *Mol Breed* 17:241–256

Seiler G, Jan C-C (2010) Basic Information. In: Hu J, Seiler G, Kole C (eds) *Genetics, genomics and breeding of sunflower*, Taylor & Francis, pp 1–50

Seiler G, Marek LF (2011) Germplasm resources for increasing the genetic diversity of global cultivated sunflower. *Helia* 34(55):1–20

Seiler GJ, Qi LL, Marek LF (2017) Utilization of sunflower crop wild relatives for cultivated sunflower improvement. *Crop Sci* 57(3):1083–1101

Seiler GJ, Rieseberg LH (1997) Systematics, origin, and germplasm resources of the wild and domesticated sunflower. In: Schneiter AA (ed) *Sunflower Technology and Production* Vol 35, Agronomy Monographs, Soil Science Society America, Inc, Madison, WI, USA Chapter 2, pp 21–65

Serieys H (2005) Identification, study, utilization in breeding programs of new CMS sources in the FAO Subnetwork. In: *Proceedings of the Sunflower Subnetwork Progress Report*, Rome, Italy, pp 47–53.

Shinozaki K, Yamaguchi-Shinozaki K (2007) Gene networks involved in drought response and tolerance. *J Exp Bot* 58(2):221–227

Shrestha A, Fendel A, Nguyen TH, Adebabay A, Kullik AS, Benndorf J, Leon J, Naz AA (2022) Natural diversity uncovers P5CS1 regulation and its role in drought stress tolerance and yield sustainability in barley. *Plant Cell Environ* 45:3523–3536

Singh BK, Ramkumar MK, Dalai M, Singh A, Solanke AU, Singh NK, Sevanthi AM (2021) Allele mining for a drought responsive gene *DRO1* determining root growth angle in donors of drought tolerance in rice (*Oryza sativa* L.). *Physiol Mol Biol Plants* 27(3):523–534

Singh BP, Jayaswal PK, Singh B, Singh PK, Kumar V, Mishra S, Singh N, Panda K, Singh NK (2015) Natural allelic diversity in *OsDREB1F* gene in the Indian wild rice germplasm led to ascertain its association with drought tolerance. *Plant Cell Rep* 34(6):993–1004

Škorić D (2009) Sunflower breeding for resistance to abiotic stresses. *Helia* 32(50):1–16

Škorić D (2012) The Genetics of Sunflower. In: Škorić D, Sakač Z (eds) *Sunflower Genetics and Breeding*, Serbian Academy of Sciences and Arts, Novi Sad, Serbia, pp 1–163

Škorić D (2016) Sunflower breeding for resistance to abiotic and biotic stresses. In: *Abiotic and Biotic Stress in Plants - Recent Advances and Future Prospects*. INTECH, Chapter 25, pp 585–635

Škorić D, Rajčan I (1992) Breeding for *Sclerotinia* tolerance in sunflower. In: *Proceedings of the 13th International Sunflower Conference*, Pisa, Italy, 7–11 September 1992, ISA, Paris, France, pp 1257–1262

Sobrado MA, Turner NC (1986) Photosynthesis, dry matter accumulation and distribution in the wild sunflower *Helianthus petiolaris* and the cultivated sunflower *Helianthus annuus* as influenced by water deficits. *Oecologia* 69:181–187

Soldatov KI (1976) Chemical mutagenesis in sunflower breeding. In: *Proceedings of the 7th International Sunflower Conference*. Krasnorar, USSR, June 27 -July 3, ISA, Paris, France, pp 352–357.

Talukder ZI, Gong L, Hulke BS, Pegadaraju V, Song Q, Schultz Q, Qi L (2014) A high-density SNP map of sunflower derived from RAD-sequencing facilitating fine-mapping of the rust resistance gene R_{12}. *PLoS One* 9(7):e98628

Talukder ZI, Ma G, Hulke BS, Jan CC, Qi L (2019) Linkage mapping and genome-wide association studies of the *Rf* gene cluster in sunflower (*Helianthus annuus* L.) and their distribution in world sunflower collections. *Front Genet* 10:216

Tan AS, Kaya Y (2019) Sunflower (*Helianthus annuus* L.) genetic resources, production and researches in Turkey. *OCL* 26:21

Tančić S, Terzić S, Dedić B, Atlagić J, Jocić S, Miklič V (2012) Disease resistance of wild sunflower species to *Macrophomina phaseolina*. In: *Proceedings of the 18th International Sunflower Conference*, Mar Del Plata, Argentina, February 27-March 1, 2012; ISA, Paris, France, pp 304–308.

Tančić-Živanov S, Dedić B, Cvejić S, Jocić S, Miklič V (2021) Sunflower genotypes tolerance to charcoal rot [*Macrophomina phaseolina* (Tassi) Goid.] under the field conditions. *Genetika* 53(3):1117–1131

Tang S, Hass CG, Knapp SJ (2006) Ty3/gypsy-like retrotransposon knockout of a 2-methyl-6-phytyl-1, 4-benzoquinone methyltransferase is non-lethal, uncovers a cryptic paralogous mutation, and produces novel tocopherol (vitamin E) profiles in sunflower. *Theor Appl Genet* 113(5):783–799

Tang S, Heesacker A, Kishore VK, Fernandez A, Sadik ES, Cole G, Knapp SJ (2003) Genetic mapping of the *Or5* gene for resistance to race E in sunflower. *Crop Sci* 43:1021–1028

Terzić S, Boniface MC, Marek L, Alvarez D, Baumann K, Gavrilova V, Joita-Pacureanu M, Sujatha M, Valkova D, Velasco L, Hulke B (2020a) Gene banks for wild and cultivated sunflower genetic resources. *OCL* 27(9):1–4

Terzić S, Dedić B, Atlagić J, Jocić S, Tančić S (2010) Screening wild sunflower species and F1 interspecific hybrids for resistance to broomrape. *Helia* 33(53):25–30

Terzić S, Zorić M, Seiler GJ (2020b) Qualitative traits in sunflower breeding: UGA-SAM1 phenotyping case study. *Crop Sci* 60(1):303–319

Till BJ, Zerr T, Comai L, Henikoff S (2006) A protocol for TILLING and Ecotilling in plants and animals. *Nat Protoc* 1(5):2465

Urie AL (1984) Inheritance of very high oleic acid content in sunflower. In: *Proceedings of the 6th Sunflower Research Workshop*, National Sunflower Association, Bismarck, ND, USA Jan 1984, pp 9–10

USDA NRCS National Plant Data Center (2006) Annual sunflower *Helianthus annuus* L. https://plantsorig.sc.egov.usda.gov/plantguide/pdf/cs_hean3.pdf

Vassilevska-Ivanova R, Shtereva L, Kraptchev B, Karceva T (2014) Response of sunflower (*Helianthus annuus* L) genotypes to PEG-mediated water stress. *Cent Eur J Biol* 9(12):1206–1214

Velasco L, Domínguez J, Fernández-Martínez JM (2004b) Registration of T589 and T2100 sunflower germplasms with modified tocopherols. *Crop Sci* 44(1):362–364

Velasco L, Fernandez-Cuesta A, Fernandez-Martinez JM (2014) Variability of seed quality traits in a collection of Spanish landraces of confectionary sunflower. *Crop Pasture Sci* 65(3):242–249

Velasco L, Pérez-Vich B, Fernández-Martínez JM (2004a) Novel variation for the tocopherol profile in a sunflower created by mutagenesis and recombination. *Plant Breed* 123(5):490–492

Velasco L, Perez-Vich B, Yassein AAM, Jan CC, Fernandez-Martinez JM (2012) Inheritance of resistance to sunflower broomrape (*Orobanche cumana* Wallr.) in an interspecific cross between *Helianthus annuus* and *Helianthus debilis* subsp. tardiflorus. *Plant Breed* 131(1):220–221

Vera-Ruiz EM, Pérez-Vich B, Fernández-Martínez JM, Velasco L (2005) Comparative genetic study of two sources of beta-tocopherol in sunflower. *Helia* 28(42):1–8

Vera-Ruiz EM, Velasco L, Leon AJ, Fernández-Martínez JM, Pérez-Vich B (2006) Genetic mapping of the *Tph1* gene controlling beta-tocopherol accumulation in sunflower seeds. *Mol Breed* 17(3):291–296

Vincourt P, As-Sadi F, Bordat A, Langlade NB, Gouzy J, Pouilly N, Lippi Y, Serre F, Godiard L, Tourvieille de Labrouhe D, Vear F (2012) Consensus mapping of major resistance genes and independent QTL for quantitative resistance to sunflower downy mildew. *Theor Appl Genet* 125(5):909–20

Vranceanu AV, Stoenescu FM (1978) Genes for pollen fertility restoration in sunflowers. *Euphytica* 27:617–627

Vranceanu AV, Tudor VA, Stoenescu FM, Pirvu N (1980) Virulence groups of *Orobanche cumana* Wallr. Differential hosts and resistance sources and genes in sunflower. In: *Proceedings of the 9th International Sunflower Conference*, Torremolinos, Spain, July 8–13, 1980. International Sunflower Association, Paris, France, pp 74–80

Wang J, Li C, Li L, Reynolds M, Mao X, Jing R (2021) Exploitation of drought tolerance related genes for crop improvement. *Int J Mol Sci* 22:10265

Warburton ML, Rauf S, Marek L, Hussain M, Ogunola O, Gonzalez J de JS (2017) The use of crop wild relatives in maize and sunflower breeding. *Crop Sci* 57(3):1227–1240

Wieckhorst S, Bachlava E, Dussle CM, Tang S, Gao W, Saski C, Knapp SJ, Schön CC, Hahn V, Bauer E (2010) Fine mapping of the sunflower resistance locus *PlARG* introduced from the wild species *Helianthus argophyllus*. *Theor Appl Genet* 121(8):1633–44

Wu B-b, Shi M-m, Poukheirandish M, Zhao Q, Wang Y, Yang C-k, Qiao L, Zhao J-j, Yan S-x, Zheng X-w, Zheng J (2022a) Allele mining of wheat ABA receptor at *TaPYL4* suggest neo-functionalization among the wheat homoeologs. *J Integr Agric* 21(8):2183–2196

Wu Y, Wang Y, Shi H, Hu H, Yi L, Hou J (2022b) Time-course transcriptome and WGCNA analysis revealed the drought response mechanism of two sunflower inbred lines. *PLoS One* 17(4):e0265447

Yang W, Feng H, Zhang X, Zhang J, Doonan JH, Batchelor WD, Xiong L, Yan J (2020) Crop phenomics and high-throughput phenotyping: past decades, current challenges, and future perspectives. *Mol Plant* 13(2):187–214

Young PA, Morris HE (1927) Plasmopara downy mildew of cultivated sunflowers. *Am J Bot* 14(9):551–553

Yu Z, Xia L, Su F, Liu W, Luo F, Han R, Mu Y, Zhang W, Wu L, Liang X, Sun N, Li L, Ma P (2022) Mining of wheat *Pm2* alleles for goal-oriented marker-assisted breeding. *Front Plant Sci* 13:912589

Yue B, Vick BA, Cai X, Hu J (2010) Genetic mapping for the *Rf1* (fertility restoration) gene in sunflower (*Helianthus annuus* L.) by SSR and TRAP markers. *Plant Breed* 129:24–28

Zhang M, Liu Z, Jan CC (2016) Molecular mapping of a rust resistance gene R_{14} in cultivated sunflower line PH3. *Mol Breed* 36(3):1–12

Zhang ZW, Ma GJ, Zhao J, Markell SJ, Qi LL (2017) Discovery and introgression of the wild sunflower-derived novel downy mildew resistance gene Pl_{19} in confection sunflower (*Helianthus annuus* L.). *Theor Appl Genet* 130(1):29–39

Zorić M, Cvejić S, Mladenović E, Jocić S, Babić Z, Marjanović Jeromela A, Miladinović D (2020) Digital image analysis using FloCIA software for ornamental sunflower ray floret color evaluation. *Front Plant Sci* 11:584822

Links: https://www.chr-khr.org/en/institute/inrae-french-national-research-institute-agriculture-food-and-environment (updated*:* August 16th, 2022)

https://www.inrae.fr/actualites/centre-ressources-biologiques-tournesol (updated*:* August 16th, 2022)

https://www.inrae.fr/en/centres/occitanie-toulouse (updated*:* August 16th, 2022)

https://www.atlasbig.com/en-us (accessed 28th July 2022)

6 Allele Mining for Crop Genomic Designing in Peanut

S. K. Bera, Kirti Rani, Jignesh H. Kamdar, M. K. Pandey, P. Janila, B. N. Motagi, Haile Desmae, Engin Yol, and Rajeev K. Varshney

6.1 INTRODUCTION

Advances in plant breeding have led to the creation of superior and high-yielding varieties of peanut, thanks to the accumulation of beneficial alleles from the vast global plant genetic resources. Even nevertheless, a sizable fraction of these advantageous/superior genes went unused since they were lost through domestication and evolution. This untapped genetic diversity found in peanuts' wild relatives and land races could be profitably used to create cultivars that are more agronomically successful. Novel alleles from crop plant relatives in the wild have been introduced into cultivated varieties (1–8), demonstrating conclusively that some alleles and their combinations have the potential to dramatically alter trait expression when introduced into cultivars by overcoming the genetic bottlenecks that previously prevented their introgression. To further improve the genetic potential of crop varieties for diverse agronomic traits, it is necessary to look again at the huge germplasm resources for novel alleles.

Allele mining can be effectively used to find desired alleles by "mining" genes of interest from various germplasm. Aware of the big potential of allele mining, many crop research organizations are coordinating allele mining efforts. In the past 10 years, enormous strides have been made toward depositing an exponential amount of sequence data into GenBank. Accelerated identification and annotation of new genes can be anticipated with the rapid accumulation of sequence and expression data in various genomic databases, enabling the creation of allele-specific markers (9). Polymerase chain reaction (PCR) techniques are developed to isolate beneficial alleles of genes from a wide range of animals based on gene and genome sequences. Direct access to important alleles that provide resistance to biotic and abiotic stressors, increased nutrient usage efficiency, increased yield, and improved quality are made possible by this capability. Similar alleles causing a particular trait and their variations in other genotypes can be found using modern genomic methods. This is also known as "allele mining" or "dissection of naturally occurring variation at candidate genes/loci." In addition to supplying fresh germplasm for introducing novel alleles to improve desired traits, the identification of allelic variants from germplasm collections also classifies the entries for the conservation of the germplasm.

The allelic variety of several peanut genomes has been demonstrated by the recently published 2.54 Gb genome of peanut, which harbors 20 pseudomolecules and 83,709 models of protein-coding genes (10,11). The community has been able to find peanut germplasm with beneficial alleles by using data mining and the available single nucleotide polymorphism (SNP) database. However, linkage dragging frequently makes it difficult to use landraces for breeding. An effective mating strategy is required to break the unfavorable connections and transform landraces into breeding-ready genetic resources in order to take advantage of the diversity. A breeding technique called multi-parent advanced generation intercross (MAGIC) is used to create highly recombined germplasm (12). Enhanced genotypic diversity and new allele rearrangements may result from the increased recombination in MAGIC populations. Such crosses may result in pre-breeding materials

DOI: 10.1201/9781003385035-6

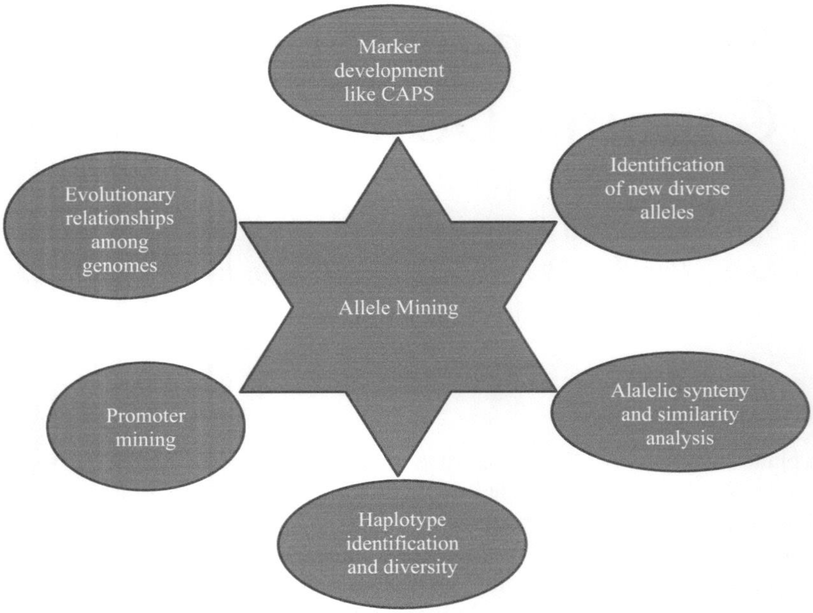

FIGURE 6.1 Application of allele mining.

with less linkage drag if elite varieties are combined with the landraces. This kind of breeding can produce a gene pool that is enriched with crucial features. We examine the idea of allele mining and its approaches in peanut, along with the opportunities and problems, after realizing the significance of allele mining in the era of genomics-driven plant breeding. The general applications of allele mining are described in Figure 6.1.

6.2 CONCEPTS AND POTENTIAL OF ALLELE MINING

Most of the agronomically important traits are governed by polygenes or complex loci. Multiple alleles from the complex locus serve as a reservoir of variation to generate functional genes. Allele mining for these genes can be an important method to identify additional alleles and new haplotypes. These allelic differences were predicted to alter the structure and/or function of the encoded protein, thus differential trait expressions. The initial investigations of allele mining concentrate primarily on the discovery of SNPs/InDels (insertion-deletions) at coding sequences or exons of the candidate gene. There are numerous instances that show how such sequence changes in genic regions might change phenotypes. However, according to recent studies, nucleotide alterations in non-coding regions (5′ UTR), such as the promoter, introns, and 3′ UTR, also significantly affect the synthesis and accumulation of transcripts, which in turn affects how traits are expressed. The A-genome mutation is an SNP, whereas three B-genome mutant alleles have been discovered, each of which caused an increase in the amount of oleic acid in peanuts through a single nucleotide substitution or MITE InDel (13–15). In the context of gene expression, mining for sequence variation, particularly in regulatory regions of genes, has gained increased significance recently. Since it is well recognized that promoter elements are crucial for the control of genes, any alterations to their sequences will have a significant impact on how genes are expressed, leading to varying trait expression. Such variable gene expression is typically brought on by sequence variation in upstream region binding sites for cis/trans-acting regulatory elements. For instance, numerous gene expression studies on the peanut genes *ubq10*, *hel1*, *ubc*, *act7*, and many others have been conducted in recent years. Table 6.1 lists the comprehensive reference genes for investigations on quantitative gene expression in peanuts.

TABLE 6.1
Reference Genes Quantitative Gene Expression Studies in Peanut

Genes Tested	Tissues	References
Hexameric polyubiquitin (ubq10)	Pod, seed, leaf, gynophore, root	(16)
Polyubiquitin10	Cotyledon, embryo axis	(17)
Ubiquitin10	32 tissues/stages/treatments	(18)
Ubiquitin	Cotyledon, embryo axis	(17)
Ubiquitin conjugating enzyme		(19)
Glyceraldehyde-3-phosphatedehydrogenase(gapdh)	Pod, seed, leaf, gynophore, root	(16)
Glucose-6-phosphate dehydrogenase	Leaf, cotyledon, stem, root, seed/stress	(19)
RNAhelicase1(hel1)	Pod, seed, leaf, gynophore, root	(16)
Nuclear helicase	32 tissues/stages/treatments	(18)
Yellow leaf-speci*fic*8 gene (yls8)	Pod, seed, leaf, gynophore, root	(16)
14-3-3protein(14-3-3)	Pod, seed, leaf, gynophore, root	(16)
60s ribosomal protein L7 (60s)	Pod, seed, leaf, gynophore, root	(16)
Ubiquitin C (ubc)	Pod, seed, leaf, gynophore, root	(16)
Elongation factor 1(ef1alpha)	Pod, seed, leaf, gynophore, root	(16)
Elongation factor1-alpha	Cotyledon, embryo axis	(17)
Elongation factor1-beta	32 tissues/stages/treatments	(18)
Elongation factor1-beta	Leaf, cotyledon, stem, root, seed/stress	(19)
Elongation factor EF-2	Cotyledon, embryo axis	(17)
Actin7(act7)	Pod, seed, leaf, gynophore, root	(16)
Actin	Cotyledon, embryo axis	(17)
Actin11	32 tissues/stages/treatments	(18)
Actin11	Leaf, cotyledon, stem, root, seed/stress	(19)
Actin2/7	32 tissues/stages/treatments	(18)
Alcohol dehydrogenase class III	Pod, seed, leaf, gynophore, root	(16)
Alcohol dehydrogenase class III	Leaf, cotyledon, stem, root, seed/stress	(19)
Alphatubulin1	Cotyledon, embryo axis	(17)
Alphatubulin	32 tissues/stages/treatments	(18)
TubulinB4	Cotyledon, embryo axis	(17)
Betatubulin	32 tissues/stages/treatments	(18)
Phospholipase D	Cotyledon, embryo axis	(17)
Malate dehydrogenase	Cotyledon, embryo axis	(17)
RibosomalproteinL24	Cotyledon, embryo axis	(17)
Cyclophilin	32 tissues/stages/treatments	(18)
Cyclophilin	Leaf, cotyledon, stem, root, seed/stress	(19)
SKP1/ask-interacting protein 16	32 tissues/stages/treatments	(18)
Metalloprotease	32 tissues/stages/treatments	(18)
Phosphoenolpyruvatecarboxylase-relatedKinase1	32 tissues/stages/treatments	(18)
TIP41-like family protein	32 tissues/stages/treatments	(18)
Hypothetical protein EG028875	32 tissues/stages/treatments	(18)
Hypothetical protein GO334386	32 tissues/stages/treatments	(18)
ATPsynthase	Leaf, cotyledon, stem, root, seed/stress	(19)
Lectin precursor	Leaf, cotyledon, stem, root, seed/stress	(19)

In order to account for the majority of the functional variations of importance in the genes, "true" allele mining should additionally analyze the non-coding and regulatory portions of the candidate genes in addition to the sequence differences in the coding regions of the key genes for agronomy traits (20) (Figure 6.2). The promoters of genes have been selected as a potential target for allele mining, also known as "promoter mining," in recognition of the significance of searching

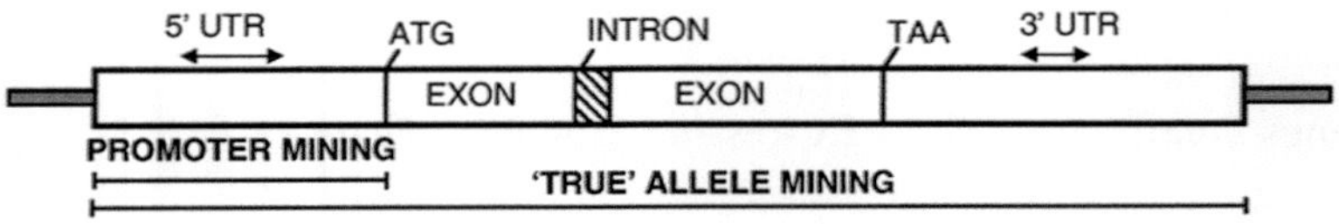

FIGURE 6.2 "True" allele mining includes coding, non-coding and regulatory regions of a gene.

for variation in regulatory areas. Promoter mining and in silico analysis were used to determine the number/frequency, location, and nucleotide variation in the transcription factor binding motif (TFBM), as well as the corresponding transcription factors, that are linked to a particular expression pattern in response to constitutive, developmental, tissue-specific, hormonal, and environmental regulation (21). Such analysis of promoter sequences also provides knowledge on the conservation of regulatory motifs specific to particular differentiated cell/tissue types. Ultimately, promoter mining would offer insights into the molecular mechanisms of gene expression and the isolation of novel and efficient promoters for use in genetic engineering applications. As global gene expression experiments to identify sets of genes with similar expression patterns are rapidly increasing, it is highly essential to incorporate approaches like in vitro and in silico promoter mining into regular allele mining programs to identify and categorize cis-regulatory elements and to understand their role in gene regulation.

6.2.1 Approaches to Allele Mining

Two major approaches are available for the identification of sequence polymorphisms for a given gene in naturally occurring populations. They are (i) modified TILLING (Targeting Induced Local Lesions in Genomes) procedures called EcoTilling and (ii) sequencing-based allele mining. A brief description of these two approaches is given in this section, and the steps involved in these approaches are illustrated in Figure 6.3.

TILLING is a technique that can identify polymorphisms (more specifically, point mutations) resulting from induced mutations in a target gene by heteroduplex analysis (22). While EcoTilling, a modified version of tilling, is used to determine the extent of natural variation in selected genes in crops, both methods are essentially the same except that the mutations are not induced artificially in EcoTilling and are detected from naturally existing alleles in the primary and secondary crop gene pools (23,24). Like TILLING, EcoTilling also relies on the enzymatic cleavage of heteroduplexed DNA (formed due to a single nucleotide mismatch in sequence between the reference and test genotype) with a single-strand-specific nuclease (i.e., Cel-1, mung bean nuclease, S1 nuclease, etc.) under specific conditions, followed by detection through Li-Corgenotypers (Li-Cor, USA). There were few reports in peanut for the determination of sequence polymorphisms with the TILLING approach. Knoll et al. (25) screened the M_2 mutant population to find possible allele changes in major allergen genes, *Ara h 1* and *Ara h 2*, and fatty acid desaturase, *AhFAD2*. They found that the mutations disrupted the start codon in *Ara h 2.02* and a premature stop codon in *Ara h 1.02*. These allergen and fatty acid desaturase genes were also used by Guo et al. (26) to find possible nucleotide variation with the use of TILLING. By screening 768 M_2 lines, different numbers of missense mutations were detected in these studied genes. In their latest study, Karaman et al. (27) developed an ethylmethane sulfonate (EMS)-induced mutant population, and M_2 lines were screened for nucleotide changes in the genes *Ara h 1.01*, *Ara h 1.02*, *ahFAD2A*, and *ahFAD2B*. Results showed that the mutation leads to a change in amino acid from glycine to aspartic acid in the *ahFAD2A* gene. These studies showed that TILLING is a powerful tool to interfere with allele function in peanuts.

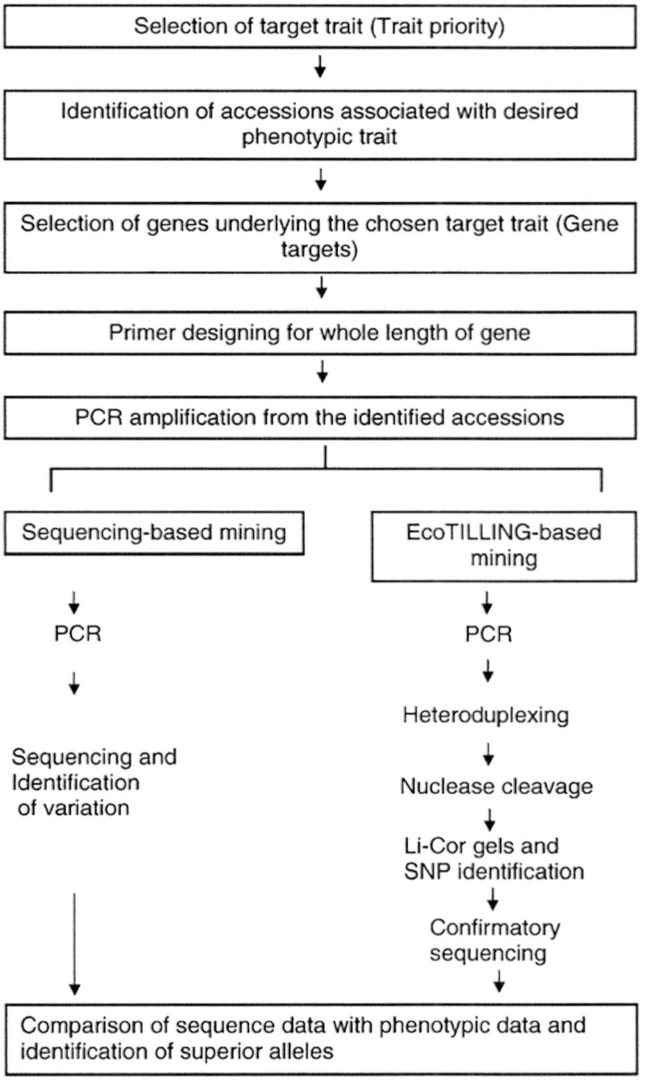

FIGURE 6.3 Steps involved in allele mining. *Source*: Kumari et al. (28).

Sequencing-based allele mining includes the amplification of alleles in diverse genotypes through PCR, DNA sequencing, and the identification of nucleotide variation. It would help to analyze individuals for haplotype structure and diversity to infer genetic association studies in plants. It does not require much sophisticated equipment or involve tedious steps but involves huge costs of sequencing.

In recent years, next-generation sequencing technologies have made immense improvements in allele mining. Though the DNA sequence reveals profound insights into genetic organization, the cost, cumbersomeness, and time were major limitations of the earlier versions of sequencers, which depended upon either chain termination sequencing or cycle sequencing. With the advent of faster sequencing platforms, the sequencing process has become more simple and cheaper. In the last few years, 'massively parallel' methods have also emerged, leading to the development of 'next generation' sequencing platforms with increased throughput and accuracy. These methods were mainly used for resequencing, alignment of the sequence data, and their comparison with the reference genome. The first of this kind was commercialized by 454 Life Sciences (29); this technique relied on pyrosequencing while eliminating the need for cloning. With this 454 sequencing platform, it was possible to produce 100 Mb of sequence with 99.5% accuracy and increased read length, averaging over 250 bases. Another massively parallel sequencer, the Illumina/Solexa genome analyzer, has been developed (30), and it is capable of sequencing one billion bases (1 Gb) of 30–40 base sequence reads in a single run in a short time period of 3–4 days. In addition to the above, Applied Biosystems Inc. (USA) has also developed a massively parallel sequencer using a supported oligonucleotide ligation and detection system (SOLiD) (31). This system, despite its shorter read length of about 25–35, can generate 2–3 Gb of sequences per run. Furthermore, Helicosbio (32) has commercialized a new massively parallel system based on single-molecule sequencing (SMS) technology. SMS technology is much faster and cheaper than second-generation sequencing technologies from Applied Biosystems and Illumina and has been described as third generation or next-to-next-generation sequencing technology (33). With the advent of high-throughput sequencing technologies, different methods have been established to genotype the mapping population of peanuts, such as restriction site-associated sequencing (RAD-seq), double digest RAD-seq, genotyping by sequencing (GBS), and high-density SNPs or InDel (34–38). The genetic diversity in Taiwanese peanut germplasm was assessed with SNPs produced by RAD-seq (39). The first genetic map based on SNPs for cultivated peanuts was constructed using ddRADseq with 1,621 SNPs (37). Jadhav et al. (40) used the GBS approach to map regions for productivity and quality traits in peanuts. The genetic map with 700 markers distributed in 2,438.1 cM was employed for the detection of quantitative trait loci (QTLs) in that study. Recently, specific length amplified fragment sequencing (SLAF-seq) technology was used to construct a high-density linkage map in peanut. Wang et al. (41) constructed a high-density genetic map that contains 3,630 SNPs on 20 linkage groups covering 2,098.14 cM and an average marker distance of 0.58 cM. They identified two stable co-located QTLs for seed and pod-related traits using the SLAF-seq approach. Goswami et al. (42) reported a simple sequence repeat (SSR) marker associated with large kernel size in interspecific peanut breeding lines. Bera et al. (43) further reported WRKY and Na+/K+ antiporter genes associated with salinity tolerance in interspecific peanut breeding lines. Bera et al. (44) first reported two major QTLs for stem rot resistance in peanuts and validated associated markers and alleles. Jasani et al. (45) reported major QTLs and associated alleles and markers for peanut bud necrosis diseases in peanuts. The recombinant inbred line (RIL) population comprising 146 lines was also sequenced with SLAF-seq and a high-resolution genetic map containing 2,266 SNPs on 20 linkage groups (LGs) with a length of 2,586.37 cM (46). The other RIL population evolved from two genotypes with contrasting seed weight, size, and shape (47). The developed genetic map was used for QTL identification of seed-related traits. A total of 27 QTLs for 100 seed weight, seed length-width, and its ratio were identified on 8 chromosomes, explaining the phenotypic variance from 0.74 to 83.23. These studies showed the effectiveness of the SLAF-seq method for allele discovery in peanuts.

6.2.2 Prerequisites for Allele Mining

Successful and efficient allele mining activity would depend mainly on the existing genetic base and the availability of genome and gene sequence information for a particular crop species. Several other important factors include the availability of efficient and reliable phenotyping techniques; genomic resources; the availability of high throughput techniques for quick and easy generation of allelic data points; cost-effective sequencing platforms; efficient bioinformatic tools for identification of nucleotide variation; and molecular marker construction technologies for marker-assisted selection (MAS). There are many criteria to be considered for initiating allele mining in any crop (Table 6.2).

TABLE 6.2
An Overview of Steps in Allele Mining

SN	Steps	Criteria for Selection/Design
1	Germplasm identification	• Reference collection representing maximum genetic diversity in a minimum possible number
2	Crop species	• Availability of genetic map, segregatingpopulations, ESTs or gene/genome sequences • Relatedness to model species for which targetgene sequence is available • Availability of comparative map/sequence collinearity • Availability of mini core collections or 200–300 reference accessions covering all varietal groups selected from mini core
3	Target trait	• Traits of interest
4	Phenotypic characterization	• Precise phenotyping using efficient protocols in suitable environmental condition
5	Candidate gene targets	• A priority gene list considering the traits targeted • Cross-species: genes with known homologues in related crops • Within species: genes for important agronomic traits. For complex traits, genes thatco-localize with reported major QTL • For crops with no full genome sequences: EST sequences and information from cDNA libraries can be used • The genes under consideration should be candidates with strong evidence, or be part of a well-understood metabolic pathway or be related by clear ortholog across species, to the extent possible
6	Primer design	• Comparative genomics-based prediction of the conserved regions that may be suitable for amplification of the homologs in related species • Covering the coding and upstream regulatory regions • Primer pairs with a minimum overlap of 100–200 bp to maximize their utility for SNP detection and sequencing
7	Sequencing analysis for detection of SNPs and InDels	• Sequencing of single and clear amplicons • Amplicons of genotypes with high trait intensity in desired direction
8	Database of candidate gene diversity	• Well classified and easily accessible database for each nucleotide polymorphism associated with trait expression • Tools for designing and modeling superior alleles for enhanced trait expression

6.3 EVOLUTION AND DISCOVERY OF NEW SUPERIOR ALLELES IN PEANUT

Mutation is considered as an evolutionary driving force that underlies existing allelic diversity in any crop species. For the creation of new alleles or causing variations in the existing allele and allelic combinations, mutations in the genic regions of the genome, either as SNPs or as InDels are important. Mutations in coding regions and/or regulatory regions may have a tremendous effect on the phenotype by altering the encoded protein structure and/or function, while those that occur in noncoding regions of a gene could often be silent without any effect on the phenotype. Despite the fact that plants propagate these altered/novel alleles (either through natural or artificial selection) based on their need in the adaptation mechanism, it can be suggested that such altered/novel alleles of key genes are preserved in many related wild species, land races, and cultivated varieties of peanut. Therefore, exploring the natural variation in important genes is highly indispensable in modern crop improvement programs.

6.4 CHARACTERIZATION OF ALLELIC DIVERSITY IN PEANUT GENE BANKS

Identification and access to allelic variation that affects the plant phenotype are of utmost importance for the utilization of genetic resources in crop improvement. Exploitation of gene banks for efficient utilization depends on the knowledge of genetic diversity in general and allelic diversity at candidate gene(s) of interest. Hence, the characterization of genetic diversity, or allelic/genic diversity, among the accessions of the collection is highly important to establish the genetic relationships among them and to assess their genetic worthiness in terms of their utility for improving a target trait. In peanut, molecular diversity studies have been conducted by various researchers to identify allelic diversity and the association of markers with biotic and abiotic resistance (48–52). Allele mining seems to be a promising, although largely untested, method to unlock the diversity in the collections of genetic resources in the world's gene banks (53). The availability of the sequence of candidate genes that underlie agronomic traits will facilitate a pragmatic gene-based use of genetic resources, which, at least in theory, can be extended to any given trait (54). Realizing the potential of allele mining in plant genetic resources (PGR) management applications, many of the international crop research institutes that maintain crop germplasm collections have initiated studies to characterize the allelic diversity of crop plants. Although many hurdles still need to be overcome, the application of allele mining strategies is expected to result in a quantum leap in the use of PGRs.

6.5 IDENTIFICATION OF NEW PEANUT HAPLOTYPES

Allele mining can be potentially employed in the identification of nucleotide variation at a genomic region (candidate gene) associated with phenotypic variation for a trait. Through this, one can evaluate the frequency, type, and extent of the occurrence of new haplotypes and the resulting phenotypic changes. In peanut, few case studies are available for analyzing the haplotype diversity and identification of new haplotypes at candidate genes (55). Knowledge of the most common haplotype changes and their frequency in populations would form the basis for association mapping studies.

6.6 DEVELOPMENT OF ALLELE-SPECIFIC MARKERS FOR MAS AND GWAS APPROACH IN PEANUT

Identification of sequence variation will pave the way to develop allele-specific marker assays for precise introgression of the identified 'superior and/or novel' alleles into a suitable genetic background. In recent years, several case studies have been reported that demonstrate the existence of sequence variation at key genes, while others have demonstrated the utility of these variations through the development of allele-specific molecular markers for MAS. For instance, in peanut, root-knot nematode (RKN) resistance markers have been developed to introgress this trait into cultivated peanut from an A-genome wild relative, *A. cardenasii* (56) and successfully utilized in MAS (57). Furthermore, AS-PCR markers linked with high O/L traits have been developed (58–61) and have been applied in multiple peanut breeding programs (62–66). Recently, allele-specific functional markers and SNP markers for rust resistance and late leaf spot (LLS) resistance were identified in peanuts using the QTL-seq method (67,68).

The genome-wide association study (GWAS) approach is widely used to detect associations between allele or genotype frequency and trait status (69). In peanuts, it was successfully applied for different traits. The peanut reference set is characterized by 50 agronomic traits and genotyping with SSR markers (70). Results showed that a total of 30 significant allele effects for these 15 traits were associated with 24 markers. Zhang et al. (71) used the GWAS approach to analyze 11 agronomic traits in 158 peanut accessions. They identified 51 SNP peaks corresponding to domestication-related traits such as seed width, pod length, seed coat color, and seed weight. In another study, 195 peanut accessions were genotyped and phenotyped, with 93 non-overlapping peak SNPs that are significantly associated with seed weight, pod weight, yield per plant, and pod branch number per plant (72). Leaf chlorophyll content is a fundamental component of photosynthesis. Zou et al. (73) detected that SNP AX-176820297 on Araip.B05 was significantly linked with this trait. A GWAS study was also conducted for the identification of biotic stress factors. Ravelombola et al. (74) reported a peanut panel consisting of 775 USDA accessions, which were phenotyped for sting nematode resistance. They found that nine SNPs were found to be associated with sting nematode resistance using single-marker regression. GWAS analysis with SNPs in 103 accessions of the U.S. mini core collection identified 90 suggestive and significant SNPs associated with angle of lateral branch with main stem, height of main stem, length of lateral branch, extent of main stem radius, and the plant type index. A consolidated list of allele mining in peanuts is given in Table 6.3.

TABLE 6.3
Allelic richness and diversity, and Grouping of Accessions Based on Allelic Diversity in Peanut

#Genotypes	#SSR	Allelic Richness	Summarized Findings	References
22 accessions including wild species	6	38 fragments, averaged 6 fragments per primer	Effective in detecting molecular variation in cultivated peanut	(75)
24 accessions	56	79 alleles, averaged 4 alleles per locus	Highlighted a simple and reliable way in obtaining Polymorphic SSR markers from cultivated peanut	(76)
188 landraces	10	89 alleles, averaged 7 alleles per locus	differentiated South American landraces than those from Africa and Asia	(77)
60 accessions	8	74 alleles, averaged 8.4 alleles per locus	Highlighted the usefulness of SSR markers for genetic diversity analysis of cultivated peanut	(78)
60 cultivated and 36 wild species accessions	12	-	More diversity amongst the Brazilian germplasm; two major groups among *A. hypogaea* accessions; high marker transferability between *Arachis* species, Ah-041 showing 100% transferability across species	(78)
48 accessions representing six botanical varieties	38	-	Differentiated six botanical types (*fastigiata, vulgaris, hypogaea, hirsuta, peruviana* and *aequatoriana*) into separate groups	(79)
22 accessions	23	135 alleles, averaged 6 alleles per locus	Few SSRs differentiated rust and/or late leaf spot resistant germplasm	(80)
46 accessions representing six botanical types	32	107 alleles, averaged 3 alleles per locus	Two distinct groups corresponding to subspecies *hypogaea* and *fastigiata;* six alleles differentiated bacterial wilt resistant and susceptible accessions	(81)
31 bacterial wilt resistant accessions	SSR: 78;AFLP126	91 loci (3.14 alleles per locus) with SSR and 72 loci (2.25allelesperlocus) with AFLP markers	14H06,7G02,3A8, 16C6,andP1M62 were effective in detecting polymorphism	(82)

(Continued)

TABLE 6.3 (*Continued*)
Allelic richness and diversity, and Grouping of Accessions Based on Allelic Diversity in Peanut

#Genotypes	#SSR	Allelic Richness	Summarized Findings	References
US peanut mini core and wild relatives	31	477 alleles, averaged 15.4 alleles per locus	13 transferable markers across all wild relative accessions	(83)
Two peanut genotypes	162	10 primer pairs of which exhibited polymorphism with 2–3 alleles, with an average of 2.2 alleles	EST could be a resource for developing SSR. The high level of transferability to wild species also implied that EST-SSR is a potentially useful marker for genetic studies in wild species.	(84)
466 accessions	26	-	Greater diversity among the Chinese mini core accessions; L2 Gangguo and ICG 12625 the most genetically diverse; among six botanical types, accessions of *fastigiata* and *hypogaea* were more diverse than other types	(85)
114Valenciapeanut accessions	52	683 alleles, averaged 13 alleles per locus	Differentiated Valencia germplasm into five clusters with two distinct major groups	(86)
292 peanut varieties	583	3,663 alleles with an average of 6.28 alleles per locus.	Study provided insights into the genetic basis of oil content and confirmed highly associated two SSR markers facilitating marker-assisted selection for developing high-oil content peanut varieties.	(87)

6.7 CONSERVED SYNTENY/EVOLUTIONARY RELATIONSHIP AMONG GENOMES/ALLELES

Using the sequence information obtained from allele mining studies, syntenic relationships can be assessed among the identified loci/genes across the species/genera. In peanut, the IPGI study provided the syntenic relations between A and B subgenomes and their sequence comparison with tetraploid peanut. Most pseudomolecules had symmetrically positioned peri-centromeres that were in accordance with cytogenetic observations (88,89). Most of the pseudomolecules have one-to-one correspondence between the two species: pairs 02, 03, 04, and 10 were collinear. While pairs 05, 06, and 09 were individually differentiated by a large inversion in one arm of the pseudomolecules. Besides, the pseudomolecules in pair 01 were differentiated by large inversions of both arms.

In contrast, chromosomes 07 and 08 might have undergone complex rearrangements that could have transported repeat-rich DNA to A07 and gene-rich DNA to A08, resulting in only one normal (upper) euchromatic arm in A07 and an abnormally small arm in A08 with low repetitive content. In accordance with cytogenetic observations (90,91), A08 could be assigned as the characteristic small "A chromosome" (cytogenetic chromosome A09).

6.8 CHALLENGES AND FUTURE PROSPECTS

Considering the huge number of accessions that are held collectively in various gene banks, genetic resource collections are deemed to harbor a wealth of undisclosed allelic variants. Now the challenge is to efficiently identify and exploit the useful variation for crop improvement. Here, we outline the difficulties associated with allele mining and possible solutions for enhancing its effectiveness.

6.8.1 Selection of Representation of Whole Genetic Diversity of Crop

The foremost challenge in unlocking the existing variation is the selection of germplasm to be 'mined'. Screening the entire collection for characterizing a gene would certainly be helpful to find rare alleles, but this is an enormous and inefficient way. Hence, we need to find a way forward from genotyping the whole composite collection to efficiently screening the accessions to discover new alleles and allele combinations while minimizing the number of accessions to be screened.

6.8.2 Effective Designing of Core/Mini Core Collection

Handling the entire germplasm is a Herculean task, be it for conventional plant breeding or for allele mining, and hence requires sampling strategies to narrow down to a manageable size while maintaining variability. Under this background, Upadhyaya et al. (92) postulated the concept of "mini core", wherein approximately 10% of the core collection is subsampled (or 1% of the entire collection) to represent global diversity in the species. The development of core and mini collections out of the entire collection has been proposed to simplify the conservation of germplasm resources and the effective utilization of the existing variation in gene banks. A huge repository of variation in the cultivated peanut is present as germplasm accessions in the gene banks. The largest collection of peanut germplasm is being held at ICRISAT, India (15,445 accessions), followed by ICAR-National Bureau of Plant Genetic Resources (ICAR-NBPGR) with 14,585 accessions; ICAR-Directorate of Groundnut Research (ICAR-DGR) in India with 9,024 accessions; 9,917 accessions at the U.S. Department of Agriculture (USDA); Oil Crops Research Institute (OCRI) of the Chinese Academy of Agricultural Sciences (CAAS) with 8,083 accessions; and 4,210 accessions in China at the Crops Research Institute of the Guangdong Academy of Agricultural Sciences. Besides, germplasm collections are also held at the North Carolina State University (NCSU) and Texas A&M University (TAMU) in the USA; in Brazil, at the Instituto Agronomico de Campinas and EMBRAPA-CENARGEN; and at the Instituto de Botánica del Nordeste (IBONE); and in Argentina, at the Instituto Nacional de Technologia Agropecuaria (INTA). The development of core collections from these huge genetic resources should be genotyped to generate data that can be used for trait association analyses. In addition to this, investigation of the phylogenetic and network traits of the collection provides data about the historical spread of peanut diversity (93).

6.8.3 Advanced High Throughput- Phenotyping Techniques

Phenotyping would be the key to success in shortlisting the genotypes for allele mining, as phenotypes are considered as the best clues for genotypes. Increased precision in efficient phenotyping techniques maximizes the chances of 'mining' the prospective genotypes. Furthermore, by developing trait-specific subsets through phenotyping experiments, the number of genotypes for 'mining' is significantly reduced. In peanuts, methods for accurate phenotyping of different traits like drought, salinity, aluminum toxicity tolerance, foliar fungal disease, bacterial disease, nematode, and insect pests were given by Janila and Nigam (94). In addition, a cost effective method for precise phenotyping for stem rot resistance was reported by Bera et al. (95).

6.8.4 Future Prospects

Due to its tremendous potential and applications, allele mining can be visualized as a vital link between the effective utilization of genetic and genomic resources in genomics-driven modern plant breeding. In order to keep pace with ever-increasing sequence data in GenBank and ever-expanding crop gene banks, it is highly domineering to evolve novel and efficient 'mining' strategies. Efforts must be initiated to develop tools and strategies for efficient handling of both genetic and genomic resources. Firstly, there is a need to develop more refined sampling strategies and tools that can support such selection for enriching useful alleles in a minimum number of genotypes with the desired levels of trait expression. More efforts are needed to develop computational tools for efficient identification of core sets from large germplasm collections that give a correct representation of all phenotypic classes (both qualitative and quantitative). Information on molecular genetic diversity should be best utilized to shortlist the genotypes while maintaining the diversity intact within the mini core sets.

However, in-depth studies are required on the number and type of markers to be used for downsizing the mini core sets. Through careful and accurate phenotyping, trait-specific subsets with all phenotypic classes can be developed, which would further result in the selection of 'key' genotypes for 'mining' the novel alleles associated with a particular trait. Secondly, several molecular, biological, and bioinformatic tools are currently available for effective mining and utilization of information from large genomic resources. Using these tools, identification of functional nucleotide polymorphisms in superior alleles and comparison of sequence variation in a range of alleles for a particular gene with the corresponding trait expression across genotypes are now possible. Furthermore, this information could be used to identify the most preferable allele with the desired levels of expression. Subsequently, improved marker designing techniques coupled with a highly powerful yet simple assay technique to discriminate even a single base pair change would be essential to transferring the identified allele into the desired background through MAS. The outcome of IPGI will make allele mining possible for many genes in peanut within the limits of genome synteny and sequence conservation. New technologies that facilitate allele discovery and integrate it into the varietal development process would greatly enhance the power of this approach. As we are rapidly entering an era of low-cost and high-throughput sequencing, routine allele mining through sequencing will soon become a universal tool for scoring variations at candidate genes between individuals and an effective and economically viable method for the identification and deployment of novel genetic resources in plant breeding. Whole genome SNP detections at key candidate regions using array chips are becoming available and are expected to further increase the speed of mutation for the detection and allow rapid identification and characterization of allelic variants. Furthermore, advanced allele mining strategies that involve the design of artificial allelic variants with more relevant functions by domain shuffling or gene shuffling among related genes offer new domains and provide powerful tools for targeted crop improvement. These days, the challenge before us is the integration of information from whole genome sequencing, genome-scale SNP assays, and targeted gene mapping studies with phenotypic trait variation for more efficient utilization of genetic and genomic resources for directed crop improvement. There is also a need to bridge the gap between channeling the information of genes and its function in the generation of new desired varieties that are an essential component of a sustainable food supply for the future. Once the genes/loci, the germplasm sources carrying trait-enhancing alleles, cataloguing of novel and superior alleles and their behavior in appropriate genetic backgrounds are identified, environment for maximum expression needs to be documented. As cost-effective, efficient tools for data mining and high-throughput sequencing platforms become available, it is certainly expected that sequencing-based allele mining will emerge as the universal method of choice for revealing natural variations, providing novel and effective alleles, and would take center stage in all crop improvement activities.

REFERENCES

(1) Tiwari S, Mishra DK, Singh A, Singh PK, Tuli R. Expression of a synthetic cry1EC gene for resistance against *Spodoptera litura* in transgenic peanut (*Arachis hypogaea* L.). *Plant Cell Rep* 2008;27(6):1017–1025.

(2) Tiwari S, Mishra DK, Chandrasekhar K, Singh PK, Tuli R. Expression of δ-endotoxin Cry1EC from an inducible promoter confers insect protection in peanut (*Arachis hypogaea* L.) plants. *Pest Manag Sci* 2011;67(2):137–145.

(3) Mehta R, Radhakrishnan T, Kumar A, Yadav R, Dobaria JR, Thirumalaisamy PP, et al. Coat protein-mediated transgenic resistance of peanut (*Arachis hypogaea* L.) to peanut stem necrosis disease through *Agrobacterium*-mediated genetic transformation. *Indian J Virol* 2013;24(2):205–213.

(4) Sarkar T, Thankappan R, Kumar A, Mishra GP, Dobaria JR. Heterologous expression of the *AtDREB1A* gene in transgenic peanut-conferred tolerance to drought and salinity stresses. *PLoS One* 2014;9(12):e110507.

(5) Sarkar T, Thankappan R, Kumar A, Mishra GP, Dobaria JR. Stress inducible expression of *AtDREB1A* transcription factor in transgenic peanut (*Arachis hypogaea* L.) conferred tolerance to soil-moisture deficit stress. *Front Plant Sci* 2016;28(7):935.

(6) Bala M, Radhakrishnan T, Kumar A, Mishra GP, Dobraia JR, Kirti PB. Overexpression of a fusion *defensin* gene from radish and fenugreek improves resistance against leaf spot diseases caused by *Cercospora arachidicola* and *Phaeoisariopsis personata* in peanut. *Turk J Biol* 2016;40(1):139–149.

(7) Patil AS, Thankappan R, Mehta R, Yadav R, Kumar A, Mishra GP, et al. Evaluation of transgenic peanut plants encoding coat protein and nucleocapsid protein genes for resistance to tobacco streak virus and peanut bud necrosis virus. *J Environ Biol* 2017; 38:187–196.

(8) Bhalani H, Thankappan R, Mishra GP, Sarkar T, Bosamia TC, Dobaria JR. Regulation of antioxidant mechanisms by *AtDREB1A* improves soil-moisture deficit stress tolerance in transgenic peanut (*Arachis hypogaea* L.). *PLoS One* 2019;14(5):e0216706.

(9) Bera SK, Rani K, Kamdar JH, Pandey MK, Desmae H, Holbrook CC, et al. Genomic designing for biotic stress resistant peanut. In: Kole C (ed) *Genomic Designing for Biotic Stress Resistant Oilseed* Crops. Springer, Cham, 2022, pp. 137–214.

(10) Varshney RK, Pandey MK, Puppala N. (2017) The peanut genome. In: Kole C (ed) *Compendium of Plant Genomes*. Springer, Cham, 2017, pp. 1–6.

(11) Zhuang W, Chen H, Yang M, Wang J, Pandey MK, Zhang C, et al. The genome of cultivated peanut provides insight into legume karyotypes, polyploid evolution and crop domestication. *Nat Genet* 2019;51(5):865–876.

(12) Gupta PK, Kulwal PL, Jaiswal V. Association mapping in crop plants: Opportunities and challenges. *Adv Genet* 2014;85:109–147.

(13) Jung S, Swift D, Sengoku E, Patel M, Teule F, Powell G, et al. The high oleate trait in the cultivated peanut (*Arachis* hypogaea L.). I. Isolation and characterization of two genes encoding microsomal oleoyl-PC desaturases. *Mol Genet Genom* 2000;263(5):796–805.

(14) Lopez Y, Nadaf HL, Smith OD, Connell JP, Reddy AS, Fritz AK. Isolation and characterization of the 12-fatty acid desaturase in peanut (*Arachis hypogaea* L.) and search for polymorphisms for the high oleate trait in Spanish market-type lines. *Theor Appl Genet* 2000;7(101):1131–1138.

(15) Patel M, Jung S, Moore K, Powell G, Ainsworth C, Abbott A. High-oleate peanut mutants result from a MITE insertion into the FAD2 gene. *Theor Appl Genet* 2004;108(8):1492–1502.

(16) Brand Y, Hovav R. Identification of suitable internal control genes for quantitative real-time PCR expression analyses in peanut (*Arachis hypogaea*). *Peanut Sci* 2010;37(1):12–19.

(17) Jiang S, Sun Y, Wang S. Selection of reference genes in peanut seed by real-time quantitative polymerase chain reaction. *Int J Food Sci Technol* 2011;46(10):2191–2196.

(18) Chi X, Hu R, Yang Q, Zhang X, Pan L, Chen N et al. (2012). Validation of reference genes for gene expression studies in peanut by quantitative real-time RT-PCR. *Mol Genet Genom* 2012;287(2):167–176.

(19) Reddy DS, Bhatnagar MP, Cindhuri KS, Sharma KK. Evaluation and validation of reference genes for normalization of quantitative real-time PCR based gene expression studies in peanut. *PLoS One* 2013;8(10):e78555.

(20) Latha R, Rubia L, Bennett J, Swaminathan MS. Allele mining for stress tolerance genes in *Oryza* species and related germplasm. *Mol Biotechnol* 2004;27(2):101–108.

(21) Veerla S, Höglund M. Analysis of promoter regions of co-expressed genes identified by microarray analysis. *BMC Bioinformatics* 2006;7(1):1–15.

(22) Till BJ, Reynolds SH, Greene EA, Codomo CA, Enns LC, Johnson JE et al. Large-scale discovery of induced point mutations with high-throughput TILLING. *Genome Res* 2003;13(3):524–530.
(23) Comai L, Young K, Till BJ, Reynolds SH, Greene EA, Codomo CA, et al. Efficient discovery of DNA polymorphisms in natural populations by Ecotilling. *Plant J* 2004;37(5):778–786.
(24) Comai L, Henikoff S. TILLING: Practical single-nucleotide mutation discovery. *Plant* 2006;45(4): 684–694.
(25) Knoll JE, Ramos ML, Zeng Y, Holbrook CC, Chow M, Chen S, et al. TILLING for allergen reduction and improvement of quality traits in peanut (*Arachis hypogaea* L.). *BMC Plant Biol* 2011;11:81
(26) Guo Y, Abernathy B, Zeng Y, Ozias-Akins P. TILLING by sequencing to identify induced mutations instress resistance genes of peanut (*Arachis hypogaea*). *BMC Genomics* 2015;16:157.
(27) Karaman K, Kizil S, Başak M, Uzun B, Yol E. Development of EMS-induced mutagenized groundnut population and discovery of point mutations in the *ahFAD2* and *Ara h 1* genes by TILLING. *J Oleo Sci* 2021;70(11):1631–1640.
(28) Kumari R, Kumar P, Sharma VK, Kumar H. Allele mining for crop improvement. *Int J Pure Appl Biosci* 2018:6(1):1456–1465.
(29) Margulies M, Egholm M, Altman WE, Attiya S, Bader JS, Bemben LA et al. Genome sequencing in microfabricated high-density picolitre reactors. *Nature* 2005;437(7057):376–380.
(30) Bennett S. Solexa ltd. *Pharmacogenomics* 2004;5(4):433–438.
(31) Hutchison CA. DNA sequencing: Bench to bedside and beyond. *Nucleic Acids Res* 2007;35(18):6227–6237.
(32) https://www.helico.bio/technology
(33) Gupta PK. Single-molecule DNA sequencing technologies for future genomics research. *Trends Biotechnol* 2008;26(11):602–611.
(34) Miller MR, Dunham JP, Amores A, Cresko WA, Johnson EA. Rapid and cost-effective polymorphism identification and genotyping using restriction site associated DNA (RAD) markers. *Genome Res* 2007;17(2):240–248.
(35) Peterson BK, Weber JN, Kay EH, Fisher HS, Hoekstra HE. Double digest RADseq: An inexpensive method for de novo SNP discovery and genotyping in model and non-model species. *PloS One* 2012;7(5):e37135.
(36) Poland JA, Brown PJ, Sorrells ME, Jannink JL. Development of high-density genetic maps for barley and wheat using a novel two-enzyme genotyping-by-sequencing approach. *PloS One* 2012;7(2):e32253.
(37) Zhou X, Xia Y, Ren X, Chen Y, Huang L, Huang S, et al. Construction of a SNP-based genetic linkage map in cultivated peanut based on large scale marker development using next-generation double-digest restriction-site-associated DNA sequencing (ddRADseq). *BMC Genomics* 2014;15(1):1–14.
(38) Han S, Yuan M, Clevenger JP, Li C, Hagan A, Zhang X et al. A SNP-based linkage map revealed QTLs for resistance to early and late leaf spot diseases in peanut (*Arachis hypogaea* L.). *Front Plant Sci* 2018;9:1012.
(39) Hsu YM, Wang SS, Tseng YC, Shin-Ruei L, Hsiang F, Wei-Chia H, et al. Assessment of genetic diversity and SNP marker development within peanut germplasm in Taiwan by RAD-seq. *Sci Rep* 2022;12:14495.
(40) Jadhav MP, Gangurde SS, Hake AA, Yadawad A, Mahadevaiah SS, Pattanashetti SK, et al. Genotyping-by-sequencing based genetic mapping identified major and consistent genomic regions for productivity and quality traits in peanut. *Front Plant Sci* 2021;12:668020.
(41) Wang, Z, Huai D, Zhang Z, Cheng K, Kang Y, Wan L, et al. Development of a high-density genetic map based on specific length amplified fragment sequencing and its application in quantitative trait loci analysis for yield-related traits in cultivated peanut. *Front Plant Sci* 2018;9:827.
(42) Goswami B, Kamdar JH, Bera SK. Molecular diversity and association of simple sequence repeat markers with kernel mass in cultivated groundnut (*Arachis hypogaea* L.). *Aust J Crop Sci* 2013;7(8):1152–1159.
(43) Bera SK, Ajay BC, Singh AL. WRKY and Na+/H+ antiporter genes conferring tolerance to salinity in interspecific derivatives of peanut (*Arachis hypogaea* L.). *Aust. J Crop Sci* 2013;7(8):1173–1180.
(44) Bera SK, Kamdar JH, Kasundra SV, Ajay BC. A novel QTL governing resistance to stem rot disease caused by Sclerotium rolfsii in peanut. *Australas Plant Pathol* 2016;45(6):637–644.
(45) Jasani MD, Kamdar JH, Bera S, Sunkad G, Bera SK. Novel and stable major QTLs conferring resistance to peanut bud necrosis disease and identification of resistant high yielding peanut breeding lines. *Euphytica* 2021;217(6):1–13.
(46) Hu XH, Zhang SZ, Miao HR, Cui FG, Shen Y, Yang WQ et al. High-density genetic map construction and identification of QTLs controlling oleic and linoleic acid in peanut using SLAF-seq and SSRs. *Sci Rep* 2018;8(1):1–10.
(47) Zhang S, Hu X, Miao H, Chu Y, Cui F, Yang W, et al. QTL identification for seed weight and size based on a high-density SLAF-seq genetic map in peanut (*Arachis hypogaea* L.). *BMC Plant Biol* 2019;19:537

(48) Mondal S, Badigannavar AM. Molecular diversity and association of SSR markers to rust and late leaf spot resistance in cultivated groundnut (*Arachis hypogaea* L.). *Plant Breed* 2010;129(1):68–71.
(49) Koilkonda P, Sato S, Tabata S, Shirasawa K, Hirakawa H, Sakai H, et al. Large-scale development of expressed sequence tag-derived simple sequence repeat markers and diversity analysis in *Arachis* spp. *Mol Breed* 2012;30(1):125–138.
(50) Khera P, Upadhyaya HD, Pandey MK, Roorkiwal M, Sriswathi M, Janila P, et al. Single nucleotide polymorphism-based genetic diversity in the reference set of peanut (*Arachis* spp.) by developing and applying cost-effective kompetitive allele specific polymerase chain reaction genotyping assays. *Plant Genome* 2013;6:plantgenome2013.06.0019.
(51) Bera SK, Kamdar JH, Maurya AK, Dash P. Molecular diversity and association of simple sequence repeat markers with bud necrosis disease in interspecific breeding lines and cultivars of peanut (*Arachis hypogaea* L.). *Aust J Crop Sci* 2014;8(5):771–781.
(52) Jiang H, Huang L, Ren X, Chen Y, Zhou X, Xia Y, et al. Diversity characterization and association analysis of agronomic traits in a Chinese peanut (*Arachis hypogaea* L.) mini-core collection. *J Integr Plant Biol* 2014;56(2):159–169.
(53) Kaur N, Street K, Mackay M, Yahiaoui N, Keller B. Allele mining and sequence diversity at the wheat powdery mildew resistance locus *Pm3*. In: *Proceedings of the 11th International Wheat Genetics Symposium*, Sydney, January 1, 2008; pp. 1–3
(54) Graner A. Barley research at IPK. In: de Vincente M C, Glaszmann J C (eds), *Molecular Markers for Allele Mining, Proceedings*. IPGRI. Rome, 2006, 96p.
(55) Li YH, Li W, Zhang C, Yang L, Chang RZ, Gaut BS, et al. Genetic diversity in domesticated soybean (*Glycine max*) and its wild progenitor (*Glycine soja*) for simple sequence repeat and single-nucleotide polymorphism loci. *New Phytol* 2010;188(1):242–253.
(56) Simpson CE. Use of wild *Arachis* species/introgression of genes into *A. hypogaea* L. *Peanut Sci* 2001;28(2):114–116.
(57) Chu Y, Wu CL, Holbrook CC, Tillman BL, Person G, Ozias-Akins P. Marker-assisted selection to pyramid nematode resistance and the high oleic trait in peanut. *Plant Genome* 2011;4(2):110–117.
(58) Chu Y, Holbrook CC, Timper P, Ozias-Akins P. Development of a PCR-based molecular marker to select for nematode resistance in peanut. *Crop Sci* 2007;47(2):841–845.
(59) Chu Y, Holbrook CC, Ozias-Akins P. Two alleles of ahFAD2B control the high oleic acid trait in cultivated peanut. *Crop Sci* 2009;49(6):2029–2036.
(60) Barkley NA, Wang ML, Pittman RN. A real-time PCR genotyping assay to detect *FAD2A* SNPs in peanuts (*Arachis hypogaea* L.). *Elect J Biotechnol* 2011;14(1):9–10.
(61) Chen Z, Wang ML, Barkley NA, Pittman RN. A simple allele-specific PCR assay for detecting *FAD2* alleles in both A and B genomes of the cultivated peanut for high-oleate trait selection. *Plant Mol Biol Rep* 2010;28(3):542–548.
(62) Janila P, Pandey MK, Shasidhar Y, Variath MT, Sriswathi M, Khera P, et al. Molecular breeding for introgression of fatty acid desaturase mutant alleles (*ahFAD2A* and *ahFAD2B*) enhances oil quality in high and low oil containing peanut genotypes. *Plant Sci* 2016;242:203–213.
(63) Bera SK, Kamdar JH, Kasundra SV, Dash P, Maurya AK, Jasani MD, et al. Improving oil quality by altering levels of fatty acids through marker-assisted selection of *ahFAD2* alleles in peanut (*Arachis hypogaea* L.). *Euphytica* 2018;214(9):1–15.
(64) Bera SK, Kamdar JH, Kasundra SV, Patel SV, Jasani MD, Maurya AK, et al. Steady expression of high oleic acid in peanut bred by marker-assisted backcrossing for fatty acid desaturase mutant alleles and its effect on seed germination along with other seedling traits. *PLoS One* 2019:14(12):e0226252.
(65) Kamdar JH, Jasani MD, Ajay BC, Rani K, Manivannan N, Vasanthi RP et al. Fatty acid desaturase-2 (*ahFAD2*) mutant alleles in peanut (*Arachis hypogaea* L.) pre-breeding lines: An insight into the source, features, discourse, and selection of novel pre-breeding lines. *Genet Resour Crop Evol* 2021;68(2):529–549.
(66) Kamdar JH, Jasani MD, Chandrashekar AB, Janila P, Pandey MK, Georrge JJ, et al. Does improved oleic acid content due to marker-assisted introgression of ahFAD2 mutant alleles in peanuts alter its mineral and vitamin composition? *Front Plant Sci* 2022;13:942617
(67) Pandey MK, Roorkiwal M, Singh VK, Ramalingam A, Kudapa H, Thudi M, et al. Emerging genomic tools for legume breeding: Current status and future prospects. *Front Plant Sci* 2016;7:455.
(68) Pandey MK, Khan AW, Singh VK, Vishwakarma MK, Shasidhar Y, Kumar V, et al. QTL-seq approach identified genomic regions and diagnostic markers for rust and late leaf spot resistance in groundnut (*Arachis hypogaea* L.). *Plant Biotechnol J* 2017; 15(8):927–941.

(69) Alseekh S, Kostova D, Bulut M, Fernie AR. Genome-wide association studies: Assessing trait characteristics in model and crop plants. *Cell Mol Life Sci* 2021;78(15):5743–5754.
(70) Pandey MK, Upadhyaya HD, Rathore A, Vadez V, Sheshshayee MS, Sriswathi M, et al. Genome wide association studies for 50 agronomic traits in peanut using the 'reference set' comprising 300 genotypes from 48 countries of the semi-arid tropics of the world. *PLoS One* 2014;9(8):e105228.
(71) Zhang H, Wang ML, Schaefer R, Dang P, Jiang T, Chen C. GWAS and co-expression network reveal ionomic variation in cultivated Peanut. *J Agric Food Chem* 2019;67:12026–12036.
(72) Wang J, Yan C, Li Y, Li C, Zhao X, Yuan, C, et al. GWAS discovery of candidate genes for yield-related traits in peanut and support from earlier QTL mapping studies. *Genes* 2019;10(10):803.
(73) Zou K, Kim KS, Kang D, Kim MC, Ha J, Moon JK, et al. Genome-wide association study of leaf chlorophyll content using high-density SNP array in peanuts (*Arachis hypogaea* L.). *Agronomy.* 2022;12:152
(74) Ravelombola W, Cason J, Tallury S, Manley A, Pham H. Genome-wide association study and genomic selection for sting nematode resistance in peanut using the USDA public data. *J Crop Improv* 2023;37(2):273–290.
(75) Hopkins MS, Casa AM, Wang T, Mitchell SE, Dean RE, Kochert GD et al. Discovery and characterization of polymorphic simple sequence repeats (SSRs) in peanut. *Crop Sci* 1999;39:1243–1247.
(76) He GH, Meng R, Newman M, Gao G, Pittman RN, Prakash CS. Microsatellites as DNA markers in cultivated peanut *Arachis hypogaea* L., *BMC Plant Biol* 2003;3:3–9.
(77) Ferguson ME, Burow MD, Schulze SR, Bramel PJ, Paterson AH, Kresovich S. Microsatellite identification and characterization in peanut *A. hypogaea* L. *Theor Appl Genet* 2004;108:1064–1070.
(78) Moretzsohn MC, Hopkins MS, Mitchell SE, Kresovich S, Valls JFM, Ferreira ME. Genetic diversity of peanut *Arachis hypogaea* L. and its wild relatives based on the analysis of hypervariable regions of the genome. *BMC Plant Biol* 2004;4:11.
(79) He GH, Meng R, Gao H, Guo B, Gao G, Newman M, et al. Simple sequence repeat markers for botanical varieties of cultivated peanut *Arachis hypogaea* L. *Euphytica* 2005;142(1-2):131–136.
(80) Mace ES, Phong DT, Upadhyaya HD, Chandra S, Crouch JH. SSR analysis of cultivated groundnut *Arachis hypogaea* L. germplasm resistant to rust and late leaf spot diseases. *Euphytica* 2006;152:317–330.
(81) Mace ES, Varshney RK, Mahalakshmi V, Seetha K, Gafoor A, Leeladevi Y, et al. *In silico* development of simple sequence repeat markers within the aeshynomenoid/dalbergoid and genistoid clades of the Leguminosae family and their transferability to *Arachis hypogaea*, groundnut. *Plant Sci* 2007;174:51–60.
(82) Jiang H, Liao B, Ren X, Lei Y, Mace E, Fu T, et al. Comparative assessment of genetic diversity of peanut (*Arachis hypogaea* L.) genotypes with various levels of resistance to bacterial wilt through SSR and AFLP analyses. *J Genet Genom* 2007;34(6), 544–554.
(83) Barkley NA, Dean R, Pittman RN, Wang ML, Holbrook CC, Pederson GA. Genetic diversity of cultivated and wild-type peanuts evaluated with M13-tailed SSR markers and sequencing. *Genetics Res* 2007;89:93–106
(84) Liang XQ, Hong YB, Chen XP, Liu HY, Zhou GY, Li SX, et al. Characterization and application of EST-SSRs in peanut (*Arachis hypogaea* L.). *Acta Agron Sin* 2009;35(2):246–254.
(85) Jiang HF, Ren XP, Zhang XJ, Huang JQ, Wang SY, Yong L, et al. Genetic diversity of peanut mini core collection detected by SSR markers. *Chin J Oil Crop Sci* 2010;32:472–478.
(86) Kottapalli P, Upadhyaya HD, Kottapalli KR, Payton P, Dwivedi S, Burow M, et al. Population structure and diversity in valencia peanut germplasm collection. *Crop Sci* 2011;51:1089–1100.
(87) Liu N, Li H, Weigang C, Bei W, Pandey MK, Luo H, et al. Dissection of the genetic basis of oil content in Chinese peanut cultivars through association mapping. *BMC Genetics* 2020;21:60.
(88) Robledo G, Seijo G. Species relationships among the wild B genome of *Arachis* species (section *Arachis*) based on FISH mapping of rDNA loci and heterochromatin detection: A new proposal for genome arrangement. *Theor Appl Genet* 2010;121(6), 1033–1046.
(89) Robledo G, Lavia GI, Seijo G. Species relations among wild *Arachis* species with the A genome as revealed by FISH mapping of rDNA loci and heterochromatin detection. *Theor Appl Genet* 2009;118(7):1295–1307.
(90) Seijo G, Lavia GI, Fernández A, Krapovickas A, Ducasse DA, Bertioli DJ, et al. Genomic relationships between the cultivated peanut (*Arachis hypogaea*, Leguminosae) and its close relatives revealed by double GISH. *Am J Bot* 2007;94(12):1963–1971.
(91) Nielen S, Campos-Fonseca F, Leal-Bertioli S, Guimarães P, Seijo G, Town C et al. FIDEL-a retrovirus-like retrotransposon and its distinct evolutionary histories in the A-and B-genome components of cultivated peanut. *Chrom Res* 2010;18(2):227–246.

(92) Upadhyaya HD, Bramel PJ, Ortiz R, Singh S. Developing a mini core of peanut for utilization of genetic resources. *Crop Sci* 2002;42(6):2150–2156.
(93) Otyama PI, Kulkarni R, Chamberlin K, Ozias-Akins P, Chu Y, Lincoln LM, et al. Genotypic characterization of the U.S. Peanut core collection. *G3 (Bethesda)* 2020;10(11):4013–4026.
(94) Pasupuleti J, Nigam SN. Phenotyping for groundnut (*Arachis hypogaea* L.) improvement. In: Panguluri SK, Kumar AA (eds) *Phenotyping for Plant Breeding*. Springer, New York, 2013, pp. 129–167.
(95) Bera SK, Kamdar JH, Kasundra SV, Darvhankar M, Jasani MD, Ajay BC, et al. An efficient phenotyping technique for wild *Arachis* species and groundnut breeding lines resistant to stem rot disease under field conditions. *Indian Phytopathol* 2016;69(4s):646–648.

7 Allelic Variation and Its Functional Characterization in Peanut

Ramesh S. Bhat, Kenta Shirasawa, Sunil S. Gangurde, and M. V. C. Gowda

7.1 INTRODUCTION

Peanut (*Arachis hypogaea* L. 2n=4x=40) is an important oilseed, legume food, and fodder crop in more than 100 countries. It is grown globally on an area of 31.6 million hectares, with a production of 53.6 million tons (https://www.fao.org/faostat/en/#data/QCL/visualize) and a productivity of 1,699 kg/ha. Globally, over half of the peanut produce goes for oil extraction, while the remaining is consumed as raw and processed food. Peanut has earned the name "poor man's almond" because of its rich nutrient contents in terms of oil, proteins, fibers, polyphenols, antioxidants, vitamins, and minerals. In addition, peanut is an excellent source of compounds like resveratrol, phenolic acids, flavonoids and phytosterols, co-enzyme Q10, and amino acids (all 20, with the highest content of arginine). With these nutrient profiles, peanuts are being considered as a functional food (Arya et al., 2016). Numerous studies have identified the value-added attributes of peanut skins and/or peanut skin extracts (Toomer, 2020) as an antioxidant, functional food ingredient, animal feed ingredient, and antimicrobial agent (Parmar et al., 2022).

Enormous efforts are being made to improve the peanut for its productivity, quality, and tolerance to biotic and abiotic stresses (Pandey et al., 2020). Genetic control of most of the yield, quality, and disease resistance-related traits has been investigated in the post-genomic era (Gangurde et al., 2019). Beneficial alleles have been discovered by linkage and association mapping for multiple traits such as seed and pod weights (Gangurde et al., 2020), seed dormancy (Kumar et al., 2020, Bomireddy et al., 2022), and stem rot resistance (Dodia et al., 2019). Progress in peanut breeding in terms of developing superior varieties was made possible by the accumulation of beneficial alleles from cultivated or wild genetic resources (see Bohra et al., 2021).

The peanut gene pool being vast and diverse, significant scope lies for studying the superior alleles that were left behind during evolution and domestication. This untapped genetic variation existing in wild relatives and landraces of peanut could be exploited for the development of agronomically superior cultivars. For instance, the resistance alleles for late leaf spot and rust resistance among most of the Indian and USA peanut germplasm were derived from GPBD 4 (Gowda et al., 2002), VG 9514 (Mondal et al., 2021), and GP NC WS 16 (SPT06-06) (Tallury et al., 2014), which in turn resulted from crosses involving the wild species *Arachis cardenasii*.

The method of studying allelic forms has undergone a paradigm shift from classical genetic screening to marker- and genome-sequence-based methods. Peanut genomics, population genomics (see Bhat et al., 2021a), epigenomics, transcriptomics, and proteomics are contributing to the identification of allele diversity in peanuts. With the surge in peanut genome sequencing projects and the comparison of multiple related individuals within and between the species of genus *Arachis*, now allele mining could be taken up with ease by understanding the structural polymorphisms. Efforts are also in progress to study pangenomes (see Zanini et al., 2022) and identify the haplotypes for

 DOI: 10.1201/9781003385035-7

efficient crop breeding (see Bhat et al., 2021b). In this regard, the current chapter reviews the latest developments in the area of allele identification, characterizing their structural and functional variations, their genetic and epigenetic roles in their evolution, and their influence on the phenotypic traits of peanuts.

7.2 EVOLUTION OF NEW ALLELES

British geneticists William Bateson and Edith Rebecca Saunders coined the term "allelomorph", which was later shortened to "allele". Alleles are the different forms of a gene that determine alternate traits or characteristics. Therefore, an allele is one pair (or more) of different forms of a given gene. Functional allelism is determined experimentally by the *cis-trans* or complementation test, based on which Seymour Benzer introduced the term *cistron* to refer to the unit of function (gene) (Benzer, 1959).

Any mutation occurring within a gene or regulatory sequence can produce a new allele. Because of the degeneracy of the genetic code, the mutant allele coding for the unaltered protein and the phenotype are called as isoalleles. Mutant alleles coding for altered gene product (protein) may show a modified phenotype, and the mutation resulting in the loss of gene product in an essential gene may be lethal.

Mutations may be either recessive or dominant. In diploids (or polyploids), the recessive mutations can be recognized only in their homozygous state, whereas the dominant mutations can be identified both in the heterozygous state and their homozygous state. Most of the mutations that have been identified and studied by geneticists are recessive in nature. It is also important to note that a mutation can continue to exist in heterozygous conditions (Branch et al., 2020). Mutations can occur in any cell cycle state of somatic and germinal cells (Schoen, Schultz, 2019). Thus, the effect of mutation and the subsequent phenotypic change depends on the nature of dominance, the cell type, and the state of the cell cycle.

Mutations involving single base pairs are referred to as point mutations, which can be found due to substitutions (change of one base pair to another), duplications, and deletions. Mutations resulting from tautomeric shifts in the DNA involve the replacement of a purine in one strand of DNA with another purine and/or the replacement of a pyrimidine in the complementary strand with another pyrimidine, which collectively are called transitions. Base-pair substitutions involving the replacement of a purine with a pyrimidine and vice versa are called transversions. Four different transitions and eight different transversions are possible in DNA. Additions and deletions of one or a few base pairs are collectively called as frameshift mutations since they alter the reading frame of a gene distal to the site of mutation.

Based on the occurrence, the mutations can be classified as spontaneous or natural mutations (those occurring without a known cause) and induced (resulting from exposure to physical or chemical mutagens). Operationally, it is difficult to discern spontaneous mutations from induced mutations. The effect of mutation on the phenotype might range from minor alterations, which can be detected only through special genetic or biochemical techniques, to gross modifications of morphology that are lethal.

Apart from mutations, the evolution of new alleles through intra- and intergenic recombination is quite frequent in many resistance genes that are arranged in multiple-copy tandem arrays (Parniske et al., 1997, Song et al., 1997). In contrast, the single-copy gene encoding the S RNase in the gametophytic SI system is highly polymorphic, suggesting that point mutations rather than recombination are likely to be involved in generating new S alleles (Coleman and Kao, 1992, Matton et al., 1995).

Plant evolution and domestication are also driven by polyploidy. This is exemplified by the fact that, though several diploid species in the genus *Arachis* were traditionally cultivated for their seeds, only the allotetraploid peanut has evolved into a domesticated crop. This was also demonstrated using the induced allotetraploid (neopolyploid) lineages obtained from crosses between the two progenitor species of peanut, *Arachis ipaensis* and *Arachis duranensis* (Leal-Bertioli et al., 2021). The neopolyploid lineages showed more variability for a few traits (fertility and seed morphology) than their progenitors and cultivated peanuts. The neopolyploid lineages also showed somatic and gametic genome instability and extensive homoeologous recombination. Studies on wheat allotetraploids showed that instability caused by homoeologous exchange was nonrandomly distributed along the chromosomes. Similar observations were also made by Chu et al. (2021) among the neopolyploid lineages of *Arachis ipaënsis*×*Arachis correntina*. The frequency of homoeologous exchanges increased with higher gene density, higher DNA similarity, a lower DNA methylation level, and higher chromatin accessibility (Zhang et al., 2020). The immediate generations after polyploidization are prone to high genetic instability, but the later generations become stabilized by the suppression of homeologous exchange and/or the selection of lines with lower levels of chromosome rearrangements. Therefore, it is likely that more unstable lineages tend to perish. Hence, it was concluded that the origin and domestication of the allotetraploid peanut involved two genetic bottlenecks: rare hybridization and polyploidization events, followed by sexual reproductive isolation from its wild diploid relatives and the survival of the very few lineages that had stronger mechanisms for limiting genomic instability. Using a large number of accessions (~249) and specific length amplified fragment (SLAF) markers, regions with high-density crossover events were identified. These regions showed a negative relationship with marker haplotypes. These regions carried those genes, mainly responding to environmental stimuli, suggesting that recombination plays a crucial role in peanut adaptation to changing environments (Wang et al., 2020a). It was also observed that intragenic homologous recombination preferentially takes place upstream and downstream of coding regions and is highly enriched within gene bodies, giving rise to novel recombinant transcripts, which in turn are predicted to generate new protein fusion variants (Zhang et al., 2020). Overall, the study showed that intragenic homologous recombination provides a mechanism for generating new chimeric transcripts containing the promoter of one parent and the coding region of another parent, as well as a mechanism for generating new protein variants due to in-frame fusion of homoeologous genes of parental species. This reshuffling within genes could give rise to new patterns of gene expression or to new protein functions. Through synteny-based comparisons of two diploid and three tetraploid peanut genomes, Shen et al. (2021) identified a high level of duplicated genes that mainly governed growth and development. These regions might contribute to the evolution of new alleles through homologous recombination. Bertioli et al. (2021) also found "invasion" of A subgenome alleles onto B subgenome chromosome", which indicates the genetic erosion of the ancestral A subgenome of cultivated peanuts following polyploidy by invasion of B subgenome alleles into the A subgenome. Regions exhibiting tetrasomic recombination were identified, and they were shown to have undergone positive selection (Clevenger et al., 2017). A substantial fraction of alleles have arisen through subgenome exchanges, and these exchanges appear to be ongoing, as there are numerous differences between accessions in the subgenome allele status at a given locus (Otyama et al., 2020). All these efforts demonstrate the evolution of new alleles through a few known mechanisms in peanut.

7.3 PEANUT GENE POOL

The genus *Arachis* of the family Fabaceae consists of nine sections (*Arachis, Erectoides, Heteranthae, Caulorrhizae, Rhizomatosae, Extranervosae, Triseminatae, Procumbentes*, and *Trierectoides*) based on morphology, cross-compatibility, viability of the hybrids, geographic distribution, and cytogenetics. Across these nine sections, 81 species have been described, of which 69 were described by Krapovickas and Gregory (1994), 11 were described by Valls and Simpson (2005), and one was described by Valls et al. (2013) and Stalker (2017). Of the 81 species, 72 are diploids ($2n=2x=20$), five are tetraploids ($2n=2x=40$), and four are aneuploids ($2n=2x=18$) (Friend et al., 2010). Two tetraploids (*A. hypogaea* and *A. monticola*) are in the section *Arachis*, and the remaining three (*A. glabrata, A. pseudovillosa*, and *A. nitida*) are in the section *Rhizomatosae*

(Krapovickas and Gregory, 1994, Valls, Simpson, 2005). Of the four aneuploids, three (*A. decora*, *A. palustris*, and *A. praecox*) belong to the section *Arachis* (Peñaloza and Valls, 1997, Lavia, 1998) and one (*A. porphyrocalyx*) belongs to the section *Erectoides* (Peñaloza and Valls, 2005).

The widely cultivated peanut or groundnut, *Arachis hypogaea,* is among the 31 species belonging to the section *Arachis*, along with the forage species *Arachis pintoi*, *Arachis glabrata*, and *A. sylvestris* (Valls and Simpson, 1994) and the ornamental species *Arachis repens* (Stalker and Simpson, 1995). The occurrence of diploid parental and wild allotetraploid species (*A. monticola*) indicates that the eastern foothills of the Andes in the region of northern Argentina and southern Bolivia could be the area of origin of the cultivated peanut *A. hypogaea* (Krapovickas and Rigoni, 1957, Kochert et al., 1996, Ravi et al., 2010). Allotetraploid *Arachis* species was then put under selection for domestication (Kochert et al., 1996), and the domesticated peanut dispersed to South and Central America, where both natural and artificial selection produced many landraces of domesticated peanut.

Several efforts were made to identify the probable ancestors of cultivated peanuts. Now there is a consensus among the researchers that the cultivated peanut has originated from a hybridization between the diploid species, *A. duranensis* (female parent with A genome under the section *Arachis*) and *A. ipaensis* (male parent with B genome under the section *Arachis*) (Kochert et al., 1996), followed by a single allotetraploidization event involving either chromosome doubling in diploid F_1 (Favero et al., 2006) or unreduced gametes in F_1 (Harlan and deWet, 1975). Recent findings indicate that the cultivated peanut originated about 0.40 million years ago (Zhuang et al., 2019, Zhuang et al., 2020), though there is another view that it originated about 10,000 years ago (Bertioli et al., 2019, 2020).

Based on several morphological differences, two subspecies (*A. hypogaea* ssp. *hypogaea* and *A. hypogaea* ssp. *fastigiata*) have been recognized in the cultivated peanut. *A. hypogaea* ssp. *hypogaea* is characterized by alternate branching, absence of flowers on the main axis, long duration (120–160 days), and presence of seed dormancy, while *A. hypogaea* ssp. *fastigiata* is recognized by sequential branching, presence of flowers on the main axis, short duration (85–130 days), and lack of seed dormancy. Based on the morphological differences, the two subspecies are subdivided into botanical varieties. *A. hypogaea* ssp. *hypogaea* is divided into var. *hypogaea* (Virginia bunch/runner) and var. *hirsuta* (Peruvian runner), whereas *A. hypogaea* ssp. *fastigiata* is classified into var. *fastigiata* (Valencia), var. *vulgaris* (Spanish bunch), var. *aequatoriana*, and var. *peruviana* (Krapovickas, 1969, 1973). *A. hypogaea* ssp. *hypogaea* is thought to have originated by a mutation from within *A. hypogaea* ssp. *fastigiata* (Singh, 1988).

The germplasm collections maintained in India (15,445 accessions), the USA (9,310 accessions), and China (7,837 accessions) make up the germplasm repository. These germplasm collections have been characterized for important agronomic traits and have different diversity panels for use in breeding. ICRISAT, the USA, and China developed core collections with 1,704, 831, and 576 accessions, respectively, from the total germplasm collection, which were further reduced to minicore collections of 184, 112, and 298 genotypes, respectively. These accessions with novel alleles, heritability, combining ability, and trait correlations have significantly contributed to the better understanding of the peanut population genomics and overall genetics needed to develop appropriate breeding strategies for target traits. Population structure, linkage disequilibrium, and marker-trait associations have been studied in these minicore accessions. Several improved lines and sources of variability have been identified or developed for various economically important traits through conventional breeding (see Desmae et al., 2019). These genetic resources have helped the advancement of peanut genomics, including genome sequencing, marker development, genetic and trait mapping, discovery of genes/variants for traits of interest, and integration of marker-assisted breeding for selected traits. These accessions are now contributing to allele mining (Lv et al., 2020, Wan et al., 2020, Peng et al., 2021).

7.4 SEQUENCING

7.4.1 Genome Sequencing

As a first step toward characterizing the genome of cultivated peanut, the genomes of the two diploid ancestors (*A. duranensis* V14167 and *A. ipaensis* K30076) of cultivated peanut were sequenced and analyzed (Bertioli et al., 2016) to overcome the challenge of assembling chromosomal pseudomolecules. Both of these accessions were collected from the most likely geographic region of origin for the cultivated peanut. In the same year, Chen et al. (2016a) sequenced *A. duranensis* (accession PI475845 from Bolivia) as well as four synthetic tetraploids and their six diploid parents [two A genomes and four B genomes, including the suspected B genome progenitor, *Arachis ipaensis*] to gain insight into peanut evolution. Based on the draft genome of *A. duranensis*, gene models with 50,324 protein-coding genes were proposed. Also, Lu et al. (2018) sequenced *A. ipaensis* and recorded ~1.39 Gb genome with 39,704 predicted protein-encoding genes.

The first reference quality assembly of the *A. monticola* (PI263393) genome was developed with a genome size of ~2.62 Gb (Yin et al., 2018). The efficiency of the current state of the strategy for *de novo* assembly of the highly complex allotetraploid species based on whole-genome shotgun sequencing, single-molecule real-time sequencing, high-throughput chromosome conformation capture technology, and BioNano optical genome maps was demonstrated. Subsequently, Yin et al. (2020) re-sequenced 17 wild diploids from AA, BB, EE, KK, and CC groups and 30 tetraploids and compared the previously sequenced genome of *A. monticola* (Yin et al., 2018).

During 2019, two reference genomes were reported: one for *A. hypogaea* ssp. *fastigiata* and the other for *A. hypogaea* ssp. *hypogaea* of the cultivated tetraploid. The International Peanut Genomic Research Initiative (IPGI)-led initiative (Bertioli et al., 2019) completed the sequencing of Tifrunner (PI644011 with registration number CV-93) (Holbrook and Culbreath, 2007), a runner-type belonging to *Arachis hypogaea* ssp. *Hypogaea*, by deploying several modern sequencing and assembly technologies such as PacBio and Hi-C data/technology. A genome of ~2.56 Gb with 20 pseudomolecules and 66,469 predicted genes was reported. Similar advanced technologies were deployed by two independent efforts in China, leading to the development of high-quality reference genome assemblies for "Shitouqi" (Zhuang et al., 2019) and "Fuhuasheng" (Chen et al., 2019a) both belonging to *A. hypogaea* ssp. *fastigiata.* The variety "Shitouqi" (zh.h0235) is a well-known Chinese landrace and breeding parent belonging to *A. hypogaea*ssp. *fastigiata* and botanical type *vulgaris* (agronomic type Spanish), while "Fuhuasheng" is a landrace from North China. For Shitouqi, a genome of ~2.54 Gb with 83,709 predicted genes across 20 pseudomolecules was reported, and the heterozygosity was very low (1/6,537 nucleotides), while a genome of ~2.55 Gb with 83,087 predicted genes across 20 pseudomolecules was reported for Fuhuasheng.

The peanut chromosomes are named from Arahy.01 to Arahy.20, where the first 10 (Arahy.1–Arahy.10) represent the A subgenome while the latter 10 (Arahy.11–Arahy.20) represent the B subgenome. As it has been reported in other recent polyploids (Chalhoub et al., 2014, Zhang et al., 2015), the number of homeologous gene pairs with expression biased toward the A subgenome was not significantly different from the number biased toward the B subgenome. No significant bias was observed for the expression of the genes between A and B subgenomes. The average length of the transcripts was 1,589.5 bp with 6.8 exons, which translate the proteins into 403 amino acids. These sizes were comparable to those of other legumes but longer than those of A and B genomes. Approximately 76.6% of predicted genes were functionally annotated. In total, 39,127 non-coding RNAs (ncRNAs) including 4,723 transfer RNAs (tRNAs), 3,107 ribosomal RNAs (rRNAs), 480 microRNAs (miRNAs), and 30,817 small nuclear RNAs have been identified. Most of the small RNA coding regions corresponded to repeat-rich regions. Among the small RNAs corresponding to gene-rich regions, those originating from the B subgenome were more common than those of the A subgenome.

Overall, 77.65% (1.97 Gb) of the genome consisted of repeat elements, of which 64.74% (1.67 Gb) comprised of retrotransposons and 4.49% (114 Mb) comprised of DNA transposons. Long terminal repeats (LTRs) are highly abundant in pericentromeric regions, whereas DNA transposons are more frequent in euchromatic arms. Most of the transposable elements (like Gypsy and unclassified LTRs) expanded after tetraploidization, especially in the A subgenome, while those of the B subgenome were derived from the progenitor B genome. However, this observation was in contrast to that of Bertioli et al. (2019), who could not find any major transpositions or insertions after tetraploidization in Tifrunner.

Five chromosomes (Arahy.01, Arahy.05, Arahy.06, Arahy.07, and Arahy.11) showed inversions. Genetic exchange between ancestral genomes could be noticed toward the ends of collinear pairs of homeologous chromosomes. Two patterns of homeologous recombination were evident; the first involved the transfer of chromosome segments between distal collinear regions of chromosomes, mostly resulting in tetrasomic genome structures, while the second involved the transfer of alleles dispersed in the chromosomes. However, the latter was strongly biased, where the transfer of alleles from the B subgenome to the A subgenome was more frequent.

The chloroplast genome of *A. hypogaea* and a chloroplastic plasmid were inherited from *A. duranensis*, the female parent of peanut (see Paterson et al., 2004). Sequencing of the chloroplast genome of *A. hypogaea* (Wang et al., 2018) reported a length of 156,354 bp (var. *hypogaea*), 156,878 bp (var. *hirsuta*), 156,718 bp (var. *fastigiata*), and 156,399 bp (var. *vulgaris*). Comparative analysis of the sequences revealed that their gene content, gene order, and GC content were highly conserved since only 46 single nucleotide polymorphisms (SNPs) and 26 insertions/deletions were identified. Most of these variations were found in the non-coding sequences, especially the trnI-GAU intronic region. Many of these features contribute to allelic variations; hence, they are important components in allele mining.

7.4.2 Epigenome Sequencing

Low DNA sequence polymorphism despite enormous phenotypic variations in peanuts indicates the possible role of epigenetic variations. Detection of epigenetic marks (along with associated expression) provides high power to identify genomic regions associated with traits or evolutionary processes such as fitness, phenotypes, and selection. In peanuts, Bertioli et al. (2019) observed lower methylation in the transcribed regions and a characteristic decline in methylation at transcription start and end sites, like in most plant genomes. Genome-wide methylation per cytosine content was higher in pericentromeric regions than the chromosome arms. Methylation was lower in the A subgenome than the B subgenome, with 76.0% and 80.5% methylation at CG sites, 61.7% and 65.1% methylation at CHG sites (where H is A, T, or C), and 5.14% and 5.51% methylation at CHH sites, respectively. All of this information provides a good resource for population epigenomics studies.

A genome-wide DNA methylation pattern and its influence on gene expression were reported across 11 diverse peanut genotypes (Bhat et al., 2019a). Bisulfite sequencing of 11 genotypes after 21 days of sowing (DAS) produced an average of 127,852,977 reads with a mapping range of 1–1,658 at each DNA-methylated site (Bhat et al., 2019a). A total of 75,973,928 sites across the genotypes belonged to the category where all the mapped reads (100%) showed cytosine methylation, while 101,137,805 sites belonged to the category where at least 50% of the mapped reads showed cytosine methylation, and 126,487,183 sites showed less than 50% of the mapped reads showing cytosine methylation. The B subgenome exhibited higher DNA methylation sites (46,294,063) than the A subgenome (30,415,166) across the genotypes. A total of 177,574 sites were found on the scaffolds. CHG (where H is A, C, or T) regions showed the highest methylation sites (30,537,376), followed by CpG (30,356,066) and CHH (15,993,361) regions across the genotypes. This observation is in line with the previous reports (Feng et al., 2010, Zemach et al., 2010) that DNA methylation in plants is found both in CpG and non-CpG (CHG and CHH, where H is A, C, or T) contexts, in contrast to mammals, where DNA methylation occurs predominantly at CpG dinucleotides.

Genotypes JL 24 and TMV 2 showed the highest (82,137,767) and the lowest methylation sites (69,044,110), respectively. Such a natural epigenetic variation was also observed among the different ecotypes of *Arabidopsis* (Vaughn et al., 2007). Bhat et al. (2019a) found that a total of 5,379,101 sites were conserved across the 11 genotypes, while the unique sites ranged from 6,575,363 (TMV 2) to 9,190,780 (JL 24). On average, inter-genic regions (70,464,637 sites) were more prone to DNA methylation than genic regions including 2 kb upstream and 2 kb downstream regions (6,422,166 sites). Within the genic regions, the introns (1,590,263) showed a greater number of DNA methylation sites than the exonic regions (971,274). The 2 kb upstream and 2 kb downstream regions had 3,860,629 DNA methylation sites, indicating a higher proportion of DNA methylation in the upstream and downstream regions than in the gene body region. The distribution of DNA methylation within the genome, especially in the promoter and gene body regions, is very important because it influences gene expression (Suzuki, Bird, 2008).

The number of genes showing at least one methylated site ranged from 51,179 (ICGV 86855) to 55,497 (ICGV 99005), indicating the variation in DNA methylation at the genic regions between the individuals (Bhat et al., 2019a). Of them, *Arahy.0DU9MH*, a 342,359 bp long gene on chromosome 11, showed the highest number of methylated sites, which ranged from 11,488 (ICGV 86855) to 14,026 (JL 24). Within *Arahy.0DU9MH*, the promoter region had 131 methylated sites, while the gene body (142 in exons and 12,573 in introns) had 12,715 sites. The expression of ~53,740 genes varied widely among the 11 genotypes. *Arahy.0DU9MH* with the highest DNA methylation sites did not show any expression at 21 DAS in the leaves of the 11 genotypes. Fifty genes with a wide range of fragments per kilobase of transcript per million mapped reads (FPKM) across the genotypes were selected and checked for DNA methylation to check their influence. Many of these genes showed a negative association between the number of DNA methylation sites and FPKM among the genotypes. *Arahy.FHUH7B* on chromosome 10 showing the highest FPKM of 54,951 had a maximum of 102 DNA methylation sites.

Fourteen C5-MTase coding genes and ten DNA demethylase coding genes identified in the diploid peanut earlier (Wang et al., 2016) were analyzed for DNA methylation and expression. A considerable variation was observed for methylation across the genes; however, not much variation was observed for methylation across the genotypes.

An attempt was made to identify the differentially DNA-methylated sites between the foliar disease-resistant (GPBD 4, VG 9514, ICGV 86855, ICGV 99005, and ICGV 86699) and susceptible (TAG 24, TMV 2, and JL 24) genotypes (Bhat et al., 2019a). Foliar disease-resistant genotypes showed significant differential DNA methylation at 766 sites, corresponding to 25 genes. Of them, two genes (*Arahy.1XYC2X* on chromosome 01 and *Arahy.00Z2SH* on chromosome 17) coding for senescence-associated protein showed differential expression, where the resistant genotypes recorded higher FPKM at their epialleles.

There is growing evidence indicating that DNA methylation influences plant phenotypes, including development and stress responses (see Kim et al., 2021). Several naturally occurring and artificially induced mutants with altered DNA methylation (i.e. epialleles) have been identified to be associated with diverse phenotypic changes in plants. The induction of methylome changes has been performed in plants by several methods, including chemical treatment, tissue culture, exogenous RNAi mediated by virus-induced gene silencing and grafting, and CRISPR–Cas9 have shown induction of methylome changes. Studies on epigenetic Recombinant Inbred Lines (epiRILs) obtained by an initial cross of isogenic parents with different DNA methylation profiles have revealed the role and significance of epigenetic alteration in identical or almost identical genetic mechanisms involved in epiallele formation, stability, and consequences of DNA methylation changes in genomic stability, transposable element activation, and phenotypic traits have been investigated (see Kim et al., 2021).

7.4.3 Transcriptome Sequencing

All RNA transcripts (transcriptomes) produced by the genome are studied (transcriptomics) for gene annotation, gene discovery, and marker development. However, population transcriptomics uses transcriptome-wide data to study variation in gene expression within and among populations to understand mechanisms underlying allelic variations, plasticity in gene expression, response to environmental change, and evolutionary change. Numerous efforts have been made in peanut to collect and study the transcriptome (see Chen and Liang, 2014). Though the initial efforts employed microarrays (Chen et al., 2012, Zhu et al., 2014), the current studies embark on RNA-Sequencing. *Arachis hypogaea* gene expression atlas (AhGEA) has been reported for the widely cultivated *A. hypogaea* ssp. *fastigiata* based on RNA-Seq data using 20 diverse tissues across five key developmental stages from an early-maturing, high-yielding, drought-tolerant variety, ICGV 91114 (Sinha et al., 2020). AhGEA sheds light on the complex regulatory networks of gravitropism and photomorphogenesis, seed development, allergens, and oil biosynthesis.

Transcriptomic analysis conducted so far could identify the differentially expressed genes for growth and development, response to biotic and abiotic stresses, and quality (see Pandey et al., 2020). Transcriptomes have been reported for seed dormancy release and germination (Xu et al., 2020), plant height (Guo et al., 2020), pod development (Yang et al., 2017), nodulation (Peng et al., 2017, Peng et al., 2020), seed coat cracking and pigmentation (Zhang et al., 2016, Huang et al., 2019, Xia et al., 2020), light-mediated embryo and pod development (Zhang et al., 2016), Ca^{2+} regulation in pod development (Yang et al., 2017), regulation of plant triacylglycerol metabolism (Zheng et al., 2017), calcium- and hormone-related responses during different stages of pod development (Hao et al., 2017), seed development (Zhang et al., 2012), seed abortion (Zhu et al., 2014), geocarpy (Chen et al., 2016b), early pod development in darkness (Xia et al., 2013), and oil accumulation and profile (Yin et al., 2013, Huai et al., 2020).

Similarly, the genes involved in cold tolerance (Jiang et al., 2020), drought tolerance (Guimarães et al., 2012, Bhogireddy et al., 2020, Zhao et al., 2020a), nematode resistance (Guimaraes et al., 2015), response to *Ralstonia solanacearum* (Chen et al., 2014), response of roots to the beneficial and pathogenic fungi (Hao et al., 2017) including late leaf spot (LLS) (Han et al., 2017, Gangurde et al., 2021) and leaf rust (Rathod et al., 2020), response to low temperature during germination among high oleate types (Chen et al., 2019b), and embryo abortion under calcium deficiency (Chen et al., 2019c) have been identified. Bosamia et al. (2020) used RNA-Seq to unravel the mechanisms of resistance to stem rot caused by *Sclerotium rolfsii* using a resistant (NRCG-CS85) and susceptible (TG37A) genotype. Several annotated transcripts are known to have a role in channeling downstream pathogen perception, like receptor-like kinases, jasmonic acid pathway enzymes, and transcription factors (TFs), including WRKY, zinc finger protein, and C2-H2 zinc finger, showed higher expression in resistant genotypes upon infection.

Gene networks regulating cadmium uptake and translocation in roots under iron deficiency (Huang et al., 2019) and differential cadmium transport and retention in roots (Yu et al., 2018) have also been identified. Transcriptome analyses have also been used for assessing genetic variability (Chopra et al., 2014, Chopra et al., 2015, Chopra et al., 2016), mining tissue-specific contigs for promoter cloning (Geng et al., 2014), and understanding alternative splicing (Ruan et al., 2018). Key genes involved in the regulation of symbiotic pathways induced by *Metarhizium anisopliae* in roots (Wang et al., 2020b) and the members of the monosaccharide transporter (MST) gene family have been identified by transcriptional profiling (Wan et al., 2020). A transcriptomic study was conducted to check the variation in the expression of genome-wide genes due to variation in DNA methylation across 11 genotypes (Bhat et al., 2019a).

In most of these studies, alternatively spiced transcripts were observed. Alernative splicing can generate more than one transcript from a single coding sequence, thereby providing post-transcriptional variation. Alternative splicing has been shown to regulate growth, development, signal transduction, flowering, circadian clock function, and the response to various environmental stresses, thus being associated with speciation (see Ruan et al., 2018).

7.4.4 Proteome

Regulations during transcription, translation, and post-translational modifications are widely spread; therefore, the mRNA content does not necessarily correspond with the protein content (Dhingra et al., 2005). Proteomics is constantly advancing to bridge the gap between DNA sequence-transcriptome-phenotype in the diverse and dynamic stages of growth and development. Since proteins influence important phenotypes and are the products of genes and epigenetic or posttranslational mechanisms, population proteomics has the potential to provide key insights into functional and metapopulation ecology, adaptation, and acclimation processes under various climate and environmental conditions. Population proteomics approaches also help identify genetic loci underlying the risk of disease and clinical biomarkers for many human disease conditions. In peanuts, the low DNA polymorphism coupled with high morphological variation might involve differences in the proteomics. Katam et al. (2014) reviewed the progress made on proteomics in peanuts, especially peanut allergens and adaptive responses to various stresses. Proteomic analyses have identified the specific proteins produced in response to high-oleic acid content in seed (Liu et al., 2020), cadmium detoxification and translocation (Yu et al., 2019), allergen production (Mamone et al., 2019), response to water stress (Kottapalli et al., 2009, Kottapalli et al., 2013, Katam et al., 2016), gynophore development (Sun et al., 2013, Zhao et al., 2015), development of aerial and subterranean pods (Zhu et al., 2013), response to salinity (Jain et al., 2006), and polyphenol content (Muralidharan et al., 2020). The advances in stress proteomics in recent years have enabled plant biologists to investigate the molecular events associated with stress adaptation in legumes, including peanuts (Rathi et al., 2016).

Post-translational modifications, which are also important for gene regulation, can be effectively studied through proteomics. Effects on protein function, cell growth and differentiation, metabolism, and so on are generally associated with modifications such as lysine acetylation, succinylation, crotonylation, malonylation, 2-hydroxyisobutyrylation, and β-hydroxybutyrylation. Post-translational modification is also one of the important targets for epigenetic variation. Recently, a comprehensive proteomics analysis of lysine crotonylation in peanut leaves was reported (Xu et al., 2021).

7.5 ALLELE MINING: APPROACHES, METHODS, AND CURRENT STATUS

Allele mining refers to the systematic identification and characterization of new allelic variants for one or more genes in large and diverse collections. With the growing evidence suggesting that searching for new allelic variants in the gene pool is a powerful strategy to increase the diversity available for plant breeding, efforts on allele mining are now being undertaken (Kumar et al., 2010, Ogawa et al., 2018, Guerra et al., 2022). Though the initial studies of allele mining focused mainly on the identification of SNPs/InDels at coding sequences or exons of the gene, the current efforts of allele mining also include nucleotide variations in non-coding regions (5′ UTR), including promoters, introns, and 3′ UTR, considering their role in gene-regulating gene expression (see Kumar et al., 2010). Promoter mining aims at identifying the nucleotide variation in transcription factor binding sites.

In peanuts, a series of reciprocal crosses were made between cultivars from different regions and HG1, a cultivar from India, to describe three loci (*Hb1*, *Hb2*, and *Hb5*) affecting plant growth habits (Ashri, 1976). Also, the allelic nature and intra- and inter-allelic relationships were recognized. Probably, the next efforts at identifying the allelic forms in the cultivated peanut came from studies on isozymes (Lu, Pickersgill, 1993, Stalker et al., 1994), random amplified polymorphic DNA (RAPD), and restriction fragment length polymorphism (RFLP) approaches (Kochert et al., 1991, Halward et al., 1992, Paik-Ro et al., 1992). The detection of DNA polymorphism in cultivated peanuts was reported for the first time by He and Prakash (1997) using DNA amplification fingerprinting (DAF) and amplified fragment length polymorphism (AFLP) approaches. Simple sequence repeat (SSR) markers (Hopkins et al., 1999, Palmieri et al., 2002, He et al., 2003, Ferguson et al., 2004, Krishna et al., 2004, Moretzsohn et al., 2004, 2005, Palmieri et al., 2005, Hong et al., 2008) were developed and tested in peanuts for detecting allelic variations. Subsequently, a large number of SSR markers were developed and/or used in peanuts (Zhang et al., 2012, Huang et al., 2016, Peng et al., 2016, Zhou et al., 2016). Cleaved amplified polymorphic sequence (CAPS) (Chu et al., 2009) and Diversity Array Technology (DArT) marker platforms were also developed (Kilian, 2008). Currently, the advent of next-generation sequencing and genotyping technologies has enabled the detection of SNPs, which have emerged as the marker of choice in crop breeding (Varshney et al., 2009). SNPs have been detected in diploids (Alves et al., 2008, Nagy et al., 2012) as well as cultivated tetraploids (Wang et al., 2013, Zhou et al., 2014, Hong et al., 2015, Shirasawa et al., 2016, Clevenger et al., 2017).

A transposable element-based marker system with a higher polymorphic rate (up to 22%) was developed (see Bhat et al., 2019b) as an improvement over other marker systems with a low polymorphic rate (5%–6%). This new marker system, named as *Arachis hypogaea* transposable element (AhTE) marker system, detects the polymorphism for the insertion of 205 bp long *Arachis hypogaea* miniature inverted-repeat transposable element (*AhMITE1*). AhTE marker system was proposed and developed (Bhat et al., 2008, Gowda et al., 2010, 2011); and subsequently, it was scaled up in peanut by Shirasawa et al. (2012a), who developed 504 AhTE markers using *AhMITE1*-enriched libraries. Later, Shirasawa et al. (2012b) developed additional 535 AhTE markers using transposon-enriched libraries of other cultivars. With the availability of the genome sequences of the diploid progenitors of peanut (Bertioli et al., 2016), efforts were made to identify the genome-wide distribution of *AhMITE1* (Gayathri et al., 2018). Since these AhTE markers were highly polymorphic and user-friendly (Kolekar et al., 2016), they were successfully used to discriminate alleles through mapping (Shirasawa et al., 2013, Kolekar et al., 2016, Jadhav et al., 2021).

EST-SSRs have been extensively used in peanut (Baozhu et al., 2009, Guangzhou, 2009, Liang et al., 2009) as an approach for the development of functional markers. Allelic forms identified through EST-SSRs truly correspond to the coding region (genes), thereby reflecting allele mining. EST-SSRs have been successfully used for trait mapping and candidate gene analysis (Nagy et al., 2012, Mondal and Badigannavar, 2018, Ahmad et al., 2020).

All these marker systems provided platforms for identifying the allelic variations in peanut based on certain sequence features (restriction site, microsatellites, TE insertion, etc.). Nagy et al. (2010) used 2,847 SSR and 380 single-strand conformational polymorphisms (SSCP) to map the *Rma* gene governing root-knot nematode resistance in peanuts. Further fine mapping could identify the genomic location and the resistant and susceptible alleles at *Rma*. Subsequently, the genetic and molecular study (Jung et al., 2000) involving cloning and sequencing suggested that a mutation in *ahFAD2A* and a significant reduction in the levels of *ahFAD2B* transcript together cause the high oleate phenotype.

Though allele mining studies have mainly targeted only one or a few genes at a time in the past in different crops, the advances in the genome-wide approach of allele mining have favored using whole genome resequencing. Various methods of sequencing [genotyping by sequencing (GBS) or sequence-based genotyping (SBG), normalized GBS (nGBS), restriction site-associated DNA sequencing (RAD-Seq), double-digest RadSeq (ddRAD), RNASeq, exome capture and sequencing, whole genome re-sequencing, etc.] could generate good-quality sequence data that can be employed for allele mining.

In peanuts, the two homeologous genes (*AhGPAT9A* and *AhGPAT9B*) coding for rate-limiting enzymes in triacylglycerol (TAG) synthesis play an important role in seed oil accumulation. These two genes cloned separately from the A and B genomes shared a similarity of 95.65% with 165 site differences (Lv et al., 2020). The overexpression or knock-down of AhGPAT9 increased or decreased the seed oil content, respectively. Allelic polymorphism analysis in 171 peanut germplasm identified 118 polymorphic sites in *AhGPAT9A* forming 64 haplotypes, while the study identified 94 polymorphic sites in *AhGPAT9B* forming 75 haplotypes.

A genome-wide search could identify a total of 76 monosaccharide transporter (*AhMST*) genes (Wan et al., 2020). Chromosome distribution pattern and synteny analysis strongly indicated that genome-wide segmental and tandem duplication contributed to the expansion of these genes. Allelic variations were identified with four SNPs in *AhMST11*, *AhMST13*, and *AhMST60*, which were significantly correlated with hundred pod weight and hundred seed weight.

Through map-based cloning and QTL-Seq approaches, Peng et al. (2021) identified a pair of homoeologous Nodulation Signaling Pathway 2 (*AhNSP2*) genes controlling nodulation in cultivated peanuts. QTL-Seq analysis identified the natural allelic polymorphisms in terms of SNPs in these genes. Considering the importance of auxin-responsive genes in plant growth and development, Zhang et al. (2022) identified 44 *AhIAA* genes in the cultivated peanut. These genes showed a diverse expression pattern. Allelic polymorphism analysis of *AhIAA-1A* and *AhIAA-1B* showed three SNP loci in the 3′ UTR of *AhIAA-1A* among the germplasm. Zhao et al. (2020b) mapped *AhTc1* by the WGRS-based QTL-Seq approach for testa color in peanut. This study also helped develop allele-specific markers, where the marker pTesta1089 was closely linked with purple testa. *AhTc1* coding for a R2R3-MYB was cloned and expressed in tobacco, where the transgenic plants had purple leaves, flowers, pods, and seeds.

Repeatome, the landscape of both tandem and dispersed repeats, was studied in *Hieracium* (Zagorski et al., 2020) and peanut (Samoluk et al., 2022) to identify the structural variations. Sequence comparisons revealed that the monomeric sequences or satDNA sequences (named AraSat1–AraSat4) differed widely for length (31–445 bp) among the wild diploid *Arachis* species (Samoluk et al., 2022). Such reports of sequence variation at tandem and dispersed repeats can also be useful in allele mining.

Our effort on genome-wide analysis could detect 4,309,724 SNPs from the WGRS data of 178 peanut accessions, along with the reference genome sequence of Tifrunner. SNPs were analyzed for their structural and functional features in order to conclude on their utility and employability in genetic and genomic studies (Bhat et al., 2022). The number of SNPs greatly varied between the genes across the chromosomes (Table 7.1) (our unpublished data). Gene *arahy.ZLMA8V* on chromosome 3 accumulated the highest number of SNPs (672), indicating a possible greater allelic variation, while 7,401 genes carried a single SNP, indicating the presence of probably only two alleles among these accessions. All 4,309,724 SNPs were checked for their impact on gene function, and when only those with moderate and high impact were considered, the gene *arahy.9NZ7QY* on chromosome 19 showed the highest (90) number of SNPs, followed by others (Table 7.2). Haplotype analysis identified 100 haplotypes within *arahy.9NZ7QY* based on the total 264 SNPs. The gene *arahy.9NZ7QY* was found to code for the disease resistance protein of the TIR-NBS-LRR class. The haplotype of GPBD 4, a foliar disease resistant elite variety, was found to differ from Tifrunner for 29 SNP sites at *arahy.9NZ7QYI* (Figure 7.1).

TABLE 7.1
Variation for the Number of Single Nucleotide Polymorphisms (SNPs) in the Genes of Peanut

SN	Chromosome	Gene with the Highest Number of SNPs	Number of Genes with the Lowest Number of SNPs (1)
1	Chr01	*arahy.UR9BCX* (138)[a]	209
2	Chr02	*arahy.KG7SDJ* (253)	215
3	Chr03	*arahy.ZLMA8V* (672)	335
4	Chr04	*arahy.MUUL4Z* (205)	244
5	Chr05	*arahy.FI8Y5W* (209)	255
6	Chr06	*arahy.2GNH2V* (333)	232
7	Chr07	*arahy.EVYE6S* (266)	200
8	Chr08	*arahy.6WB007* (197)	205
9	Chr09	*arahy.UY05SB* (570)	277
10	Chr10	*arahy.QG99LK* (155)	279
11	Chr11	*arahy.NU1CTT* (440)	503
12	Chr12	*arahy.5BTH38* (172)	395
13	Chr13	*arahy.PP5TI0* (155)	582
14	Chr14	*arahy.3TG68A* (264)	442
15	Chr15	*arahy.GH1081* (190)	448
16	Chr16	*arahy.DHQL63* (376)	486
17	Chr17	*arahy.64IA1E* (163)	549
18	Chr18	*arahy.3E1ZKR* (251)	489
19	Chr19	*arahy.9NZ7QY* (264)	535
20	Chr20	*arahy.V6JT5K* (243)	521

[a] Values in the parenthesis indicate the number of SNPs in the respective gene.

TABLE 7.2
Allelic Variation among the Top 10 Genes with High and Moderate Impact SNPs in the Peanut Genome

SN	Gene ID	Chromosome	Length	Total SNPs	Intronic SNPs	Synonymous SNPs	Non-synonymous SNPs	Low Impact SNPs	Moderate Impact SNPs	High Impact SNPs
1	*arahy.9NZ7QY*	19	8,377	264	130	39	87	43	87	3
2	*arahy.BB34IZ*	09	6,623	187	62	37	78	41	80	0
3	*arahy.UI155H*	19	4,234	203	106	28	62	34	62	1
4	*arahy.64IA1E*	17	8,550	163	79	13	53	22	55	0
5	*arahy.TKN2M5*	12	5,533	86	14	18	53	19	53	0
6	*arahy.C6L0PV*	12	19,539	112	29	18	50	20	50	1
7	*arahy.370D23*	09	3,090	87	0	38	48	38	48	1
8	*arahy.MZJT69*	19	7,756	186	43	31	48	41	49	0
9	*arahy.3RLS73*	08	19,842	88	18	18	46	19	46	0
10	*arahy.DHQL63*	16	24,119	376	291	27	43	40	44	1

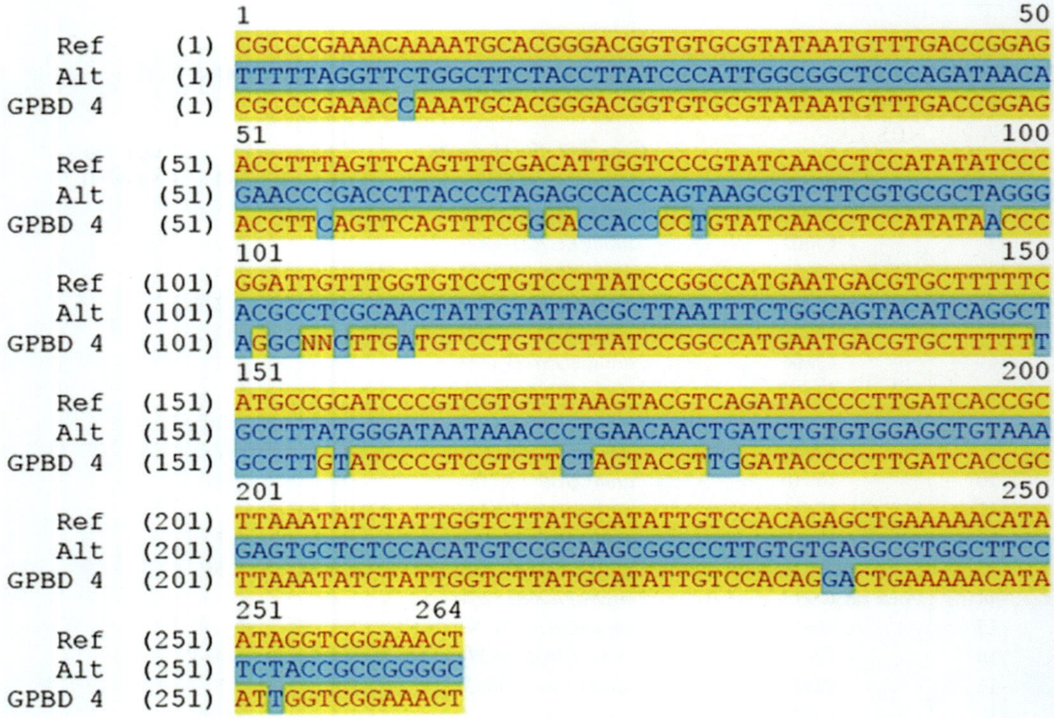

FIGURE 7.1 Single nucleotide polymorphisms at 264 loci in the region of *arahy.9NZ7QY* gene in GPBD 4 against the reference Tifrunner and alternate alleles.

7.6 CHARACTERIZATION OF ALLELIC DIVERSITY

A high degree of genomic variation exists between the individuals of most of the species (see Hu et al., 2020). Therefore, a true representation of the diversity within a species can be understood through the study of genomic variation in populations. Thus, the approach of genomics gained a new dimension of pan-genomics to represent the genomic diversity of a species in terms of core genes and variable genes. Furthermore, the pan-genome was extended to super-pangenome (Khan et al., 2020) to represent the diversity of all species within a genus. Genetic variation existing at the population level, as revealed by partial or complete genome analysis, has now turned population genetics into population genomics. This helps in understanding the genomic basis of fitness, local adaptation, and phenotypes by unraveling the locus-specific effects from genome-wide effects (Cortinovis et al., 2020). Population genomics also helps in understanding genome differentiation and identifying domestication-related genes, which might form the basis for future genomics-enabled breeding (Li et al., 2020). Also, population genomics helps identify the genome-wide effects contributing to various phenotypic traits. With the taxonomic richness, population genomics was reported for peanuts by Bhat et al. (2021a).

Gene- and allele-based diversity analysis can also be performed in addition to whole genome-based analysis. In peanuts, several genes and gene families have been described. They include *ARF* coding genes (Tang et al., 2021), *FAD2* gene family (Wang et al., 2015, Chi et al., 2017), *WRKY* transcription factor coding genes (Zhao et al., 2020a), basic/helix-loop-helix (*bHLH*) genes (Gao et al., 2017), *LEA* genes (Huang et al., 2022, Wan et al., 2021), small auxin-upregulated RNAs (*SAURs*) genes (Liu et al., 2022a), lipoxygenases (*LOXs*) genes (Mou et al., 2022), *XTHs* (xyloglucan endotransglucosylases/hydrolases) genes (Zhu et al., 2022), thioredoxins (*Trx*) genes (Li et al., 2022), sugars will eventually be exported transporter (*SWEET*) protein coding genes (Chu et al., 2020), NBS-LRR genes (Deng et al., 2018, Ratnaparkhe et al., 2011), *NAC* genes (Yuan et al., 2020), putative cytosine-5 DNA methyltransferase (C5-MTases) and demethylase genes (Wang et al., 2016), and stilbenoidprenyl transferase gene (Yang, 2016).

Number, distribution, size variation, exon-intron structure, promoter analysis, expression, and phylogeny have been reported for most of these gene families. However, the allelic variations were reported for *AhGPAT9* (Lv et al., 2020), *AhMST* (Wan et al., 2020), *AhNSP2* (Peng et al. 2021), *AhIAA* Zhang et al. (2022), and *AhTc1* (Zhao et al. 2020b), as described in Section 7.5.

A recent method of targeted mutagenesis through genome editing and base editing (Przybyla and Gilbert, 2021, Singh and Shekhawat, 2021) allows for the generation of novel alleles. Direct and irreversible mutations through non-homologous end joining of double-stranded breaks are generated by CRISPR–Cas9 to get an altered phenotype. Recently developed Cas9 variants, novel RNA-guided nucleases and base-editing systems, and DNA-free CRISPR–Cas9 (Metje-Sprink et al., 2018) delivery methods now provide great opportunities for inducing mutations (see Yin et al., 2017, Atkins and Voytas, 2020, Porto et al., 2020).

Recent studies with gene editing in several polyploid crops have demonstrated facile editing of some or all of the genes targeted for modification on homeologous chromosomes; otherwise, the polyploids were considered to be difficult for gene editing (Weeks, 2017). Gene editing and base editing methods exclusively in peanut have been recently documented (Neelakandan et al., 2022a, b, c). Targeted mutagenesis in the conserved coding regions of *AhFAD2* genes was undertaken using TALENs (Wen et al., 2018). Mutation frequencies among *AhFAD2* mutant lines were significantly correlated with oleic acid accumulation. Using CRISPR/Cas9 activity, three mutations were generated: G448A in *AhFAD2A*, 441_442insA, and G451T in *AhFAD2B*, leading to high oleic acid content in peanuts (Yuan et al., 2019). The function of NFR genes involved in responding to rhizobia was explored using CRISPR/Cas9 tool and hairy root transformation in peanuts. Knock-out mutants validated the function of *AhNFR5* genes in nodule formation (Shu et al., 2020). Protoplast isolation and transformation methods using sgRNAs for an allergen gene, Ara h 2 could induce several edits in the target gene (Biswas et al., 2022). Gene editing methods for heat tolerance in oilseeds including peanut have been discussed by Ahmad et al. (2021). Overall, these innovative editing systems can create precise mutations, and leave no transgene footprint, thereby holding potential promise for generating novel alleles.

7.7 INFLUENCE OF ALLELIC FORMS ON GENE EXPRESSION AND PHENOTYPE

Allele-specific expression is commonly observed for many genes, which can lead to phenotypic diversity and evolution. Mechanisms regulating allele-specific regulation are to be understood in depth, as they have been done in apples (Tian et al., 2022). Using RNA-Seq, allele-specific regulation was observed for 2,091 genes. In comparison to the haploid genome, the genomic sequences of the two alleles were distinguished for 817 genes showing allele-specific expression. Altered allele-specific expression in 354 of these genes was associated with the presence of a transposable element.

In peanuts, RNA-Seq analysis was performed for drought-tolerant (C-76) and drought-susceptible (Val-C) genotypes grown under stress and non-stress conditions (Bhogireddy et al., 2020). The transcriptome analysis detected 4,508 differentially expressed genes (DEGs), which showed 514 SNPs. Allele-specific primers were designed for 23 genes to validate allelic variations in C-76 and Val-C using qRT-PCR. Homozygous and heterozygous alleles were distinguished based on the variations obtained in the cycle threshold (Ct) values. Of the 23 genes, allele-specific expression was observed for two genes (*Araip.3WN1Q* and *Aradu.03ENG*).

Variation in the expression of *ahFAD2A*gene forms, which determine oleic acid accumulation in peanuts was studied (Huang et al., 2012). Initially, allelic-specific primers were used to check allelic variations of *ahFAD2A* among 52 peanut varieties. PCR results revealed the presence of two alleles of *ahFAD2A*. The two alleles differed for the G/A SNP at the 448 bp site of *ahFAD2A*. Expression of these two alleles was checked using real-time PCR at different developmental stages of the seed. The expression level of the mutant allele from Yuhua 9416 was slightly higher than that of the wild-type allele from Yuanza 9102 during the early to middle developmental stages (17–38 days). But a rapid decrease in the expression level was observed for the mutant allele as compared to the wild type at the late developmental stage (after 45 days). The genotypes with the mutant allele showed a higher oleic acid content.

A pilot study of RNA-Seq on an advanced RIL population in tetraploid peanuts identified 92 paramutation-like genes, which showed a allele-specific expression pattern among the RILs (Huang et al., 2020). Based on the results, it was speculated that most of the paramutagenic allele was expressed at lower levels than the paramutable allele.

Designing future crops through genomic breeding strategies tends to optimize crop genomes by accumulating beneficial alleles and purging out the deleterious alleles. Hence, locus identification, allele mining, haplotyping (Bhat et al., 2021b), and exploiting the allelic variation present at all associated loci, followed by crossing the genotypes in a designed way, is the roadmap for future breeding (Varshney et al., 2021). In this regard, the efforts on haplotyping the *AhGPAT9* gene in peanuts can provide a useful clue. The two homeologous genes of *AhGPAT9* coding for an enzyme in triacylglycerol (TAG) synthesis playing an important role in seed oil content were subjected to haplotyping in 171 peanut germplasm (Lv et al., 2020). In total, 64 haplotypes (a1 to a64) in *AhGPAT9A* and 75 haplotypes (b1 to b75) in *AhGPAT9B* were identified. Furthermore, haplotype analysis showed that a5, b57, b30, and b35 were elite haplotypes associated with high oil content, whereas a7, a14, a48, b51, and b54 were associated with low oil content types.

Linkage disequilibrium (LD) mapping has been successfully used in peanuts to identify the haplotypes and genetic signatures for a few traits (Pandey et al., 2014, Clevenger et al., 2017, Otyama et al., 2019, Liu et al., 2022b). Significant differentiation patterns of haplotypes were observed across the improved cultivars and landraces for the candidate genes for seed weight and plant architecture (Liu et al., 2022b).

Multi-parent populations such as nested association mapping (NAM) and multi-parent advanced generation inter-cross (MAGIC) have been developed to diversify the peanut genetic pool. NAM and MAGIC are the next-generation mapping populations and the chief source of novel alleles for future plant breeding. Recently, the alleles for seed weight were identified on chromosome A05/B05 using a nested association mapping approach (Gangurde et al., 2020). Such populations are widely used in cereal and legume crops.

7.8 CONCLUSION

Allele mining has been systematically implemented for the loci controlling frost tolerance in barley (Guerra et al., 2022), blast-resistance gene *Pi9* in rice (Zhou et al., 2020), rice yellow mottle virus resistance genes in rice (Pidon et al., 2020), forage quality and response to environmental stresses in grasses (Hanley et al., 2021), *TaGRF-2D* gene in *Triticum aestivum* and *Aegilops tauschii* (Kroupin et al., 2020), leukocyte antigen class II in East Asian dogs (Kang et al., 2021), fiber elongation in cotton (Ma et al., 2022), leaf rust resistance in the Vavilov wheat collection (Riaz et al., 2018), genes controlling the timing of spring bud flush of tea (Wang et al., 2019), and plant architecture and yield-related agronomic traits in wheat (Guo et al., 2019). With the application of omics for both cultivated and wild species, peanut has reached a stage where allele mining can be taken up efficiently. Numerous gene families have been identified and characterized, which form the target for allele mining. Candidate genes and alleles associated with various traits have been identified using QTL mapping and genome-wide association studies (GWAS) (Liu et al., 2022b). Identification of novel and superior alleles would help design better genotypes. The use of wild species in peanut breeding provides a great scope for introducing novel alleles for trait improvement. Genetic exchanges between the subgenomes enable broadening the variability, as has been shown for fruit type (Suassuna et al., 2020). Overall, allele mining in peanuts has been progressing to identify novel alleles and haplotypes for their employment in peanut breeding.

REFERENCES

Ahmad M, Waraich EA, Skalicky M, Hussain S, Zulfiqar U, et al. Adaptation strategies to improve the resistance of oilseed crops to heat stress under a changing climate: An overview. *Front Plant Sci*. 2021;12:767150. doi: 10.3389/fpls.2021.767150.

Ahmad S, Nawade B, Sangh C, Mishra GP, Bosamia TC, et al. Identification of novel QTLs for late leaf spot resistance and validation of a major rust QTL in peanut (*Arachis hypogaea* L.). *3 Biotech*. 2020;10(10):1–13.

Alves DM, Pereira RW, Leal-Bertioli SC, Moretzsohn MC, Guimaraes PM, et al. Development and use of single nucleotide polymorphism markers for candidate resistance genes in wild peanuts (*Arachis* spp.). *Genet Mol Res*. 2008;7(3):631–42.

Arya SS, Salve AR, Chauhan S. Peanuts as functional food: A review. *J Food Sci Technol*. 2016;53(1):31–41.

Ashri A. Plasmon divergence in peanuts (*Arachis hypogaea*): A third plasmon and locus affecting growth habit. *Theor Appl Genet*. 1976;48(1):17–21.

Atkins PA, Voytas DF. Overcoming bottlenecks in plant gene editing. *Curr Opin Plant Biol*. 2020;54:79–84.

Baozhu G, Xiaoping C, Yanbin H, Xuanqiang L, Phat D, et al. Analysis of gene expression profiles in leaf tissues of cultivated peanuts and development of EST-SSR markers and gene discovery. *Int J Plant Genomics*. 2009;1–14. doi: 10.1155/2009/715605.

Benzer S. On the topology of the genetic fine structure. *Proc Natl Acad Sci*. 1959;45(11):1607–20.

Bertioli DJ, Abernathy B, Seijo G, Clevenger J, Cannon SB. Evaluating two different models of peanut's origin. *Nat Genet*. 2020;52:557–9.

Bertioli DJ, Cannon SB, Froenicke L, Huang G, Farmer AD, et al. The genome sequences of *Arachis duranensis* and *Arachis ipaensis*, the diploid ancestors of cultivated peanut. *Nat Genet*. 2016;48(4):438–46. doi: 10.1038/ng.3517.

Bertioli DJ, Clevenger J, Godoy IJ, Stalker H, Wood S, et al. Legacy genetics of *Arachis cardenasii* in the peanut crop shows the profound benefits of international seed exchange. *Proc Natl Acad Sci*. 2021;118(38):e2104899118.

Bertioli DJ, Jenkins J, Clevenger J, Dudchenko O, Gao D, et al. The genome sequence of segmental allotetraploid peanut *Arachis hypogaea*. *Nat Genet*. 2019;51(5):877–84. doi: 10.1038/s41588-019-0405-z.

Bhat JA, Yu D, Bohra A, Ganie SA, Varshney RK. Features and applications of haplotypes in crop breeding. *Commun Biol*. 2021b;4(1):1–12.

Bhat RS, Patil VU, Chandrashekar TM, Sujay V, Gowda MVC, et al. Recovering flanking sequence tags of miniature inverted-repeat transposable element by thermal asymmetric interlaced-PCR in peanut. *Curr Sci*. 2008;95(4):452–3.

Bhat RS, Rockey J, Shirasawa K, Tilak IS, Brijesh Patil MP, et al. DNA methylation and expression analyses reveal epialleles for the foliar disease resistance genes in peanut (*Arachis hypogaea* L.). *BMC Res Notes*. 2019a;13(1):20. doi: 10.1186/s13104-020-4883-y.

Bhat RS, Shirasawa K, Chavadi SD. Genome-wide structural and functional features of single nucleotide polymorphisms revealed from the whole genome resequencing of 179 accessions of *Arachis*. *Physiol Plant*. 2022;174(1):e13623. doi: https://doi.org/10.1111/ppl.13623.

Bhat RS, Shirasawa K, Monden Y, Yamashita H, Tahara M. Developing transposable element marker system for molecular breeding. In: Jain M, Garg R, editors. *Legume Genomics: Methods and Protocols*. New York: Humana;2019b. pp. 233–52.

Bhat RS, Shirasawa K, Sharma V, Isobe S, Hirakawa H, et al. Population genomics of peanut. In: Rajora OP, editor. *Population Genomics*. Switzerland AG 2021: Springer Nature;2021a. pp. 793–835.

Bhogireddy S, Xavier A, Garg V, Layland N, Arias R, et al. Genome-wide transcriptome and physiological analyses provide new insights into peanut drought response mechanisms. *Sci Rep*. 2020;10(1):4071. doi: 10.1038/s41598-020-60187-z.

Biswas S, Wahl NJ, Thomson MJ, Cason JM, McCutchen BF, et al. Optimization of protoplast isolation and transformation for a pilot study of genome editing in peanut by targeting the allergen gene Ara h 2. *Int J Mol Sci*. 2022;23(2):837. doi: 10.3390/ijms23020837.

Bohra A, Kilian B, Sivakumar S, Caccamo M, McCouch SR, et al. Reap the crop wild relatives for breeding future crops. *Trends Biotechnol.* 2021. doi: 10.1016/j.tibtech.2021.08.009.

Bomireddy D, Gangurde SS, Variath MT, Janila P, Manohar SS, et al. Discovery of major quantitative trait loci and candidate genes for fresh seed dormancy in groundnut. *Agronomy*. 2022;12(2):404.

Bosamia TC, Dodia SM, Mishra GP, Ahmad S, Joshi B, et al. Unraveling the mechanisms of resistance to Sclerotium rolfsii in peanut (*Arachis hypogaea* L.) using comparative RNA-Seq analysis of resistant and susceptible genotypes. *PLoS One*. 2020;15(8):e0236823.

Branch W, Tallury S, Clevenger J, Schwartz B, Hanna W. Inheritance of a novel heterozygous peanut mutant, 5-small leaflet. *Peanut Sci.* 2020;47(1):33–7.

Chalhoub B, Denoeud F, Liu S, Parkin IA, Tang H, et al. Early allopolyploid evolution in the post-Neolithic Brassica napus oilseed genome. *Science*. 2014;345(6199):950–3.

Chen C, Cao Q, Jiang Q, Li J, Yu R, et al. Comparative transcriptome analysis reveals gene network regulating cadmium uptake and translocation in peanut roots under iron deficiency. *BMC Plant Biol*. 2019c;19(1):35.

Chen H, Yang Q, Chen K, Zhao S, Zhang C, et al. Integrated microRNA and transcriptome profiling reveals a miRNA-mediated regulatory network of embryo abortion under calcium deficiency in peanut (*Arachis hypogaea* L.). *BMC Genom*. 2019b;20(1):1–17.

Chen X, Hong Y, Zhang E, Liu H, Zhou G, et al. Comparison of gene expression profiles in cultivated peanut (*Arachis hypogaea*) under strong artificial selection. *Plant Breed*. 2012;131(5):620–30.

Chen X, Li H, Pandey MK, Yang Q, Wang X, et al. Draft genome of the peanut A-genome progenitor (*Arachis duranensis*) provides insights into geocarpy, oil biosynthesis, and allergens. *Proc Natl Acad Sci*. 2016a;113(24):6785–90.

Chen X, Liang X. Peanut transcriptomics. In: Mallikarjuna N, Varshney RK, editors. *Genetics, Genomics and Breeding of Peanuts*. CRC Press, Taylor & Francis Group; Boca Raton, FL; 2014. pp. 139–60.

Chen X, Lu Q, Liu H, Zhang J, Hong Y, et al. Sequencing of cultivated peanut, *Arachis hypogaea*, yields insights into genome evolution and oil improvement. *Mol Plant*. 2019a;12(7):920–34. doi: 10.1016/j.molp.2019.03.005.

Chen X, Yang Q, Li H, Li H, Hong Y, et al. Transcriptome-wide sequencing provides insights into geocarpy in peanut (*Arachis hypogaea* L.). *Plant Biotechnol J*. 2016b;14(5):1215–24.

Chen Y, Ren X, Zhou X, Huang L, Yan L, et al. Dynamics in the resistant and susceptible peanut (*Arachis hypogaea* L.) root transcriptome on infection with the *Ralstonia solanacearum*. *BMC Genom*. 2014;15(1):1078.

Chi X, Zhang Z, Chen N, Zhang X, Wang M, et al. Isolation and functional analysis of fatty acid desaturase genes from peanut (*Arachis* hypogaea L.). *PLoS One*. 2017;12(12):e0189759. doi: 10.1371/journal.pone.0189759.

Chopra R, Burow G, Farmer A, Mudge J, Simpson CE, et al. Comparisons of *de novo* transcriptome assemblers in diploid and polyploid species using peanut (*Arachis* spp.) RNA-Seq data. *PLoS One*. 2014;9(12):e115055.

Chopra R, Burow G, Farmer A, Mudge J, Simpson CE, et al. Next-generation transcriptome sequencing, SNP discovery and validation in four market classes of peanut, *Arachis hypogaea* L. *Mol Genet Genomics*. 2015;290(3):1169–80.

Chopra R, Burow G, Simpson CE, Chagoya J, Mudge J, et al. Transcriptome sequencing of diverse peanut (*Arachis*) wild species and the cultivated species reveals a wealth of untapped genetic variability. *G3: Genes, Genomes, Genetics*. 2016;6(12):3825–36.

Chu HD, Le QTN, Hoang YHT, Pham TP, La HV, et al. Identification, structural characterization, and in silico expression analysis of the sucrose transporter 'SWEET'gene family in peanut (*Arachis hypogaea*). *Vietnam J Sci Technol*. 2020;62(3):62–7.

Chu Y, Bertioli D, Levinson CM, Stalker HT, Holbrook CC, et al. Homoeologous recombination is recurrent in the nascent synthetic allotetraploid *Arachis ipaënsis*×*Arachis correntina* 4x and its derivatives. G3. 2021;11(4):jkab066.

Chu Y, Holbrook CC, Ozias-Akins P. Two alleles of *ahFAD2B* control the high oleic acid trait in cultivated peanut. *Crop Sci*. 2009;49(6):2029–36.

Clevenger J, Chu Y, Chavarro C, Agarwal G, Bertioli DJ, et al. Genome-wide SNP genotyping resolves signatures of selection and tetrasomic recombination in peanut. *Mol Plant*. 2017;10(2):309–22. doi: 10.1016/j.molp.2016.11.015.

Coleman CE, Kao T-H. The flanking regions of two *Petunia inflata* S alleles are heterogeneous and contain repetitive sequences. *Plant Mol Biol*. 1992;18(4):725–37.

Cortinovis G, Frascarelli G, Di Vittori V, Papa R. Current state and perspectives in population genomics of the common bean. *Plants (Basel)*. 2020;9(3):330. doi: 10.3390/plants9030330.

Deng Y, Chen H, Zhang C, Cai T, Zhang B, et al. Evolution and characterisation of the Ah RAF 4 NB-ARC gene family induced by Aspergillus flavus inoculation and abiotic stresses in peanut. *Plant Biol.* 2018;20(4):737–50.

Desmae H, Janila P, Okori P, Pandey MK, Motagi BN, et al. Genetics, genomics and breeding of groundnut (*Arachis* hypogaea L.). *Plant Breed.* 2019;138(4):425–44. doi: 10.1111/pbr.12645.

Dhingra V, Gupta M, Andacht T, Fu ZF. New frontiers in proteomics research: a perspective. *Int J Pharm.* 2005;299(1–2):1–18.

Dodia SM, Joshi B, Gangurde SS, Thirumalaisamy PP, Mishra GP, et al. Genotyping-by-sequencing based genetic mapping reveals large number of epistatic interactions for stem rot resistance in groundnut. *Theor Appl Genet.* 2019;132(4):1001–16.

Favero AP, Simpson CE, Valls JFM, Vello NA. Study of the evolution of cultivated peanut through crossability studies among *Arachis ipaensis*, *A. duranensis*, and *A. hypogaea*. *Crop Sci.* 2006;46(4):1546–52. doi: 10.2135/cropsci2005.09-0331.

Feng S, Cokus SJ, Zhang X, Chen P-Y, Bostick M, et al. Conservation and divergence of methylation patterning in plants and animals. *Proc Natl Acad Sci.* 2010;107(19):8689–94.

Ferguson M, Burow M, Schulze S, Bramel P, Paterson A, et al. Microsatellite identification and characterization in peanut (*A.* hypogaea L.). *Theor Appl Genet.* 2004;108(6):1064–70.

Friend S, Quandt D, Tallury S, Stalker H, Hilu K. Species, genomes, and section relationships in the genus *Arachis* (Fabaceae): A molecular phylogeny. *Plant Syst EVol.* 2010;290(1–4):185–99.

Gangurde SS, Kumar R, Pandey AK, Burrow M, Laze HE, et al. Climate-smart groundnuts for achieving high productivity and improved quality: Current status, challenges, and opportunities. In: Kole C, editor. *Genomic Designing of Climate-Smart Oilseed Crops*. New York: Springer Nature;2019. pp. 133–72. https://link.springer.com/chapter/10.1007/978-3-319-93536-2_3

Gangurde SS, Nayak SN, Joshi P, Purohit S, Sudini HK, et al. Comparative transcriptome analysis identified candidate genes for late leaf spot resistance and cause of defoliation in groundnut. *Int J Mol Sci.* 2021;22(9):4491.

Gangurde SS, Wang H, Yaduru S, Pandey MK, Fountain JC, et al. Nested-association mapping (NAM)-based genetic dissection uncovers candidate genes for seed and pod weights in peanut (*Arachis hypogaea*). *Plant Biotechnol J.* 2020;18(6):1457–71.

Gao C, Sun J, Wang C, Dong Y, Xiao S, et al. Genome-wide analysis of basic/helix-loop-helix gene family in peanut and assessment of its roles in pod development. *PLoS One*. 2017;12(7):e0181843.

Gayathri M, Shirasawa K, Varshney RK, Pandey MK, Bhat RS. Development of new *AhMITE1* markers through genome-wide analysis in peanut (*Arachis hypogaea* L.). *BMC Res Notes*. 2018;11(1):10. doi: 10.1186/s13104-017-3121-8.

Geng L, Duan X, Liang C, Shu C, Song F, et al. Mining tissue-specific contigs from peanut (*Arachis hypogaea* L.) for promoter cloning by deep transcriptome sequencing. *Plant Cell Physiol.* 2014;55(10):1793–801.

Gowda MVC, Bhat RS, Motagi BN, Sujay V, Varshakumari, et al. Association of high-frequency origin of late leaf spot resistant mutants with *AhMITE1* transposition in peanut. *Plant Breed.* 2010;129(5):567–9. doi: 10.1111/j.1439-0523.2009.01704.x.

Gowda MVC, Bhat RS, Sujay V, Kusuma P, Varshakumari, et al. Characterization of *AhMITE1* transposition and its association with the mutational and evolutionary origin of botanical types in peanut (*Arachis* spp.). *Plant Syst E.* 2011;291(3–4):153–8. doi: 10.1007/s00606-010-0373-3.

Gowda MVC, Motagi BN, Naidu GK, Diddimani SB, Sheshagiri R. GPBD 4: A spanish bunch groundnut genotype resistant to rust and late leaf spot. *Int Arachis News Lett.* 2002;22:29–32.

Guangzhou C. Utility of EST-derived SSR in cultivated peanut (*Arachis hypogaea* L.) and *Arachis* wild species. *BMC Plant Biol.* 2009;9:35.

Guerra D, Morcia C, Badeck F, Rizza F, Delbono S, et al. Extensive allele mining discovers novel genetic diversity in the loci controlling frost tolerance in barley. *Theor Appl Genet.* 2022;135(2):553–69.

Guimarães PM, Brasileiro AC, Morgante CV, Martins AC, Pappas G, et al. Global transcriptome analysis of two wild relatives of peanut under drought and fungi infection. *BMC Genom.* 2012;13(1):387.

Guimaraes PM, Guimaraes LA, Morgante CV, Silva Jr OB, Araujo ACG, et al. Root transcriptome analysis of wild peanut reveals candidate genes for nematode resistance. *PLoS One*. 2015;10(10):e0140937.

Guo F, Ma J, Hou L, Shi S, Sun J, et al. Transcriptome profiling provides insights into molecular mechanism in peanut semi-dwarf mutant. *BMC Genom.* 2020;21(1):211. doi: 10.1186/s12864-020-6614-0.

Guo H, Xiong H, Xie Y, Zhao L, Gu J, et al. Functional mutation allele mining of plant architecture and yield-related agronomic traits and characterization of their effects in wheat. *BMC Genet.* 2019;20(1):1–15.

Halward TM, Stalker TH, Larue EA, Kochert G. Use of single-primer DNA amplifications in genetic studies of peanut (*Arachis* hypogaea L.). *Plant Mol Biol.* 1992;18(2):315–25.

Han S, Liu H, Yan M, Qi F, Wang Y, et al. Differential gene expression in leaf tissues between mutant and wild-type genotypes response to late leaf spot in peanut (*Arachis* hypogaea L.). *PLoS One*. 2017;12(8):e0183428.

Hanley SJ, Pellny TK, de Vega JJ, Castiblanco V, Arango J, et al. Allele mining in diverse accessions of tropical grasses to improve forage quality and reduce environmental impact. *Ann Bot*. 2021;128(5):627–37.

Hao K, Wang F, Nong X, McNeill MR, Liu S, et al. Response of peanut *Arachis hypogaea* roots to the presence of beneficial and pathogenic fungi by transcriptome analysis. *Sci Rep*. 2017;7(1):1–15.

Harlan JR, deWet JMJ. On Ö. Winge and a prayer: the origins of polyploidy. *Bot Rev*. 1975;41(4):361–90.

He G, Meng R, Newman M, Gao G, Pittman RN, et al. Microsatellites as DNA markers in cultivated peanut (*Arachis hypogaea* L.). *BMC Plant Biol*. 2003;3(1):1–6.

He G, Prakash C. Identification of polymorphic DNA markers in cultivated peanut (*Arachis hypogaea* L.). *Euphytica*. 1997;97(2):143–9.

Holbrook CC, Culbreath AK. Registration of 'Tifrunner' peanut. *J Plant Regist*. 2007;1(2):124.

Hong Y, Liang X-Q, Chen X-P, Liu H-Y, Zhou G-Y, et al. Construction of genetic linkage map based on SSR markers in peanut (*Arachis* hypogaea L.). *Agril Sci China*. 2008;7(8):915–21.

Hong Y, Pandey MK, Liu Y, Chen X, Liu H, et al. Identification and evaluation of single-nucleotide polymorphisms in allotetraploid peanut (*Arachis hypogaea* L.) based on amplicon sequencing combined with high resolution melting (HRM) analysis. *Front Plant Sci*. 2015;6:1068. doi: 10.3389/fpls.2015.01068.

Hopkins MS, Casa AM, Wang T, Mitchell SE, Dean RE, et al. Discovery and characterization of polymorphic simple sequence repeats (SSRs) in peanut. *Crop Sci*. 1999;39(4):1243–7.

Hu Z, Wei C, Li Z. Computational strategies for eukaryotic pangenome analyses. *The Pangenome*. 2020:293–307.

Huai D, Xue X, Li Y, Wang P, Li J, et al. Genome-wide identification of peanut KCS genes reveals that AhKCS1 and AhKCS28 are involved in regulating VLCFA contents in seeds. *Front Plant Sci*. 2020;11:406. doi: 10.3389/fpls.2020.00406.

Huang B, Zhang X, Miao L, Gao W, Han S, et al. Allelic expression variation of ahFAD2A and its relationship with oleic acid accumulation in peanut. *Acta Agronomica Sinica*. 2012;38(10):1752–9.

Huang J, Xing M, Li Y, Cheng F, Gu H, et al. Comparative transcriptome analysis of the skin-specific accumulation of anthocyanins in black peanut (*Arachis* hypogaea L.). *J Agric Food Chem.* 2019;67(4):1312–24.

Huang L, Liu X, Pandey MK, Ren X, Chen H, et al. Genome-wide expression quantitative trait locus analysis in a recombinant inbred line population for trait dissection in peanut. *Plant Biotechnol J*. 2020;18(3):779–90. doi: 10.1111/pbi.13246.

Huang L, Wu B, Zhao J, Li H, Chen W, et al. Characterization and transferable utility of microsatellite markers in the wild and cultivated *Arachis* species. *PLoS One*. 2016;11(5):e0156633.

Huang R, Xiao D, Wang X, Zhan J, Wang A, et al. Genome-wide identification, evolutionary and expression analyses of LEA gene family in peanut (*Arachis hypogaea* L.). *BMC Plant Biol*. 2022;22(1):1–22.

Jadhav MP, Gangurde SS, Hake AA, Yadwad A, Supriya SM, et al. Genotyping-by-sequencing based genetic mapping identified major and consistent genomic regions for productivity and quality traits in peanut. *Front Plant Sci*. 2021;12:2034. doi: 10.3389/fpls.2021.668020.

Jain S, Srivastava S, Sarin NB, Kav NN. Proteomics reveals elevated levels of PR 10 proteins in saline-tolerant peanut (*Arachis hypogaea*) calli. *Plant Physiol Biochem*. 2006;44(4):253–259.

Jiang C, Zhang H, Ren J, Dong J, Zhao X, et al. Comparative transcriptome-based mining and expression profiling of transcription factors related to cold tolerance in peanut. *Int J Mol Sci*. 2020;21(6):1921.

Jung S, Swift D, Sengoku E, Patel M, Teulé F, et al. The high oleate trait in the cultivated peanut [*Arachis hypogaea* L.]. I. Isolation and characterization of two genes encoding microsomal oleoyl-PC desaturases. *Mol Gen Genet*. 2000;263(5):796–805.

Kang M, Ahn B, Youk S, Cho H-s, Choi M, et al. High allelic diversity of dog leukocyte antigen class II in East Asian dogs: Identification of new alleles and haplotypes. *J Mamm E* 2021;28(3):773–84.

Katam R, Gottschalk V, Survajahala P, Brewster G, Paxton P, et al. Advances in proteomics research for peanut genetics and breeding. In: Mallikarjuna N, Varshney RK, editors. *Genetics, Genomics and Breeding of Peanuts*. Boca Raton, FL: CRC Press;2014. pp. 161–77.

Katam R, Sakata K, Suravajhala P, Pechan T, Kambiranda DM, et al. Comparative leaf proteomics of drought-tolerant and-susceptible peanut in response to water stress. *J Proteomics*. 2016;143:209–26.

Khan AW, Garg V, Roorkiwal M, Golicz AA, Edwards D, et al. Super-pangenome by integrating the wild side of a species for accelerated crop improvement. *Trends Plant Sci*. 2020;25(2):148–58. doi: 10.1016/j.tplants.2019.10.012.

Kilian A. DArT-based whole genome profiling and novel information technologies in support system of modern breeding of groundnut. 3rd International Conference for Peanut Genomics and Biotechnology on "Advances in *Arachis* through Genomics and Biotechnology" (AAGB). ICRISAT, Patancheru, India 2008. p. 2.

Kim EY, Kim KD, Cho J. Harnessing epigenetic variability for crop improvement: current status and future prospects. *Genes & Genomics*. 2021;44:259–66.

Kochert G, Halward T, Branch WD, Simpson CE. RFLP variability in peanut (*Arachis hypogaea* L.) cultivars and wild species. *Theor Appl Genet*. 1991;81(5):565–70.

Kochert G, Stalker HT, Gimenes M, Galgaro L, Lopes CR, et al. RFLP and cytogenetic evidence on the origin and evolution of allotetraploid domesticated peanut, *Arachis hypogaea* (*Leguminosae*). *Am J Bot*. 1996;83(10):1282–91.

Kolekar RM, Sujay V, Shirasawa K, Sukruth M, Khedikar YP, et al. QTL mapping for late leaf spot and rust resistance using an improved genetic map and extensive phenotypic data on a recombinant inbred line population in peanut (*Arachis* hypogaea L.). *Euphytica*. 2016;209(1):147–56. doi: 10.1007/s10681-016-1651-0.

Kottapalli KR, Rakwal R, Shibato J, Burow G, Tissue D, et al. Physiology and proteomics of the water-deficit stress response in three contrasting peanut genotypes. *Plant Cell Environ*. 2009;32(4):380–407.

Kottapalli KR, Zabet-Moghaddam M, Rowland D, Faircloth W, Mirzaei M, et al. Shotgun label-free quantitative proteomics of water-deficit-stressed midmature peanut (Arachis hypogaea L.) seed. *J Proteome Res*. 2013;12(11):5048–57.

Krapovickas A. The origin, variability and spread of the groundnut (*Arachis hypogaea*). In: Ucko RJ, Dimbledy WC, editors. *The Domestication and Exploitation of Plant and Animals*. London: Greald Duckworth Co. Ltd.;1969. pp. 427–41.

Krapovickas A. Evolution of the genus *Arachis*. In: Moav R, editor. *Agricultural Genetics-Selected Topics*. New York: John Wiley & Sons;1973. pp. 135–51.

Krapovickas A, Gregory WC. Taxonomía del género *Arachis* (Leguminosae). *Bonplandia*. 1994;8:1–186.

Krapovickas A, Rigoni VA. Nuevas especies de *Arachis vinculadas* al problema del origen del mani. *Darwiniana*. 1957;11:431–55.

Krishna GK, Zhang J, Burow M, Pittman RN, Delikostadinov SG, et al. Genetic diversity analysis in Valencia peanut (*Arachis hypogaea* L.) using microsatellite markers. *Cell Mol Biol Lett*. 2004;9:685–97.

Kroupin PY, Chernook AG, Bazhenov MS, Karlov GI, Goncharov NP, et al. Allele mining of TaGRF-2D gene 5′-UTR in *Triticum aestivum* and *Aegilops tauschii* genotypes. *PLoS One*. 2020;15(4):e0231704.

Kumar GR, Sakthivel K, Sundaram RM, Neeraja CN, Balachandran SM, et al. Allele mining in crops: Prospects and potentials. *Biotechnol Adv*. 2010;28(4):451–61.

Kumar R, Janila P, Vishwakarma MK, Khan AW, Manohar SS, et al. Whole-genome resequencing-based QTL-seq identified candidate genes and molecular markers for fresh seed dormancy in groundnut. *Plant Biotechnol J*. 2020;18(4):992–1003.

Lavia GI. Karyotypes of *Arachis palustris* and *A. praecox* (section *Arachis*), two species with basic chromosome number x= 9. *Cytologia*. 1998;63(2):177–81.

Leal-Bertioli S, Nascimento EF, Chavarro M, Custódio AR, Hopkins MS, et al. Spontaneous generation of diversity in *Arachis* neopolyploids (*A. ipaënsis* x *A. duranensis*) 4x replays the early stages of peanut evolution. *G3 (Bethesda)*. 2021;11(11):jkab289. doi: 10.1093/g3journal/jkab289.

Li X, Su G, Ntambiyukuri A, Tong B, Zhan J, et al. Genome-wide identification and expression analysis of the AhTrx family genes in peanut. *Biol Plant*. 2022;66:112–22.

Li Y, Zhu FL, Zheng XW, Hu ML, Dong C, et al. Comparative population genomics reveals genetic divergence and selection in lotus, *Nelumbo nucifera*. *BMC Genom*. 2020;21(1):146. doi: 10.1186/s12864-019-6376-8.

Liang X, Chen X, Hong Y, Liu H, Zhou G, et al. Utility of EST-derived SSR in cultivated peanut (*Arachis hypogaea* L.) and *Arachis* wild species. *BMC Plant Biol*. 2009;9(1):35.

Liu H, Hong Y, Lu Q, Li H, Gu J, et al. Integrated analysis of comparative lipidomics and proteomics reveals the dynamic changes of lipid molecular species in high-oleic acid peanut seed. *J Agric Food Chem*. 2020;68(1):426–38. doi: 10.1021/acs.jafc.9b04179.

Liu Y, Shao L, Zhou J, Li R, Pandey MK, et al. Genomic insights into the genetic signatures of selection and seed trait loci in cultivated peanut. *J Adv Res*. 2022b:https://doi.org/10.1016/j.jare.2022.01.016.

Liu Y, Xiao L, Chi J, Li R, Han Y, et al. Genome-wide identification and expression of SAUR gene family in peanut (*Arachis hypogaea* L.) and functional identification of AhSAUR3 in drought tolerance. *BMC Plant Biol*. 2022a;22(1):1–13.

Lu J, Pickersgill B. Isozyme variation and species relationships in peanut and its wild relatives (*Arachis* L.-Leguminosae). *Theor Appl Genet*. 1993;85(5):550–60.

Lu Q, Li H, Hong Y, Zhang G, Wen S, et al. Genome sequencing and analysis of the peanut B-genome progenitor (*Arachis ipaensis*). *Front Plant Sci*. 2018;9:604.

Lv Y, Zhang X, Luo L, Yang H, Li P, et al. Characterization of glycerol-3-phosphate acyltransferase 9 (AhGPAT9) genes, their allelic polymorphism and association with oil content in peanut (*Arachis* hypogaea L.). *Sci Rep*. 2020;10(1):1–15.

Ma J, Jiang Y, Pei W, Wu M, Ma Q, et al. Expressed-genes and their new alleles identification during fibre elongation reveal the genetic factors underlying improvements of fibre length in cotton. *Plant Biotechnol J*. 2022. doi: 10.1111/pbi.13874.

Mamone G, Di Stasio L, De Caro S, Picariello G, Nicolai MA, et al. Comprehensive analysis of the peanut allergome combining 2-DE gel-based and gel-free proteomics. *Food Res Int*. 2019;116:1059–65.

Matton DP, Mau S-L, Okamoto S, Clarke AE, Newbigin E. The S-locus of Nicotiana alata: genomic organization and sequence analysis of two S-RNase alleles. *Plant Mol Biol*. 1995;28(5):847–58.

Metje-Sprink J, Menz J, Modrzejewski D, Sprink T. DNA-free genome editing: Past, present and future. *Front Plant Sci*. 2018;9:1957. doi: 10.3389/fpls.2018.01957.

Mondal S, Badigannavar AM. Mapping of a dominant rust resistance gene revealed two R genes around the major Rust_QTL in cultivated peanut (*Arachis* hypogaea L.). *Theor Appl Genet*. 2018;131:1671–81.

Mondal S, Motagi BN, Badigannavar AM. Identification of major consensus and epistatic QTLs for late leaf spot resistance in a cultivated groundnut breeding line VG 9514. *Euphytica*. 2021;217(9):1–14.

Moretzsohn M, Hopkins M, Mitchell S, Kresovich S, Valls J, et al. Genetic diversity of peanut (*Arachis hypogaea* L.) and its wild relatives based on the analysis of hypervariable regions of the genome. *BMC Plant Biol*. 2004;4(1):11.

Moretzsohn MC, Leoi L, Proite K, Guimaraes PM, Leal-Bertioli SCM, et al. A microsatellite-based, gene-rich linkage map for the AA genome of *Arachis* (Fabaceae). *Theor Appl Genet*. 2005;111(6):1060–71.

Mou Y, Sun Q, Yuan C, Zhao X, Wang J, et al. Identification of the LOX gene family in peanut and functional characterization of AhLOX29 in drought tolerance. *Front Plant Sci*. 2022;13:832785.

Muralidharan S, Poon YY, Wright GC, Haynes PA, Lee NA. Quantitative proteomics analysis of high and low polyphenol expressing recombinant inbred lines (RILs) of peanut (*Arachis* hypogaea L.). *Food Chem*. 2020;334:127517. doi: 10.1016/j.foodchem.2020.127517.

Nagy ED, Chu Y, Guo Y, Khanal S, Tang S, et al. Recombination is suppressed in an alien introgression in peanut harboring Rma, a dominant root-knot nematode resistance gene. *Mol Breed*. 2010;26:357–70.

Nagy ED, Guo Y, Tang S, Bowers JE, Okashah RA, et al. A high-density genetic map of *Arachis duranensis*, a diploid ancestor of cultivated peanut. *BMC Genom*. 2012;13(1):469.

Neelakandan AK, Subedi B, Traore SM, Binagwa P, Wright DA, et al. Base editing in peanut using CRISPR/nCas9. *Front Genome Ed*. 2022a;4:901444. doi: 10.3389/fgeed.2022.901444.

Neelakandan AK, Wright DA, Traore SM, Chen X, Spalding MH, et al. CRISPR/Cas9 based site-specific modification of *FAD2* cis-regulatory motifs in peanut (*Arachis* hypogaea L). *Front Genet*. 2022b;13:849961. doi: 10.3389/fgene.2022.849961.

Neelakandan AK, Wright DA, Traore SM, Ma X, Subedi B, et al. Application of CRISPR/Cas9 system for efficient gene editing in peanut. *Plants (Basel)*. 2022c;11(10). doi: 10.3390/plants11101361.

Ogawa D, Yamamoto E, Ohtani T, Kanno N, Tsunematsu H, et al. Haplotype-based allele mining in the Japan-MAGIC rice population. *Sci Rep*. 2018;8(1):1–11.

Otyama PI, Kulkarni R, Chamberlin K, Ozias-Akins P, Chu Y, et al. Genotypic characterization of the US peanut core collection. *G3: Genes, Genomes, Genetics*. 2020;10(11):4013–26.

Otyama PI, Wilkey A, Kulkarni R, Assefa T, Chu Y, et al. Evaluation of linkage disequilibrium, population structure, and genetic diversity in the U.S. peanut mini core collection. *BMC Genom*. 2019;20(1):481. doi: 10.1186/s12864-019-5824-9.

Paik-Ro OG, Smith RL, Knauft DA. Restriction fragment length polymorphism evaluation of six peanut species within the *Arachis* section. *Theor Appl Genet*. 1992;84(1):201–8.

Palmieri D, Bechara M, Curi R, Gimenes M, Lopes C. Novel polymorphic microsatellite markers in section Caulorrhizae (*Arachis*, Fabaceae). *Mol Ecol Notes*. 2005;5(1):77–9.

Palmieri DA, Hoshino A, Bravo J, Lopes C, Gimenes M. Isolation and characterization of microsatellite loci from the forage species *Arachis pintoi* (Genus *Arachis*). *Mol Ecol Notes*. 2002;2(4):551–3.

Pandey MK, Pandey AK, Kumar R, Nwosu CV, Guo B, et al. Translational genomics for achieving higher genetic gains in post-genome era in groundnut. *Theor Appl Genet*. 2020;133(5):1679–702. doi: 10.1007/s00122-020-03592-2.

Pandey MK, Upadhyaya HD, Rathore A, Vadez V, Sheshshaye MS, et al. Genomewide association studies for 50 agronomic traits in peanut using the reference set comprising 300 genotypes from 48 countries of semi-arid tropics of the world. *PLoS One*. 2014;9(8):e105228. doi: 10.1371/journal.pone.0105228.

Parmar S, Sharma V, Deekshitha B, Pooja S, Joshi P, et al. Recent advances in genetics, genomics and breeding for nutritional quality in groundnut. In: Gosal S, Wani SH, editors. *Accelerated Plant Breeding, Volume 4 Oil Crops*. Cham, Switzerland: Springer Nature;2022. p. 111–37.

Parniske M, Hammond-Kosack KE, Golstein C, Thomas CM, Jones DA, et al. Novel disease resistance specificities result from sequence exchange between tandemly repeated genes at the Cf-4/9 locus of tomato. *Cell*. 1997;91(6):821–32.

Paterson AH, Stalker HT, Gallo-Meagher M, Burrow M, Dwivedi SL, et al. Genomics and genetic enhancement of peanut. *Legume Crop Genomics*. 2004:97–109.

Peñaloza APS, Valls JFM. Contagen do número cromossômico en assos de *Arachis* decora (Legumonsae). In: Vega RFA, Bovi MLA, Betti JA, Voltan RBQ, editors. *Simpósio Latino Slericanno de Recursos Genéticos Vegetais*. IAC/Embrapa-Cenargen, Campanias;1997. p. 21.

Peñaloza APS, Valls JFM. Chromosome number and satellited chromosome morphology of eleven species of *Arachis* (Leguminosae). Bonpladia 15:65–72. *Bonplandia*. 2005;14(15):65–72.

Peng Z, Chen H, Tan L, Shu H, Varshney RK, et al. Natural polymorphisms in a pair of NSP2 homoeologs can cause loss of nodulation in peanut. *J Exp Bot*. 2021;72(4):1104–18. doi: 10.1093/jxb/eraa505.

Peng Z, Gallo M, Tillman BL, Rowland D, Wang J. Molecular marker development from transcript sequences and germplasm evaluation for cultivated peanut (*Arachis* hypogaea L.). *Mol Genet Genomics*. 2016;291(1): 363–81. doi: 10.1007/s00438-015-1115-6.

Peng Z, Liu F, Wang L, Zhou H, Paudel D, et al. Transcriptome profiles reveal gene regulation of peanut (*Arachis* hypogaea L.) nodulation. *Sci Rep*. 2017;7(1):1–12.

Peng Z, Zhao Z, Clevenger JP, Chu Y, Paudel D, et al. Comparison of SNP calling pipelines and NGS platforms to predict the genomic regions harboring candidate genes for nodulation in cultivated peanut. *Front Genet*. 2020;11:222. doi: 10.3389/fgene.2020.00222.

Pidon H, Chéron S, Ghesquière A, Albar L. Allele mining unlocks the identification of RYMV resistance genes and alleles in African cultivated rice. *BMC Plant Biol*. 2020;20(1):1–14.

Porto EM, Komor AC, Slaymaker IM, Yeo GW. Base editing: advances and therapeutic opportunities. *Nature Reviews Drug Discovery*. 2020;19(12):839–59.

Przybyla L, Gilbert LA. A new era in functional genomics screens. *Nat Rev Genet*. 2021:1–15.

Rathi D, Gayen D, Gayali S, Chakraborty S, Chakraborty N. Legume proteomics: progress, prospects, and challenges. *Proteomics*. 2016;16(2):310–27.

Rathod V, Hamid R, Tomar RS, Patel R, Padhiyar S, et al. Comparative RNA-Seq profiling of a resistant and susceptible peanut (*Arachis hypogaea*) genotypes in response to leaf rust infection caused by *Puccinia arachidis*. *3 Biotech*. 2020;10(6):284. doi: 10.1007/s13205-020-02270-w.

Ratnaparkhe MB, Wang X, Li J, Compton RO, Rainville LK, et al. Comparative analysis of peanut NBS-LRR gene clusters suggests evolutionary innovation among duplicated domains and erosion of gene microsynteny. *New Phytol*. 2011;192(1):164–78.

Ravi K, Hari U, Sangam D, David H, Varshney RK. Genetic relationships among seven sections of genus *Arachis* studied by using SSR markers. *BMC Plant Biol*. 2010;10(1):15.

Riaz A, Athiyannan N, Periyannan SK, Afanasenko O, Mitrofanova OP, et al. Unlocking new alleles for leaf rust resistance in the Vavilov wheat collection. *Theor Appl Genet*. 2018;131(1):127–44.

Ruan J, Guo F, Wang Y, Li X, Wan S, et al. Transcriptome analysis of alternative splicing in peanut (*Arachis hypogaea* L.). *BMC Plant Biol*. 2018;18(1):139.

Samoluk SS, Vaio M, Ortíz AM, Chalup LM, Robledo G, et al. Comparative repeatome analysis reveals new evidence on genome evolution in wild diploid *Arachis* (Fabaceae) species. *Planta*. 2022;256(3):1–18.

Schoen DJ, Schultz ST. Somatic mutation and evolution in plants. *Annu Rev Ecol Evol Syst*. 2019;50:49–73.

Shen S, Li Y, Wang J, Wei C, Wang Z, et al. Illegitimate recombination between duplicated genes generated from recursive polyploidizations accelerated the divergence of the genus *Arachis*. *Genes*. 2021;12(12):1944.

Shirasawa K, Bertioli DJ, Varshney RK, Moretzsohn MC, Leal-Bertioli SC, et al. Integrated consensus map of cultivated peanut and wild relatives reveals structures of the A and B genomes of *Arachis* and divergence of the legume genomes. *DNA Res*. 2013;20(2):173–84. doi: 10.1093/dnares/dss042.

Shirasawa K, Hirakawa H, Tabata S, Hasegawa M, Kiyoshima H, et al. Characterization of active miniature inverted-repeat transposable elements in the peanut genome. *Theor Appl Genet*. 2012a;124(8):1429–38. doi: 10.1007/s00122-012-1798-6.

Shirasawa K, Koilkonda P, Aoki K, Hirakawa H, Tabata S, et al. *In silico* polymorphism analysis for the development of simple sequence repeat and transposon markers and construction of linkage map in cultivated peanut. *BMC Plant Biol*. 2012b;12(1):80. doi: 10.1186/1471-2229-12-80.

Shirasawa K, Kuwata C, Watanabe M, Fukami M, Hirakawa H, et al. Target amplicon sequencing for genotyping genome-wide single nucleotide polymorphisms identified by whole-genome resequencing in peanut. *Plant Genome*. 2016. doi: 10.3835/plantgenome2016.06.0052.

Shu H, Luo Z, Peng Z, Wang J. The application of CRISPR/Cas9 in hairy roots to explore the functions of *AhNFR1* and *AhNFR5* genes during peanut nodulation. *BMC Plant Biol*. 2020;20(1):1–15.

Singh AK. Putative genome donors of *Arachis hypogaea* (*Fabaceae*), evidence from crosses with synthetic amphidiploids. *Plant Syst E*. 1988;160(3):143–51.

Singh BD, Shekhawat NS. *Gene Editing: Principles and Applications*. Jodhpur, India: Scientific Publishers;2021.

Sinha P, Bajaj P, Pazhamala LT, Nayak SN, Pandey MK, et al. *Arachis hypogea* gene expression atlas (AhGEA) for *fastigiata* subspecies of cultivated groundnut to accelerate functional and translational genomics applications. *Plant Biotechnol J*. 2020;18:2187–200.

Song W-Y, Pi L-Y, Wang G-L, Gardner J, Holsten T, et al. Evolution of the rice Xa21 disease resistance gene family. *Plant Cell*. 1997;9(8):1279–87.

Stalker HT. Utilizing wild species for peanut improvement. *Crop Sci*. 2017;57(3):1102–20.

Stalker HT, Phillips TD, Murphy JP, Jones TM. Variation of isozyme patterns among *Arachis* species. *Theor Appl Genet*. 1994;87(6):746–55.

Stalker HT, Simpson CE. Genetic resources in *Arachis*. In: He P, Stalker HT, editors. *Advances in Peanut Science*. Stillwatter, OK: APRES Inc.;1995. pp. 14–53.

Suassuna T, Suassuna N, Martins K, Matos R, Heuert J, et al. Broadening the variability for peanut breeding with a wild species-derived induced allotetraploid. *Agronomy*. 2020;10(12):1917.

Sun Y, Wang Q, Li Z, Hou L, Dai S, et al. Comparative proteomics of peanut gynophore development under dark and mechanical stimulation. *J Proteome Res*. 2013;12(12):5502–11.

Suzuki MM, Bird A. DNA methylation landscapes: provocative insights from epigenomics. *Nat Rev Genet*. 2008;9(6):465.

Tallury S, Isleib T, Copeland S, Rosas-Anderson P, Balota M, et al. Registration of two multiple disease resistant peanut germplasm lines derived from *Arachis cardenasii* Krapov. & WC Gregory, GKP 10017. *J Plant Regist*. 2014;8(1):86–9.

Tang Y, Du G, Xiang J, Hu C, Li X, et al. Genome-wide identification of auxin response factor (ARF) gene family and the miR160-ARF18-mediated response to salt stress in peanut (*Arachis* hypogaea L.). *Genomics*. 2021;114(1):171–84.

Tian Y, Thrimawithana A, Ding T, Guo J, Gleave A, et al. Transposon insertions regulate genome-wide allele-specific expression and underpin flower colour variations in apple (Malus spp.). *Plant Biotechnol J*. 2022;20(7):1285–97. doi: 10.1111/pbi.13806.

Toomer OT. A comprehensive review of the value-added uses of peanut (*Arachis hypogaea*) skins and by-products. *Crit Rev Food Sci Nutr*. 2020;60(2):341–50. doi: 10.1080/10408398.2018.1538101.

Valls J, Simpson C. Taxonomy, natural distribution, and attributes of *Arachis*. In: Kerridge PC, Hardy B, editors. *Biology and Agronomy of Forage Arachis*. San Diego, CA: CIAT;1994. pp. 1–18.

Valls JF, Simpson CE. New species of *Arachis* (*Leguminosae*) from Brazil, Paraguay and Bolivia. *Bonplandia*. 2005;14:35–63.

Valls JFM, Da Costa LC, Custodio AR. A novel trifoliolate species of *Arachis* (*Fabaceae*) and further comments on the taxonomic section Trierectoides. *Bonplandia*. 2013:91–7.

Varshney RK, Bohra A, Yu J, Graner A, Zhang Q, et al. Designing future crops: genomics-assisted breeding comes of age. Trends Plant Sci. 2021;26(6):631–49.

Varshney RK, Nayak SN, May GD, Jackson SA. Next-generation sequencing technologies and their implications for crop genetics and breeding. *Trends Biotechnol*. 2009;27(9):522–30.

Vaughn MW, Tanurdžić M, Lippman Z, Jiang H, Carrasquillo R, et al. Epigenetic natural variation in *Arabidopsis* thaliana. *PLoS One*. 2007;5(7):e174.

Wan L, Ren W, Miao H, Zhang J, Fang J. Genome-wide identification, expression, and association analysis of the monosaccharide transporter (MST) gene family in peanut (*Arachis hypogaea* L.). *3 Biotech*. 2020;10(3):130. doi: 10.1007/s13205-020-2123-8.

Wan Q, Luo L, Zhang X, Lv Y, Zhu S, et al. Genome-wide identification and abiotic stress response pattern analysis of NF-Y gene family in peanut (*Arachis* hypogaea L.). *Trop Plant Biol*. 2021;14:329–344.

Wang F, Nong X, Hao K, Cai N, Wang G, et al. Identification of the key genes involved in the regulation of symbiotic pathways induced by Metarhizium anisopliae in peanut (*Arachis hypogaea*) roots. *3 Biotech*. 2020b;10(3):124. doi: 10.1007/s13205-020-2105-x.

Wang J, Li C, Yan C, Zhao X, Shan S. A comparative analysis of the complete chloroplast genome sequences of four peanut botanical varieties. *PeerJ*. 2018;6:e5349.

Wang ML, Chen CY, Tonnis B, Barkley NA, Pinnow DL, et al. Oil, fatty acid, flavonoid and resveratrol content variability and *FAD2A* functional SNP genotypes in the U.S. peanut mini-core collection. *J Agric Food Chem.* 2013;61(11):2875–82. doi: 10.1021/jf305208e.

Wang P, Gao C, Bian X, Zhao S, Zhao C, et al. Genome-wide identification and comparative analysis of cytosine-5 DNA methyltransferase and demethylase families in wild and cultivated peanut. *Front Plant Sci.* 2016;7:7. doi: 10.3389/fpls.2016.00007.

Wang RJ, Gao XF, Yang J, Kong XR. Genome-wide association study to identify favorable SNP allelic variations and candidate genes that control the timing of Spring bud flush of Tea (*Camellia sinensis*) using SLAF-seq. *J Agric Food Chem.* 2019;67(37):10380–91.

Wang X, Xu P, Ren Y, Yin L, Li S, et al. Genome-wide identification of meiotic recombination hot spots detected by SLAF in peanut (*Arachis* hypogaea L.). *Sci Rep.* 2020a;10(1):1–11.

Wang Y, Zhang X, Zhao Y, Prakash C, He G, et al. Insights into the novel members of the *FAD2* gene family involved in high-oleate fluxes in peanut. *Genome.* 2015;58(8):375–83.

Weeks DP. Gene editing in polyploid crops: wheat, camelina, canola, potato, cotton, peanut, sugar cane, and citrus. In: Donald P. Weeks, Bing Yang, editors. *Progress in Molecular Biology and Translational Science.* Elsevier, Academic Press, Cambridge:2017. pp. 65–80.

Wen S, Liu H, Li X, Chen X, Hong Y, et al. TALEN-mediated targeted mutagenesis of fatty acid desaturase 2 (FAD2) in peanut (*Arachis hypogaea* L.) promotes the accumulation of oleic acid. *Plant Mol Biol.* 2018;97(1–2):177–85.

Xia H, Zhao C, Hou L, Li A, Zhao S, et al. Transcriptome profiling of peanut gynophores revealed global reprogramming of gene expression during early pod development in darkness. *BMC Genom.* 2013;14(1):517.

Xia H, Zhu L, Zhao C, Li K, Shang C, et al. Comparative transcriptome analysis of anthocyanin synthesis in black and pink peanut. *Plant Signal Behav.* 2020;15(2):1721044. doi: 10.1080/15592324.2020.1721044.

Xu M, Luo J, Li Y, Shen L, Zhang X, et al. First comprehensive proteomics analysis of lysine crotonylation in leaves of peanut (*Arachis* hypogaea L.). *Proteomics.* 2021;21(5):2000156.

Xu P, Tang G, Cui W, Chen G, Ma CL, et al. Transcriptional differences in peanut (*Arachis hypogaea* L.) seeds at the freshly harvested, after-ripening and newly germinated seed stages: Insights into the regulatory networks of seed dormancy release and germination. *PLoS One.* 2020;15(1):e0219413. doi: 10.1371/journal.pone.0219413.

Yang S, Li L, Zhang J, Geng Y, Guo F, et al. Transcriptome and differential expression profiling analysis of the mechanism of Ca2+ regulation in peanut (*Arachis* hypogaea) pod development. *Front Plant Sci.* 2017;8:1609.

Yang T. *Molecular Cloning and Characterization of a Stilbenoid Prenyltransferase Gene Family in Peanut.* Arkansas State University, 2016.

Yin D, Ji C, Ma X, Li H, Zhang W, et al. Genome of an allotetraploid wild peanut *Arachis monticola*: a *de novo* assembly. *GigaScience.* 2018;7(6):giy066.

Yin D, Ji C, Song Q, Zhang W, Zhang X, et al. Comparison of *Arachis monticola* with diploid and cultivated tetraploid genomes reveals asymmetric subgenome evolution and improvement of peanut. *Adv Sci.* 2020;7(4):1901672. doi: 10.1002/advs.201901672.

Yin D, Wang Y, Zhang X, Li H, Lu X, et al. De novo assembly of the peanut (*Arachis hypogaea* L.) seed transcriptome revealed candidate unigenes for oil accumulation pathways. *PLoS One.* 2013;8(9):e73767.

Yin K, Gao C, Qiu J-L. Progress and prospects in plant genome editing. *Nat Plants.* 2017;3(8):1–6.

Yu R, Jiang Q, Xv C, Li L, Bu S, et al. Comparative proteomics analysis of peanut roots reveals differential mechanisms of cadmium detoxification and translocation between two cultivars differing in cadmium accumulation. *BMC Plant Biol.* 2019;19(1):137.

Yu R, Ma Y, Li Y, Li X, Liu C, et al. Comparative transcriptome analysis revealed key factors for differential cadmium transport and retention in roots of two contrasting peanut cultivars. *BMC Genom.* 2018;19(1):938.

Yuan C, Li C, Lu X, Zhao X, Yan C, et al. Comprehensive genomic characterization of NAC transcription factor family and their response to salt and drought stress in peanut. *BMC Plant Biol.* 2020;20(1):1–21.

Yuan M, Zhu J, Gong L, He L, Lee C, et al. Mutagenesis of *FAD2* genes in peanut with CRISPR/Cas9 based gene editing. *BMC Biotechnol.* 2019;19(1):24.

Zagorski D, Hartmann M, Bertrand YJ, Paštová L, Slavíková R, et al. Characterization and dynamics of repeatomes in closely related species of *Hieracium* (*Asteraceae*) and their synthetic and apomictic hybrids. *Front Plant Sci.* 2020;11:591053. doi: 10.3389/fpls.2020.591053

Zanini SF, Bayer PE, Wells R, Snowdon RJ, Batley J, et al. Pangenomics in crop improvement-From coding structural variations to finding regulatory variants with pangenome graphs. *Plant Genome.* 2022;15(1): e20177.

Zemach A, McDaniel IE, Silva P, Zilberman D. Genome-wide evolutionary analysis of eukaryotic DNA methylation. *Science*. 2010;328(5980):916–9.

Zhang J, Liang S, Duan J, Wang J, Chen S, et al. De novo assembly and characterisation of the transcriptome during seed development, and generation of genic-SSR markers in peanut (*Arachis hypogaea* L.). *BMC Genom*. 2012b;13:90. doi: 10.1186/1471-2164-13-90.

Zhang T, Hu Y, Jiang W, Fang L, Guan X, et al. Sequencing of allotetraploid cotton (*Gossypium hirsutum* L. acc. TM-1) provides a resource for fiber improvement. *Nat Biotechnol*. 2015;33(5):531–7.

Zhang X, Zhang K, Luo L, Lv Y, Li Y, et al. Identification of peanut Aux/IAA genes and functional prediction during seed development and maturation. *Plants*. 2022;11(4):472.

Zhang Y, Wang P, Xia H, Zhao C, Hou L, et al. Comparative transcriptome analysis of basal and zygote-located tip regions of peanut ovaries provides insight into the mechanism of light regulation in peanut embryo and pod development. *BMC Genom.* 2016;17(1):606.

Zhang Z, Gou X, Xun H, Bian Y, Ma X, et al. Homoeologous exchanges occur through intragenic recombination generating novel transcripts and proteins in wheat and other polyploids. *Proceedings of the National Academy of Sciences*. 2020;117(25):14561–71.

Zhao C, Zhao S, Hou L, Xia H, Wang J, et al. Proteomics analysis reveals differentially activated pathways that operate in peanut gynophores at different developmental stages. *BMC Plant Biol*. 2015;15(1):188.

Zhao N, He M, Li L, Cui S, Hou M, et al. Identification and expression analysis of WRKY gene family under drought stress in peanut (*Arachis* hypogaea L.). *PLoS One*. 2020a;15(4):e0231396. doi: 10.1371/journal.pone.0231396.

Zhao Y, Ma J, Li M, Deng L, Li G, et al. Whole-genome resequencing-based QTL-seq identified *AhTc1* gene encoding a R2R3-MYB transcription factor controlling peanut purple testa colour. *Plant Biotechnol J*. 2020b;18(1):96–105. doi: 10.1111/pbi.13175.

Zheng L, Shockey J, Guo F, Shi L, Li X, et al. Discovery of a new mechanism for regulation of plant triacylglycerol metabolism: The peanut diacylglycerol acyltransferase-1 gene family transcriptome is highly enriched in alternative splicing variants. *J Plant Physiol*. 2017;219:62–70.

Zhou X, Dong Y, Zhao J, Huang L, Ren X, et al. Genomic survey sequencing for development and validation of single-locus SSR markers in peanut (*Arachis hypogaea* L.). *BMC Genom*. 2016;17(1):420.

Zhou X, Xia Y, Ren X, Chen Y, Huang L, et al. Construction of a SNP-based genetic linkage map in cultivated peanut based on large scale marker development using next-generation double-digest restriction-site-associated DNA sequencing (ddRADseq). *BMC Genom*. 2014;15:351. doi: 10.1186/1471-2164-15-351.

Zhou Y, Lei F, Wang Q, He W, Yuan B, et al. Identification of novel alleles of the rice blast-resistance gene Pi9 through sequence-based allele mining. *Rice*. 2020;13(1):1–15.

Zhu J, Tang G, Xu P, Li G, Ma C, et al. Genome-wide identification of xyloglucan endotransglucosylase/hydrolase gene family members in peanut and their expression profiles during seed germination. *PeerJ*. 2022;10:e13428.

Zhu W, Chen X, Li H, Zhu F, Hong Y, et al. Comparative transcriptome analysis of aerial and subterranean pods development provides insights into seed abortion in peanut. *Plant Mol Biol*. 2014;85(4–5):395–409.

Zhu W, Zhang E, Li H, Chen X, Zhu F, et al. Comparative proteomics analysis of developing peanut aerial and subterranean pods identifies pod swelling related proteins. *J Proteomics*. 2013;91:172–87.

Zhuang W, Chen H, Yang M, Wang J, Pandey MK, et al. The genome of cultivated peanut provides insight into legume karyotypes, polyploid evolution and crop domestication. *Nat Genet*. 2019;51(5):865–76.

Zhuang W, Wang X, Paterson AH, Chen H, Yang M, et al. Reply to: Evaluating two different models of peanut's origin. *Nat Genet*. 2020;52:560–3.

8 Integration of Genomics-Assisted Allele Mining Approaches for Improvement of Flax

Current Advancements and Future Perspective

Shashank Kumar Yadav, Mamta Singh, Dhammaprakash Pandhari Wankhede, Devender Singh, and Vikender Kaur

8.1 INTRODUCTION

Flax (*Linum usitatissimum* L.) is a self-pollinated annual crop with a genome size of around 370 Mb. It belongs to the family Linaceae, comprising 22 genera and approximately 300 species, although only *L. usitatissimum* is commercially important (Hickey, 1988). Flax is contemplated as a "founder crop" providing basal material for food, paints, automobiles, biofuel, soap, varnishes, polymers, textiles, medicine, and other industrial uses (Przybylski, 2005; Diederichsen and Ulrich, 2009; Shim et al., 2014). Nutritionally flaxseed is rich in oil with the highest proportion of omega-3 fatty acids (alpha-linolenic acid; 55%–57%), dietary fibers (27.3%), protein (18.29%), and lignans (294–700 mg/100 g on a dry weight basis) (Singh et al., 2011; Goyal et al., 2014; Kajla et al., 2015). Canada is the leading producer and exporter of flax (~23% of world production), followed by Kazakhstan and Russia (FAOSTAT, 2022; Yadav et al., 2022). Presently, it is cultivated over more than 24.37 lakh ha, with a production of 22 lakh tons and a productivity of 903 kg/ha (FAOSTAT, 2022; Figure 8.1). However, various abiotic and biotic stresses are major constraints for sustainable

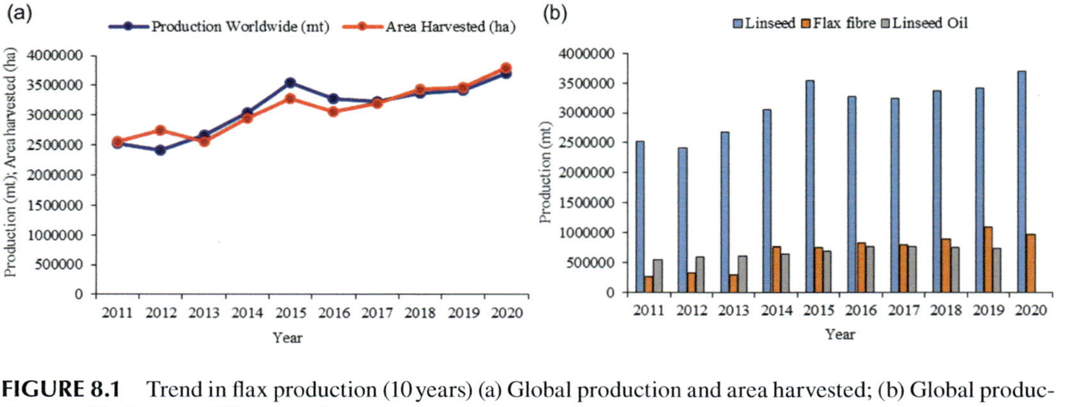

FIGURE 8.1 Trend in flax production (10 years) (a) Global production and area harvested; (b) Global production specified as per different end uses.

DOI: 10.1201/9781003385035-8

production in all important flax-producing areas. Furthermore, climatic fluctuations certainly put more pressure on the overall vegetative and reproductive growth of the plant, primarily in the seed filling/dough stage, resulting in considerable yield losses. Drought, high temperature, and soil salinity are the main abiotic stresses, resulting in numerous physiological, morphological, and metabolic alterations ultimately leading to yield reduction (Fischer and Maurer, 1978; Ludlow and Muchow, 1990). Among biotic stress, insect pests like linseed bud fly (*Dasyneura lini* Barnes), defoliators like semilooper (*Plusia orichalcea* Fabr), Lucern caterpillar (*Spodoptera exigua* Hubn.), tobacco caterpillar (*Spodoptera litura* Fabr.), and Bihar hairy caterpillar (*Spilarctia obliqua* Walk.) appear as sporadic and region-specific pests with heavy incidence in some cropping seasons, resulting in 40%–60% crop loss globally (Mukherji et al., 1999). Genetic enhancement of varieties remains the most compelling, sustainable, and ecologically synergistic approach to overcoming these steadily evolving threats and improving stress tolerance and yield potential (Kaur et al., 2017, 2018). The yield gains in flax had been very low and slow (0.5 tons/ha since the past 50 years) through conventional breeding. Therefore, genome-wide availability of molecular markers and QTLs/QTNs/SNPs/candidate genes for economically important traits is expected to play a pivotal role in further genetic gains in flax through markers/genomics-assisted and/or haplotype-based breeding. The diverse genetic pool conserved in global genebanks can support breeding activities by providing promising germplasm as sources of novel alleles to meet these challenges for sustainable flax production. Knowledge pertaining to the extent of genetic variability is imperative for association mapping of flax collections and to expedite the process of breeding through marker-assisted selection (MAS) (Flint-Garcia et al., 2005; Rajwade et al., 2010; Kale et al., 2012; Chandrawati et al., 2014; Pali et al., 2014, 2015). Until now, genetic markers such as amplified fragment length polymorphism (AFLP), inter-simple sequence repeat (ISSR), random amplified polymorphic DNA (RAPD), and expressed sequence tags-simple sequence repeats (EST-SSR) have been extensively used for understanding genetic diversity and population structure in flax (Fu et al., 2002a, b, 2003a, b, c; Cloutier et al., 2009, 2011, 2012a, b; Rajwade et al., 2010; Uysal et al., 2010; Saroha et al., 2022b). After the availability of the whole genome sequence of flax (Wang et al., 2012) and reduced representation sequencing information, numerous studies have led to the identification of a number of SNPs linked to major agronomic and seed quality traits (Deng, 2013; Soto-Cerda et al., 2013, 2014, 2018, 2021; Xie et al., 2018; You et al., 2018b; Saroha et al., 2022a), fiber content (Xie et al., 2018), and biotic and abiotic stress resistance (He et al., 2019; Sertse et al., 2021; Li et al., 2022; You et al., 2022) in flax. The present chapter is intended to briefly describe various aspects of allele mining, strategic methodologies/techniques employed, and their applications in genomics-assisted improvement of flax.

8.2 GENOMICS TOOLS AND RESOURCES FOR GENETIC ENHANCEMENT OF FLAX

Over the past decade, genomics research in flax has grown by leaps and bounds owing much to the availability of high-throughput whole-genome sequencing (Wang et al., 2012), re-sequencing (Guo et al., 2020), and reduced representation sequencing such as genotyping-by-sequencing (GBS) (Saroha et al., 2022a). These approaches have translated genotypic-phenotypic data into highly applicable trait-specific information for flax improvement. Primarily, these traits can be summarized either for value enrichment in the form of better oil content and fatty acid composition in linseed, fiber content, and quality in fiber flax (Diederichsen and Ulrich, 2009), or for the identification and functional characterization of genes to achieve resistance to various biotic and abiotic stresses. Molecular or genome-based markers (allele-specific) have evidently been established as the most promising tool for assessing huge genetic variability both within and between populations by investigating a large number of loci distributed throughout the genome (Powell et al., 1995; Spooner et al., 2005). The dissection of novel variants in candidate gene(s)/loci, referred to as allele mining, has a key role in targeted trait improvement as well as the categorization of germplasm for conservation.

8.3 EVOLUTION OF NOVEL ALLELES

Recombination events and mutations are the primary propulsive forces underlying genetic diversity across plant species. Majorly, the resultant allelic variants are deleterious, but in generic terms, they prove to be beneficial (0.1%) for better adaptability and survival (Singh, 2005). Identification and trait-responsive characterization of existing alleles/allelic combinations, and novel alleles in the genic and regulatory regions of the genome as single nucleotide polymorphism (SNP) or as insertion and deletion (InDel) is imperative for trait-tailored precision breeding. These genomic changes or mutations decisively regulate cellular metabolism, leading to an alteration in the dynamics of proteins and resultant phenotypic changes (Doebley and Lukens, 1998; Buckler and Thornsberry, 2002). Naturally occurring allelic variability modulates the expression of a gene(s), causing trait-specific quantitative variation, whereas induced mutations either eliminate or cause an exponential reduction in the functionality of a gene (Tanksley, 1993; Yano and Sasaki, 1997; Mackay, 2001).

8.4 APPROACHES FOR ALLELE MINING

Three key approaches are being employed for discerning sequence polymorphism in naturally occurring populations, *viz.*, (i) targeting induced local lesions in genomes (modified TILLinG), also known as Eco-TILLinG-based allele mining; (ii) sequencing-based allele mining; and (iii) association mapping-based allele mining. A brief description of these approaches with respect to their application in flax has been summarized in the subsequent sections.

8.4.1 Eco-TILLinG or Modified TILLinG

TILLinG is a reverse genetics approach for the identification of single base-pair allelic variation in the target gene, more specifically through artificially induced point mutations (Comai et al., 2004), while the Eco-TILLinG technique detects natural mutations/variations in genes. The main steps involve DNA isolation and pooling from reference and test genotypes, cleavage by nuclease, discernment in polyacrylamide gels using Li-Cor-based genotypers, and confirmation through sequencing (Raghavan et al., 2007). It allows rapid discernment across genetic variants in many individuals in a cost-effective manner, as only one individual from each haplotype group needs to be sequenced. TILLinG utilizes chemical mutagens like ethyl methane sulfonate (EMS) and azacitidine to induce random mutations in the genome through mutagenizing seeds, causing G/C-to-A/T point mutations. The resulting M1 plants are selfed, and subsequent M2 individuals are polymerase chain reaction (PCR) analyzed using infrared responsive (IR) dye-labeled primers for the formation of mismatches or heteroduplexes, which represent naturally occurring SNPs and induced SNPs. Single-strand-specific nucleases (S1 nuclease, T4 endonuclease VII, and Cel-1) are used for digesting mismatched base pairs, which are analyzed through denaturing polyacrylamide gel electrophoresis. Green and Marshall (1984) developed EMS-mediated mutant linseed lines with reduced α-Linolenic acid (ALA) content (<29%). A further reduction in ALA content to ~2% was reported by Green (1986) and Rowland (1991). The *fatty acid desaturase* 3 (FAD3) genes, *lufad3a* and *lufad3b*, had point mutations leading to premature termination of translation in one of the characterized EMS mutant lines resulting in non-functional FAD3 enzymatic activity (Vrinten et al., 2005). Moreover, flaxseed EMS mutagenized population has been successfully phenotyped for the identification of a number of lignin biosynthetic genes; *Coumarate-3-Hydroxylase* (*C3H*) and *Cinnamyl Alcohol Dehydrogenase* (*CAD*) gene mutants: *c3h* and *cad* (Chantreau et al., 2013).

8.4.2 Sequencing-Based Allele Mining

This technique involves PCR amplification of alleles across diverse genotypes/population followed by identification of allelic variability by DNA sequencing for the elucidation of haplotype structure and diversity analysis. Sequencing-based allele mining requires simple instrumentation and technical know-how, although it involves huge operational costs. On the other hand, Eco-Tilling is cost-efficient but requires technical expertise and a sophisticated, tedious methodology. However, the advent and advancement of NGS-based technologies with low cost and good sequence quality are being strived for. This will facilitate large-scale genotyping of genetic resources, leading to faster generation of allelic data.

8.4.2.1 Next-Generation Sequencing

Over the past decade, many high-throughput (HT) sequencing or NGS platforms have been devised, leading to the generation of genome sequence data more swiftly, inexpensively, and efficaciously compared to preceding sequencing approaches (Gupta, 2008; Hudson, 2008; Mardis, 2008; Varshney et al., 2009; Kilian and Graner, 2012). The accuracy and detection power of NGS-based TILLinG have an edge over other established methods for the identification of rare alleles in populations (Tsai et al., 2011). Several modern large-scale NGS-based sequencing platforms, including Illumina/GA, Roche/GS FLX, supported oligonucleotide ligation and detection system (SOLiD) (Applied Biosystems), cPAL sequencing (Shendure, 2008; Metzker, 2010), and the HeliScope Single Molecule Sequencer: Helicos Heliscope (http://seqll.com/) are commercially available depending upon the sequencing read length. Earliest amongst these types was 454 Life Sciences, based on pyro-sequencing and producing 100 Mb of sequence with 99.5% accuracy. Another extensively used sequencer, Illumina/Solexa, can sequence one billion bases (1 Gb) of 30–40 base sequence reads per run in a short duration of 3–4 days. SOLiD can generate 2–3 Gb of sequence in a single run, despite shorter read lengths of 25–35 (Hutchison, 2007). With the availability of flax genome sequence data, NGS has witnessed a major thrust. The genome of flax (cultivar CDC Bethune) was sequenced *de novo* employing a whole-genome shotgun (WGS) approach on the Illumina platform using paired-end and mate-pair libraries. The resulting assembly of 116,602 contigs and 88,384 scaffolds covered 81% of the 370 Mb of the flax genome (Wang et al., 2012). Lately in 2018, chromosome-level assembly for 15 chromosome pairs of CDC Bethune was achieved using BioNano genome optical, BAC-based physical and genetic mapping (You et al., 2018a). de Santana Lopes et al. (2018) sequenced the plastid genome of *L. usitatissimum*, representing 156,721 bp of DNA-ring molecule, leading to the identification of 109 unique genes and 2 pseudo-genes in the genome. Furthermore, scaffold-level genome assemblies of flaxseed cultivar Longya-10, fiber cultivar Heiya-14, and pale flax (*L. bienne*) using llumina sequencing and genetic mapping were reported by Zhang et al. (2020). More recently, the genome of the fiber flax cultivar Atlant (Dmitriev et al., 2021) and the chromosome-scale high-quality genome sequence of another fiber flax cultivar YY5 were reported, which substantially improved the earlier assembly (Sa et al., 2021). Owing to the availability and advancement of whole genome sequence data, NGS could be extensively explored for harnessing, identification, and elucidation of genetic and functional variability of candidate genes/alleles and QTL mapping to aid studies related to population genetics, germplasm conservation, and molecular ecology (Kilian and Graner, 2012). Using genomic and transcriptomic assemblies, polymorphic variability in gene expression across different agro-climatic and geographical origins has been assessed in different species of the genus *Linum* and cultivars of *L. usitatissimum* (Bolsheva et al., 2019; Dmitriev et al., 2019, 2020a, b; Novakovskiy et al., 2019; Wu et al., 2019; Ali et al., 2020; Guo et al., 2020). Another more recent technology, single-molecule sequencing, termed as third-generation sequencing, is much faster and more economical compared to second-generation sequencing technologies offered by Applied Biosystems and Illumina; however, presently, it is in a nascent stage and is evolving rapidly around modern technological tools.

8.4.3 Association Mapping-Based Allele Mining

Association mapping, also known as "linkage disequilibrium mapping", is the key to the identification of specific functional genetic variants (loci, alleles) linked to phenotypic differences in a trait (Flint-Garcia et al., 2003; Yu and Buckler, 2006; Zhu et al., 2008). It takes advantage of historic linkage disequilibrium to link phenotypes to genotypes and uncover genetic associations. It has an advantage over bi-parental mapping, where high-density linkage maps are constructed through limited parent hybridization for localization of genes or QTLs for target traits in specific linkage segments. The efficacy and precision for QTL detection are superior in association-based mapping because of the utilization of a diverse germplasm population with genetic variation (Gupta et al., 2005). Information thus generated can be utilized by geneticists and breeders for producing progenies from pre-designated crosses, leading to inbreds and recombinant-inbred lines (RILs) for genomics-assisted crop improvement.

8.5 TOOLS REQUIRED FOR ALLELE MINING

After the generation and acquisition of humongous genomic data, sorting and analyzing it into useful and easily interpretable form becomes imperative. An indispensable part of allele mining is the identification of polymorphisms in the test population through comparative analysis of sequences from the bulk data generated. Therefore, several bioinformatics tools are available for handling complex nucleotide data, structural and functional prediction of complex biomolecules, transcription factor binding sites, *in silico* putative functional classification of plant promoters (Yadav et al., 2020), and identification of polymorphisms in genomic and protein sequences resulting in an altered phenotype. These tools also simplify the identification of sequence polymorphism and access whether the concerned variation translates to variation in transcription factor binding motifs (TFBMs). The majority of these computational-based prediction models rely on algorithms, which are largely developed on the basis of pre-characterized sequences available in databases that need to be validated through *in vivo* or *in vitro* experimental approaches.

8.6 APPLICATIONS OF ALLELE MINING

Allele mining plays a key role in the identification and functional characterization of novel genes and superior alleles. It is instrumental in unraveling molecular mechanisms underlying novel trait variations in relation to changes in their corresponding nucleotide sequence. Furthermore, the rate of genetic drift, divergence of candidate gene(s), and allelic synteny with members of close and distinctly related species can be studied. Allele mining also plays a key role in molecular discrimination among related species, the development of allele-specific molecular markers, and thereby facilitating the introgression of novel alleles through MAS or genome editing through clustered regularly interspaced short palindromic repeats (CRISPR) and CRISPR-associated protein 9 (CRISPR-Cas9) technology.

8.6.1 Diversity Assessment

Several national and international institutes and genebanks are maintaining crop germplasm collections, which can be used to unravel the genetic diversity of crop plants and gain insights into the allelic richness of conserved material for efficient utilization. Diversity in flax germplasm maintained at Plant Gene Resources of Canada (PGRC) has been delineated for distinct fiber and oilseed type morphotypes to develop a global core germplasm set for association mapping and genetic improvement based on the 414 microsatellite or simple sequence repeat (SSR) marker (Soto-Cerda et al., 2013). You et al. (2017) assessed diversity analysis for agronomic, seed quality, fiber and disease resistance in the PGRC flax core collection. Genetic diversity analysis of the 220 Indian and a few exotic linseed accessions, particularly for flowering time and maturity-related traits, was carried out based upon the polymorphism using SSR markers (Saroha et al., 2022b). These studies play an important role in flax genetic improvement programs through the identification of underlying marker alleles specific to certain agronomically important traits.

8.6.2 Identification of Novel Haplotypes

Variations in the genic regions can be associated with phenotypic variation for a particular trait which can be harnessed for evaluating the frequency, type, and degree of occurrence of new haplotypes resulting in altered phenotypes. Information on common haplotype variants and their frequency in populations forms the basis for association mapping studies. Interactions between haplotypes can significantly regulate numerous agronomic traits. Furthermore, knowledge of haplotypic associations with complex traits can be exploited for genomic selection (GS) and breeding. Many case studies are available for analyzing the haplotype diversity at candidate genes, for example, α-ALA content in linseed (Thambugala et al., 2013; Rajwade et al., 2016). Similarly, six *desaturase* gene (*FAD*) haplotypes were identified in 10 linseed varieties revealing two to nine novel alleles for each *FAD* gene resulting in 1–4 protein isoforms. Most of these changes were non-synonymous in nature. Likewise, Guo et al. (2020) re-sequenced 200 flax cultivars and generated a genome variation map comprising 674,074 SNPs. These SNP variations were used for categorizing the selected cultivars into two groups, including haplotypes (functional alleles) with longer seed length, width, and more thousand seed weight, and the other group having haplotypes (nonfunctional alleles) with small seed size and less thousand seed weight. Haplotypic variants for *Melampsora lini* avirulence (*Avr*) genes have been extensively used for the characterization of rust resistance in flax (Dodds et al., 2004, 2006; Catanzariti et al., 2006).

8.6.3 Development of Allele-Specific Markers

Different genetic markers have been developed based on various polymorphism-detecting techniques such as southern blotting-nucleic acid hybridization, PCR-based sequencing, RFLP, AFLP, RAPD, SSR, SNPs, sequence-tagged sites (STS), cleaved amplified polymorphic sequences (CAPS), etc. (Collard et al., 2005). These markers are being extensively used for gene cloning, genetic variability analysis, genotyping, deciphering of QTLs, identification of disease resistance genes, phylogenetic studies, linkage mapping, etc.

8.6.3.1 RAPD and AFLP Markers

The preliminary reports on *Linum* molecular markers were based on RAPD and AFLP technologies for molecular characterization and genetic variability analysis in flax genotypes, revealing a high degree of polymorphism (Aldrich and Cullis, 1993; Spielmeyer et al., 1998a; Stegnii et al., 2000; Fu et al., 2002a, b, 2003a, b, c; Fu, 2005; Diederichsen and Fu, 2006; Cullis, 2007; Ottai et al., 2012; Yurkevich et al., 2013; Chandrawati et al., 2014, 2017; Iqbal et al., 2014; Bolsheva et al., 2015; Nagabhushanam et al., 2021). AFLP markers have been employed for the identification of dominant polymorphisms and linkage map construction (Oh et al., 2000; Everaert et al., 2001; Schulman, 2007). Two QTLs were identified through AFLP-based independent linkage mapping for resistance to *Fusarium* wilt, the most devastating fungal disease affecting flax. The construction of the linkage map was accomplished through the mapping of populations from doubled-haploid (DH) lines from a cross between the polyembryonic, low-linolenic-acid genotype (CRZY8/RA91), and the Australian cultivar 'Glenelg' (Spielmeyer et al., 1998b). Several attempts at refining the AFLP technique in flax were made (van Treuren, 2001); however, broader applications are still limited.

8.6.3.2 SSR Markers

SSR markers are created by the amplification of a locus with a variable number of short tandem repeats, hence producing amplicons of different lengths. SSRs have been found to be ubiquitous across all eukaryotes and are widely used owing to co-dominance, multi-allelism, heritability, genome-wide distribution, polymorphism, and ease of genotyping (Weber and May, 1989; Powell et al., 1996). However, the generation of SSR markers requires the technical expertise of *de-novo* development through the construction and sequencing of genomic libraries. Earlier degenerate primers with SSR tails (Fisher et al., 1996) were used in flax (Wiesner et al., 2001), but this approach had limited applicability as it generated few markers. Subsequently, SSR-enriched libraries were used for the identification of flax-specific SSRs (Roose-amsaleg et al., 2006; Deng et al., 2008; Bickel et al., 2011; Kale et al., 2012); however, this method too was time- and labour-intensive and resulted in a small number of unique and polymorphic SSRs. Later flax genome sequences and data from large-scale expressed sequence tag (EST) collections of 274,278 ESTs generated from libraries from 14 different tissues were mined to identify SSRs (Cloutier et al., 2009, 2012a; Soto-Cerda et al., 2011a, b; Venglat et al., 2011). Overall, the mining of ESTs and genomic survey sequences (GSSs) resulted in 88% of the SSR markers available to date for flax in public databases. The summary of SSR markers developed in flax has been briefly recapitulated under Table 8.1 with further additions after Cloutier et al. (2019). These markers were further used to characterize segregating populations (Cloutier et al., 2011; Kumar et al., 2015), germplasm collections (Soto-Cerda et al., 2013), and other *Linum* species (Fu and Peterson, 2010; Soto-Cerda et al., 2011b, 2014). Likewise, ISSR markers were applied for the evaluation of genetic diversity in flax genotypes (Wiesnerová and Wiesner, 2004; Rajwade et al., 2010; Uysal et al., 2010; Pali et al., 2015). Genetic diversity studies using retro transposon-based SSAP markers are, however, rare (Melnikova et al., 2014). Asgarinia et al. (2013) constructed a linkage map based on 143 SSR markers in the F_2 population of 300 individuals made from a cross between NorMan and the Linda cultivars. More recently, a considerably larger number of SNP markers have been available, which is fortifying genome-wide association studies.

8.6.3.3 QTLs and SNPs-Based Markers

Most of the agronomically important traits, such as yield, quality, and tolerance to biotic and abiotic stress, are polygenic traits and are often designated as quantitative traits. A total of 313 QTLs for 31 quantitative traits have been reported in chromosome-scale pseudo-molecules in flax that have been associated with seed quality traits (oil and amino acid content), fiber content, flowering time, and resistance to various biotic and abiotic stresses. SNPs are bi-allelic, single genetic code variations

TABLE 8.1
Summary of SSR Markers Developed in Flax Using Different Molecular Approaches

SSR Source	Approach	No. of Germplasm Accessions/ Genotypes Evaluated	Polymorphic Primer Combinations	Loci Identified	Loci per Primer Pair	References
Genomic	Degenerate primers	8	10	–	–	Wiesner et al. (2001)
	SSR-enriched libraries	93	23	28	1.22	Roose-amsaleg et al. (2006)
	SSR-enriched libraries	8	35	37	1.06	Deng et al. (2010)
	SSR-enriched libraries	8	38	38	1.00	Deng et al. (2011)
	SSR-enriched libraries	27	9	–	–	Kale et al. (2012)
	SSR-enriched libraries	19	42	42	1.00	Bickel et al. (2011)
	Sequence data mining	60	150	408	1.00	Soto-Cerda et al. (2012)
	Sequence data mining	34	89	29	–	Saroha et al. (2022b)
ESTs	Sequence data mining	23	248	275	1.11	Cloutier et al. (2009)
	Sequence data mining	61	23	23	1.00	Soto-Cerda et al. (2011b)
	Sequence data mining	16	145	149	1.03	Cloutier et al. (2012a)
BESs	Sequence data mining	16	673	720	1.07	Cloutier et al. (2012a)

distributed throughout the genome and a powerful tool for the analysis of allelic polymorphism in a population. The abundance of these SNPs makes them the most favorable markers for many association studies and the construction of high-density genetic maps. Fu and Peterson (2012) discovered 1,067 SNPs in the flax genome using pyro-sequencing. Furthermore, Kumar et al. (2012) discovered 55,465 SNPs using the Illumina platform, of which 96.8% were validated in the biparental population. Using the NGS approach, 407 genotypes in the PGRC core collection were re-sequenced, leading to the identification of 1,775,461 high-quality SNPs (You et al., 2016; He et al., 2019). Kumar et al. (2015) employed 329 SNP markers and 362 SSR markers for the construction of a genetic map in an RIL population derived from a cross between Canadian varieties CDC Bethune and Macbeth. Likewise, Saroha et al. (2022a) identified 335 unique QTNs in an association mapping panel of 131 accessions for flowering time, maturity, and plant height across five environments using the Multi-Locus Genome Wide Association Study (ML-GWAS). These studies vividly substantiate the existence of a large amount of unexplored allelic diversity and provide tremendous scope for the identification and implementation of genetic variability for trait-tailored enhancement in flax (Table 8.2).

TABLE 8.2
List of QTLs/SNPs Identified in Flax for Various Agro-Morphological, Seed Quality Traits, Biotic and Abiotic Stresses

Trait	QTN/QTL	Candidate Gene/ Marker	Function	Reference
Agro-morphological traits	*QPh.1*	*LUSc_149_01*	Plant height	Chandrawati and Yadav (2017)
	QPh.2	*LU_2497*		
	Chr1	Scaffold 156	Plant height	Wu et al. (2018)
	Chr8	Scaffold 2786	Stem length	
	uq.C1–1	*Lus10012562, Lus10016188*	Total length	Zhang et al. (2018)
	uq.C8–1	*Lus10011957*		
	uq.C2–2	*Lus10014561, Lus10014535, Lus10030692*	Plant height	
	uq.C7–1	–		
	uq.C4–1	*Lus10041481*		
	uq.C5–1	*Lus10007137*		
	uq.C6–1	*Lus10034082, Lus10017587, Lus10019435*		
	uq.C8–2	–		
	uq.C8–3	*Lus10012249, Lus10012252*		
	uq.C11–1	*Lus10038920*		
	uq.C12–1	*Lus10036228, Lus10036227, Lus10036229, Lus10036247, Lus10023392*		
	uq.C14–1	*Lus10015608, Lus10015614*		
	PH population	*Lus10027782, Lus10016188,*		
	uq.C1–1	*Lus10027887*		
	uq.C3–1	–		
	uq.C9–1	*Lus10002031, Lus10015623*		
	uq.C11–1	*Lus10038920*		
	uq.C12–1	*Lus10031416*		
	uq.C13–1	*Lus10026439, Lus10026441*		
	QBp.1	*LU_3017*	Branches/plant	Chandrawati and Yadav (2017)

(*Continued*)

TABLE 8.2 (*Continued*)
List of QTLs/SNPs Identified in Flax for Various Agro-Morphological, Seed Quality Traits, Biotic and Abiotic Stresses

Trait	QTN/QTL	Candidate Gene/ Marker	Function	Reference
	_	*Lus10021620*	Seed and flower colour *Flavonoid 3'5' Hydroxylase* (*F3'5'H*) gene	Sudarshan et al. (2017)
	QDm.BM.crc-LG4	*Lu2031*	Days to maturity (DM)	Kumar et al. (2015)
	QCw.BM.crc-LG4	*Lu2031*	Cell wall	
	QSw.BM.crc-LG4	*Lu2031*	Straw weight	Kumar et al. (2015)
Yield and related traits	*Chr10*	Scaffold 319	Seed yield	Wu et al. (2018)
	Chr12	Scaffold 117 and Scaffold 27		
	Chr5, Chr15	Scaffold 33, Scaffold355	Stem yield	
	QYld.BM.crc-LG4	*Lu2031*	Yield	Kumar et al. (2015)
	Qcp.nbri.2.1	*LUSc_270_01*	Capsule/plant	Singh et al. (2021)
	Qcp.nbri.6.1	*LUSc_629_4*		
	Qcp.nbri.7.1	*LUSc_257_01*		
	Qcp.nbri.14.1	*LU_3043*		
	Qcp.nbri.2.1	*LUSc_270_01*		
	Qcp.nbri.6.1	*LUSc_629_4*		
	Qcp.nbri.7.1	*LUSc_257_01*		
	Qcp.nbri.14.1	*LU_3043*		
	Qcwp.nbri.7.1	*LU_a58*	Capsule/Seed weight/plant	
	Qcwp.nbri.9.1	*LU_934*		
	QTsw.BM. crc-LG15	*Lu2010a/Lu2001*	Thousand seed weight	Kumar et al. (2015)
	QSpb.BM.crc-LG4	*Lu2031*	Seeds per boll	
	QCp.1	*LUSc_182_01*	Capsules/plant	Chandrawati and Yadav (2017)
	QCp.2	*LU_344*		
	QCp.3	*LU_2972*		
	QSpc.1	*LUC_7894041_1*	Seeds/capsule	
	QSpc.2	*LU_2223*		
	QSw.1	*LU_2161*	Seed weight	
	Chr1	Scaffold156	Fiber yield	Wu et al. (2018)
	Chr11	Scaffold127		
	Chr5	Scaffold680	Fiber content	
	Chr11	Scaffold376		
Protein content	*QPro.BM. crc-LG11*	*Lu11–21716266/Lu52*	Seed protein content	Kumar et al. (2015)
Oil content and quality	*QOc.1*	*LU_1151*	Oil content	Chandrawati and Yadav (2017)
	QOc.2	*LU_209*		
	QOil.BM.crc-LG8	*Lu8–22516618/Lu3189*	Oil content	Kumar et al. (2015)
	QPal.BM.crc-LG7	*Lu402/Lu7–1820805*	Palmitic acid content	
	QSte.BM.crc-LG1	*Lu2183a/Lu1–2670961*	Stearic acid content	
	QSte.BM.crc-LG3	*Lu3–8415336/Lu2164*		
	QSte.BM.crc-LG11	*Lu2128/Lu11–19000928*		

(*Continued*)

TABLE 8.2 (*Continued*)
List of QTLs/SNPs Identified in Flax for Various Agro-Morphological, Seed Quality Traits, Biotic and Abiotic Stresses

Trait	QTN/QTL	Candidate Gene/ Marker	Function	Reference
	QOle.BM.crc-LG3-1	*Lu3–3979616/Lu3–5950394*	Oleic acid content	
	QOle.BM.crc-LG3-2	*Lu658/Lu3150*		
	QOle.BM.crc-LG5	*Lu5–9728492*		
	QLin.BM.crc-LG5	*Lu5–9728492*	Linolenic acid content	
	QLio.BM.crc-LG3	*Lu3–3979616/Lu3–5950394*	Linoleic acid content	
	QLio.BM.crc-LG6	*Lu2545*		
Biotic stress	*QTL45/ Lu9–6270376*	*Lus10031043* (leucine-rich repeat receptor kinase)	Bacterial pathogen associated molecular pattern (PAMP) receptor	He et al. (2019)
		Lus10031058 (elongation factor)	Effector triggered immunity	
	afB13	–	*Fusarium* wilt	You and Cloutier (2020)
	Qabr.nbri.14.1	*Lu3043*	*Alternaria* blight resistance	Singh et al. (2021)
	Qabr.nbri.14.2	*Lu207F198*		
	QPM-crc-LG1 (Lu2698-Lu2712)	–	Powdery mildew resistance	You and Cloutier (2020)
	QPM-crc-LG7 (Lu2810-Lu2832)	–		Asgarinia et al. (2013)
	QPM-crc-LG9 (Lu1125a-Lu932)	–		
Abiotic stress	*Lus-20209630*	*Lus10039723 (IAA amido synthetase GH3.6)*	Response to stress and root development (Root length)	Soto-Cerda et al. (2020)
		Lus10039747 (Diacylglycserol kinase 5)	Cold and drought stress tolerance	
	Lu6–19733117	*Lus10021019 (Allene oxide synthase 3)*	Stomatal closure and drought tolerance	
		Lus10020997 (S/T Protein kinase SRK2E)	Response to water deprivation, regulation of stomatal closure	
	Lu6–15961789	*Lus10016017* (Catalase isozyme C)	Promotes drought stress tolerance and response to water deprivation (Root volume)	
	Lu5–4774423	*Lus10034840* (Calcium transporting ATPase 9, plasma membrane type)	Pollen development (Root surface area stability	

(*Continued*)

TABLE 8.2 (*Continued*)
List of QTLs/SNPs Identified in Flax for Various Agro-Morphological, Seed Quality Traits, Biotic and Abiotic Stresses

Trait	QTN/QTL	Candidate Gene/ Marker	Function	Reference
	Lu6–15939492	*Lus10016017* (Catalase isozyme C)	Response to water deprivation, promotes drought stress tolerance and recovery	
	Chr5:1375386	*Lus10029690/1* (Cellulose synthase interactive 3)	Flax fiber, stress tolerance index	Sertse et al. (2021)
	Lus10029692 (AFI)	Xylem development, stress tolerance		
	Chr8:2514743	*Lus10025166* (PIN-LIKES 3)	Plant height, drought tolerance	
		Lus10025172 (Set domain protein 25)	Flowering time	
	Chr14:205508	*Lus10009472* (Agamous-like 12)	Drought tolerance, root growth	
		Lus10009476 (C-terminally encoded peptide receptor 2,)	Biomass, N uptake	
		Lus10009480 (Wax inducer 1)	Cell wall structure	
		Lus10009481 (Agamous-like MADS-box protein AGL11)	Plant height	
	Chr9:4203006	*Lus10040333* (3-ketoacyl-CoA synthase 19)	Straw weight under drought stress (biomass)	
		Lus10040335 (Ankyrin repeat-containing protein ITN1)	Salt and drought susceptibility index (DSI), biomass related traits	
	Chr8:16534117	*Lus10004554* (Poltergeist like 1)	Root and Shoot development	
	Chr12:6352775	*Lus10016846* (Two-component response regulator ARR1-related)	Shoot development, drought tolerance	
		Lus10016831 (Early growth response gene 1)	Seed development, drought tolerance	

(*Continued*)

TABLE 8.2 (*Continued*)
List of QTLs/SNPs Identified in Flax for Various Agro-Morphological, Seed Quality Traits, Biotic and Abiotic Stresses

Trait	QTN/QTL	Candidate Gene/ Marker	Function	Reference
	Chr12:20557728	*Lus10031398* (Inositol Monophosphatase 1)	Drought tolerance	
		Chr11:3972867	*Lus10042229* (CBL-interacting protein kinases)	
		Lus10042231 (Translocon at the inner envelope membrane of chloroplasts 110)	Heat shock, DSI	
	Chr1:7029139	*Lus10029127* (Kelch repeat F-box)	Ovule development, stress tolerance index	
		Lus10029115 (Ribosomal pentatricopeptide repeat protein 4)	Seed development, stress tolerance	
	Chr12:10910146	*Lus10030137* (Nuclear factor Y subunit A1)	Seed development, drought stress tolerance	
		Lus10030142 (Nuclear pore anchor, Translocated promoter region)	Flowering, auxin signalling	
	Chr9:15446958	*Lus10021766* Mitogen-activated protein kinase kinase kinase 5	Seeds per boll (DSI)	
	Lu6–17376408	*Lus10019811* (Probable Cinnamyl alcohol dehydrogenase 1)	Stress tolerance index (Drought tolerance)	Soto-Cerda et al. (2020)
		Lus10019781 (L-Ascorbate peroxidase)	Enhanced salt tolerance, drought and cold tolerance	
	Lu14–23517150	*Lus10014978* (Aquaporin PIP2-2)	Drought tolerance	
	Lus.o.m. Scaffold1.14	*Cytochrome P450*	Responsive to salinity stress	Zhao et al. (2022)
	Lus.o.m. scaffold 91.141	*WD40*		
	qRSL2.3	*Lus10013205*	Salinity tolerance	Li et al. (2022)
	Chr2:23123754	*Lus10013240* (xyloglucan endotransglucosylase/ hydrolase protein 27)	Canopy temperature under drought stress	Sertse et al. (2021)
	Chr3:9279281	*Lus10019365* (Stromal cell derived factor 2)		
	Chr9:18937269	*Lus10024816* (cytochrome p450, family 81, subfamily d, polypeptide 8)		

8.7 CHALLENGES

A large number of germplasm accessions in various genebanks worldwide hold enormous potential for undisclosed allelic variations. However, the prime challenge is the efficient identification of novel alleles and the refined utilization of allelic variants for crop improvement. The challenges in allele mining and different approaches for improving efficiency are discussed in the following sections.

8.7.1 Selection of Genotypes and Development of Core and Mini-Core Collections

Screening the huge genebank collection may lead to the exploration of novel useful alleles; however, it will be a herculean task involving both time and a high degree of technical expertise. Therefore, an efficient protocol for characterizing gene(s) and screening the collection that represents diversity in the true sense while minimizing the number of accessions to be screened is a prerequisite. The development of core and mini collections from entire germplasm collections has been advocated for the conservation and efficient utilization of existing diversity across global genebanks. A core collection is a subset of a broader germplasm collection that represents the genetic variability of the species chosen for evaluation (Brown, 1989). For the designation of core collections, several crop specific descriptors have been identified to record variability and published (Mahajan et al., 2000; Nôžková et al., 2016). The size of the core set is largely determined by the global amount of resources available for collection. Principally, 10% or 1% of the entire collection (in case of a large collection) of the crop accessions comprises the core and mini-core collections, respectively, representing the variability of the entire collection has been proposed considering the neutral allele theory (Brown, 1989; Upadhyaya and Ortiz, 2001). Several computational tools like 'PowerCore' (Kim et al., 2007) and Core Hunter 3 (De Beukelaer and Davenport, 2018; De Beukelaer et al., 2018) have been extensively used for developing core collections that capture allelic richness and reduce redundancy. Genetic markers and qualitative trait-based reference collections or sets (subset of mini core) constitute the source materials for discerning the naturally existing allelic variation in candidate genes through allele mining.

8.7.2 Precision Phenotyping

Genome sequencing and molecular-assisted breeding techniques have accentuated the genotyping of a large set of population; however, phenotyping remains the rate-limiting step towards trait-specific crop improvement. High-throughput real-time phenotyping techniques for various traits like morphology, growth, pest resistance, abiotic stress tolerance, and biochemical profile involve mechanized data collection and automation of data analysis.

8.7.3 Managing Genetic Resources

The generation of a large amount of genomic data has also led to genetic redundancy within the crop gene pool thereby posing a challenge for effective management and utilization of resources. Therefore, the sorting and analysis of data through various bioinformatics tools for the selection of desirable alleles remains a challenging task. Furthermore, with the rapid accumulation of nucleotide and transcript data, the development of computational tools for efficient analysis and interpretation of sequence-specific diversity needs to be kept on par.

8.7.4 Delimitations and Characterization of Regulatory Region

Promoter or *cis* regulatory regions are the sites or loci that are generally spread upstream from the start codon. The sequence variability in these regulatory loci plays a detrimental role in the spatiotemporal expression of a gene, thereby instrumental in shaping the external as well as internal phenotype.

Demarcation of the promoter region, even with the availability of genomic sequence, is a strenuous task owing to a lack of consensus over the length or size of the regulatory region. The span of these regions can be downstream in the coding sequence of the gene, spanning a window ranging from as low as 100 bp to –3,000 bp. However, high densities of regulatory elements have been reported to exist between –100 and –600 bp (Park et al., 2002; Molina and Grotewold, 2005; Veerla and Hoglund, 2006). Deciphering regulatory variants based on transcript abundance between individuals or subsets is governed either by trans-acting factors or environmental cues (Dai et al., 2007). Furthermore, an insertion or deletion (InDel) in the regulatory sequence might translate to the loss or gain of *cis* elements, which might lead to an altered phenotype. The functionality of the promoter region can be bi-directional, so ensuring the orientation will be arduous even with the most advanced software tools. Furthermore, the intricacy of analyzing these promoters is accentuated owing to the fact that transcription factors (TFs) primarily regulate multiple gene(s) (Hehl and Wingender, 2001). Thus, promoter mining remains a tedious task for precise positional identification of regulatory regions, and the need to devise bioinformatic software that can precisely predict the core promoter region based on the representation/over-representation of the conserved regulatory motifs is emphasized.

8.8 FUTURE PERSPECTIVES

Allele mining holds tremendous potential for efficient utilization of genomic resources for value enrichment and stress tolerance in flax. It has been successfully applied to unravel the genetic diversity of flax, allowing the construction of high-density genetic maps and precise QTL mapping for important agronomic and seed quality traits in the recent past. However, further efforts to materialize tools and strategies for the refinement of both genetic and genomic resources for the characterization of novel alleles need to be envisaged. A vast amount of genome-wide SNPs and the available association data can be employed for integration with phenotypic variation, although precision phenotyping of trait-tailored subsets is essential to unraveling the novel alleles. With the cost-effective and high-throughput sequencing platforms at disposal, the allele mining approach will further evolve as the reliable and sorted method for elucidation and characterization of natural variations towards flax improvement.

REFERENCES

Aldrich J, Cullis CA (1993) RAPD analysis in flax: Optimization of yield and reproducibility using klen Taq 1 DNA polymerase, chelex 100, and gel purification of genomic DNA. *Plant Mol Biol Rep* 11:128–141

Ali E, Saand MA, Khan AR, Shah JM, Feng S, Ming C, et al. (2020) Genome-wide identification and expression analysis of detoxification efflux carriers (*DTX*) genes family under abiotic stresses in flax. *Physiol Plant* 171(4):483–501

Asgarinia P, Cloutier S, Duguid S, Rashid K, Mirlohi AF, Banik M, et al. (2013) Mapping quantitative trait loci for powdery mildew resistance in flax (*Linum usitatissimum* L.). *Crop Sci* 53:2462–2472

Bickel CL, Gadani S, Lukacs M, Cullis CA (2011) SSR markers developed for genetic mapping in flax (*Linum usitatissimum* L.). *Res Rep Biol* 2:23–29

Bolsheva NL, Melnikova NV, Kirov IV, Dmitriev AA, Krasnov GS, Amosova AV, et al. (2019) Characterization of repeated DNA sequences in genomes of blue-flowered flax. *BMC Evol Biol* 19(1):49

Bolsheva NL, Zelenin AV, Nosova IV, Amosova AV, Samatadze TE, Yurkevich OY, et al. (2015) The diversity of karyotypes and genomes within section *Syllinum* of the Genus *Linum* (*Linaceae*) revealed by molecular cytogenetic markers and RAPD analysis. *PloS ONE* 10(4):1–15

Brown AHD (1989) Core collections: A practical approach to genetic resources management. *Genome* 31(2):818–824

Buckler ES, 4th, Thornsberry JM (2002) Plant molecular diversity and applications to genomics. *Curr. Opin. Plant Biol* 5(2):107–111.

Catanzariti AM, Dodds PN, Lawrence GJ, Ayliffe MA, Ellis JG (2006) Haustorially expressed secreted proteins from flax rust are highly enriched for a virulence elicitors. *Plant Cell* 18(1):243–256

Chandrawati, Maurya R, Singh PK, Ranade SA, Yadav HK (2014) Diversity analysis in Indian genotypes of linseed (*Linum usitatissimum* L.) using AFLP markers. *Gene* 549(1):171–178

Chandrawati, Singh N, Kumar R, Kumar S, Singh PK, Yadav VK, et al. (2017) Genetic diversity, population structure and association analysis in linseed (*Linum usitatissimum* L.). *Physiol Mol Biol Plants* 23(1):207–219

Chandrawati, Yadav HK (2017) Development of linkage map and mapping of QTLs for oil content and yield attributes in linseed (*Linum usitatissimum* L.). *Euphytica* 213:258

Chantreau M, Grec S, Gutierrez L, Dalmais M, Pineau C, Demailly H, et al. (2013) PT-Flax (phenotyping and TILLinG of flax): Development of a flax (*Linum usitatissimum* L.) mutant population and TILLinG platform for forward and reverse genetics. *BMC Plant Biol* 13:159

Cloutier S, Miranda E, Ward K, Radovanovic N, Reimer E, Walichnowski A, et al. (2012a) Simple sequence repeat marker development from bacterial artificial chromosome end sequences and expressed sequence tags of flax (*Linum usitatissimum* L.). *Theor Appl Genet* 125(4):685–694.

Cloutier S, Niu Z, Datla R, Duguid S (2009) EST-derived microsatellites from flax (*Linum usitatissimum* L.). *Theor Appl Genet* 119:53–63.

Cloutier S, Ragupathy R, Miranda E, Radovanovic N, Reimer E, Walichnowski A, et al. (2012b) Integrated consensus genetic and physical maps of flax (*Linum usitatissimum* L.). *Theor Appl Genet* 125(8):1783–1795.

Cloutier S, Ragupathy R, Niu Z, Duguid S (2011) SSR-based linkage map of flax (*Linum usitatissimum* L.) and mapping of QTLs underlying fatty acid composition traits. *Mol Breed* 28:437–451.

Cloutier S, You FM, Soto-Cerda BJ (2019) *Linum* Genetic Markers, Maps, and QTL Discovery. In: Cullis CA (ed) *Genetics and Genomics of Linum*, Springer Nature Switzerland, 23:97–117.

Collard BCY, Jahufer MZZ, Brouwer JB, Pang ECK (2005) An introduction to markers, quantitative trait loci (QTL) mapping and marker-assisted selection for crop improvement: The basic concepts. *Euphytica* 142:169–196.

Comai L, Young K, Till BJ, Reynolds SH, Greene EA, Codomo CA, et al. (2004) Efficient discovery of DNA polymorphisms in natural populations by EcoTilling. *Plant J* 37(5):778–786.

Cullis CA (2007) Flax. In: Kole C (eds) *Oilseeds. Genome Mapping and Molecular Breeding in Plants*, vol 2. Springer, Berlin, Heidelberg, 2:275–295.

Dai LY, Liu XL, Xiao YH, Wang GL (2007) Recent advances in cloning and characterization of disease resistance genes in rice. *J Integr Plant Biol* 49(1):112–119

De Beukelaer H, Davenport GF (2018) Core Hunter: Multi-Purpose Core Subset Selection. R package version 3.2.1.

De Beukelaer H, Davenport GF, Fack V (2018) Core Hunter 3: Lexible core subset selection. *BMC Bioinform* 19:203.

de Santana Lopes A, Pacheco TG, Santos K, Vieira L, Guerra MP, Nodari RO, et al. (2018) The *Linum usitatissimum* L. plastome reveals atypical structural evolution, new editing sites, and the phylogenetic position of *Linaceae* within *Malpighiales*. *Plant Cell Rep* 37(2):307–328.

Deng X (2013). Development of molecular markers and association analysis of yield related traits in flax (*Linum* usitatissimum L). Ph.D thesis. *Chin. Acad. Agric. Sci.*

Deng X, Chen XB, Long SH, Wang XC, GAO Y, He DF, et al. (2008) Microsatellite marker enrichment with magnetic beads in flax. *Acta Agron Sin* 34(12):2099–2105.

Deng X, Long SH, He DF, Li X, Wang YF, Liu J, et al. (2010) Development and characterization of polymorphic microsatellite markers in *Linum* usitatissimum. *J. Plant Stud* 123(1):119–123.

Deng X, Long SH, He DF, Li X, Wang YF, Hao DM, et al. (2011) Isolation and characterization of polymorphic microsatellite markers from flax (*Linum usitatissimum* L.). *Afr J Biotechnol* 10(5):734–739. Diederichsen A, Fu YB (2006). Phenotypic and molecular (RAPD) differentiation of four infraspecific groups of cultivated Flax (*Linum usitatissimum* L. subsp. usitatissimum). *Genet Resour Crop E* 53:77–90.

Diederichsen A, Ulrich A (2009) Variability in stem fibre content and its association with other characteristics in 1177 flax (Linum *usitatissimum* L.) genebank accessions. *Ind Crops Prod* 30(1):33–39.

Dmitriev AA, Kezimana P, Rozhmina TA, Zhuchenko AA, Povkhova LV, Pushkova EN, et al. (2020a). Genetic diversity of SAD and FAD genes responsible for the fatty acid composition in flax cultivars and lines. *BMC Plant Biol* 20:301.

Dmitriev AA, Krasnov GS, Rozhmina TA, Zyablitsin AV, Snezhkina AV, Fedorova MS, et al. (2019) Flax (*Linum usitatissimum* L.) response to non-optimal soil acidity and zinc deficiency. *BMC Plant Biol* 19:54.

Dmitriev AA, Novakovskiy RO, Pushkova EN, Rozhmina TA, Zhuchenko AA, Bolsheva NL, et al. (2020b) Transcriptomes of different tissues of flax (*Linum usitatissimum* L.) cultivars with diverse characteristics. *Front Genet* 11:565146.

Dmitriev AA, Pushkova EN, Novakovskiy RO, Beniaminov AD, Rozhmina TA, Zhuchenko AA et al. (2021) Genome sequencing of fiber flax cultivar Atlant using oxford nanopore and illumina platforms. *Front Genet* 11:590282.

Dodds PN, Lawrence GJ, Catanzariti AM, Ayliffe MA, Ellis JG (2004) The *Melampsora lini AvrL567* avirulence genes are expressed in haustoria and their products are recognized inside plant cells. *Plant Cell* 16:755–768.

Dodds PN, Lawrence GJ, Catanzariti AM, Teh T, Wang CI, Ayliffe MA, et al. (2006) Direct protein interaction underlies gene-for-gene specificity and coevolution of the flax resistance genes and flax rust avirulence genes. *Proc Natl Acad Sci USA* 103:8888–8893.

Doebley J, Lukens L (1998) Transcriptional regulators and the evolution of plant form. *Plant Cell* 10(7):1075–1082.

Everaert I, De Riek J, De Loose M, Van Waes J, Van Bockstaele E (2001) Most similar variety grouping for distinctness evaluation of flax and linseed (*Linum usitatissimum* L.) varieties by means of AFLP and morphological data. *Plant Var Seeds* 14:69–87.

FAOSTAT (2022) Available at: https://www.fao.org/faostat/en/#data/QCL (Accessed August 5, 2022).

Fischer R., Maurer R (1978) Drought resistance in spring Wheat cultivars, 1. Grain yield responses. *Aust J Agric Res* 29:897–912.

Fisher PJ, Gardner RC, Richardson TE (1996) Single locus microsatellites isolated using 5' anchored PCR. *Nucleic Acids Res* 24(21):4369–4371.

Flint-Garcia SA, Thornsberry JM, Buckler ES 4th (2003) Structure of linkage disequilibrium in plants. *Annu Rev Plant Biol* 54:357–374.

Flint-Garcia SA, Thuillet AC, Yu J., Pressoir, G., Romero SM., Mitchell SE, et al. (2005) Maize association population: A high-resolution platform for quantitative trait locus dissection. *Plant J* 44(6):1054–1064.

Fu YB (2005) Geographic patterns of RAPD variation in cultivated flax. *Crop Sci* 45(3):1084–1091.

Fu YB, Diederichsen A, Richards KW, Peterson G (2002a) Genetic diversity within a range of cultivars and landraces of flax (*Linum usitatissimum* L.) as revealed by RAPDs. *Genet Resour Crop E* 49:167–174.

Fu YB, Guérin S, Peterson G, Carlson JE, Richards KW (2003b) Assessment of bulking strategies for RAPD analyses of flax germplasm. *Genet Resour Crop E* 50:743–746.

Fu YB, Guérin S, Peterson GW, Diederichsen A, Rowland GG, Richards KW (2003c) RAPD analysis of genetic variability of regenerated seeds in the Canadian flax cultivar CDC Normandy. *Seed Sci Technol* 31:207–211.

Fu YB, Peterson GW (2010) Characterization of expressed sequence tag-derived simple sequence repeat markers for 17 *Linum* species. *Botany* 88:537–543.

Fu YB, Peterson GW (2012) Developing genomic resources in two *Linum* species via 454 pyrosequencing and genomic reduction. *Mol Ecol Resour* 12(3):492–500.

Fu YB, Peterson GW, Diederichsen A, Richards KW (2002b) RAPD analysis of genetic relationships of seven flax species in the genus *Linum* L. *Genet Resour Crop E* 49:253–259.

Fu YB, Rowland GG, Duguid SD, Richards KW (2003a) RAPD analysis of 54 North American flax cultivars. *Crop Sci* 43:1510–1515.

Goyal A, Sharma V, Upadhyay N, Gill S, Sihag M (2014) Flax and flaxseed oil: An ancient medicine & modern functional food. *J Food Sci Technol* 51:1633–1653.

Green AG (1986) A mutant genotype of flax (*Linum usitatissimum* L.) containing very low levels of linolenic acid in its seed oil. *Can J Plant Sci* 66:499–503.

Green AG, Marshall AR (1984) Isolation of induced mutants in linseed (*Linum usitatissimum* L.) having reduced linolenic acid content. *Euphytica* 33:321–328.

Guo D, Jiang H, Yan W, Yang L, Ye J, Wang Y et al. (2020) Resequencing 200 flax cultivated accessions identifies candidate genes related to seed size and weight and reveals signatures of artificial selection. *Front Plant Sci* 10:1682.

Gupta PK (2008) Single-molecule DNA sequencing technologies for future genomics research. *Trends Biotechnol* 26(11):602–611.

Gupta PK, Rustgi S, Kulwal PL (2005) Linkage disequilibrium and association studies in higher plants: Present status and future prospects. *Plant Mol Biol* 57(4):461–485.

He L, Xiao J, Rashid KY, Yao Z, Li P, Jia G, et al. (2019) Genome-wide association studies for Pasmo resistance in flax (*Linum usitatissimum* L.). *Front Plant Sci* 9:1982.

Hehl R, Wingender E (2001) Database-assisted promoter analysis. *Trends Plant Sci* 6(6):251–255.

Hickey M (1988) *100 Families of Flowering Plants*. 2nd ed. Cambridge University Press, Cambridge.

Hudson ME (2008) Sequencing breakthroughs for genomic ecology and evolutionary biology. *Mol Ecol Resour* 8(1):3–17.

Hutchison CA 3rd (2007) DNA sequencing: Bench to bedside and beyond. *Nucleic Acids Res* 35(18):6227–6237.

Iqbal A, Nawaz S, Babar AD, Bukhari SA, Amjad MS, Khan MA et al. (2014) Estimation of genetic variability in Linseed (*Linum usitatissimum* L.) using molecular marker: Molecular characterization of linseed genotypes. *Int J Agric Sci* 4(4):151–156.

Kajla P, Sharma A, Sood DR (2015) Flaxseed-a potential functional food source. *J Food Sci Technol* 52:1857–1871.

Kale SM, Pardeshi VC, Kadoo NY, Ghorpade PB, Jana MM, Gupta VS (2012) Development of genomic simple sequence repeat markers for linseed using next-generation sequencing technology. *Mol Breed* 30:597–606.

Kaur V, Kumar S, Yadav R, Wankhede DP, Aravind J, Radhamani J, et al. (2018) Analysis of genetic diversity in Indian and exotic linseed germplasm and identification of trait specific superior accessions. *J Environ Biol* 39:702–709.

Kaur V, Yadav R, Wankhede DP (2017) Linseed (*Linum usitatissimum* L.) genetic resources for climate change intervention and its future breeding. *J Appl Nat Sci* 9(2):1112–1118.

Kilian B, Graner A (2012) NGS technologies for analyzing germplasm diversity in genebanks. *Brief Funct Genom* 11(1):38–50.

Kim KW, Chung HK, Cho GT, Ma KH, Chandrabalan D, Gwag JG, et al. (2007) Power Core: A program applying the advanced M strategy with a heuristic search for establishing core sets. *Bioinformatics (Oxford, England)* 23(16):2155–2162.

Kumar S, Kerkhi SA, Gangwar LK, Chand P, Kumar M (2012) Improvement in the genetic architecture through study of variability, heritability and genetic advance in linseed crop (*Linum usitatissimum* L.). *Agric Sci Eng.* (ICASE2012):170.

Kumar S, You FM, Duguid S, Booker H, Rowland G, Cloutier S (2015) QTL for fatty acid composition and yield in linseed (*Linum usitatissimum* L.). *Theor Appl Genet* 128(5):965–984.

Li X, Guo D, Xue M, Li G, Yan Q, Jiang H, et al. (2022) Genome-wide association study of salt tolerance at the seed germination stage in flax (*Linum usitatissimum* L.). *Genes* 13(3):486.

Ludlow MM, Muchow RC (1990) A critical evaluation of traits for improving crop yields in water-limited environments. *Adv Agron* 43:107–153.

Mackay TF (2001) Quantitative trait loci in *Drosophila*. *Nat Rev Genet* 2(1):11–20.

Mahajan RK, Sapra RL, Srivastava U, Singh M, Sharma GD (2000) *Minimal Descriptors (for Characterization and Evaluation) of Agri-Horticultural Crops (Part I)*. National Bureau of Plant Genetic Resources, New Delhi.

Mardis ER (2008) Next-generation DNA sequencing methods. *Annu Rev Genomics Hum Genet* 9:387–402.

Melnikova NV, Kudryavtseva AV, Zelenin AV, Lakunina VA, Yurkevich OY, Speranskaya AS, et al. (2014) Retrotransposon-based molecular markers for analysis of genetic diversity within the Genus Linum. *Biomed Res Int* 2014:231589.

Metzker ML (2010) Sequencing technologies - the next generation. *Nat Rev Genet* 11(1) 31–46.

Molina C, Grotewold E (2005) Genome wide analysis of Arabidopsis core promoters. *BMC Genom* 6:25.

Mukherji KG, Dubey OP, Upadhyaya R.K., (1999) Insect pests of linseed. In: Upadhyaya RK (ed.), *IPM System in Agriculture (Oilseeds Crops)*. Aditya Book Pvt. Ltd, India, 5:250–263.

Nagabhushanam B, Mir MI, Nagaraju M, Sujatha E, Devi BR, Kumar BK (2021) Genetic diversity analysis of Linseed (*Linum usitatissimum* L.) accessions using RAPD Markers. *Emir J Food Agric* 33(7):589–599.

Novakovskiy RO, Povkhova LV, Krasnov GS, Rozhmina TA, Zhuchenko AA, Kudryavtseva LP, et al. (2019) The cinnamyl alcohol dehydrogenase gene family is involved in the response to *Fusarium oxysporumin* resistant and susceptible flax genotypes. *Vavilov J Genet Breed* 23:896–901.

Nôžková J, Pavelek M, Bjelková M, Brutch NB, Tejklová E, Porokhovinova E, et al. (2016) Descriptor list for flax (*Linum usitatissimum* L.). Slovak University of Agriculture in Nitra 552-1484-9.

Oh T, Gorman M, Cullis C (2000) RFLP and RAPD mapping in flax (*Linum usitatissimum*). *Theor Appl Genet* 101:590–593.

Ottai MES, Al-Kordy MAA, Hussein RM, Hassanein MS (2012) Genetic diversity among Romanian fiber flax varieties under Egyptian conditions. *Aust J Basic Appl Sci* 6:162–168.

Pali V, Mehta N, Verulkar SB, Xalxo MS, Saxena RR (2015) Molecular diversity in Flax (*Linum usitatissimum* L.) as revealed by DNA based markers. *Int J Plant Res* 28:157–165.

Pali V, Verma SK, Xalxo MS, Saxena RR, Mehta N, Verulkar SB (2014) Identification of microsatellite markers for fingerprinting popular Indian flax (*Linum usitatissimum* L.) cultivars and their utilization in seed genetic purity assessments. *Aust J Crop Sci* 8:119–126.

Park PJ, Butte AJ, Kohane IS (2002) Comparing expression profiles of genes with similar promoter regions. *Bioinformatics (Oxford, England)* 18(12):1576–1584.

Powell W, Machray GC, Provan J (1996) Polymorphism revealed by simple sequence repeats. *Trends Plant Sci* 1(7):215–222.

Powell W, Morgante M, Andre C, McNicol JW, Machray GC, Doyle JJ, et al. (1995) Hypervariable microsatellites provide a general source of polymorphic DNA markers for the chloroplast genome. *Curr Biol* 5(9):1023–1029.

Przybylski R (2005) Flax oil and high linolenic oils. In: Shahidi F (ed) *Bailey's Industrial Oil and Fat Products*, Wiley, Hoboken, NJ, pp. 281–302.

Raghavan C, Naredo MEB, Wang H, Atienza G, Liu B, Qiu F, et al. (2007) Rapid method for detecting SNPs on agarose gels and its application in candidate gene mapping. *Mol Breed* 19:87–101.

Rajwade AV, Arora RS, Kadoo NY, Harsulkar AM, Ghorpade PB, Gupta VS (2010) Relatedness of Indian flax genotypes (*Linum usitatissimum* L.): An inter-simple sequence repeat (ISSR) primer assay. *Mol Biotechnol* 45:161–170.

Rajwade AV, Joshi RS, Kadoo NY, Gupta VS (2016) Sequence characterization and in silico structure prediction of fatty acid desaturases in linseed varieties with differential fatty acid composition. *J Sci Food Agric* 96(15):4896–4906.

Roose-amsaleg C, Cariou-pham E, Vautrin D, Tavernier R, Solignac M (2006) Polymorphic microsatellite loci in *Linum usitatissimum*. *Mol Ecol Notes* 6:796–799.

Rowland GG (1991) An EMS-induced low linolenic acid mutant in McGregor flax (*Linum usitatissimum* L.). *Can J Plant Sci* 71:393–396.

Sa R, Yi L, Siqin B, An M, Bao H, Song X et al. (2021) Chromosome-level genome assembly and annotation of the fiber flax (*Linum usitatissimum*) genome. *Front Genet* 12,735690.

Saroha A, Pal D, Gomashe SS, Akash, Kaur V, Ujjainwal S, et al. (2022a). Identification of QTNs associated with flowering time, maturity, and plant height traits in *Linum usitatissimum* L. using genome-wide association study. *Front Genet* 13:811924.

Saroha A, Pal D, Kaur V, Kumar S, Bartwal A, Aravind J, et al. (2022b) Agro-morphological variability and genetic diversity in linseed (*Linum usitatissimum* L.) germplasm accessions with emphasis on flowering and maturity time. *Genet Resour Crop E* 69:315–333. Schulman AH (2007) Molecular markers to assess genetic diversity. *Euphytica* 158:313–321.

Sertse D, You FM, Ravichandran S, Soto-Cerda BJ, Duguid S, Cloutier S (2021) Loci harboring genes with important role in drought and related abiotic stress responses in flax revealed by multiple GWAS models. *Theor Appl Genet* 134(1):191–212.

Shendure J, Ji H (2008) Next-generation DNA sequencing. *Nat Biotechnol* 26(10) 1135–1145.

Shim YY, Gui B, Arnison PG, Wang Y, Reaney MJT (2014) Flaxseed (*Linum usitatissimum* L.) bioactive compounds and peptide nomenclature: A review. *Trends Food Sci Technol* 38(1):5–20.

Singh BD (2005) A text book of plant breeding. Kalyani Publishers. p. 171–2.

Singh KK, Mridula D, Rehal J, Barnwal P (2011) Flaxseed: A potential source of food, feed and fiber. *Crit Rev Food Sci Nutr* 51(3):210–222.

Singh N, Kumar R, Kumar S, Singh PK, Yadav HK (2021) Mapping QTLs for *Alternaria* blight in Linseed (*Linum usitatissimum* L.). *3 Biotech* 11(2):91.

Soto-Cerda BJ, Aravena G, Cloutier S (2021) Genetic dissection of flowering time in flax (*Linum usitatissimum* L.) through single and multi-locus genome-wide association studies. *Mol. Genet. Genomics* 296(4):877–891.

Soto-Cerda BJ, Carrasco RA, Aravena GA, Urbina HA, Navarro CS, (2011a). Identifying novel polymorphic microsatellites from cultivated flax (*Linum usitatissimum* L.) following data mining. *Plant Mol Biol Rep* 29:753–759.

Soto-Cerda BJ, Cloutier S, Gajardo HA, Aravena G, Quian R, You FM (2020) Drought response of flax accessions and identification of quantitative trait nucleotides (QTNs) governing agronomic and root traits by genome-wide association analysis. *Mol Breed* 40:15.

Soto-Cerda BJ, Cloutier S, Quian R, Gajardo HA, Olivos M, You FM (2018). Genome-wide association analysis of mucilage and hull content in flax (*Linum usitatissimum L.) seeds. Int J Mol Sci* 19:2870.

Soto-Cerda BJ, Diederichsen A, Duguid S, Booker H, Rowland G, Cloutier S (2014) The potential of pale flax as a source of useful genetic variation for cultivated flax revealed through molecular diversity and association analyses. *Mol Breed* 34(4):2091–2107.

Soto-Cerda BJ, Diederichsen A, Ragupathy R, Cloutier S (2013) Genetic characterization of a core collection of flax (*Linum usitatissimum* L.) suitable for association mapping studies and evidence of divergent selection between fiber and linseed types. *BMC Plant Biol* 13:78.

Soto-Cerda BJ, Maureira-Butler I, Muñoz G, Rupayan A, Cloutier S (2012) SSR-based population structure, molecular diversity and linkage disequilibrium analysis of a collection of flax (*Linum usitatissimum* L.) varying for mucilage seed-coat content. *Mol Breed* 30:875–888.

Soto-Cerda BJ, Saavedra HU, Navarro C, Ortega P (2011b) Characterization of novel genic SSR markers in *Linum usitatissimum* (L.) and their transferability across eleven *Linum* species. *Electron J Biotechnol* 4:4–4.

Spielmeyer W, Green AG, Bittisnich D, Mendham N, Lagudah ES (1998a) Identification of quantitative trait loci contributing to *Fusarium* wilt resistance on an AFLP linkage map of flax (*Linum usitatissimum*). *Theor Appl Genet* 97:633–641.

Spielmeyer W, Lagudah E, Mendham N, Green AG (1998b) Inheritance of resistance to flax wilt (*Fusarium oxysporum* f. sp. lini Schlecht) in a doubled haploid population of *Linum usitatissimum* L. *Euphytica* 101:287–291.

Spooner D, Treuren R Van, De Vicente MC (2005) *Molecular Markers for Genebank Management.* IPGRI Technical Bulletin no. 10. IPGRI, Rome, Italy.

Stegnii VN, Chudinova YV, Salina EA (2000) RAPD analysis of flax (*Linum usitatissimum* L.) varieties and hybrids of various productivity. *Russ J Genet* 36:1149–1152.

Sudarshan GP, Kulkarni M, Akhov L, Ashe P, Shaterian H, Cloutier S, et al. (2017) QTL mapping and molecular characterization of the classical D locus controlling seed and flower color in *Linum usitatissimum* (flax). *Sci Rep* 7:15751.

Tanksley SD (1993) Mapping polygenes. *Annu Rev Genet* 27:205–233.

Thambugala D, Duguid S, Loewen E, Rowland G, Booker H, You FM, et al. (2013) Genetic variation of six desaturase genes in flax and their impact on fatty acid composition. *Theor Appl Genet* 126(10):2627–2641.

Tsai H, Howell T, Nitcher R, Missirian V, Watson B, Ngo KJ, et al. (2011) Discovery of rare mutations in populations: TILLING by sequencing. *Plant Physiol.* 156(3):1257–1268.

Upadhyaya HD, Ortiz R (2001) A mini core subset for capturing diversity and promoting utilization of chickpea genetic resources in crop improvement. *Theor Appl Genet* 102(8):1292–1298.

Uysal H, Fu YB, Kurt O. Peterson GW, Diederichsen A, Kusters P (2010) Genetic diversity of cultivated flax (Linum *usitatissimum* L.) and its wild progenitor pale flax (*Linum bienne* Mill.) as revealed by ISSR markers. *Genet Resour Crop E* 57:1109–1119.

van Treuren R (2001) Efficiency of reduced primer selectivity and bulked DNA analysis for the rapid detection of AFLP† polymorphisms in a range of crop species. *Euphytica* 117:27–37.

Varshney RK, Nayak SN, May GD, Jackson SA (2009) Next-generation sequencing technologies and their implications for crop genetics and breeding. *Trends Biotechnol* 27(9):522–530.

Veerla S, Höglund M (2006) Analysis of promoter regions of co-expressed genes identified by microarray analysis. *BMC Bioinform* 7:384.

Venglat P, Xiang D, Qiu S, Stone SL, Tibiche C, Cram D et al. (2011) Gene expression analysis of flax seed development. *BMC Plant Biol* 11:74.

Vrinten P, Hu Z, Munchinsky MA, Rowland G, Qiu X (2005) Two FAD3 desaturase genes control the level of linolenic acid in flaxseed. *Plant Physiol* 139:79–87.

Wang Z, Hobson N, Galindo L, Zhu S, Shi D, McDill J, et al. (2012) The genome of flax (*Linum usitatissimum*) assembled de novo from short shotgun sequence reads. *Plant J* 72(3):461–473.

Weber JL, May PE (1989) Abundant class of human DNA polymorphisms which can be typed using the polymerase chain reaction. *Am J Hum Genet* 44(3):388–396.

Wiesner I, Wiesnerová D, Tejklová E (2001) Effect of anchor and core sequence in microsatellite primers on flax fingerprinting patterns. *J Agric Sci* 137(1):37–44.

Wiesnerová D, Wiesner I (2004) ISSR-based clustering of cultivated flax germplasm is statistically correlated to thousand seed mass. *Mol. Biotechnol* 26(3):207–214.

Wu J, Zhao Q, Zhang L, Li S, Ma Y, Pan L, et al. (2018) QTL mapping of fiber-related traits based on a high-density genetic map in flax (*Linum usitatissimum* L.). *Front Plant Sci* 9:885.

Wu W, Nemri A, Blackman LM, Catanzariti AM, Sperschneider J, Lawrence GJ, et al. (2019) Flax rust infection transcriptomics reveals a transcriptional profile that may be indicative for rust *Avr* genes. *PLoS One* 14(12):e0226106.

Xie D, Dai Z, Yang Z, Sun J, Zhao D, Yang X, et al. (2018). Genome-wide association study identifying candidate genes influencing important agronomic traits of flax (*Linum usitatissimum* L.) using SLAF-seq. *Front Plant Sci* 8:2232.

Yadav B, Kaur V, Narayan OP, Yadav SK, Kumar A, Wankhede DP (2022) Integrated omics approaches for flax improvement under abiotic and biotic stress: Current status and future prospects. *Front Plant Sci* 13:931275.

Yadav SK, Santosh Kumar VV, Verma RK, Yadav P, Saroha A, Wankhede DP, et al. (2020) Genome-wide identification and characterization of ABA receptor PYL gene family in rice. *BMC Genom* 21:676.

Yano M, Sasaki T (1997) Genetic and molecular dissection of quantitative traits in rice. *Plant Mol Biol* 35(1–2):145–153.

You FM, Booker HM, Duguid SD, Jia G, Cloutier S (2016) Accuracy of genomic selection in biparental populations of flax (*Linum usitatissimum* L.). *Crop J* 4(4):290–303.

You FM, Cloutier S (2020) Mapping quantitative trait loci onto chromosome-scale pseudomolecules in flax. *Methods Protoc* 3(2):28.

You FM, Jia G, Xiao J, Duguid SD, Rashid KY, Booker HM, et al. (2017) Genetic variability of 27 traits in a core collection of flax (*Linum usitatissimum* L.). *Front Plant Sci* 8:1636.

You FM, Rashid KY, Zheng C, Khan N, Li P, Xiao J, He L, Yao Z, Cloutier S. (2022) Insights into the genetic architecture and genomic prediction of powdery mildew resistance in flax (*Linum usitatissimum* L.). *Int J Mol Sci*. 23(9):4960.

You FM, Xiao J, Li P, Yao Z, Jia G, He L, et al. (2018a) Chromosome-scale pseudomolecules refined by optical, physical and genetic maps in flax. *Plant J*. 95(2):371–384.

You FM, Xiao J, Li P, Yao Z, Jia G, He L, et al. (2018b) Genome-wide association study and selection signatures detect genomic regions associated with seed yield and oil quality in flax. *Int J Mol Sci* 19:2303.

Yu J, Buckler ES (2006) Genetic association mapping and genome organization of maize. *Curr. Opin. Biotechnol* 17(2):155–160.

Yurkevich OY, Naumenko-Svetlova AA, Bolsheva NL, Samatadze TE, Rachinskaya OA, Kudryavtseva AV, et al. (2013) Investigation of genome polymorphism and seed coat anatomy of species of section *Adenolinum* from the genus *Linum*. *Genet Resour Crop E* 60:661–676.

Zhang J, Long Y, Wang L, Dand Z, Zhang T, Song X, et al. (2018) Consensus genetic linkage map construction and QTL mapping for plant height-related traits in linseed flax (*Linum usitatissimum* L.). *BMC Plant Biol* 18(160):1–12.

Zhang J, Qi Y, Wang L, Wang L, Yan X, Dang Z, et al. (2020) Genomic comparison and population diversity analysis provide insights into the domestication and improvement of flax. *iScience* 23:100967.

Zhao W, Zhang Y, Zhang J, Qi Y, Wang L, Dang Z, et al. (2022) QTL mapping by whole genome re-sequencing and analysis of candidate genes for salt tolerance in linseed (*Linum usitatissimum* L.). *Oil Crop Sci* 7(2):80–85.

Zhu C, Gore M, Buckler ES, Yu J (2008) Status and prospects of association mapping in plants. *Plant Genome* 1(1):5–20.

9 Molecular Breeding to Design Ideal Sesame Crop

Birgul Guden, Moin Qureshi, Linhai Wang, Huan Li, Boubacar Sinare, Bulent Uzun, and Engin Yol

9.1 INTRODUCTION

Sesame is an important oilseed crop that belongs to the family, Pedaliaceae. The crop grows well and gives high yields in both tropical and temperate climates (Morris 2009). The total cultivated area of sesame in the world was about 13.96 million hectares, with an annual production of 6.8 million tons, according to the Food and Agriculture Organization (FAO) (2020). Although this crop has nutritional and social-economic importance, genetic and breeding efforts in sesame have been limited and slow (Ashri 2007). One of the key points in sustainable sesame improvement programs is accumulating beneficial alleles from existing plant genetic diversity and wild relatives to develop new sesame cultivars.

With the advent of DNA based technologies, novel and superior alleles of agronomically important genes from a broad range of crop species have been identified and used to organize appropriately for the development of improved varieties based on genomic data (Kumar et al. 2010). Allele mining can be effectively used to find high-quality alleles by "mining" genes of interest from different genetic resources. Aware of the enormous potential of allele mining, many international crop research institutions are coordinating allele mining efforts. Although it has been widely used especially for major oil crops like soybean and sunflower, there has been limited effort for sesame. Therefore, investigating allelic variation in candidate genes controlling key agronomic traits will help to improve sesame studies.

From these perspectives, we have designed this chapter with the main topics of utilization of wild species with interspecific crosses, marker-assisted selection, and genomic approaches for designing new sesame crops that have high oil/yield/antioxidants/oleic-linoleic acids and resistance to important biotic-abiotic stresses.

9.2 CROP DESCRIPTION, PRODUCTION, AND DOMESTICATION

Sesame is annual in habit, self-pollinated, and has indeterminate growth (Yol and Uzun 2012). It has fairly long branches, while certain nonbranching/shy branching forms are also observed (Tripathy et al. 2019). The crop is normally 60–120 cm in height; however, the plants grow more rapidly and can reach a height of 2 m in regularly irrigated fields (Ashri 2007). Plant stems carry ridges; the leaves are sessile at the bottom and petiolate from the middle to the top, with whole edges; occasionally with lobed leaves (Bedigian 2010). The flowers are generally white, pink, purple, and also shades of these colors, which appear at various stages of development (Kafiriti and Mponda 2010), and the fruit is a dehiscent capsule that shatters near the harvest (Qureshi et al. 2022). Sesame is an important nutritional source, including 50% oil and 25% protein. After oil extraction, the remaining residue also contains high-quality protein suitable for animal feeding (Hwang 2005). Oil content in sesame seed reaches up to 62.7% (Uzun et al. 2008), which mainly consists of five major fatty acids: oleic, linoleic, stearic, palmitic, and arachidic acids. The predominant fatty acids are oleic and linoleic (about 80%) in oil, indicating its quality together with unique antioxidants (Uzun et al. 2007; Yol et al. 2015). Sesame seed has several health-promoting properties, some of which have been related to lignans (sesamin, sesamolinol, sesaminol, and sesamolin), a group of nonglycerol

DOI: 10.1201/9781003385035-9

fraction chemicals that contribute to its oxidative stability and antioxidative function (Moazzami and Kamal-Eldin 2006). With these benefits, its seeds have long been used in human nutrition, cosmetics, perfumery, soaps, insecticides, anti-aging treatments, and medicine.

Although sesame is cultivated largely in both the tropics and subtropical regions of Asia and Africa, several varieties of sesame may be cultivated in temperate climates free of frost throughout the growing season (Hegde 2012). It is cultivated around the globe in over 70 nations, of which 26 are situated in Africa and 24 in Asia (Najeeb et al. 2012). Asia was leading in sesame output with 51.3% of global production, but later in the year 2015, sesame production on the African continent surged substantially. According to FAO (2019), the global output of sesame topped 5.5 million tons in 2017, of which roughly 40% was in Asia and 57% was produced in Africa. The main reasons for this huge increase were the African government's initiative to support mechanical and technological developments to improve agronomy and harvesting processes, resulting in higher production (Rahman et al. 2020). According to the latest reports by FAO (2020), Sudan is at the top of the sesame seed output globally, with 22.41% of the worldwide production, followed by Myanmar, Tanzania, India, and Nigeria, accounting for 10.88%, 10.44%, 9.68%, and 7.21%, respectively. Sesame production predictions imply a consistent expansion of global sesame output. The estimated findings imply that the output of sesame would reach 9.26 million tons per year by the year 2040, a rise of roughly 67% over the present production levels (Rahman et al. 2020). Most of the sesame-producing nations in Africa export sesame beyond the continent. Tanzania is the world's third-biggest producer of sesame seeds, but it's also one of the continent's major exporters, according to the latest available statistics from FAO. Although Asia-Pacific is the greatest producer and consumer of sesame seeds, China owns the largest proportion of the sesame import trade. However, Japan is the second-largest sesame importer and accounts for 10% of the world's sesame imports (Rahman et al. 2020). With increased health awareness and a change in dietary patterns among European consumers, Europe is developing as a new market for sesame seeds.

The exact origin of cultivated sesame (*Sesamum indicum* L.) has been a major contradictory subject for a long time, with the African or Indian subcontinent as the two suggested origins. Earlier, the crop was believed to have originated in Africa, where a high number of wild species of the genus Sesamum were discovered (Bedigian and Harlan 1986). Since then, the continent has been pointed out as a primary center of origin for sesame diversity. The presence of wide genetic diversity in Africa led to this subcontinent being considered the primary center of origin (Bedigian 2003). A description of the major evolution and distribution of the genus Sesamum has been documented in Africa and is believed to have spread from African to Asian and South American countries (Najeeb et al. 2012). Later, some of the described species were found to be synonyms. On the other hand, archeological evidence documented in Pakistan at Harappa in the Indus valley shows India to be the place of earliest domestication (Najeeb et al. 2012). Studies of inter-specific hybridization and molecular analysis supported and confirmed its domestication in the Indian subcontinent (Bedigian 2004, 2010). The different species have then spread from Asia to other places in the world, such as Africa, the Far East, China, and Americas, along the trade routes (Bedigian and Harlan 1986). At present, sesame is widely grown as an oilseed crop in Asian countries like India, China, and Korea, as well as on the American and African continents. Though archeological evidence was found in India, most of the genetic diversity of traditional sesame landraces and related wild species of cultivated sesame was found in Africa (Ashri 1998). The diversity of cultivar types and wild species was reported to be unexplored in East Africa (Nyongesa et al. 2013). The studies performed on the diversity of the cultivated sesame revealed a huge genetic distance between the cultivated sesame (Basak et al. 2019) and some wild species (Nyongesa et al. 2013). The genus *Sesamum* and related *Pedaliaceae* were reported to have exceptionally wide adaptability and environmental flexibility, which stand for their successful distribution in various places (Bedigian 2010). Species of *Sesamum* exhibit vigor development in previously undeveloped land and flourish in nutritionally deficient soils and sites where moisture is in short supply. Throughout the ages, sesamum seed dispersal has been reported to be correlated to the habitat of each species: For example, in *Sesamum alatum*, dispersal is widely distributed in tropical Africa in the driest habitats thanks to its winged seed, which enables dissemination to suitable locations. *S. angustifolium* and *Sesamum calycinum* have tiny, lightweight seeds that are more easily dispersed in savannah grasslands (Bedigian 2010).

9.3 SESAME-PRODUCING CHALLENGES

Sesame is able to thrive in a tough climate and requires minimal irrigation, fertilizer, and manure without the utilization of pesticides, owing to its high levels of inherent resistance to various biotic and abiotic stresses. However, the world's population is growing every day, and climate change is affecting agricultural output, negatively impacting the economic and cultural surroundings of the oilseed industry, including sesame (Yadav et al. 2022). Traditional techniques and non-resistant cultivars have an impact on sesame yield as well. Abiotic stresses like drought, waterlogging, salt, and high temperatures have reduced sesame yield in certain locations (Dossa et al. 2019b). Despite sesame being drought-resistant, prolonged and acute drought stress during the vegetative stage has negative impact on different growth stages, for example, flowering, capsule and seed production, seed quantity, and oil accumulation (Dossa et al. 2019a). Also, sesame cannot withstand salt. It is particularly susceptible to high levels of calcium and sodium chloride ions in soil solutions (Myint et al. 2020).

Sesame is sensitive to waterlogging and freezing, which affects sustainable production specifically (You et al. 2018). Waterlogging is the most severe challenge to sesame production in East and South Asia, particularly in China and Korea (Dossa et al. 2019c). Sesame grain development and yield decline after 2–3 days of waterlogging when the crop is cultivated on poorly drained soils (Sarkar et al. 2016). Prolonged exposure to flooding has a highly detrimental effect on the maturation, chlorosis, growth, and yield traits of sesame (Mensah et al. 2006). The sensitivity to waterlogging during the reproductive stage in sesame was reported by Saha et al. (2010), who obtained a lower seed output by roughly 44%.

Besides these abiotic challenges, sesame fields are also experiencing a substantial decline in the average loss of 25% of potential global output caused by biotic stresses (Myint et al. 2020), which comprises live organisms such as fungus, viruses, bacteria, insects, and nematodes. In sesame, most biotic stress is largely caused by two catastrophic diseases, i.e., phyllody and dry root rot caused by phytoplasma and fungi, respectively (Singh et al. 2020). Phyllody is one of the destructive biotic stress factors in sesame; most cultivars are susceptible to this disease (Ustun et al. 2017). Abnormal development of green pigmentation on flowers, witches' broom, and limited fertile seeds are the common symptoms of this disease (Rao and Nabi 2015). Additionally, blight, charcoal rot, stem anthracnose, mildew, and phyllody are other major diseases in sesame. Insect pests such as leaf roller, capsule borer, sphinx moth, aphids, and gall midge also have negative effects on sesame (Dilipsundar et al. 2019).

The growth habit of sesame is generally indeterminate, and the seed-bearing capsules rupture upon harvest (Langham and Wiemers 2002). These two traits render the crop inappropriate for mechanical harvesting and restrict its economic viability by restricting the producing regions to nations that have access to inexpensive labor in many areas (Qureshi et al. 2022). Prior to threshing, up to 50% of sesame seeds can be lost due to shattering (Weiss 1971).

9.4 BREEDING APPROACHES FOR SESAME IMPROVEMENT

Breeding activities in sesame are targeted at decreasing both biotic and abiotic stress as well as improving the oil content, quality, and seed yield. In the past, sesame breeding was based on the exploitation of naturally existing varieties or on creating variation by inter- and intra-specific crosses or mutations (Savant 2016). So far, there is no cultivar that is totally resistant to diseases like bacterial leaf spot, Fusarium wilt, and phyllody (Belay 2018). Additionally, to minimize diseases in sesame, mutation breeding is applied, which involves producing a range of acceptable morphological and physiological mutants in disease-prone regions. Similarly, gamma rays have also been used to obtain mutant lines that are resistant to phytophthora (Islam et al. 2016). Sesame is characterized as a largely self-pollinated crop, as well as having a variety of natural phenotypic polymorphisms. Hence, the selection approaches applied for breeding the self-pollinated presence of a number of these opposing features are equally successful in sesame (Degefa 2019). The presence of genetic variation is a requirement for hybridization and the accumulation of favorable alleles into a single genetic background via selection (Singh et al. 2022). Domestication, plant introduction, and pure line selection are the primary breeding procedures relevant to existing genetic variability (Begna 2021). Crossing and selection methods, pedigree, mass, and single seed descent, are important tools for widening diversity, which is important in sesame breeding (Singh et al. 2022). Despite all these available conventional breeding approaches, ideal sesame crops have not been developed.

As sesame is the principal crop in undeveloped nations, major efforts for sesame development have been conducted primarily through the conventional plant breeding approach. A conventional breeding procedure includes crossing genotypes, followed by a selection of the best progenies from the several segregation outcomes (Kumar 1999). Indeed, such a technique is laborious and time-consuming, taking numerous crossings and generations, and precise phenotypic selection together with the linkage drag may make it more difficult to attain the desired objective. Therefore, the use of modern plant breeding can be more revolutionary than continuing to use conventional breeding methods. Modern plant breeding methods such as proteomics, genomics, molecular marker-assisted selection, bioinformatics tools, and genetic engineering have been used on key crops including rice, rapeseed-mustard, wheat, sorghum, maize, and potato (Singh et al. 2022).

The utilization of new biotechnology approaches should be integrated into sesame breeding programs, especially taking into consideration the changing climatic conditions. Using molecular approaches, the selection of plants with agronomic traits and molecular markers for abiotic stress factors should be more successful and quicker in breeding studies. The molecular breeding efforts in sesame started late, with just one genetic map released and no information on quantitative trait locus (QTL) mapping until 2013 (Dossa et al. 2017b). However, significant progress has been achieved in sesame breeding efforts over the past decade to apply modern molecular biology methods, such as molecular markers, linkage mapping, and genome sequencing (Tripathy et al. 2019). Molecular marker technologies, among other modern tools, have considerably expedited conventional sesame breeding efforts in several crop species, improving genetic gain and decreasing breeding cycles (Myint et al. 2020). Along with these molecular marker technologies, the development of allele-mining-specialized molecular markers through allele mining is another new method followed in advanced breeding programs. Allele mining is an approach that tries to dissect naturally occurring allelic variations or suitable alleles for candidate genes impacting critical agronomic parameters that have the potential for crop improvement. It might help in the discovery of superior alleles, inter- and intra-species similarity analysis, the discovery of new haplotypes, and the development of functional molecular markers for marker-assisted selection (Sharma et al. 2014).

9.5 WILD RELATIVES AND OPPORTUNITIES FOR SESAME BREEDING

The cultivated sesame (*Sesamum indicum* L.) is a self-pollinated diploid species with 2n=26 chromosomes and low percentages of cross-pollination were reported (Ashri 1998). The genus consists of three cytogenetic groups that differentiate from each other by their chromosomal numbers. *S. indicum* 2n=26, along with its wild species *S. alatum, S. capense*, and *S. malabaricum*, the group 2n=32 consists of *S. prostratum, S. laciniatum, S. angolense, S. angustifolium*, and *S. latifolium*, while *S. radiatum* belongs to the group 2n=64 (Nimmakayala et al. 2011). There are also wild types whose chromosome numbers have not been identified (Table 9.1) (IPGRI and NBPGR 2004). There is limited cross-compatibility among the species from the different groups, creating difficulties in transferring desirable traits from wild species. Therefore, it has been difficult to transfer desirable characteristics such as drought tolerance, pest resistance, and resistance to diseases from wild relatives into cultivated sesame (Ashri 1998). Chromosomes of *S. indicum* are small, with variable length and X=13, resulting from the probable polyploidy process of ancestral

TABLE 9.1
The List of Sesamum Species

Chromosome Number (2n)	Species	Distribution	Specific Traits
26	S. alatum Thonn.	Nigeria, Sudan, Mozambique	Resistant to phyllody and antigastra
	S. malabaricum Burm.	India (Malabar)	Resistant to water logging, donor for cytoplasmic male sterility
	S. indicum L.	Tropical to temperate zone	High quality oil and seed
	S. capense Burm. f.	South Africa	
32	S. angolense Welw.	Angola	
	S. angustifolium Engl.	Congo, Mozambique, Uganda	
	S. laciniatum Willd.	India (Deccan hills), Africa	Resistant to antigastra and drought. Non-shattering capsules
	S. latifolium Gillett	East Africa	
	S. prostratum Retz	India, Africa	Resistant to antigastra, and salinity. Non-shattering capsules
64	S. radiatum Schum. & Thonn.	Africa, Upper Guinea, Sri Lanka	Resistant to drought
Unknown	S. abbreviatum Merxm		
	S. calycinum Welw. ssp.calycinum		
	S. calycinum Welw. ssp. Pseudoangolense Seidenst ex. Ihlenf.		
	S. calycinum Welw. ssp. baumii (Stapf) Seidenst. ex. Ihlenf.	Angola	
	S. capense Burm. f. ssp. lepidotum Schinz	Africa	
	S. marlothii Engl	S.W. Africa, East Indies	
	S. parviflorum Grabow- Seidensticker		
	S. pedalioides Hiern	Angola	
	S. rigidum Peyr. ssp. rigidum	Angola	
	S. rigidum ssp merenksyanum Ihlenf. & Seidenst.		
	S. schinzianum Aschers. ex. Schinz	S.W. Africa (Damaraland), East Indies	
	S. triphyllum Welw. ex. Aschers.	Africa, East Indies	
	S. triphyllum Welw. ex. Aschers. var. grandiflorum (Schinz) Merxm.		

Source: IPGRI and NBPGR (2004), Descriptors for Sesame (Sesamum spp.), IPGRI, Rome; NBPGR, New Delhi. Nimmakayala et al. (2011)

species. Successive hybridization of cultivated *S. indium* and the closely related species in the same cytogenic group was observed with maternal effect in some crosses. In general, for the different cytogenetic groups, inter-specific hybridization within each group was reported with success, but nuclear-cytoplasmic interactions seem to be the major characteristic of the whole genus (Bedigian 2015). At present, there is limited knowledge regarding the relationships between the species across the three cytogenic groups, and further studies using suitable DNA tools are required for efficient exploitation of the potential of the wild species. This would open a path of hybridization and introgression for large genetic diversity.

The international gene bank collections have a large number of closely related sesame crop species, which provides an opportunity for the introgression of agronomical traits into the cultivated sesame. Numerous attempts at crosses among *Sesamum* species by researchers across the world yielded various results in the introgression of desirable genes from various wild species into the cultivated species (Nimmakayala et al. 2011). So far, limited success stories reporting desirable gene transfer from wild species to the cultivated one in sesame exist. In the same cytogenic group, hybridization between *S. indicum* and *S. malabaricum* produced fertile F_1s (Nimmakayala 1997), while crosses in which *S. malabaricum* is the female and *S. indicum* is used as a male provided the F_1 hybrids with male sterility (Prabakaran 1996). The male sterile plants were obtained by Kavitha (1998) after backcrossing F_1 hybrids of *S. malabaricum* and *S. indicum*, showing a dominant gene was responsible for fertility restoration. Crosses of the cultivated *S. indicum* with *S. capense*, all belonging to the same cytogenic group (2n=26), produced a normal fruit set with non-viable seeds (Nayar and Mehra 1970), while reciprocal crosses of the cultivated species with its closed related species, *S. alatum*, were reported to produce some non-viable seeds, underscoring the wide genetic distance between the two species, although they belonged to the same cytogenetic group (Kobayashi 1991). Ram et al. (2006) reported crossability barriers for *S. alatum*×*S. indicum*. Later, crossing *S. alatum* with other cultivars exposed the chemical products such as gibberellic acid, naphthalene, and kinetin to produce viable capsules (Rajeswari et al. 2010). In the 2n=32 cytogenetic group, fertile F_1 hybrids and the F_2 (second filial generation) were reported in interspecific crossing between *S. prostratum* and *S. laciniatum* (Kobayashi 1991). In the 2n=64 chromosome groups, hybridization of *S. radiatum* (as *S. occidentale*) with *S. schinzianum* was found to be fully fertile, and they were assessed as one biological species by Hiremath and Patil (2002). The species *S. radiatum* and *S. laciniatum* have different chromosome numbers and have resistance to drought (Table 9.1). Subramanian (1972) made hybridization between two species; however, non-viable and shriveled seeds were obtained. Similarly, the cross between *S. occidentale* and *S. laciniatum* produced non-viable seeds. *S. prostratum* showed resistance to *Antigastra catalaunalis* and salinity with non-shattering capsules (Nimmakayala et al. 2011). Cultivated sesame is non-tolerant to salinity; therefore, the inclusion of related genes responsible for salinity tolerance in *S. prostratum* by crossings would be beneficial. Ramanathan (1950) and Kedharnath et al. (1959) acquired viable seeds – F1 sterile – amphidiploid fertile and doubling the chromosome number of the F1 with mutagen after this cross (Joshi 1961). The cultivated species was crossed with *S. radiatum*, 2n=64, resulting in a limited number of viable capsules, with mostly inviable seeds (Kobayashi 1991). For this crossing, an embryo culture protocol was developed by Lee et al. (1991) and then used to rescue embryos of hybrids, although no viable seeds were produced in the reciprocal. The embryos of *S. schinzianum*×*S. indicum* were also cultured by Qu et al. (1994) with the three-stage embryo culture protocol.

9.6 DEVELOPMENT AND USE OF MOLECULAR MARKERS FOR SESAME IMPROVEMENT

In classical plant breeding, identification of cultivars with the desired traits generally involves morphological and visual selection. Through advancements in molecular biology, molecular markers that can identify variation in a gene/DNA sequence with a specific chromosomal position and their effects on the phenotype have revolutionized in plant breeding systems (Cahill and Schmidt 2004). Molecular markers are derived from sequence differences that emerge because of mutations, including insertions, deletions, inversions, duplications, and translocations, that are closely associated with the quantitative trait locus (QTL) (Nadeem et al. 2018). Molecular markers are important tools to find differences among the diversity, and they provide tremendous advantages in all aspects of plant breeding: (i) can determine variations in genomic basis between individuals by enhancing selection efficiency at early phases of growth; (ii) offer objective data that is not affected by environmental conditions or management practices (Mondini et al. 2009); (iii) can be obtained from any tissue at any development stage of plants (Francia et al. 2005); and (iv) can enable us to screen a great number of plants with limited time and resources (Liu et al. 2012).

Different types of molecular markers have been developed and successfully implemented in a variety of crops for most of the agriculturally important traits such as biotic and abiotic stress factors, quality, and yield (Francia et al. 2005). Especially for commercially important crops, these markers have been used as a key for plant breeders who utilize marker-assisted selection for crop development. Different markers have been used to select resistant/susceptible lines against important biotic/abiotic factors in industrial crops such as tomato (yellow leaf curl virus) (Nevame et al. 2018), wheat (leaf and stem rust) (Genievskaya et al. 2020), maize (downy mildew) (Kim et al. 2020), and rice (drought stress) (Mardani et al. 2013). In recent years, they have been largely used in sesame breeding studies such as genetic diversity, germplasm evaluation, and high-throughput analysis like QTL mapping and genome-wide association studies.

The sesame genome information provides the necessary data for gene and allele discovery for important sesame traits. Many candidate genotypes, genes, QTLs, and alleles associated with those traits were collected and discovered as great-demand research topics, including plant architecture, yield, quality, disease resistance, abiotic stress tolerance, and so on.

9.6.1 Genetic Diversity

Genotype-specific amplifications produced by various markers give important information for further classification (Ali et al. 2007). Different genetic diversity studies were conducted with the use of molecular markers to understand the population structure of sesame. The diversity studies based on molecular markers were started with random amplified polymorphic DNA (RAPD) and amplified fragment length polymorphism (AFLP) markers. Bhat et al. (1999) analyzed 58 accessions of sesame genotypes with RAPD markers. They found a high level of genetic diversity, and genetic diversity was greater in the collections from the Indian subcontinent as compared to the exotics in that study. Ali et al. (2007) used AFLP markers to identify the genetic architecture of the collection, including 96 sesame varieties. They constructed the unweighted pair group method with an arithmetic mean (UPGMA) tree separated into two major groups, which mostly consist of Eastern Asian and South Asian origins. Abdellatef et al. (2008) used RAPDs on 10 sesame germplasms collected from different regions of Sudan. They obtained a low level of genetic similarity in the collection. Cho et al. (2011) developed and used 16 SSR (simple sequence repeat) markers to exploit the genetic diversity of 150 sesame genotypes collected from 22 countries. Cluster analyses were performed for accession and population, and the mean genetic similarity coefficient was 0.45 for accession and 0.52 for population in that study. The cultivar and wild types were analyzed by inter-simple sequence repeat (ISSR) markers to understand genetic variation (Nyongesa et al. 2013). Forty-six accessions of sesame were divided into six clusters; however, the clustering did not indicate a clear division among sesame accessions based on their origins. The genotypes, Indian-1 and Indian-2, showed a good amount of genetic divergence in that study; therefore, they should be the parents to obtain heterosis. The other study used ISSR markers for diversity: 128 sesame genotypes were handled to uncover the population structure of the collection (Abate et al. 2015). The polymorphic information content (PIC) value ranged between 0.26 and 0.76, showing the high informativeness of the selected primers in this study. A total of 60 sesame genotypes were evaluated with 8 polymorphic expressed sequence tag (EST)-SSR markers by Pandey et al. (2015), who detected 27 alleles, averaging 3.37 alleles per locus, with the number of alleles ranging from 2 to 6. Dossa et al. (2016) utilized simple sequence repeat markers (SSRs) to determine the genetic relatedness of 96 sesame germplasm from 22 countries and identified that the diversity detected in African germplasm was lower than in Asian germplasm. Wu et al. (2014b) analyzed genetic diversity with 88 polymorphic markers (insertion-deletion (InDel) and SSR) in 130 Chinese sesame accessions and determined 325 alleles, including some unique alleles. Stavridou et al. (2021) characterized 37 landraces from 10 geographical areas using 28 qualitative morpho-physiological descriptors with 7 expressed sequence tag-derived simple sequence repeat (EST-SSR) markers and identified different clusters in the grouping. Adu-Gyamfi et al. (2019) evaluated sesame collection with agronomic and molecular markers. In a molecular approach, they found that 21 out of the 38 SSR markers were polymorphic, the heterozygosity was 0.52, and the average allele per locus was 19.52. With next-generation sequencing technology, higher numbers of markers were identified for genetic diversity and relationships. Cui et al. (2017) analyzed 366 sesame genotypes with the use of 89,924 single-nucleotide polymorphisms (SNPs). They found that the genetic structure analysis showed that sesame genotypes were structured primarily on the basis of their geographic collection. The other comprehensive diversity study was conducted by Basak et al. (2019), who used 5,292 high-quality SNPs discovered by a genome-reducing approach. They found that there was no clear division evident among the sesame genotypes with regard to their continental locations. Similarly, Laurentin and Karlovsky (2006) found no association between genetic diversity and genotype origin. These studies showed that molecular markers are highly important tools to exploit the genetic structure of collections in sesame.

9.6.2 Wildish Traits

One of the main reasons for low seed yield in sesame is some wildish traits, including capsule shattering (Uzun et al. 2004) and indeterminate growth habit (Uzun and Cagirgan 2006). Only a few studies using molecular markers were conducted on these traits. Capsule shattering causes high yield losses is one of the important problems for sesame as an industrial crop. Uzun et al. (2003) identified a closely linked AFLP marker for the closed capsule in the segregated population obtained by crossing the non-fragmented mutant line (cc3) and the Turkish variety Muganlı-57 were developed. This marker later was converted to a sequence characterized amplified region (SCAR) marker significantly advances the reproducibility and reliability of PCR analysis. The Si-SR-32-19 marker was produced by conversion from the AFLP to the SCAR marker in mapping population obtained by crossing between Cplus1 (shattering-resistant) and Kuaox25 (shattering-susceptible) (Phumichai et al. 2017). The development of sequencing technologies has made it possible to identify candidate genes and it facilitated the identification of functional molecular markers derived from sequence polymorphisms discovered within the functional gene(s) associated with phenotypic traits (Kage et al. 2016). Yol et al. (2021) identified a functional In-Del marker with the use of double digest restriction-site associated DNA sequencing (ddRAD-seq) in the position of chromosome 8/S8–5062843 that included six coding sequences (CDSs) and was associated with the *KAN*1 isoform *X*1 (*S. indicum*) affect shattering trait in sesame. Parallel with the current study, a loss of function of the KAN1-like protein was also detected in a curly leaf, non-shattering sesame mutant by Zhang et al. (2018a), who named this gene *SiCl1*.

The other limitation in sesame production is the indeterminate growth habit that causes non-synchronous maturity and incompatibility with combined harvesting (Uzun and Cagirgan 2006). Firstly, Uzun and Cagirgan (2009) determined two ISSR markers associated with the *dt* gene in the F_2 segregating population derived from crossing between the mutant line (dt1) and indeterminate (cv. Muganl 57) lines. Zhang et al. (2016) also conducted mapping analysis and exploited the *SiDt* gene associated with this trait. The target SNP, SiDt27-1, and the determinacy gene were located in Scaffold 00170 of the Yuzhi 11 reference genome. Zhang et al. (2018b) constructed a high-density genetic map using a genotyping by sequencing (GBS) approach. The researchers developed the SSR marker ZMM2498, which is linked to the determinate trait as a *dt* gene in linkage group (LG) 09. Furthermore, Dutta et al. (2022a) identified a *dt* gene-based EST-SSR marker and validated it in an interspecific hybrid sesame population (derived from *S. indicum* and *S. prostratum*).

9.6.3 Yield-Related Traits

One of the most important objectives of sesame breeding is to increase yield. However, it is a time-consuming and low-efficiency process as it is affected by many complex and environmentally sensitive factors (Wu et al. 2014a; Gadri et al. 2020). It is known that integrating traditional breeding strategies with molecular approaches is an effective way of improving yield (Guo et al. 2019). Having knowledge of the genetic structure and detecting important QTLs of yield-related characteristics is important in sesame breeding studies to improve high-yielding cultivars. Wu et al. (2014a) constructed a high-density genetic map of 224 recombinant inbred lines and identified 21 seed yield-related QTLs. Mei et al. (2021) determined a total of 46 QTLs for yield-related traits. The number of seeds per capsule and seed weight is also known as major contributors to seed yield to a large extent (Emamgholizadeh et al. 2015). In a study conducted by Wei et al. (2016), it was determined that two markers, ID71 and ZZM452, were stably associated with the number of seeds per capsule and thousand seed weight, respectively. Du et al. (2019) captured two QTLs for a thousand seed weights. Although branching habits significantly affect grain yield (Baydar 2005; Teichmann and Muhr 2015), very few genetic mapping studies have been performed. Mei et al. (2017) mapped genes for basal branching habit (*SiBH*) and flowers per leaf axil (*SiFA*) on LG5 and LG11, respectively, and identified three (Marker58311, Marker34507, and Marker36337) markers linked to the *SiFA* gene.

Plant height is positively correlated with seed yield. However, higher heights in sesame are not very suitable due to damage caused by strong winds or hurricanes, which create more challenges for mechanized harvest (Langham and Wiemers 2002; Zhang et al. 2019a). Recently, scientists created a dwarf mutant dw607 (type dwf1) with ethyl methane-sulfonate (EMS) mutagenesis, where plant height is reduced by more than 40% and internode length is reduced by approximately 50% compared to the wild type (Wang et al. 2017b). Wei et al. (2015) discovered two promising candidate genes at the locus of flowering time (*SiDOG1* and *SiIAA14*) and two other candidate genes at the locus of plant height (*SiDFL1* and *SiILR1*), which had a significant correlation with oilseed yield. Miao et al. (2020) developed a functional marker, SiSNPdwf1, from the *C9.scaffold*2.572 region that was assumed to be the target gene associated with the short internode length. Leaf size is a very important portion of plant architecture (Jiang et al. 2013). Due to the shading effect, the leaf shape/size serve to capture light energy and synthesize organic compounds, thus affecting the yield by changing the number of capsules produced (Langham and Wiemers 2002). Zhang et al. (2018a) determined the gene (*SiCL*1) that controls leaf curling and capsule indehiscence by the pooled bulked segregant analysis (BSA) performed by resequencing the genome of 130 F_2 individuals derived from crossing the cl1 (curly leaf shape)and USA (0)–26 (normal leaf shape). The authors developed the SiCLInDel1 marker from the region in the *SiCL*1 gene where the 20 nucleic acids were mutagenized into 6 nucleic acids. Furthermore, Sheng et al. (2021) combined the QTL-seq approach with SSR marker mapping in a recombinant inbred line (RIL) derived from a cross between sesame varieties Zhongzhi No. 13 (with big leaves) and ZZM2289 (with small leaves) and determined a major pleiotropic locus, *qLS*15-1, for both leaf length and width.

9.6.4 Quality Traits

Sesame is a widely accepted oilseed crop worldwide with its nutritional and therapeutic benefits (Wan et al. 2015). In order to exploit the genetic basis of quality traits, different studies have been conducted (Magni et al. 2010; Wei et al. 2013). An association mapping of seed oil-related traits and protein content using 79 (SSRs, sequence-related amplified polymorphisms (SRAPs), and AFLPs) markers was conducted in 216 Chinese sesame accessions. However, only one environment-stable marker, M15E10–3, was determined for oil content (Wei et al. 2013). Li et al. (2014) evaluated 369 sesame accessions in 5 environments using 112 SSRs, and a total of 19 and 24 SSRs were found for oil and protein content, respectively. Moreover, 36 candidate genes associated with the lipid pathway, including fatty acid elongation genes and a gene encoding *Stearoyl-ACP Desaturase*, were determined. Furthermore, Chen et al. (2014) searched for the sequence divergency in the coding region of the fatty acid desaturase (*FAD*) gene between wild and cultivated sesame (high fatty acid composition). Researchers determined some SNPs that led to amino acid changes located in the enzyme active site between the wild and cultivated forms. Wu et al. (2017) conducted a study to detect QTLs for quality traits using a mapping population containing 224 RILs. The highest phenotypic variation was observed as 10.1% for oil content LG1 in marker intervals SBN2389–SBN2485. The LG5 has the highest phenotypic variation ratios of 10.9% and 18.6% for protein and sesamin content, respectively, in that study. In one of the latest QTL studies, Xu et al. (2021) identified loci governing the main sesame lignans, sesamin and sesamolin variations in RILs under two locations. They captured 26 QTLs on the linkage map including 424 SSR markers and found 16 and 10 loci associated with sesamin and sesamolin variation, respectively. Two regions, *qSmin_11.1* and *qSmol_11.1*, showed very high phenotypic variation detected in both environments as 67.69% and 46.05% for sesamin and sesamolin, respectively.

9.6.5 Resistance to Biotic and Abiotic Stress Factors

Plants are exposed to a range of factors during their growth cycles, including biotic stressors like herbivore assault and pathogen infection, as well as abiotic stressors like heat and cold, drought, lack of nutrients, increased salt levels, and toxic metals (Younis et al. 2020). Sesame usually suffers from various bacterial, fungal, and viral diseases, and their causative agents include *Cercospora sesami, Alternaria sesami, Cylindosporium sesami, Macrophomina phaseoli, Phytophthora parasitica, Oidim* spp., *Erysiphe cirhoracearum*, *Phytophthora nicotianae* var. sesame, and *Helminthosporim sesamine* (Min and Toyota 2019). However, there were rare reports on the genes or QTLs for sesame disease resistance. Although the development of cultivars with resistant genes is the best idea, limited research has been conducted. The marker associated with powdery mildew, which is one of the most devastating diseases of sesame and causes significant yield loss, was developed by Kabi et al. (2021). This marker is the RGA (resistance gene analog) marker (h2_13m22a) tightly linked with the powdery mildew resistance gene developed from segregating generations of crosses involving Rama (susceptible) and VRI-1 (resistant) genotypes. Rao et al. (2013) used 30 accessions and determined two RAPD markers (OPAE 6 and OPM 12) to map disease tolerance gene(s) associated with powdery mildew. They were tightly linked to the powdery mildew tolerance gene *Pm* 1. Another important damaging disease is charcoal rot, caused by *Macrophomina phaseolina* (Tassi) Goid. M., inducing a 30% reduction or complete loss of sesame crops (Wang et al. 2017a). The 548 lines generated from the cross between ZZM2748 (susceptible) and Zhongzhi No. 13 (resistance) were tested in four environments and constructed an SSR-based genetic map (Wang et al. 2017a). Among the 14 QTLs for coal rot disease, the locus with the highest phenotype contribution was determined to be *qCRR12.2*, which contains 15 genes. *Fusarium oxysporum* sesame is one of the important diseases causing quantitative and qualitative losses in the yield of sesame (Shabana et al. 2014), and Anter and Ghada (2021) identified five RAPD and two ISSR markers linked to *fusarium* resistance.

Drought, waterlogging, and salt are major abiotic stressors that limit sesame production. Wang et al. (2016) figured out a comprehensive transcriptome study on two tolerant and susceptible genotypes and revealed 66 genes that may be candidates for improving sesame tolerance to waterlogging. Liang et al. (2021) detected 34 QTLs on 10 chromosomes. Among them, 13 stable QTLs were revealed in two independent experiments, and eight of them were associated with traits under water stress conditions. Li et al. (2018) identified two common QTLs for drought and salt responses located in linkage groups 5 and 7, respectively. Zhang et al. (2019b) used two contrasting sesame genotypes (salt-tolerant – WZM3063) (salt-sensitive – ZZM4028) in transcriptome analysis and revealed strongly induced salt-responsive genes in sesame mainly related to amino acid metabolism, carbohydrate metabolism, biosynthesis of secondary metabolites, plant hormone signal transduction, and oxidation-reduction process.

9.7 COMBINING OF DESIRED TRAITS VIA MOLECULAR MARKERS

In recent years, it has been known that lots of functional QTLs and genes have been identified for desirable agronomic traits with genomic advances. The gene/QTL-related markers are therefore important tools to develop an ideal sesame crop that has higher quantity and quality. In this section, we summarized the important molecular markers for important traits for sesame breeders (Table 9.2).

Plant height is an important selection criterion for machine harvesting. Langham and Wiemers (2002) reported that 150 cm and lower is suitable for all harvest practices, which also contribute

TABLE 9.2
Key Markers to Develop Cultivars in Breeding Studies

Traits	Marker Systems	Marker/Gen Name	Authors
Oil Content		*SIN_1016759 (PPO)*	Wei et al. (2015)
Determinate Growth	SSR	*ZMM2498*	Zhang et al. (2018a)
Capsule Size	SNP	*SIN_1011000, SIN_1010995 and SIN_1010983*	Zhou et al. (2018)
Capsule Shattering	In-Del	*S8_5062843*	Yol et al. (2021)
Seeds per Capsule	In-Del	*ID71*	Wei et al. (2016)
Thousand Seed Weight	SSR	*ZZM452*	
Plant Height	SNP	*SiSNPdwf1*	Miao et al. (2020)

positively to lodging resistance (Ashikari et al. 1999). The 'Yuzhi Dw607' has a high yield level and is also a short internode length mutant variety (Zhang et al. 2019a). This mutant trait can be selected by the functional marker *SiSNPdwf1* (Miao et al. 2020). Similarly, an effective SSR (ZMM2498) for a determinate growth habit trait was discovered by Zhang et al. (2018b). Although determinate lines had a lower seed yield compared to cultivars, they had important traits such as synchronous flowering, lodging resistance, uniformity, and early maturing (Uzun and Cagirgan 2006). These two markers should therefore be used especially for crosses with dwarf types. Branching is also significant for higher yields in sesame because stems carry flowers and capsules, and branching was identified as one of the important parts of an ideal sesame cultivar (Baydar et al. 2005). However, too much branching is also an undesirable feature. During harvest, higher and taller branching can cause capsule losses and/or seed cracking (Langham and Wiemers 2002). Mei et al. (2017) handled branching traits in sesame and classified genotypes: plants with branches ≤2 were classified as "unbranched type" and ≥3 as "branched type,". After segregation and molecular analysis, showed that the basal branching habit was controlled by one single gene (designated as *SiBH*). Three markers, Marker129539, Marker41538, and Marker31462, were found to be tightly linked, and they can be used to select branching types. The trait of higher numbers of capsules per axil and capsule density are desirable traits that positively affect yield (Baydar et al. 1999; Sarwar et al. 2007; Yol and Uzun 2011). Generally, tri-capsule types produce more capsules compared to mono-capsule types, but having more capsules isn't adequate for the production of high yields (Baydar 2005). Wei et al. (2015) found the gene (*SiACS*) controlling capsule per leaf axil, and therefore related markers should be integrated into breeding studies. Long and dense capsule characteristics have a great impact on seed yield in sesame by enabling plants to produce more capsules and seeds (Bisht et al. 2004; Sarwar et al. 2007). Zhou et al. (2018) conducted GWAS (genome-wide association study) analysis and found three SNP markers related to capsule size. They should be integrated into the breeding program because there is a positive correlation between seed yield and longer capsules of sesame (Ogbonna and Ukaan 2012). The traits of 1,000 seed weight and the number of seeds are important traits contributing to seed yield. Wei et al. (2016) identified marker ID71 as stably associated with the number of seeds per capsule and ZZM452 as associated with the thousand-seed weight. These two traits can be selected with the use of polymerase chain reaction (PCR)-based applications. The trait of capsule shattering at harvest is one of the most important factors preventing the commercial value of the plant from being revealed. The non-shattering and/or semi-shattering capsule seems to be a better option for mechanized harvesting (Yol and Uzun 2019). However, non-shattering capsule mutants have low yields, which restricts the utilization of these genotypes in sustainable sesame cultivation (Qureshi et al. 2022). The developed InDel marker (Yol et al. 2021) should be useful to separate shattering and non-shattering types. The oil content has a polygenic nature, and a major locus was discovered for this trait that contained the candidate gene SiPPO (*SIN_1016759*) (Wei et al. 2015). The potential markers should be validated in different genetic resources to identify markers that have higher phenotypic variation. It should be very useful for future breeding studies.

9.8 GENOMIC APPROACHES FOR DISCOVERING NEW ALLELES

9.8.1 Sesame Genome

The genome size of sesame was different in several reports. Zhang et al. (2013a) reported that "Yuzhi 11" was about 369 Mb. Wang et al. (2014) reported "Zhongzhi No. 13" was about 357 Mb; Wei et al. (2017) reported 350 Mb for two other sesame varieties; and Sabag et al. (2021) reported a sesame about 375 Mb. Of those, "Zhongzhi No. 13" was first fully published and opened the genome assembly and annotation information, and 27,148 protein-coding genes were predicted (Table 9.3). With the five sesame genomes, including two landraces (*S. indicum* cv. Baizhima and Mishuozhima) and three modern cultivars (*S. indicum* var. Zhongzhi13, Yuzhi11, and Swetha), Yu et al. (2019) constructed a sesame pan-genome of 554.05 Mb. The pan-genome contained 26,472 orthologous gene clusters; 15,409 (58.21%) of them were core (present across all five sesame genomes), whereas the remaining 41.79% (11,063) clusters and the 15,890 variety-specific genes were dispensable. These genomes and pan-genomes constituted a resource for further gene identification and sesame crop improvement.

9.8.2 Genetic Linkage Map

Genetic linkage mapping is an important genetic tool to tag molecular markers and identify QTL/gene location. The first genetic map for sesame was constructed by Wei et al. (2009) using EST-SSRs, AFLPs, and random selective amplification of microsatellite polymorphic loci (RSAMPLs), which were mapped in 30 linkage groups covering 936.72 cM with an average marker spacing of 4.93 cM. The SSR, AFLP, and RSAMPL were also used to construct a high-density map, and 724 polymorphic markers were grouped into 14 linkage groups with a total length of 1216.00 cM (Zhang et al. 2013b). Wu et al. (2014a) developed a high-density genetic linkage map including 1,230 markers using the RIL population through restriction site-associated DNA sequencing (RAD-Seq). Along with the published sesame genome, more and more high-density genetic linkage maps of sesame have been developed. Wang et al. (2017a) developed 7,357 SSR markers from the sesame genome and transcriptomes, and then constructed a genetic map containing 424 novel polymorphic markers. To identify the determinacy gene, Zhang et al. (2018b) used the genotyping by sequencing (GBS) method for high-throughput SNP identification in the F_2 population. This genetic map length was 1086.403 cM in, which was composed of 13,679 SNPs. Another sesame genetic map was constructed with the SLAF-seq (specific-locus amplified fragment sequencing) approach by Du et al. (2019), who obtained a genetic map of 2128.51 cM in length with an average distance of 0.99 cM between adjacent. Most recently, many high-density genetic linkage maps of sesame using bin markers and SNPs were constructed by Liang et al. (2021), Asekova et al. (2021), Yol et al. (2021), and Li et al. (2021). These maps provided the foundation and convenience for the discovery of sesame genes and QTLs. For example, Wu et al. (2014a) detected 13 QTLs on 7 LGs and 17 QTLs on 10 LGs for seven-grain yield-related traits by the multiple interval mapping (MIM) and the mixed linear composite interval mapping (MCIM), respectively. Wang et al. (2017a) discovered 14 QTLs responsible for sesame charcoal rot disease resistance.

TABLE 9.3
The Genome Sequencing Information of Several Sesame Varieties

Cultivar	Total Length (Mb)	Contig N50 (kb)	Scaffold N50(Mb)	Coding Genes	Average Gene Length (bp)	Reference
Zhongzhi 13	274	52,2	2.10	27148	3171	Wang et al. (2014)
Baizhima	267	47,3	–	31558	3673	Yu et al. (2019)
Mishouzhima	254	47,9	–	30995	3700	Yu et al. (2019)
Swetha	340	11,5	0.02	41859	4032	Yu et al. (2019)
Yuzhi11	211	17,9	0.32	26022	3623	Yu et al. (2019)
Zhongzhi 13	273	53,1	20.26	36189	3718	Yu et al. (2019)

9.8.3 Genome Resequencing and GWAS Used in Sesame Allele Discovery

The genome-wide association study (GWAS) approach is widely used to detect relationships between traits and molecular markers or genes in a studied population based on linkage disequilibrium (Li et al. 2018). It provides more accurate positioning and mapping without the requirement for a linkage group (Flint-Garcia et al. 2005; Huang and Han 2014). This approach finds the association between SNPs and selected agronomic traits with the advent of next-generation genotyping techniques (Berhe et al. 2021). It has been successfully applied to several crops such as alfalfa, *Arabidopsis thaliana*, soybean, maize, rice, *Brassica napus*, etc. (Kumar et al. 2015; Mao et al. 2015; Clauw et al. 2016; Xu et al. 2016; Liu and Yu 2017; Zeng et al. 2017).

The completion of a high-quality reference genome sequence in sesame provides an opportunity to comprehensively investigate the genetic architecture of sesame traits. One of the most significant agronomic characteristics of sesame is the color of the seed coat, which is linked to biochemical processes involved in the metabolism of proteins and oils, antioxidant content, and disease resistance (Zhang et al. 2013b). Wang et al. (2014) reported the resequencing of 29 sesame accessions and then found adenylate isopentenyl transferase (ITP) genes may function in determining the number of flowers per leaf axil of sesame by mediating zeatin biosynthesis. Wei et al. (2015) sequenced 705 diverse sesame varieties, constructed a haplotype map of the sesame genome, and then identified 549 associated loci for 56 agronomic traits in four environments. In a study of 705 sesame accessions in four environments using GWAS, 13 important associations were determined for oilseed components (oil, protein, sesamin, sesamolin, saturated fatty acid (SFA), unsaturated fatty acid (USFA), and the ratio SFA/USFA). The genes related to fatty acid biosynthesis (*CXE17* and *GDSL*-like lipase), flower lip color (*SiGL3*), yield (*SiACS*), and oil content (*SiPPO*) were identified in this study. Dossa et al. (2019a) handled drought tolerance in sesame with a GWAS approach. They investigated 400 sesame genotypes and found a total of 140 reliable and stable QTLs, and the genes *SiABI4*, *SiTTM3*, *SiGOLS1*, *SiNIMIN1*, and *SiSAM* were detected as potential genes to modulate drought tolerance. Li et al. (2018) reported that a large-scale GWAS study was performed on a total of 490 sesame accessions and detected 132 and 120 significant SNPs, then resolved to 9 and 15 QTLs for drought and salt stresses, respectively. The genes *SiCCD8*, SiGPAT3, *SiCYP76C7*, *SiGRF5*, *SiGDH2*, *SiRABA1D*, *SiEMF1*, and *SiGRV2* were discovered as potential genes related to drought stress; On the other hand *SiGPAT3*, *SiNAC43*, *SiGDH2*, *SiLHCB6*, *SiMLP31*, *SiPOD*, *SiWRKY14*, *SiXXT5*, *SiXTH15*, *SiG6PD1*, *SiHSFA1*, *SiUDG*, *SiCP24*, *SiDUF538*, and *SiCC-NBS-LRR* were also identified as potential genes for salt tolerance regulation in that study. He et al. (2019) utilized GWAS to identify alleles that contribute to vitamin E content variation in sesame and observed a significantly associated locus on chromosome 8 (*LG08_6621957*). Cui et al. (2021) re-sequenced 366 sesame germplasm lines and performed GWAS on seed coat color in 12 environments, and then identified a total of 224 significant single nucleotide polymorphisms. Tesfaye et al. (2022) determined 21 significant marker-trait associations for yield-related traits and reported that they determined the first QTLs for flowering time and maturity in sesame. The presence of E vitamin is a significant biological property of sesame which comes in eight different forms: four tocopherols (α, β, γ, and δ) and four tocotrienols (α, β, γ, and δ) (Jiang et al. 2001). He et al. (2019) conducted a GWAS utilizing 5,962 genome-wide markers in 96 core sesame accessions and determined a region on chromosome 8 (LG08_6621957) including 10 SNPs that are associated with γ-tocopherol. The *SIPPO* (*SIN* 10167599) gene encodes a putative polyphenol oxidase (PPO), which contributes to the oxidation phase in the manufacture of proanthocyanidin, lignin, and melanin, and creates black colors via the browning reaction in plants (Yu 2013). Zhou et al. (2018) identified several QTLs associated with seed-related traits. A total of 646 loci were significantly associated with the 39 traits ($p < 10{-}7$) and resolved to 547 QTLs. Significantly, they identified novel candidate genes, such as *SiLPT3* and *SiACS8*, which may control capsule length and capsule number traits. Sabag et al. (2021) used a panel of 148 sesame genotypes, which they analyzed using 20,294 single-nucleotide polymorphism markers. They defined a major genomic region on LG2 that was associated with flowering dates and yield-related traits. Dossa et al. (2020) exploited the genetic basis of seven root-related traits and uncovered 409 significant signals and 19 quantitative trait loci containing 32 candidate genes. A peak SNP significantly associated with root number and root dry weight traits was located in the promoter of the gene named 'Big Root Biomass' which modulates sesame root biomass through the auxin pathway. Sesame oil comprises important nutraceuticals such as UFA, lignans, vitamin E, and phytosterols (Wang et al. 2022). Phytosterol is one of the crucial dietary constituents and has multiple uses in areas such as food, cosmetics, and medicine (Sharma et al. 2021). Wang et al. (2022) evaluated 402 sesame accessions for phytosterol content in two environments. They determined 33 significant associated SNP loci and 37 candidate genes for the four traits (campesterol, stigmasterol, β-sitosterol, and Δ5-avenasterol) performed with GWAS. Among SNPs, Ch6–39270 and Ch11–142842 were stably linked with campesterol and stigmasterol content.

9.8.4 Transcriptomic Approach

The term transcriptome refers to the complete set of transcripts in a cell and their quantity in specific developmental stages (Wang et al. 2009). Transcriptome profiling can be identified by RNA-sequencing (RNA-seq) which uses next-generation sequencing technology (Behera et al. 2021). It provides an effective approach to the characterization of the transcriptomic profile of plants for gene discovery and analyzing differential gene expression of mRNAs (messenger RNA) (Jain 2012). The transcriptome sequencing of a sesame plant was conducted for different purposes. Although sesame is an important oilseed crop, the development of molecular markers is limited compared to other important oilseed crops such as soybean and sunflower. From this point, Zhang et al. (2012) conducted an RNA-Seq study to develop genic SSR markers. The 24 samples analyzed in RNA-seq experiments and 42,566 uni-transcripts were assembled from more than 260 million filtered reads. They detected 7,324 SSRs (SSRs ≥15 bp) and 4,440 SSRs (SSRs ≥18 bp); a total of 2,164 SSR primer pairs were identified in the 4,440 SSR-containing sequences (≥18 bp). This showed that transcriptomic data should be a highly valuable source for developing marker systems. Sesame is highly sensitive to waterlogging stress, and generally, plants hardly survive over 36 h in these stress conditions. The study was conducted by Wang et al. (2012a) to identify related genes for waterlogging stress with the use of RNA-seq data from the roots of two contrasting sesame cultivars. They found significant changes in the expression of genes for the pathways, including photosynthesis, cysteine and methionine metabolism, flavone and flavonol biosynthesis, steroid biosynthesis, glutathione metabolism, phenylpropanoid biosynthesis, spliceosome, and circadian rhythm. In another study about waterlogging stress response, resistant and tolerant genotypes were exposed to 0, 3, 9, and 15 hours of waterlogging as well as 20 hours post-drainage (Wang et al. 2016). Results showed that a total of 1,379 genes were significantly induced and commonly expressed in sesame under waterlogging conditions. The 1,281 genes were newly reported in that study. Dossa et al. (2017a) examined drought-tolerant and sensitive genotypes based on transcriptomes. The remarkable result of this study was that they found unique differentially expressed genes in drought-tolerant genotypes. These genes should be considered for drought tolerance in ongoing sesame studies. Roots are the sensors of water deficit, and a high number of genes are differentially expressed under drought conditions in roots (Takahashi et al. 2018). To identify these genes in sesame roots, an RNA-seq study was conducted to understand transcriptional changes in the roots of drought-tolerant (TEX-1) and drought-sensitive (VEN-1) sesame genotypes (Song et al. 2020). Transcription-factor enrichment analysis showed that multiple members of the WRKY, bZIP, MYB, and NAC families were highly expressed under osmatic stress for the roots of TEX-1. In a different study, the root diversity was studied with the use of 40 diverse sesame varieties grown in soil and hydroponics systems (Su et al. 2019). Two genotypes, including G546 (big root phenotype) and G259 (small root phenotype), were compared in this research, and a total of 2,897 differentially expressed genes were detected, and they were enriched in phenylpropanoid biosynthesis, diarylheptanoid and gingerol biosynthesis, starch and sucrose metabolism, and flavonoid biosynthesis, showing the importance of these pathways for root development. Soil salinity is one of the major inhibitors of sustainable crop production. To reveal the crucial biological pathways to salt stress in sesame, Zhang et al. (2019c) compared two contrasting sesame genotypes concerning salt tolerance. They monitored that transcriptome analysis revealed strongly induced salt-responsive genes in sesame, mainly related to plant hormone signal transduction, amino acid metabolism, oxidation-reduction process, and carbohydrate metabolism. Sesame is an important oil source, and one of the important breeding objectives is increasing oil content. The genes expressed in low and high oil content genotypes were handled by Wang et al. (2019), who used 22 seed and carpel transcriptomes from 3 varieties of sesame with high and low oil content. A total of 23 candidate genes were identified and predicted to be beneficial for higher oil accumulation. It is highly important to understand pathways related to increasing oil content. Seed coat color is linked to important quality traits in sesame (Wang et al. 2012b; Zhang et al. 2018c) and highly affects consumer preferences. Wang et al. (2020) investigated the developing seed transcriptomes of two varieties of sesame, "Zhongfengzhi No. 1" (white seed) and "Zhongzhi No. 33" (black seed). The differentially expressed genes were mainly enriched in 37 metabolic pathways between black and white sesame, especially flavonoid biosynthesis and biosynthesis of secondary metabolites, which were found to be dominant. When comparing different development stages, this work showed the significant role of 8–14 days post-anthesis in black pigment biosynthesis and accumulation. Leaf size traits are also strongly influenced by environmental changes, and they are a crucial component of sesame for plant structure and affect yield potential (Sheng et al. 2021). The differentially expressed genes between big and small leaves were investigated using RNA-seq by Sheng et al. (2021), who detected 12 expressed lower and 14 expressed higher genes in the large leaf parent at experimental stages. Among them, three candidate genes, *SIN_1004875*, *SIN_1004882*, and *SIN_1004883*, were associated with leaf growth and development in sesame. Therefore, these regions should be assessed to select bigger-leaf sesame plants in breeding studies. In a recent transcriptomic study, Dutta et al. (2022b) examined the transcriptome of developing seeds of the wild type, *S. mulayanum*, and cultivar. They carried out a transcriptomic study on the seeds at 10 and 30 days after pollination and investigated differentially expressed transcripts. The significant result of this study was that oil containing cultivated sesame revealed high expression of the key lipid biosynthesis genes.

9.8.5 Sesame Genome and Gene Information Online Databases

There were at least five sesames that had been sequenced and assembled. A list of online genetic and genome databases was constructed based on the study of the sesame genome and its variations. The first version of the sesame genome, Sinbase 1.0 (http://orci-genomics.org/Sinbase/), was released in 2015, just following the first published sesame genome, "Zhongzhi No.13". Recently, Sinbase 1.0 was upgraded to version 2.0 (http://www.sesame-bioinfo.org/Sinbase2.0) (Wang et al. 2021). A comprehensive database for sesame function genes and loci named SesameFG (http://www.ncgr.ac.cn/SesameFG) was constructed in 2017, which collected the genes and loci related to sesame yield, yield-related, disease resistance, oil quality, growth cycle-related, flowering date, waterlogging, lipid metabolism, and other morphological traits (Wei et al. 2017). Up to date, at least six sesame online genetic and genome databases have been constructed and released, which provide useful platforms and promote the exploration of sesame genes and loci (Table 9.4).

9.9 CONCLUSION

Sesame breeding studies have been boosted by the utilization of genetic resources and recent developments in the genomic era. About 35,000 accessions and wild types have been collected and deposited worldwide. Molecular biology studies have been intensively conducted in the last decade, and lots of QTLs, genes, and markers have been explored. Germplasm collections contain lots of superior alleles that can be uncovered by developed genomic data and will be applied in crop breeding to reach the ideal sesame crop. On the other side, to make more attractive sesame functional genomics studies, developed databases have been updated with sesame genetic and phenotypic data. This combined data should be effectively utilized in discovering the new QTLs/genes of significant agronomic/quality traits in sesame.

TABLE 9.4
Several Online Genetic and Genome Databases in Sesame

No.	Sesame Online Database and Their Webpages	Reference
1	Sinbase 1.0 http://ocri.genomics.org/Sinbase/	Wang et al. (2015)
2	Sinbase2.0 http://www.sesame-bioinfo.org/Sinbase2.	Wang et al. (2021)
3	SesameFG http://www.ncgr.ac.cn/SesameFG	Wei et al. (2017)
4	Sesame HapMap http://202.127.18.228/SesameHapMap	Wei et al. (2015)
5	PMDBase http://www.sesamebioinfo.org/PMDBase	Yu et al. (2017)
6	Sesame Pan-genome database. http://www.sesame-bioinfo.org/pan-genome/.	Yu et al. (2019)

REFERENCES

Abate M, Mekbib F, Ayana A, Nigussie M (2015) Assessment of genetic diversity in Ethiopian sesame (*Sesamum indicum* L.) germplasm using ISSR markers. *Biotechnol J Int* **8**(4):1–13

Abdellatef E, Sirelkhatem R, Mohamed AMM, Radwan KH, Khalafalla MM (2008) Study of genetic diversity in Sudanese sesame (*Sesamum indicum* L.) germplasm using random amplified polymorphic DNA (RAPD) markers. *Afr J Biotechnol* **7**:4423–4427

Adu-Gyamfi R, Prempeh R, Zakaria I (2019) Diversity assessment of some sesame (*Sesamum indicum* L.) genotypes cultivated in Northern Ghana using morphological and simple sequence repeat (SSR) markers. *Adv Agric* 2019:6067891

Ali GM, Yasumoto S, Seki-Katsuka M (2007) Assessment of genetic diversity in sesame (*Sesamum indicum* L.) detected by amplified fragment length polymorphism markers. *Electron J Biotechnol* **10**(1):12–23

Anter AS, Ghada SM (2021) Possibility of combining high yield and resistance to fusarium wilt disease using molecular markers in 4 élite sesame lines. *Turkish J Field Crop* **26**(1):129–138

Asekova S, Oh E, Kulkarni KP, Siddique MI, Lee M et al (2021) An integrated approach of QTL mapping and genome-wide association analysis identifies candidate genes for Phytophthora blight resistance in sesame (*Sesamum* indicum L.). *Front Plant Sci* **12**:604709

Ashikari M, Wu J, Yano M, Sasaki T, Yoshimura A (1999) Rice gibberellin-insensitive dwarf mutant gene Dwarf 1 encodes the alpha-subunit of GTP-binding protein. *P Natl Acad Sci USA* **96**(18):10284–10289

Ashri A (1998) Sesame breeding. *Plant Breed Rev* **16**:179–228

Ashri A (2007) Sesame (*Sesamum* indicum L.). In: Singh RJ (ed) *Genetics Resources, Chromosome Engineering and Crop Improvement*, Vol. 4, Oilseed Crops. CRC Press, BocaRaton, pp 231–289

Basak M, Uzun B, Yol E (2019) Genetic diversity and population structure of the Mediterranean sesame core collection with use of genome-wide SNPs developed by double digest RAD-Seq. *PLoS One* **14**:e0223757

Baydar H (2005) Breeding for the improvement of the ideal plant type of sesame. *Plant Breed* **124**(3):263–267

Baydar H, Marquard R, Turgut I (1999) Pure line selection for improved yield, oil content and different fatty acid composition of sesame, Sesamum indicum L. *Plant Breed* **118**:462–464

Bedigian D (2003) Evolution of sesame revisited: domestication, diversity and prospects. *Genet Resour Crop E* **50**(7):779–787

Bedigian D (2004) History and lore of sesame in Southwest Asia. *Econ Bot* **58**(3):329–353

Bedigian D (2010) Cultivated sesame and wild relatives in the genus Sesamum L. In: Bedigian D (ed) *Sesame: The Genus Sesamum*, CRC Press, Taylor & Francis Group, Boca Raton, pp 33–77

Bedigian D (2015) Systematics and evolution in Sesamum <producer>L. (Pedaliaceae</producer>), part 1: evidence regarding the origin of sesame and its closest relatives. *Webbia* **70**(1):1–42

Bedigian D, Harlan JR (1986) Evidence for cultivation of sesame in the ancient world. *Econ Bot* **40**(2):137–154

Begna T (2021) Conventional breeding methods widely used to improve self-pollinated crops. *Int J Res Agric Sci* **7**(1):1–16

Behera S, Voshall A, Moriyama EN (2021) Plant transcriptome assembly: Review and benchmarking. In: Helder IN (ed) *Bioinformatics*, Exon Publications, Brisbane, 33877767

Belay Y (2018) Screening of fusarium wilt, bacterial blight and phyllody diseases resistant sesame genotypes in sesame growing areas of Northern Ethiopia. *J Agric Ecol Res Int* **15**(2):1–12

Berhe M, Dossa K, You J, Mboup PA, Diallo IN et al (2021) Genome-wide association study and its applications in the non-model crop *Sesamum indicum*. *BMC Plant Biol* **21**:283

Bhat VK, Babrekar PP, Lakhanpaul S (1999) Study of genetic diversity in Indian and exotic sesame (*Sesamum indicum* L.) germplasm using random amplified polymorphic DNA (RAPD) markers. *Euphytica* **110**:21–34

Bisht IS, Bhat KV, Lakhanpaul S, Biswas BK, Pandiyan M et al (2004) Broadening the genetic base of sesame (*Sesamum* indicum L.) through germplasm enhancement. *Plant Genet Resour* **2**:143–151.

Cahill DJ, Schmidt DH (2004) Use of marker assisted selection in a product development breeding program. In: Fischer T, Turner N, Angus J, Mcintyre L, Robertson M, Borrell A, Lloyd D (eds) New Directions for a Diverse Planet. Proceedings of the 4th International Crop Science Congress, Brisbane, 26 Sept–1 Oct 2004:133

Chen Z, Tonnis B, Morris B, Wang RB, Zhang AL et al (2014) Variation in seed fatty acid composition and sequence divergence in the FAD2 gene coding region between wild and cultivated sesame. *J Agric Food Chem* **62**(48):11706–11710

Cho Y-I, Park J-H, Lee C-W, Ra W-H, Chung J-W et al (2011) Evaluation of the genetic diversity and population structure of sesame (*Sesamum* indicum L.) using microsatellite markers. *Genes Genom* **33**(2):187–195

Clauw P, Coppens F, Korte A, Herman D, Slabbinck B et al (2016) Leaf growth response to mild drought: natural variation in Arabidopsis sheds light on trait architecture. *Plant Cell* **28**(10):2417–2434

Cui C, Liu Y, Liu Y, Cui X, Sun Z et al (2021) Genome-wide association study of seed coat color in sesame (*Sesamum indicum* L.). *PLoS* **16**(5):e0251526

Cui C, Mei H, Liu Y, Zhang H, Zheng Y (2017) Genetic diversity, population structure, and linkage disequilibriumof an association-mapping panel revealed by genome-wide SNP markers in sesame. *Front Plant Sci* **8**:1189

Degefa FT (2019) Review on sesame (*Sesamum indicum* L.) breeding in Ethiopia. *Biol Agr Healthc* **9**:39–45

Dilipsundar N, Chitra N, Gowtham V (2019) Checklist of insect pests of sesame. *Indian J Entomol* **81**(4):928–944

Dossa K, Diouf D, Wang L, Wei X, Zhang Y et al (2017b) The emerging oilseed crop Sesamum indicum enters the “omics” era. *Front Plant Sci* **8**:1154

Dossa K, Li, D, Wang L, Zheng X, Liu A et al (2017a) Transcriptomic, biochemical and physio-anatomical investigations shed more light on responses to drought stress in two contrasting sesame genotypes. *Sci Rep* **7**:8755

Dossa K, Li D, Zhou R, Yu J, Wang L et al (2019a) The genetic basis of drought tolerance in the high oil crop Sesamum indicum. *Plant Biotechnol J* **17**(9):1788–803

Dossa K, Mmadi MA, Zhou R, Zhang T, Su R et al (2019b) Depicting the core transcriptome modulating multiple abiotic stresses responses in sesame (*Sesamum* indicum L.). *Int J Mol Sci* **20**(16):3930

Dossa K, Wei X, Zhang Y, Fonceka D, Wenjuan Y et al (2016) Analysis of genetic diversity and population structure of sesame samples from Africa and Asia as major centers of its cultivation. *Genes* **7**(4):14

Dossa K, You J, Wang L, Zhang Y, Li D et al (2019c) Transcriptomic profiling of sesame during waterlogging and recovery. *Sci Data* **6**:204.

Dossa K, Zhou R, Li D, Liu A, Qin L et al (2020) A novel motif in the5′-UTR of an orphan gene ‘Big Root Biomass’ modulates root biomass in sesame. *Plant Biotechnol J* **19**(5):1065–1079

Du H, Zhang, H, Wei L, Li C, Duan Y et al (2019) A high-density genetic map constructed using specific length amplified fragment (SLAF) sequencing and QTL mapping of seed-related traits in sesame (*Sesamum indicum* L.). *BMC Plant Biol* **19**:588

Dutta D, Banerjee S, Pal M, Gangopadhyay G (2022a) Validation of determinate (dt) gene-based DNA marker in inter-specific hybrid sesame and in-silico analysis of the predicted dt protein structures. *Physiol Mol Biol Plants* **28**(1):139–152

Dutta D, Harper A, Gangopadhyay G (2022b) Transcriptomic analysis of high oil-yielding cultivated white sesame and low oil-yielding wild black sesame seeds reveal differentially expressed genes for oil and seed coat colour. *Nucleus* **65**:151–164

Emamgholizadeh S, Parsaeian M, Baradaran M (2015) Seed yield prediction of sesame using artificial neural network. *Eur J Agron* **68**:89–96

FAO (2019 & 2020) FAOSTAT. FAO, Rome, Italy. https://www.fao.org/faostat/en/#data/QCL. Accessed 13 July 2022

Flint-Garcia SA, Thuillet AC, Yu J, Pressoir G, Romero SM et al (2005) Maize association population: a high-resolution platform for quantitative trait locus dissection. *Plant J* **44**(6):1054–1064

Francia E, Tacconi G, Crosatti C, Barabaschi D, Bulgarelli D et al (2005) Marker assisted selection in crop plants. *Plant Cell Tissue Organ Cult* **82**:317–342

Gadri Y, Williams LE, Peleg Z (2020) Tradeoffs between yield components promote crop stability in sesame. *Plant Sci* **295**:110105

Genievskaya Y, Abugalieva S, Rsaliyev A, Yskakova G, Turuspekov Y (2020) QTL mapping for seedling and adult plant pesistance to leaf and stem rusts in Pamyati Azieva x Paragon mapping population of bread wheat. *Agronomy* **10**:1285

Guo T, Yu H, Qiu J, Li J, Han B et al (2019) Advances in rice genetics and breeding by molecular design in China. *Sci Sin Vitae* **49**:1185–1212

He Q, Xu F, Min MH, Chu SH, Kim KW et al (2019) Genome-wide association study of vitamin E using genotyping by sequencing in sesame (*Sesamum indicum*). *Genes Genom* **41**(9):1085–1093

Hegde DM (2012) Sesame. In: Peter KV (ed) *Handbook of Herbs and Spices*. 2nd ed. Woodhead Publishing, Cambridge, UK; pp 449–486

Hiremath SC, Patil CG (2002) Genome relations among octaploid species of *Sesamum* L. (Pedaliaceae). *Cytologia* **67**:403–409.

Huang X, Han B (2014) Natural variations and genome-wide association studies in crop plants. *Annu Rev Plant Biol* **65**:531–551

Hwang LS (2005) Sesame oil. In: Shahidi F (ed) *Bailey's Industrial Oil and Fat Products*. 6th ed. John Wiley & Sons, New Jersey, pp 537–576

IPGRI and NBPGR (2004) *Descriptors for sesame (Sesamum spp.)*. International Plant Genetic Resources Institute, Rome, Italy; and National Bureau of Plant Genetic Resources, New Delhi, India

Islam F, Gill RA, Ali B, Farooq MA, Xu L et al (2016) Sesame. In: Gupta SK (ed) *Breeding Oilseed Crops for Sustainable Production*. Academic Press, San Diego, CA, pp 135–147

Jain M (2012) Next-generation sequencing technologies for gene expression profiling in plants. *Brief Funct Genomics* **11**:63–70

Jiang L, Liu X, Xiong G, Liu H, Chen F et al (2013) DWARF 53 acts as a repressor of strigolactone signalling in rice. *Nature* **504**:401–405

Jiang Q, Christen S, Shigenaga MK, Ames BN (2001) γ-Tocopherol, the major form of vitamin E in the US diet, deserves more attention. *Am J Clin Nutr* **74**(6):714–722

Joshi AB (1961) *Sesamum*. Indian Central Oilseeds Committee, Hyderabad, India

Kabi M, Baisakh B, Das TR, Tripathy SK, Dash M (2021) Inheritance pattern and validation of RGA marker for powdery mildew resistance in sesame. *Indian J Genet* **81**(2):331–333

Kafiriti E, Mponda O (2010) Growth and Production of Sesame. Soils, plant growth and crop production, Encyclopedia of life support systems (EOLSS), United Kingdom

Kage U, Kumar A, Dhokane D, Karre S, Kushalappa AC (2016) Functional molecular markers for crop improvement. *Crit Rev Biotechnol* **36**(5):917–930

Kavitha K (1998) Characterisation of newly developed cytoplasmic genic male sterility lines of Sesamum (*Sesamum indicum* L.) and evaluation of heterotic expression. Dissertation, Tamil Nadu Agriculture University

Kedharnath S, Ramanujam S, Joshi AB (1959) Chromosome pairing in two sesquiploid hybrids and its bearing on genome relationship in the genus Sesamum. *Indian J Genet* **19**:201–206

Kim HC, Kim K-H, Song K, Kim JY, Lee BM (2020) Identification and validation of candidate genes conferring resistance to downy mildew in maize (*Zea* mays L.). *Genes* **11**(2):191

Kobayashi T (1991) Cytogenetics of sesame (*Sesamum indicum*). In: Tsuchiya T, Gupta PK (eds) *Chromosome Engineering in Plants; Genetics, Breeding, Evolution*. Elsevier, Amsterdam, pp 581–592

Kumar GR, Sakthivel K, Sundaram RM, Neeraja CN, Balachandran SM et al (2010) Allele mining in crops: prospects and potentials. *Biotechnol Adv* **28**(4):451–461

Kumar LS (1999) DNA markers in plant improvement: an overview. *Biotechnol Adv* **17**(2-3):143–182

Kumar V, Singh A, Mithra SA, Krishnamurthy SL, Parida SK et al (2015) Genome-wide association mapping of salinity tolerance in rice (*Oryza sativa*). *DNA Res* **22**(2):133–145

Langham DR, Wiemers T (2002) Progress in mechanizing sesame in the US through breeding. In: Janickand J, Whipkey A (eds) *Trends in New Crops and New Uses*. American Society for Horticultural Science Press, Virginia, pp 157–173

Laurentin HE, Karlovsky P (2006) Genetic relationship and diversity in a sesame (*Sesamum indicum* L.) germplasm collection using amplified fragment length polymorphism (AFLP). *BMC Genet* 7-10:16483380

Lee BH, Lee JI, Seong NS, Kang CW (1991) Studies on inter-specific hybridization in sesame. II. Development of hybrid plants from *S. indicum* x *S. radiatum*. *Kor J Breed* **23**:42-47 (in Korean)

Li C, Duan Y, Miao H, Ju M, Wei L et al (2021) Identification of candidate genes regulating the seed coat color trait in sesame (*Sesamum indicum* L.) using an integrated approach of QTL mapping and transcriptome analysis. *Front Genet* **12**:700469

Li C, Miao H, Wei L, Zhang T, Han X et al (2014) Association mapping of seed oil and protein in *Sesamum* indicum L. using SSR markers. *Plos ONE* **9**(8):e105757

Li D, Dossa K, Zhang Y, Wei X, Wang L et al (2018) GWAS uncovers differential genetic bases for drought and salt tolerances in sesame at the germination stage. *Genes (Basel)* **9**(2):87

Liang J, Sun J, Ye Y, Yan X, Yan T et al (2021) QTL mapping of PEG-induced drought tolerance at the early seedling stage in sesame using whole genome re-sequencing. *Plos ONE* **16**(2):e0247681

Liu XP, Yu LX (2017) Genome-wide association mapping of loci associated with plant growth and forage production under salt stress in alfalfa (*Medicago* sativa L.). *Front Plant Sci* **8**:853

Liu Y, He Z, Appels R, Xia X (2012) Functional markers in wheat: current status and future prospects. *Theor Appl Genet* **125**(1):1–10

Magni C, Ballabio C, Restani P, Fuggetta D, Alessandri C et al (2010) Molecular insight into IgE-mediated reactions to sesame (*Sesamum indicum* L.) seed proteins. *Ann Allerg Asthma Im* **105**(6):458–464

Mao H, Wang H, Liu S, Li Z, Yang X et al (2015) A transposable element in a NAC gene is associated with drought tolerance in maize seedlings. *Nat Commun* **6**:8326

Mardani Z, Rabie B, Sabour H, Sabour A (2013) Mapping of QTLs for germination characteristics under non-stress and drought stress in rice. *Rice Sci* **20**(6):391–399

Mei H, Liu Y, Cui C, Hu C, Xie F et al (2021) QTL mapping of yield-related traits in sesame. *Mol Breed* **41**:43

Mei H, Liu Y, Du Z, Wu K, Cui C et al (2017) High-density genetic map construction and gene mapping of basal branching habit and flowers per leaf axil in sesame. *Front Plant Sci* **8**:636

Mensah JK, Obadoni BO, Eruotor PG, Onome-Irieguna F (2006) Simulated flooding and drought effects on germination, growth, and yield parameters of sesame (*Sesamum indicum* L.) *Afr J Biotechnol* **5**:1249–1253

Miao H, Li C, Duan Y, Wei L, Ju M (2020) Identification of a Sidwf1 gene controlling short internode length trait in the sesame dwarf mutant dw607. *Theor Appl Genet* **133**(1):73–86

Min YY, Toyota K (2019) Occurrence of different kinds of diseases in sesame cultivation in Myanmar and their impact to sesame yield. *J Exp Agric Int* **38**(4):1–9

Moazzami AA, Kamal-Eldin A (2006) Sesame seed is a rich source of dietary lignans. *J Am Oil Chem Soc* **83**(8):719–723

Mondini L, Noorani A, Pagnotta MA (2009) Assessing plant genetic diversity by molecular tools. *Diversity* **1**(1):19–35

Morris JB (2009) Characterization of sesame (*Sesamum indicum* L.) germplasm regenerated in Georgia, USA. *Genet Resour Crop E* **56**:925–936

Myint D, Gilani SA, Kawase M, Watanabe KN (2020) Sustainable sesame (*Sesamum indicum* L.) production through improved technology: an overview of production, challenges, and opportunities in Myanmar. *Sustainability* **12**(9):3515

Nadeem MA, Nawaz MA, Shahid MQ, Doğan Y, Comertpay G et al (2018) DNA molecular markers in plant breeding: current status and recent advancements in genomic selection and genome editing. *Biotechnol Biotechnol Equip* **32**(2):261–285

Najeeb U, Mirza MY, Jilani G, Mubashir AK, Zhou WJ (2012) Sesame. In: Gupta SK (ed) *Technological Innovations in Major World Oil Crops*, Vol. 1. Springer, New York, pp 131–145

Nayar NM, Mehra KL (1970) Sesame: its uses, botany, cytogenetics and origin. *Econ Bot* **24**:20–31

Nevame AYM, Xia L, Nchongboh CG, Hasan MM, Alam MA et al (2018) Development of a new molecular marker for the resistance to tomato yellow leaf curl virus. *Biomed Res Int* 2018:8120281

Nimmakayala P (1997) *Interspecific Crosses in Sesame*. Report submitted to Directorate of Oilseeds Research, Hyderabad, India

Nimmakayala P, Perumal R, Mulpuri S, Reddy UK (2011) Sesamum. In: Kole C (ed) *Wild Crop Relatives: Genomic and Breeding Resources Oilseeds*. Springer-Verlag Press, Berlin pp 261–273

Nyongesa BO, Were BA, Gudu S, Dangasuk OG, Onkware, AO (2013) Genetic diversity in cultivated sesame (*Sesamum indicum* L.) and related wild species in East Africa. *J Crop Sci Biotechnol* **16**:9–15

Ogbonna PE, Ukaan SI (2012) Yield evaluation of 13 sesame (*Sesamum indicum* L.) accessions in the derived savannah agro-ecology of south-eastern Nigeria. *Afr J Agric Res* **7**:5772–5778

Pandey SK, Das A, Rai P, Dasgupta T (2015) Morphological and genetic diversity assessment of sesame (*Sesamum indicum* L.) accessions differing in origin. *Physiol Mol Biol Plants* **21**(4):519–529

Phumichai C, Matthayatthaworn W, Chuenpom N, Wongkaew A (2017) Identification of a SCAR marker linked to a shattering resistance trait in sesame. *Turk J Field Crops* **22**:258–265

Prabakaran AJ (1996) Genetic diversity of wild sesame from southern India. *Plant Genet Resour News* 1106:44–46
Qu Z, Wu XY, Xia FJ (1994) Embryo rescue and plant regeneration in inter-specific hybridization between *Sesamum* indicum L., and S. schinzianum (Schinz) Aschers. *Oil Crops China* **16**:33–35 (in Chinese)
Qureshi M, Langham D, Lucas SJ, Uzun B, Yol E (2022) Breeding history for shattering trait in sesame: classic to genomic approach. *Mol Biol Rep* **49**:7185–7194
Rahman A, Bhattarai S, Akbar D, Thomson M, Trotter T et al. (2020) Market analysis of sesame seed. https://crcna.com.au/ Accessed 22 July 2022
Rajeswari S, Thiruvengadam V, Ramaswamy NM (2010) Production of interspecific hybrids between *Sesamum alatum* Thonn and *Sesamum indicum* L. through ovule culture and screening for phyllody disease resistance. *S Afr J Bot* **76**(2):252–258
Ram SG, Sundaravelpandian K, Kumar M, Vinod KK, Kannan Bapu JR et al. (2006) Pollen-pistil interaction in the inter-specific crosses of Sesamum sp. *Euphytica* **152**:379–385
Ramanathan K (1950) A note on the interspecific hybridization in *Sesamum*. *Madras Agric J* **37**:397–400
Rao GP, Nabi SU (2015) Overview on a century progress in research on sesame phyllody disease. *Phytopathogenic Mollicutes* **5**(2):74–83
Rao PVR, Prasuna K, Anuradha G, Srividhya A, Reddy VLN (2013) Molecular mapping and tagging of powdery mildew tolerance gene(s) in sesame (*Sesamum indicum*). *Indian J Agric Sci* **83**(6):605–610
Sabag I, Morota G, Peleg Z (2021) Genome-wide association analysis uncovers the genetic architecture of tradeoff between flowering date and yield components in sesame. *BMC Plant Biol* **21**:549
Saha RR, Ahmed B, Aziz MA, Hossain MA (2010) Screening of sesame genotypes for water logging tolerance. *Bangladesh Agron J* **13**(1):83–93
Sarkar PK, Khatun A, Singha A (2016) Effect of duration of water-logging on crop stand and yield of sesame. *Int J Innov Appl Res* **14**(1):1–6
Sarwar G, Haq MA, Chaudhry MB, Rabbani I (2007) Evaluation of early and high yielding mutants of sesame (*Sesamum indicum* L.) for different genetic parameters. *J Agric Res* **45**:125–133
Savant KD (2016) Induction of male sterility in sesame (Sesamum indicum L.). *J Plant Breed Genet* **4**(1):7–10
Shabana R, Abd El-Mohsen AA, Khalifa MMA, Saber AA (2014) Quantification of resistance of F6 sesame elite lines against Charcoal-rot and Fusarium wilt diseases. *Adv Agric Biolo* **1**:144–150
Sharma MK, Singh S, Sengar RS (2014) Allele mining and its applications. *Biotech Today* **4**(1):34–37
Sharma N, Tan MA, An SSA (2021) Phytosterols: potential metabolic modulators in neurodegenerative diseases. *Int J Mol Sci* **22**:12255
Sheng C, Song S, Zhou R, Li D, Gao Y et al (2021) QTL-Seq and transcriptome analysis disclose major QTL and candidate genes controlling leaf size in sesame (*Sesamum* indicum L.). *Front Plant Sci* **12**:580846
Singh M, Chahar S, Avtar R, Singh A, Kumar N (2022) Advances in classical and molecular breeding in sesame (*Sesamum* indicum L.). In: Gosal SS, Wani SH (eds) *Accelerated Plant Breeding*, Vol. 4. Springer, Cham, pp 491–529
Singh VB, Singh AK, Srivastava JN, Singh SK (2020) Important diseases of Sesamum (*Sesamum indicum* L.) and their management. In: Srivastava JN, Singh AK (eds) *Diseases of Field Crops: Diagnosis and Management*, Apple Academic Press, Canada, pp 259–285
Song Q, Joshi M, Wang S, Johnson CD, Joshi V (2020) Comparative analysis of root transcriptome profiles of sesame (*Sesamum indicum* L.) in response to osmotic stress. *J Plant Growth Regul* **40**:1787–1801
Stavridou E, Lagiotis G, Kalaitzidou P, Grigoriadis I, Bosmali I (2021) Characterization of the genetic diversity present in a diverse sesame landrace collection based on phenotypic traits and EST-SSR markers coupled with an HRM Analysis. *Plants* **10**(4):656
Su R, Zhou R, Mmadi MA, Li D, Qin L (2019) Root diversity in sesame (*Sesamum indicum* L.): insights into the morphological, anatomical and gene expression profiles. *Planta* **250**:1461–1474
Subramanian M (1972) Cytogenetical studies on interspecific hybrids in *Sesamum*. Dissertation, Tamil Nadu Agriculture University
Takahashi F, Kuromori T, Sato H, Shinozaki K (2018) Regulatory gene networks in drought stress responses and resistance in plants. *Adv Exp Med Biol* **1081**:189–214
Teichmann T, Muhr M (2015) Shaping plant architecture. *Front Plant Sci* **6**:233
Tesfaye T, Tesfaye K, Keneni G, Alemu T, Alemu A (2022) Genome-wide association study for yield-related traits in sesame (*Sesamum indicum*). *Plant Breed* **141**(2):246–256
Tripathy SK, Kar J, Sahu D (2019) Advances in sesame (*Sesamum* indicum L.) breeding. In: Al-Khayri JM, Jain SM, Johnson DV (eds) *Advances in Plant Breeding Strategies: Industrial and Food Crops*. Springer, Cham pp 577–635

Ustun R, Yol E, Ikten C, Catal M, Uzun B (2017) Screening, selection and real-time qPCR validation for phytoplasma resistance in sesame (*Sesamum* indicum L.). *Euphytica* **213**:159
Uzun B, Arslan C, Furat S (2008) Variation in fatty acid compositions oil content and oil yield in a germplasm collection of sesame (*Sesamum* indicum L.). *J Am Oil Chem Soc* **85**:1135–1142
Uzun B, Arslan C, Karhan M, Toker C (2007) Fat and fatty acids of the white lupin (*Lupinus albus* L.) in comparison to sesame (*Sesamum indicum* L.). *Food Chem* **102**:45–49
Uzun B, Cagirgan MI (2006) Comparison of determinate and indeterminate lines of sesame for agronomic traits. *Field Crops Res* **96**:13–18
Uzun B, Cagirgan MI (2009) Identification of molecular markers linked to determinate growth habit in sesame. *Euphytica* **166**(3):379–384
Uzun B, Lee D, Donini P, Cagirgan MI (2003) Identification of a molecular marker linked to the closed capsule mutant trait in sesame using AFLP. *Plant Breed* **122**:95–97
Uzun B, Ozbas MO, Canci H, Cagirgan MI (2004) Heterosis for agronomic traits in sesame hybrids of cultivars x closed capsule mutants. *Acta Agric Scand B Soil Plant Sci* **54**:108–112
Wan Y, Li H, Fu G, Chen X, Chen F, Xie M (2015) The relationship of antioxidant components and antioxidant activity of sesame seed oil. *J Sci Food Agric* **95**:2571–2578
Wang H, Zhang H, Ma Q, Wei L, Ju M et al (2017b) Optimization of EMS mutagenesis condition and screening of mutants in sesame. *J Henan Agri Sci* **46**(1):36–41
Wang L, Han X, Zhang Y, Li D, Wei X et al (2014) Deep resequencing reveals allelic variation in *Sesamum indicum*. *BMC Plant Biol* **14**(1):1–11
Wang L, Li D, Zhang Y, Gao Y, Yu J et al (2016) Tolerant and susceptible sesame genotypes reveal waterlogging stress response patterns. *Plos ONE* **11**(3):e0149912
Wang L, Segla S, Dossou K, Wei X, Zhang Y et al (2020) Transcriptome dynamics during black and white sesame (*Sesamum indicum* L.) seed development and identification of candidate genes associated with black pigmentation. *Genes* **11**(12):1399
Wang L, Yu J, Zhang Y, You J, Zhang X et al (2021) Sinbase 2.0: an updated database to study multi-omics in *Sesamum indicum*. *Plants* **10**(2):272
Wang L, Zhang Y, Li D, Dossa K, Wang ML et al (2019) Gene expression profiles that shape high and low oil content sesames. *BMC Genet* **20**:45
Wang L, Zhang Y, Li P, Wang X, Zhang W et al (2012b) HPLC Analysis of seed sesamin and sesamolin variation in a sesame germplasm collection in China. *J Am Oil Chem Soc* **89**:1011–1020
Wang L, Zhang Y, Qi X, Li D, Wei W et al (2012a) Global gene expression responses to waterlogging in roots of sesame (*Sesamum indicum* L.). *Acta Physiol Plant* **34**:2241–2249
Wang L, Zhang Y, Zhu X, Zhu X, Li D (2017a) Development of an SSR-based genetic map in sesame and identification of quantitative trait loci associated with charcoal rot resistance. *Sci Rep* **7**:8349
Wang Z, Gerstein M, Snyder M (2009) RNA-Seq: a revolutionary tool for transcriptomics. *Nat Rev Genet* **10**(1):57–63
Wang L, Yu J, Li D, Zhang X (2015) Sinbase: an integrated database to study genomics, genetics and comparative genomics in *Sesamum indicum*. *Plant and Cell Physiol* **56**(1): e2–e2
Wang Z, Zhou Q, Dossou SSK, Zhou R, Zhao Y et al (2022) Genome-wide association study uncovers loci and candidate genes underlying phytosterol variation in sesame (*Sesamum* indicum L.). *Agriculture* **12**:392
Wei LB, Zhang HY, Zheng YZ, Miao HM, Zhang TZ et al (2009) A genetic linkage map construction for sesame (*Sesamum* indicum L.). *Genes Genom* **31**:199–208
Wei W, Zhang Y, Lv H, Li D, Wang L et al (2013) Association analysis for quality traits in a diverse panel of Chinese sesame (*Sesamum* indicum L.) germplasm. *J Integr Plant Biol* **55**(8):745–758
Wei W, Zhang Y, Wang L, Li D, Gao Y et al (2016) Genetic diversity, population structure, and association mapping of 10 agronomic traits in sesame. *Crop Sci* **56**:331–343
Wei X, Gong H, Yu J, Liu P, Wang L et al (2017) SesameFG: an integrated database for the functional genomics of sesame. *Sci Rep* **7**(1):1–10
Wei X, Liu K, Zhang Y, Feng Q, Wang L et al (2015) Genetic discovery for oil production and quality in sesame. *Nat Commun* **6**(1):1–10
Weiss EA (1971) *Castor, Sesame and Safflower*. Barnes and Noble Inc, New York
Wu K, Liu H, Yang M, Tao Y, Ma H et al (2014a) High-density genetic map construction and QTLs analysis of grain yield-related traits in sesame (*Sesamum indicum* L.) based on RAD-Seq technology. *BMC Plant Biol* **14**:274
Wu K, Wu WX, Yang MM, Liu, HY et al (2017) QTL Mapping for oil, protein and sesamin contents in seeds of white sesame. *Acta Agron Sin* **43**(7):1003–1011

Wu K, Yang M, Liu H, Tao Y, Mei J et al (2014b) Genetic analysis and molecular characterization of Chinese sesame (*Sesamum indicum* L.) cultivars using insertion-deletion (InDel) and simple sequence repeat (SSR) markers. *BMC Genet* **15**(1):1–15

Xu F, Zhou R, Dossou SSK, Song S, Wang L (2021) Fine mapping of a major pleiotropic QTL associated with sesamin and sesamolin variation in sesame (*Sesamum* indicum L.). *Plants* **10**:1343

Xu L, Hu K, Zhang Z, Guan C, Chen S et al (2016) Genome-wide association study reveals the genetic architecture of flowering time in rapeseed (*Brassica napus* L.). *DNA Res* **23**(1):43–52

Yadav R, Kalia S, Rangan P, Pradheep K, Rao GP et al (2022) Current research trends and prospects for yield and quality improvement in sesame, an important oilseed crop. *Front Plant Sci* **13**:863521

Yol E, Basak M, Kizil S, Lucas S, Uzun B (2021) A high-density SNP genetic map construction using ddRAD-Seq and mapping of capsule shattering trait in sesame. *Front Plant Sci* **12**:679659

Yol E, Toker R, Golukcu M, Uzun B (2015) Oil content and fatty acid characteristics in Mediterranean sesame core collection. *Crop Sci* **55**:2177–2185

Yol E, Uzun B (2011) Inheritance of number of capsules per leaf axil and hairiness on stem, leaf and capsule of sesame (Sesamum indicum L.). *Aust J Crop Sci* **5**(1):78–81

Yol E, Uzun B (2012) Geographical patterns of sesame (*Sesamum indicum* L.) accessions grown under Mediterranean environmental conditions, and establishment of a core collection. *Crop Sci* **52**:2206–2214

Yol E, Uzun B (2019) Inheritance of indehiscent capsule character, heritability and genetic advance analyses in the segregation generations of dehiscent x indehiscent capsules in sesame. *Tarim Bilim Derg* **25**:79–85

You J, Wang Y, Zhang Y, Dossa K, Li D et al (2018) Genome-wide identification and expression analyses of genes involved in raffinose accumulation in sesame. *Sci Rep* **8**(1):1–11

Younis A, Ramzan F, Ramzan Y, Zulfiqar F, Ahsan M et al (2020) Molecular markers improve abiotic stress tolerance in crops: a review. *Plants* **9**(10):1374

Yu CY (2013) Molecular mechanism of manipulating seed coat coloration in oilseed Brassica species. *J Appl Genet* **54**(2):135–145

Yu J, Dossa K, Wang L, Zhang Y, Wei X, Liao B, Zhang X (2017) PMDBase: a database for studying microsatellite DNA and marker development in plants. *Nucleic Acids Res* **45**(D1):D1046–D1053

Yu J, Golicz AA, Lu K, Dossa K, Zhang Y et al (2019) Insight into the evolution and functional characteristics of the pan-genome assembly from sesame landraces and modern cultivars. *Plant Biotechnol J* **17**(5):881–892

Zeng A, Chen P, Korth K, Hancock F, Pereira A et al (2017) Genome-wide association study (GWAS) of salt tolerance in worldwide soybean germplasm lines. *Mol. Breed* **37**(3):1–14

Zhang H, Miao H, Ju M (2019a) Potential for adaption to climate change through genomic breeding in sesame. In: Kole C (ed) *Genomic Designing of Climate-smart Oilseed Crops.* Springer Press, USA, pp 371–440

Zhang H, Miao H, Li C, Wei L, Duan Y et al (2016) Ultra-dense SNP genetic map construction and identification of *SiDt* gene controlling the determinate growth habit in *Sesamum indicum* L. *Sci Rep* **6**:31556

Zhang H, Miao H, Wang L, Qu L, Liu H et al (2013a). Genome sequencing of the important oilseed crop *Sesamum* indicum L. *Genome Biol* **14**(1):1–9

Zhang H, Miao H, Wei L, Li C, Duan Y et al (2018a) Identification of a *SiCL1* gene controlling leaf curling and capsule indehiscence in sesame via cross-population association mapping and genomic variants screening. *BMC Plant Biol* **18**:296

Zhang H, Miao H, Wei L, Li C, Zhao R et al (2013b) Genetic analysis and QTL mapping of seed coat color in sesame (Sesamum indicum L.). *Plos One* **8**:e63898

Zhang H, Wei L, Miao H, Zhang T, Wang C (2012) Development and validation of genic-SSR markers in sesame by RNA-seq. *BMC Genom* **13**:316

Zhang X, Wei X, Niang M, Liu P, Zhang Y et al (2018c) Near-infrared reflectance spectroscopy reveals wide variation in major components of sesame seeds from Africa and Asia. *Crop J* **6**:202–206

Zhang Y, Li D, Zhou R, Wang X, Dossa K et al (2019b) Transcriptome and metabolome analyses of two contrasting sesame genotypes reveal the crucial biological pathways involved in rapid adaptive response to salt stress. *BMC Plant Biol* **19**:66

Zhang Y, Wang L, Gao Y, Li D, Yu J et al (2018b) Genetic dissection and fine mapping of a novel dt gene associated with determinate growth habit in sesame. *BMC Genet* **19**:38

Zhang Y, Wei M, Liu A, Zhou R, Li D et al (2019c) Comparative proteomic analysis of two sesame genotypes with contrasting salinity tolerance in response to salt stress. *J Proteomics* **201**:73–83

Zhou R, Dossa K, Li D, Yu J, You J et al (2018) Genome-wide association studies of 39 seed yield-related traits in sesame (*Sesamum indicum* L.). *Int J Mol Sci* **19**(9):2794

10 Genetic Diversity, Gene Discovery, and Allele Mining in Mustard (*Brassica juncea* L. Czern.)

Chhaya Atri, Javed Akhatar, Harjeevan Kaur, Simarjeet Kaur Sra, Anna Goyal, and S.S. Banga

10.1 INTRODUCTION

Mustard (*Brassica juncea* L. Czern) belongs to the family *Brassicaceae* in angiosperms. It is one of the most important oilseed crops in the world. Mustard is an allopolyploid (AABB; 2n=36) that may have evolved as a result of multiple independent hybridizations between *B. rapa* (AA) and *B. nigra* (BB) with spontaneous chromosome doubling (U 1935; Kaur et al. 2014). Detailed genetic analysis employing nuclear or chloroplast DNA markers has suggested a polyphyletic or even monophyletic ancestry (Song et al. 1988; Chen et al. 2013; Kaur et al. 2014). Oilseed form of mustard is the predominant crop in the Indian subcontinent. But both oilseed and condiment types are cultivated in the drier environments of Europe, North America, and Australia. Leaf mustards are abundant in China and Southeast Asia. China also cultivates root mustard in Mongolia and the northeast (NE) of the country. Ancient farmers may have used domestication-related diversifying selection to create variants for different food use purposes (Banga and Banga 2016; Yang et al. 2016; Yang et al. 2018). An alternative perspective explored the role of polyploidization in the evolution of crop agronomic variation (Prakash and Hinata 1980). Vavilov (1951) proposed Central Asia (Afghanistan and neighboring regions) as the primary origin of species, with Asia Minor, central/western China, and eastern India as secondary centers of diversity. Nevertheless, a number of scientists believe that the species originated in the Middle East, where its progenitor species were sympatric (Burkill 1930; Olsson 1960; Vaughan et al. 1963; Axelsson et al. 2000). However, the existence of high allelic diversity in land races from Afghanistan bordering Pakistan (Banga and Banga 2016) necessitates a re-examination of Vavilov's assertion that Afghanistan and its environs were the primary center of origin (1949). Kang et al. (2021) used a joint analysis of chloroplast, mitochondrial, and nuclear genome data to propose that *B. juncea* might have actually originated from a single hybridization event in West Asia, followed by at least three independent domestication events. Spontaneous gene mutations and introgressions during the crop's migration eastward may have also contributed to its current level of morphotype diversity. Until recently, it was believed that natural hybridizations between parental genome donor species were unidirectional, with *B. rapa* serving as the cytoplasmic donor for *B. juncea* (Uchimiya and Wildman 1978; Banga et al. 1983; Erickson et al. 1983; Palmer et al. 1983; Warwick and Black 1991; Pradhan et al. 1992). However, the characterization of larger sets of germplasm using microsatellite markers specific to cytoplasmatic organelles has suggested the existence of two major plasma types in *B. juncea* (Kaur et al. 2014) and that the cytoplasm of some ancient East European land races was found to be phylogenetically close to that of *B. nigra*. The strategy of replacing ineffective alleles at bottleneck loci with more productive ones using pedigree and performance-based selection has helped to increase the productivity potential of this crop. However, the repeated use of elite cultivars in mustard breeding programs has led to the loss of rare alleles from breeding germplasm. Such founding events, in

DOI: 10.1201/9781003385035-10

conjunction with domestication-related effects, have produced many genetic bottlenecks that have led to a stagnating response to selection. Identification of novel alleles from crop land races or their wild relatives and substitution of a better combination of alleles at bottleneck loci can enhance genetic advance and contribute to the development of varieties with increased productivity and environmental resilience. Self-pollinated crops, such as mustard, are prone to the loss of alleles other than those that contribute to the crop's reproductive fitness in the homozygous state. This hypothesis suggests that no single piece of germplasm will contain all of the advantageous alleles currently present in a species. In addition, different breeding objectives and adaptations to diverse environments also result in a highly variable distribution of alleles among populations. Therefore, it is necessary to mine germplasm resources for novel alleles, as a breeder's selection can work only on the alleles present in a given breeding resource. Thus, overcoming the genetic bottlenecks through the introduction of more effective alleles or combinations thereof can potentially induce substantial changes in trait expression when transferred to an appropriate genetic background. When necessary, alien introgressions should be utilized to strategically introduce new genes into the crop gene pools. A significant proportion of the introgressed variant is normally harmful or neutral. However, alien introgression can alter the genetic composition of a species and contribute to the expansion of its germplasm base. Polyploidization following the resynthesis of a natural allopolyploid can also be extremely efficient at generating allelic variation. Polyploidy may also benefit from allelic activation, where transcripts silenced in one lineage are activated by the transcriptional machinery of the genome from a different lineage.

Several molecular techniques, including restriction fragment length polymorphism (RFLP), amplified fragment length polymorphism (AFLP), and SNP markers, have been utilized to characterize allelic variations in many *Brassica* species. Multiple genomic databases (e.g. https://www.ncbi.nlm.nih.gov/genome/; http://mips.helmholtz-muenchen.de/plant/genomes.jsp; http://pgrc.ipk-gatersleben.de/cr-est/; etc.) now store vast quantities of sequence and expression data that can be used to identify allelic variation. Several methods have been employed for the efficient extraction of alleles. These include targeted allele mining, mining for CGs, mining for the entire genome, mining for association studies, and mining for linkage disequilibrium (LD). This chapter examines the research pertaining to "allele mining," or the dissection of naturally occurring variation at CGs in *B. juncea*. The majority of studies explored for this review involved identifying genomic regions associated with the traits of interest through mapping QTL or genome-wide association studies (GWAS). Such studies have enabled direct access to the alleles for improved nutrient use efficiency, seed yield, and seed quality, besides the alleles that confer resistance to biotic and abiotic stresses. Not only does the identification of allelic variants from germplasm collections provide novel alleles for targeted trait improvement, but it also helps shortlist the germplasms for conservation programs (Mohd Saad et al. 2021). There are currently three chromosome-scale genome assemblies for stem mustard (Yang et al. 2016), yellow mustard (Kang et al. 2021), and Indian oilseed mustard (Paritosh et al. 2021). Kang et al. (2021) provided additional evidence that *B. juncea* has a greater nucleotide diversity than the B subgenome (= 2.05013 vs. 1.45103). LD also decayed more rapidly in the A subgenome compared to the B subgenome, indicating that the A subgenome experienced a much higher level of genetic recombination. Yang et al. (2021) could identify substantial structural differences between the oilseed and vegetable forms of *B. juncea* through direct genome comparisons. It was discovered that chromosomes A03, A05, A07, and A09 of the A subgenome and chromosomes B01, B02, B04, B05, and B07 of the B subgenome contained extensive inversions. A significant fragment loss was also observed on chromosome A06. The B subgenome appeared to have been subjected to greater selective pressures than the A subgenome, as shown by higher fixation index (FST) and estimates in the B subgenome, also in comparison to the A subgenome.

10.2 EXPLORING THE GENETIC DIVERSITY OF MUSTARD

Genetic diversity is a fundamental resource that determines the efficacy of selection programs and the performance of inter-variety hybrids. It can be defined as the combination of allelic variation in a species genes and phenotypes. In addition to observed and predicted heterozygosities, they are routinely measured in terms of genotypic or allele frequencies (Petit et al. 1998). Molecular markers cover both coding and non-coding regions of the genome and reflect variation at the DNA level. Documenting the genetic diversity of a species is essential for analyzing the distinction between cultivars, land races, and crop wild relatives, as well as knowing their evolutionary history. Such studies have utilized random amplified polymorphic DNA (RAPD) (Rabbani et al. 1998; Khan et al. 2009), AFLP (Burton et al. 2004), and simple sequence repeat (SSR) markers (Xu et al. 2010b) to differentiate germplasm resources from distinct eco-geographies. Chen et al. (2013) utilized SSR-derived polymorphism to establish two germplasm groups. Group I was composed of mustard accessions from central and western India and eastern China. Group II consisted of germplasm lines from the northern and eastern regions of India, the central and western regions of China, Europe, and Australia. Compared to the Indian lines, Group 1 Chinese germplasm lines exhibited a higher frequency of alleles and exhibited a high level of genetic diversity. The outcomes of the studies by Kaur et al. (2014) supported the existence of two groups through a population structure analysis of 122 germplasm lines based on the polymorphism produced by nuclear markers. The first group consisted of East European and Indian germplasm lines. Germplasm accessions from India, China, and Australia contributed to the second group. In a separate study, 125 landraces of *B. juncea*, *B. rapa*, and *B. nigra* were gathered from India's ancient trade routes. The study also included historical strains of *B. juncea*, *B. rapa*, and *B. nigra* from Pakistan, Kashmir, and Punjab in northwest (NW) and northeastern India. These accessions were genotyped using SSRs developed for the chloroplast and nuclear genomes using sequence data from the genomic regions closely associated with flowering and seed dispersal. Land races from NW India and NE India formed two distinct groups. NW germplasms were more closely related to *B. nigra* and *B. tournefortii*, another species of the B lineage. The NE group, including some *B. juncea* accessions was placed closer to *B. rapa* (Banga and Banga 2016). SSR genotyping of 145 accessions (Sharma et al. 2022) has suggested a high level of genetic diversity [0.03 (SJ1668I) to 0.75 (EJU4)]. Different clustering and population structure analyses demarcated the test germplasm into two primary clusters and three subpopulations with varying degrees of admixing. Singh et al. (2022) discovered 36 distinct SSR alleles that could be used to differentiate 27 mustard varieties. The above-cited studies have identified essential sources of allelic diversity in released cultivars, land races, and crop wild relatives. Documentation and preservation of this diversity are crucial for effective crop improvement programs as well as in reducing or preventing the loss of variation, an indispensable breeding resource. Additionally, it is essential to identify the genes underlying selective sweeps and desirable alleles in order to ensure continuous breeding gains. The process of identifying allelic variation can now be optimized as high-quality nuclear and organellar genome assemblies are available for *B. juncea* and its ancestor species. Whole-genome resequencing across germplasms and the construction of pan-genomes are now helping to overcome the limitations of the single reference genome(s) and laying the groundwork for future research, especially utilizing structural and presence-absence variations for GWAS.

10.3 GENE DISCOVERY AND ALLELE MINING

The genetic and environmental bottlenecks of a biological entity define its phenotypic performance. Genetic bottlenecks can theoretically be mitigated by substituting desirable alleles at bottleneck loci, whereas environmental bottlenecks are largely uncontrollable components of phenotypic variation. Recent advances in molecular marker and next-generation sequencing technologies have improved our understanding of the genetic factors underlying the phenotypic variations of a vast array of agronomic and seed quality traits. The combination of QTL mapping, GWAS, transcriptomics, and reverse genetics has facilitated the identification of many loci that are able to explain observable phenotypic variances for agronomic and seed quality traits, besides resistance to biotic and abiotic stresses. Early successes in the construction of linkage maps using a variety of marker systems were instrumental in conducting the QTL mapping analysis (see Yadava and Ramchiary 2022 for details). However, most earlier maps were limited by the small number of markers included and the insufficient coverage of 18 *B. juncea* chromosomes. Paritosh et al. (2014) made a very significant contribution to the development of high-density linkage maps by using RNA-seq of two *B. juncea* genotypes to develop SNP markers for applications in both genome-wide and localized fine mapping. A large number of QTLs are now available, which explains a substantial proportion of the observed phenotypic variations. The use of genome-wide markers for the pre-selection of elite genotypes is gradually replacing the application of a small number of gene-based markers in breeding programs, thereby enhancing selection gains and accelerating plant breeding activities.

10.3.1 Grain Yields and Their Constituents

Grain yield is a complex trait with many contributing components like plant height, number of branches, number of siliques, silique size, seed size, and harvest index. Additionally, phenological features such as days to flowering, duration of flowering, and maturity are also strongly associated with seed yield. Early genetic research had highlighted the significance of additive and non-additive gene action, as well as epistasis (Banga et al. 2015). Genotype-environment (GE) interactions continue to be a significant concern. These substantially impair the response to selection in early generations (Gunasekera et al. 2006). It is also virtually impossible to generalize genetic inferences due to polyploidy, parental heterogeneities, and admixed gene pools. Studies involving QTL mapping and GWAS have discovered multiple QTLs for each trait, indicating the involvement of multiple genes or genomic regions in defining trait heredity. This information is extremely valuable for using marker-assisted breeding to accelerate and optimize crop improvement programs. QTL mapping studies focusing on flowering and other yield-related traits have led to the identification of QTLs for plant height, main shoot length, pods on the main shoot, seed size, and seed yield (Ramchiary et al. 2007b; Yadava et al. 2012; Dhaka et al. 2017b). Most QTLs discovered were clustered on chromosomes A07 and A10. A meta-analysis of 187 identified QTLs led to their amalgamation into 20 meta-QTLs (MQTLs). These include 54 QTLs in LG-A10 which formed a single yield-related meta-QTL. Mahmood et al. (2005) identified a number of significant and stable QTLs for numerous mustard yield components; the majority of these compensated each other through a shared developmental pathway. Seed size is a very important factor for mustard productivity. The growth of female gametophytes in terms of the size of the embryo, endosperm, and integument helps determine the seed weight in dicots. Cell number and cell size are equally important (Sundaresan 2005). Dhaka et al. (2017) identified six consensus QTLs (C-QTLs) in A03, A07, A10, and B03 linkage groups (LG). Eleven genes (*ATGRF3, CKX1, HAIKU1, TTG2, ARR1, SGT, BZR1, DA2, MYB56,* and *TT4*) known to affect seed weight in *Arabidopsis thaliana* were also annotated. Li et al. (2016) identified QTLs associated with tuber mustard stem expansion traits, which include stem weight, stem diameter, and stem length. GWAS has allowed the discovery of marker-trait associations (MTAs) for grain yield, seed size, secondary branches, plant height, root diameter, and root length (Akhatar and Banga 2015). A single marker, SB1822-1, was repeatedly detected for seed size and grain yield, and its position on chromosome B03 was ascertained to be 17.5 cM. Under normal moisture conditions, marker SB3872–3 had a significant effect on seed size (R^2=15.16%), and it was located at 60.9 cM on chromosome B5. The allele SB1822-1 had an average positive effect on grain yield and seed size. SNPs associated with plant height, main shoot length, seed size, and seed yield were also reported under the conditions of terminal heat stress (Sandhu et al. 2019). The majority of these MTAs were found on chromosome A09. Akhatar et al. (2021) also identified 282 significant MTAs for mustard flowering, maturity, and plant height. Annotating genomic regions within 50 kb of the peak SNPs allowed for the prediction of 30 CGs involved in the metabolic pathways to light perception, circadian rhythms, floral meristem identity, flowering regulation, gibberellic acid pathways, and plant development. More than one copy each of *AGL24*, *AP1*, *FVE*, *FRI*, *GID1A*, and *GNC* was detected. *FLC* and *CO* genes were predicted on chromosomes A02 and B08, respectively. *CDF1*, *CO*, *FLC*, *AGL24*, GNC, and *FAF2* contributed to the observed phenotypic variation for the plant height. Mathur et al. (2022) have identified *ARF2* and *SWEET15* homologs as gene candidates for the seed size. Interestingly, these genes were also identified in a putative selective sweep region of *B. juncea*

(Kang et al. 2021). Kumar et al. 2022 predicted a large number of CGs following a literature survey of published QTLs associated with the yield's component characters. Mining the five most robust and stable yield-related MQTL regions allowed the discovery of gene clusters including protein kinases, UDP-glucuronosyl/UDP-glucosyltransferase, proteinase inhibitor I3, glycoside hydrolase, and many more. Further literature review and functional annotation of corresponding *A. thaliana* orthologs, facilitated the same authors to identify 15 CGs associated with 15 yield-related and 46 quality traits.

10.3.2 Seed Quality Characteristics

Oil, protein, erucic acid, and meal GSLs are the primary targets of *B. juncea* plant breeding improvements. Complex quantitative inheritance and environmental influences have made simultaneous improvement of these interrelated seed quality traits difficult to achieve (Banga et al. 2015). Numerous studies have been conducted to comprehend the genetics of seed quality traits (Akhatar et al. 2020). Increasing mustard's oil content has been a primary breeding objective. Various breeding techniques have led to a 38%–42% oil content in current cultivar yields. Genetic studies are suggestive of a complex heritability involving a large number of additive-effect genes, in addition to maternal and environmental influences, (Banga and Banga 2009; Weselake et al. 2009; Wang et al. 2010). As the genes controlling oil biosynthesis are strongly influenced by environmental factors, we can identify QTLs using different mapping populations (Cheung et al. 1998b; Mahmood et al. 2006; Ramchiary et al. 2007b; Yadava et al. 2012) or by utilizing GWAS (Akhatar et al. 2020). QTL mapping studies were largely conducted using genic SSRs (Dhaka et al. 2017a), genomic SSRs (Lakshmi Padmaja et al. 2005; Kim et al. 2009; Xu et al. 2010b), RFLP and AFLP markers (Pradhan et al. 2003). Singh et al. (2013) identified QTLs for oil content (14), erucic acid (2), eicosenoic acid (2), linolenic acid (3), linoleic acid (3), and palmitic acid (2). The QTLs identified for the presence of erucic, eicosapentaenoic, and linoleic acids were clustered in a specific genomic region on LG2 (Panjabi et al. 2008). Rout et al. 2018 conducted QTL metanalysis in multiple biparental populations, developed in the background of high and low (2%) erucic acid genotypes. Meta-analysis of QTLs mapped in high erucic acid populations led to the identification of nine significant common QTLs. The majority of oil QTLs co-localized with the erucic acid loci on LG A08 and B07. In contrast, the meta-analysis of ZE populations identified five QTLs on LGs A03 (Oil-A3-1), A08 (Oil-A8-1), B01 (Oil-B1-1), B04 (Oil-B4-1), and B06 (Oil-B6-1) that were designated as minor QTLs in populations with high erucic acid. Six novel QTLs were identified in populations derived from genotypes low in erucic acid. Akhatar et al. (2020) conducted GWAS to identify a number of genes related to oil and fatty acid biosynthesis in both A-(A04, A06, A09) and B-(B05, B06, B08) genome chromosomes. *CER26, the GDSL lipase gene*, and *LACS5*, which is a floral tissue-specific gene involved in oil biosynthesis (Shockey et al. 2002). *AD6* (on chromosome B05), *delta9 desaturase* (*9 DS*), *ADS1* on chromosome B06, and the gene encoding LTPs on chromosome B08 were also identified. Kumar et al. (2022) compiled 613 QTLs associated with quality traits of mustard through a bibliographic search. Of these, 410 QTLs could be projected onto the consensus map, allowing 324 QTLs to be merged into 51 MQTL, while 18 QTLs remained as singletons. The remaining two QTLs fell outside the MQTL supporting intervals. There were 28 MQTLs in subgenome A and 23 in subgenome B out of the 51 predicted. One QTL hotspot was identified in each of chromosomes A2, A3, A5, A7, B4, and B7. The same authors also validated MQTLs. Quality MQTLs harbored many genes encoding acyl-CoA oxidases, enoyl-acyl-CoA oxidases, enoyl-CoA hydratase/isomerases, FAD-linked oxidases, fatty acid hydroxylase, fatty acyl-coenzyme A reductases, GDSL lipase/esterase, oleosins, etc.

The seeds of *B. juncea* contain an abundance of protein, whose content is negatively correlated with fat concentration (Grami and Stefansson 1977). Two distinct types of seed storage proteins are: legumin-type globulins (11S, 12S, or cruciferin) and napin-type albumins (2S, or napins) (Wanasundara 2011). Oleosins, also known as oil body proteins, are the primary structural proteins. Other types of proteins include: metabolic proteins such as lipid transfer proteins (LTP), protease inhibitors (Ceciliani et al. 1994), Ca2+-dependent calmodulin binding proteins (Neumann et al. 1996), and dehydrins (Svensson et al. 2000). The availability of an in-depth knowledge base for protein structure, structure-function relationships, and genetic control is necessary for optimal protein utilization and the development of new food and feed applications. Association mapping analysis (Akhatar et al. 2020) helped to identify 23 MTAs on chromosomes A04, A06, A09, and B03. These MTAs accounted for between 14.09% and 23.21% of the phenotypic variation observed for the trait. Seven genes were predicted to be within 1.08–9.24 kb of their respective peak SNPs based on functional annotation. These included: *PASTICCINO-1* (A04), *AtP4H3* (A06), *ASN1, GTR2,* and *PSP1* (A09, B03).

10.3.2.1 Fatty Acids

The primary objective of oil quality alterations has been the modification of fatty acid content and composition. The crop improvement programs have largely targeted the development of varieties with high or low erucic acid contents (Kaur et al. 2020) to suit industrial applications or human consumption. The characterization of the complex biochemical metabolic reactions involved in fatty acid biosynthesis (Barker et al. 2007) indicated that *FAE1* was essential for the synthesis of erucic acid (James and Dooner 1990). *B. juncea* has two *FAE* orthologs, *BjunA.FAE1.1* and *BjunB.FAE1.2*. The early studies involving the examination of segregating generations developed from crosses between genotypes with high and low levels of erucic acid revealed the role of two genes whose effects were additive but unequal (Kirk and Hurlstone 1983; Bhat et al. 2002; Gupta et al. 2004). In subsequent analysis, simple molecular markers (e.g., RAPD, AFLP, RFLP, etc.) were identified for QTLs associated with erucic acid, oleic acid, and linolenic acid (Cheung et al. 1998b; Mahmood et al. 2003; Lionneton et al. 2004). Some of the identified QTLs accounted for more than 85% of the observed phenotypic variation for these fatty acids. Identified fatty acid QTLs act in opposition to one another, which is a natural consequence of fatty acids sharing a common biosynthetic pathway. Nearly half of the oleic acid is converted to erucic acid as a result of fatty acid elongation. On the other hand, incremental desaturation leads to the conversion of a smaller proportion of oleic acid into linoleic and linolenic acids. Gupta et al. (2004) cloned and sequenced two *FAE1* genes (*FAE1.1* and *FAE1.2*) from mustard genotypes with high and low erucic acid content. The sequence comparisons of the *FAE1* genes between the high-erucic acid cultivar Varuna and the low-erucic acid cultivar Heera revealed seven SNPs, four in *FAE1.1* and three in *FAE1.2*. These two genes were mapped to two distinct intervals on the genetic map of *B. juncea*, using the single-nucleotide primer extension (SNuPE) method for SNP genotyping. Further analysis of forty BC_5F_2 descendants with the seven SNPs in the *FAE1* genes facilitated the prediction of expected phenotypes for the descendants. In addition, it was demonstrated that zero-erucic types carried recessive allele(s) at both loci (E1E1E2E2), whereas high-erucic types revealed E1E1E2E2 genotypes. The oil produced by the heterozygote (E1e1E2e2) contained an intermediate level of erucic acid. Ramchiary et al. (2007) opined that the rise in molecular mass during the extension of oleic acid (C18:1) to erucic acid was the reason for a higher seed oil content in erucic acid-rich genotypes (C22:1) in comparison to the earlier low-erucic acid *Brassica* genotypes, which had low seed oil content (Ecke et al. 1995; Burns et al. 2003). Low and high erucic acid alleles of *FAE1.1* and *FAE1.2* genes in *B. juncea* were also distinguished using cleaved amplified polymorphic sequences (CAPS) as well (Saini et al. 2016). The same group discovered polymorphisms in the promoter regions of *FAE1.1* and *FAE1.2* as a 28-bp deletion in *FAE1.1* and a 340-bp insertion of a transposon-like element in *FAE1.2* in the low erucic acid mustard genotype (Saini et al. 2019). Sequence variability in the promoter regions was used to develop markers which could distinguish between genotypes with low and high levels of erucic acid in *B. juncea* and its ancestral diploids. Gill et al. 2021 crafted Kompetitive allele-specific PCR (KASPar) assays by comparing the sequences of *FAE1.1* and *FAE1.2* genes from high and low erucic acid (2%) genotypes of *B. juncea* (AABB) and its two progenitors, *B. rapa* (AA) and *B. nigra* (AA) (BB). These comparisons resulted in the discovery of SNPs at positions 8 and 3, respectively. Three SNPs for *FAE1.1* and one for *FAE1.2* led to missense mutations, precipitating amino acid modifications and inactivation of the KCS enzyme. SNPs at positions 735 and 1,476 for genes *FAE1.1* and *FAE1.2* were later mobilized into highly reproducible and breeder friendly KASPar assays capable of tracking recessive alleles in their heterozygous state.

10.3.2.2 Meal Glucosinolates

Only *Brassicaceae* plants contain GSLs (Zhang and Hamauzu, 2004). GSLs (-thioglucoside-N-hydroxysulfates) exist in aliphatic, aromatic, and indole forms (Clarke 2010). Significant amounts of aliphatic GSLs are present in *B. juncea* (Yadav and Rana 2018). Although GSLs possess anticancer, antibacterial, antifungal, and anti-oxidant properties (Latté et al. 2011), they can have a negative effect on thyroid function if a seed cake rich in GSLs is fed to dairy cattle in large quantities (Walker and Booth 2001). Breeding processes are now geared toward genetically reducing seed GSLs in seed meal to support animal feeding and maintaining high GSL levels in leafy tissues to prevent damage by herbivores and pathogenic microorganisms. *Brassica* GSLs are synthesized according to a complex network of regulatory mechanisms that are subjected to powerful environmental influences (Becker and Léon 1988; Hirani et al. 2016). The majority of genes associated with the biosynthetic pathway of aliphatic GSLs are known in *A. thaliana* (Sonderby et al. 2010). Somewhat less is known for the chain-elongated homophenylalanine-derived aromatic GSLs (Bhandariet al. 2015), as genes governing aromatic GSL biosynthesis are yet to be fully identified. Numerous studies have been performed to comprehend the genetics underlying the variations in GSL contents. There are also reports indicating negative associations between a number of yield component QTLs and GSL QTLs (Ramchiary et al. 2007b; Ramchiary et al. 2007a). Panjabi et al. (2008) have reported SNP markers in close proximity to GSL-regulating loci (LGs A02, A03, A09, and B01). Several QTLs for seed GSLs are now available (Ramchiary et al. 2007a; Panjabi et al. 2008; Bisht et al. 2009) and Sequence Characterized Amplified Region (SCAR) markers (Ripley and Roslinsky 2005). It is also known that *BjuB.CYP79F1* controls the biosynthesis of the propyl fraction of aliphatic GSLs in mustard (Sharma et al. 2016). Akhatar et al. (2020) used GWAS to predict the association of eight genes on chromosomes A04, A05, A06, B03, B04, and B06; which are involved in different steps of the biosynthesis pathway for GSL. These include *HB16* and *SK1*, which code for the enzyme shikimate kinase (SK). *CM1* (chromosomes A06 and B04) is involved in phenylalanine and tyrosine biosynthesis, and *JMT* (B03) is vital for the induction of indole GSLs in *Arabidopsis* by jasmonate (Ku et al. 2016). *LINC4*-B06 encodes branched-chain amino acid transaminase, which converts branched-chain amino acids and -keto-glutarate to branched-chain-keto acids and glutamate. The chromosome B06-predicted *CYP81G1* is involved in the metabolism of indole GSLs. *MYB44* (B06) regulates the expression of the majority of GSL biosynthesis genes in collaboration with *EIN2* (Brader et al. 2001). The CGs *BjuA.GSL-ELONG.a*, *BjuA.GSL-ELONG.c*, *BjuA.GSL-ELONG.d*, *BjuA.Myb28.a*, *BjuA.GSL-ALK.a*, and *BjuA.GSL-ALK.b* have been mapped via co-mapping with previously identified GSL QTLs (Ramchiary et al. 2007). The contribution of these genes to various steps of the GSL biosynthesis pathway was also validated through functional analysis based on allelic variation and phenotypic data. Pushpa et al. (2016) revalidated six GSLs markers in a recombinant inbred line (RIL) population created by crossing a low-erucic acid and a double-low mustard variety. Yang et al. (2021) found major MTAs in two regions of chromosomes A2 and A9. Both regions contain orthologs of *HAG1* (*MYB28*), *BjuVA02G46870.1* and *BjuVA09G07110.1*, which are key regulatory genes that govern the biosynthesis of aliphatic GSLs in seed. *MYB28* (the homolog of *BjuVA02G46870.1*) appeared to have an important role as its silencing reduced the GSL synthesis in the oilseed *B. juncea* (Augustine et al. 2013).

10.3.2.3 Seed Coat Color

The integuments and chalaza tissues of the ovule differentiate to form the seed coat after fertilization, and the brown pigmentation of this seed coat results from the accumulation of oxidized flavonoids (proanthocyanidin) in the inner integuments (Lepiniec et al. 2006). Mutations in pathway genes that inhibit the biosynthesis of the flavonoid pro-anthocyanidin can lead to a yellow hue in the seeds. Compared to brown seed coat color, yellow seed coat color is linked to a lower fibre content and, as a consequence, higher oil and protein contents (Woods 1980; Banga 1996; Rahman and McVetty 2011). The genetics of brown seed coat color have been widely investigated. The outcomes range from monogenic (Xu et al. 2010a; Huang et al. 2012), digenic (Vera et al. 1979;

Vera and Woods 1982; Yan et al. 2009), to multigenic (Xu et al. 2010a; Huang et al. 2012) inheritance. However, the consensus centers around the involvement of two dominant duplicate genes (R1 and R2) in the expression of the seed coat color. The presence of either of these two genes was sufficient for a brown seed coat to develop. Only when both of these genes (rl proanthocyanin r2 r2) are in a homozygous recessive state do yellow seeds develop. There have also been numerous attempts to identify molecular markers associated with this trait. *Bjcl* and *Bjc2* co-segregated with AFLP markers E3M3 7 and E8M7 4 on LG3 and LG6, respectively, as ascertained by AFLP-based linkage mapping (Lionneton et al. 2002). Padmaja et al. 2005 had identified three SSR marker loci (Ra2-A11, Na10-A08, and Ni4-F11) that were strongly associated with seed coat color. Subsequent genetic mapping identified Ra2-A11 and Na10-A08 on LG1 and marker Ni4-F11 in LG2 of the *B. juncea* linkage map (Pradhan et al. 2003). Negi et al. (2000) converted seed coat color related AFLP markers into SCAR markers. Mahmood et al. (2005) described three QTLs (LGs A06, A10, and B04) for seed coat color that were stable across environments. These QTLs accounted for 43%, 31%, and 16% of the total phenotypic variation, respectively. Huang et al. (2011) identified four closely linked AFLP and SSR markers that demonstrated high collinearity with *A. thaliana* chromosome 3. The homologue identification by comparative genomics and subsequent fine mapping analysis (Huang et al. 2012) assisted in the identification of 14 AFLP and 2 SSR markers for close association with the brown seed locus. The SSR marker CB1022, which displayed codominant inheritance, could distinguish different genotypes within a population. Padmaja et al. (2014) mapped two homoeologous flavonoid genes, *BjuA.TT8* and *BjuB.TT8*, to LGs A09 and B03, respectively. Two intron polymorphism markers were reported to flank the yellow seed color gene at respective distances of 0.1 and 0.3 cM (Huang et al. 2016). *Bra036827*, *Bra036828*, and *Bra036829* were subsequently indicated as sources of three IP markers (IP1, IP2, and IP3), and these markers co-segregated with the target gene(s). The genomic region confirming *Bra036828's* association with flavonoid biosynthesis and its high similarity to the *TRANSPARENT TESTA 6* gene is now well known. It is possible that *Bra036828* is the gene responsible for yellow seed color. Patel et al. (2022) cloned and sequenced brown-seed (DRMRIJ-31) and yellow-seed (RLC-3) *Bju.TT8* homologs. Sequencing analysis of *Bju.TT8* homologs revealed that in RLC-3, gene *Bju.ATT8* contained a 1,279-bp insertion in the 7th exon, whereas gene *Bju.BTT8* comprised a SNP (CT) in the seventh exon. These two mutations resulted in the development of yellow seed coats. The authors could develop highly reliable functional markers for both *Bju.TT8* homologs.

10.3.3 Resistance to Diseases and Insect-Pests

10.3.3.1 White Rust

Albugo candida, a fungus pathogen, is responsible for white rust, which is one of the most destructive mustard diseases. Early genetic studies had revealed a monogenic dominant pattern of inheritance for resistance against white rust race-2 (Tiwari et al. 1988; Cheung et al. 1998; Sachan et al. 2000). Cheung et al. (1998) also found a dominant RFLP marker (X140a) that co-segregated with the *Acr* locus. The same locus was also tagged using different marker systems including RFLP (Cheung et al. 1998) and RAPD (Prabhu et al. 1998), which also detected two flanking RAPD markers that fell within 10 cM of the target locus. Somers et al. (2002) also identified AFLP markers for a novel white rust resistance locus (*Ac2V1*). In a step forward, Varshney et al. (2004) designed and validated a CAPS marker using two different F_2 populations. Panjabi-Massand et al. (2010) identified two independent loci conferring resistance to *Albugo candida* race 2V in the east European mustard germplasm lines Heera and Donskaja IV.

The white rust resistance-conferring locus *AcB1-A4.1* was located on LG A4 in Heera, whereas *AcB1-A5.1* was present at LG A5 in Donskaja-IV. These genes accounted for 76% and 70%, respectively, of the total phenotypic variance explained. Comparative mapping subsequently showed that the *AcB1-A4.1* locus of *B. juncea* was homologous to the ancestral "S block" of *A. thaliana* chromosome

5 (Panjabi et al. 2008). This QTL interval was further fine-mapped on *the B. juncea* linkage map using 54 IP primer sequences derived from the S block of *A. thaliana. The AcBl-A4.1* region was also fine-mapped using eight IP markers based on IP primer sequences from the *A. thaliana S* block. Markers *AT5g41560* and *AT5g41970* were present within a 1.2 cM interval on the Varuna×Heera map. These markers are perfectly co-segregated with the resistant phenotype. The QTL region of *AcBl-A5.1* exhibited homology with the "J" block of *A. thaliana* chromosome 2 (Panjabi et al. 2008). Co-segregation analysis with four IP primer sequences from the J block of *A. thaliana* allowed identification of *AT2G34510* and *AT2G36360* as flanking markers for the *AcBl-A5.1* gene. *AT2G36360* appeared to be strongly associated with resistance loci. Additional genomic studies conducted by Arora et al. (2019) revealed the presence of a single CC-NB-LRR protein-encoding R gene in Donskaja-IV (*BjuWRR1*), conferring resistance to different pathogen isolates. Bhayana et al. (2020) identified another locus conferring resistance to *A. candida* isolate AcB1 at LG A6. Another novel resistance gene was mapped between 63.0 cM and 70.8 cM on the LG A06 for the Chinese tuber mustard genotype, Tumida (Yang et al. 2016). A CC-NBS-LRR (CNL) type R gene, *BjuA046215*, is the most likely candidate protein-encoding gene in the mapped region (Paritosh et al. 2021).

10.3.3.2 Sclerotinia Stem Rot

A high level of resistance to this disease is not present in the crop's primary gene pool (Barbetti et al. 2014). However, wild *Brassicaceae* species like *Erucastrum gallicum, Capsella bursa-pastoris, E. cardaminoides, Diplotaxis tenuisiliqua, B. fruticulosa,* and *B. oleracea* have been reported to possess high levels of resistance to this disease (Garg et al. 2010; Mei et al. 2011). SSR- and SNP-based molecular markers have been used to characterize the *B. juncea–B. fruticulosa* and *B. juncea–Erucastrum cardaminoides* ILs (Rana et al. 2017; Rana et al. 2019). Genotyping-By-Sequencing (GBS)-GWAS using *B. juncea–B. fruticulosa* ILs allowed identification of 49 MTAs corresponding to 13 loci on chromosomes A01, A03, A04, A05, A08, A09, and B05 (Atri et al. 2019). The annotation of the genomic region found associated with sclerotinia resistance helped to predict 20 CGs belonging to major disease resistance protein families, namely, TIR-NBS-LRR, chitinase, malectin/receptor-like protein kinase, defensin-like (DEFL), desulfoglucosinolate sulfotransferase protein, and lipoxygenase. The majority of significant SNPs were confirmed by comparing the genome sequences (21×) of five advanced generation lines selected for *Sclerotinia* resistance to corresponding sequences from three susceptible *B. juncea* germplasm lines. Similar studies were undertaken for *B. juncea–Erucastrum cardaminoides* ILs. These helped to detect SNPs associated with resistance on chromosomes A03, A06, and B03 (Rana et al. 2019). The annotation of related genome space emphasized the importance of signal transduction pathways, hypersensitive responses, and the production of anti-fungal proteins and metabolites for this resistance source. Multiple resistance genes were found clustered in small segments of chromosomes A03 and B03.

10.3.3.3 Alternaria Blight

Alternaria brassiceae causes Alternaria blight, which is a significant disease of *B. juncea.* There is no credible source of resistance in the primary gene pool of *B. juncea*, so attempts have been made to introgress resistance from diverged *Brassica*ceae species, namely *Diplotaxis erucoides* and *Sinapis alba*, using embryo culture and protoplast fusion, respectively (Kumari 2018; Kumari and Bhat 2019). Mapping with donor genome-specific SSRs led to the identification of three QTLs responsible for *A. Brassicae* resistance in *S. alba–B. juncea* introgression lines (ILs) (Singh 2021).

10.3.3.4 Aphids

These cause severe productivity losses in mustard, and the current level of resistance in mustard cultivars is insufficient to prevent or resist aphid infestation. It is known that *B. juncea–B. fruticulosa* ILs vary in their resistance to mustard aphids (Kumar et al. 2011; Atri et al. 2012; Kumar and Banga 2017; Agrawal et al. 2021). The wild *Brassica*ceae species *B. fruticulosa* is well-known for its aphid resistance (Kumar et al. 2011). SNP genotyping and genome-wide association analysis (Atri et al. 2023, in preparation) have helped to predict genes associated with signaling hormones such as Brassinosteroid-signaling kinase (*BSK7*-B01), JA (*4CLL2*-B03), ET (*ERF1*-(B07), and redox *CAT3* (B03) or defensive enzymes such as *PRX64, PRX71* (B06), and *PRX69* (A02) for their role in conferring aphid resistance associated with the uptake and recycling of cell wall sugars (*STP7*-B01). Genes associated with local and systemic defensive responses, were also predicted. Many of the predicted genes could be validated through associated transcriptomics.

10.4 GENOME-SEQUENCE-DEPENDENT ALLELE MINING

To make this chapter more informative, we have also summarized the results of sequence-based allele mining for the genes predicted by our group from various association mapping studies (Tables 10.1–10.5). Genome sequences from 92 *B. juncea* core germplasm lines or alien introgression sets, wherever applicable, were used for this purpose. The forward and reverse FASTQ reads of genotypes were then mapped to their respective genomic reference sequences to generate a variant call format file (VCF file). The FASTA sequences of the predicted genes were first obtained from the National Center for Biotechnology Information (NBCI) database. Afterwards, a local blast run was conducted using in-house genomic datasets as sequence databases. Gene coordinates were determined using the Nucleotide Basic Local Alignment Search Tool (BLASTn). Only the top blast hits with the highest coverage and percentage of identity were selected. SNPs and INDELs were then identified within selected gene coordinates from the VCF file to their respective genomic datasets.

10.4.1 Morphological Traits

As shown in Table 10.1, multiple SNPs and INDELs were identified and the CGs predicted for the morphological traits of flowering, plant height, pod length, and seeds per pod. We also included the flowering genes identified in prior research, including *FLM, CKB4, AGL6, SOC1, COP1, GID1A, FKF1, ARP6, FD, GI, AGL24, AP2, GA1, FAF2, CRY1, JMJ14, ATJ3, MAF5, FT, HAM1,* and *FRI.* These flowering genes were found to contain a total of 140 SNPs and 47 INDELs. The greatest number of SNPs (35) were observed in five genes on chromosome A05, including *CKB4* (PVE=9.79), *AGL6* (PVE=20.63), *SOC1* (PVE=18.57), *COP1* (PVE=18.94), and *GID1A* (PVE=16.51), followed by chromosome A02 (27). On chromosomes A08 (*ARP6* and *FD*) and B03 (*FAF2, CRY1, JMJ14*), each

TABLE 10.1
Summary of Allele Mining for Morphological Traits in *B. juncea*

Trait	Chr	Position(s)	Gene	AT ID	Reference	SNPs	INDELs
Flowering	A02	37090239-3082	*FLM*	*AT1G77080*	Akhatar et al. (2021)	27	2
	A05	2937409-45567	*CKB4*	*AT2G44680*		5	4
	A05	3354215	*AGL6*	*AT2G45650*		1	0
	A05	3354890-7582	*SOC1*	*AT2G45660*	Zhao et al. (2019),	2	0
	A05	7544327-6996	*COP1*	*AT2G32950*	Akhatar et al. (2021)	25	7
	A05, B02	30987755-390279, 62297219-20	*GID1A*	*AT3G05120*	Akhatar et al. (2021)	2, 1	9,0
	A07	1786248	*FKF1*	*AT1G68050*	Zhao et al. (2019),	1	0
	A08	52619-842	*ARP6*	*AT3G33520*	Akhatar et al. (2021)	10	0
	A08	18952302-5620	*FD*	*AT4G35900*		15	4
	A09	39780608–959	*GI*	*AT1G22770*	Akhatar et al. (2021)	3	0
	B02	59088464, 64497529-9514	*AGL24*	*AT4G24540*	Zhao et al. (2019), Akhatar et al. (2021)	1, 2	0,6
	B02	66542016-66535	*AP2*	*AT4G36920*		5	0
	B02, B08	46682290, 14118234	*GA1*	*AT4G02780*		1, 1	0,0

(*Continued*)

TABLE 10.1 (*Continued*)
Summary of Allele Mining for Morphological Traits in *B. juncea*

Trait	Chr	Position(s)	Gene	AT ID	Reference	SNPs	INDELs
	B03	115277	*FAF2*	*AT1G03170*	Akhatar et al. (2021)	1	0
	B03	338948-346586	*CRY1*	*AT4G08920*	Zhao et al. (2019), Akhatar et al. (2021)	20	10
	B03	10133662-4104	*JMJ14*	*AT4G20400*	Akhatar et al. (2021)	4	0
	B05	49722910	*ATJ3*	*AT3G44110*		1	0
	B06	8515215-24113	*MAF5*	*AT5G65080*			5
	B06, A07	6716803-8200, 33488659	*FT*	*AT1G65480*	Zhao et al. (2019), Akhatar et al. (2021)	5, 1	0,0
	B07	11563802-4171	*HAM1*	*AT5G64610*	Akhatar et al. (2021)	4	0
	B08, A10	16066635, 6054217-43	*FRI*	*AT4G00650*	Zhao et al. (2019), Akhatar et al. (2021)	1, 1	0,0
Plant Height	A02	1762477	*FLC*	*AT5G10140*	Akhatar et al. (2021)	1	0
	A02	36486270-7130	*CDF1*	*AT5G62430*		4	0
	A09, B04	10079474, 16336556–1785	*FVE*	*AT2G19520*		1, 6	0,16
	B02	64496077-6528	*AGL24*	*AT4G24540*		8	0
	B08	9748710-37	*CO*	*AT5G15840*		1	0
	B08, A10	16066635, 6054217-43	*FRI*	*AT4G00650*	Zhao et al. (2019), Akhatar et al. (2021)	1, 1	0,0
Pod Length	B06	11913579-680	*NST1*	*AT2G46770*	Kaur et al. (2020b)	6	0
Seeds per pod	B02	4867097-166	*SUR1*	*AT2G20610*		8	0

contained 25 SNPs, as shown in Table 10.1. SNPs and INDELs were identified on multiple chromosomes at loci *GID1A, AGL24, GA1, FT,* and *FRI.* In addition, *CRY1* (10), *GID1A* (9), and *COP1* (9) contained the greatest number of INDELs (7). We identified 23 SNPs and 16 INDELs in six genes – *FLC, CDF1, FVE, AGL24, CO,* and *FRI* – that contribute to plant height (Akhatar et al. 2021). *AGL24* had the most SNPs (8) on chromosome B02 (64496077–6528), whereas *FVE* had the most INDELs (16) on chromosome B04 (position 16336556–1785). For pod length, we identified six SNPs in *NST1* at positions 11913579–666 on chromosome B06 (Kaur et al. 2021). Lastly, for the trait seeds/pod, we were able to identify 8 SNPs on *SUR1* at positions 4867097–166 on chromosome B02 (Kaur et al. 2021).

10.4.2 Quality Traits

We performed allele mining for the six genes predicted to be associated with oil content in *B. juncea.* This led to the identification of a total of 11 SNPs and 3 INDELs. Two SNPs were located on chromosome A06 and one on chromosome A09. INDELs were also detected in *CER26, LACS5,* and *DIR1* (Table 10.2). Two genes, *FAE1.1* and *FAE1.2*, had eight and three SNPs, respectively, for fatty acid.

10.4.3 Shattering Resistance

We investigated allele mining in nine genes (*CUC3/NAC368, AFO/FIL, FUL1, ERF4, MAN7, STM, NST2, LEUNIG,* and *RAP2.4*) for shattering resistance (rupture energy) in *B. juncea.* According to Table 10.3, a total of 26 SNPs and 6 INDELs were identified, the majority of which were on the A-genome. At position 12315809–41 on chromosome A05, *ERF4* had the greatest number of SNPs (8). *RAP2.4*, located on chromosome B05, possessed six SNPs and two INDELs at positions 60868573–825. At positions 15096881–7017 on chromosome A02, SNPs and Indels were found in *CUC3/NAC368.* As depicted in Table 10.3, SNPs and INDELs were identified in STM at two distinct positions.

TABLE 10.2
Summary of Allele Mining for Quality Traits in *B. juncea*

Trait	Chr	Position(s)	Gene	AT ID	Reference	SNPs	INDELs
Oil	A04	6252658	*CER26*	*AT4G13840*	Akhatar et al. (2020)	1	1
	A06, A09	26456373-6466, 14737522	*GDSL esterase/lipase*	*AT2G03980*		2, 1	0
	A09	24082148-3336	*LACS5*	*AT4G11030*		2	1
	B05	5272794-2795	*FAD6*	*AT4G30950*		1	0
	B06	19315729	*ADS1*	*AT1G06080*		1	0
	B08	34645905-6006	*DIR1*	*AT4G12510*		3	1
Fatty acid			*FAE1.1*		Gill et al. (2021)	8	–
			FAE1.2			3	–

TABLE 10.3
Summary of Allele Mining for Shattering Resistance in *B. juncea*

Trait	Chr	Position(s)	Gene	AT ID	Reference	SNPs	INDELs
Rupture energy	A02	15096881-7017	*CUC3/NAC368*	*AT1G76420*	Kaur et al. (2020b)	2	2
	A03	13393564	*AFO/FIL*	*AT2G45190*		1	0
	A03	27065868	*FUL1*	*AT5G60910*		1	0
	A05	12315809-41	*ERF4*	*AT3G15210*		8	0
	A09	6549794	*MAN7*	*AT5G66460*		1	0
	A09	13105514, 13128924-9099	*STM*	*AT1G62360*		1, 4	0, 1
	A09	46982925-55	*NST2*	*AT3G61910*		1	1
	B05	4478265	*LEUNIG*	*AT4G32551*		1	0
	B05	60868573-825	*RAP2.4*	*AT1G78080*		6	2

10.4.4 Nitrogen Use Efficiency (NUE)

Gene sequences from *NRT1.1, AT3G09430, NRT1.8, AAP1, RSL4, ATG5, AAH, BFN1, RHD6, NRT3.1, DJ1C, RLF, ATG3, GS1, NPF3.1, AT1G47670*, and *CAT9* were queried to identify alleles predicted to be associated with NUE (Gupta et al. 2021). 38 SNPs and 8 INDELs were identified for these genes (Table 10.4). *NPF3.1* had the highest number of SNPs and INDELs at position 54719294–29595 on chromosome B05 (3). *ATG5* (A10) had eight SNPs at positions 14342041–233. Indels were found in four genes: *NRT1.8, RSL4, NPF3.1,* and *ATG3* (Table 10.4).

10.4.5 Biotic Stresses

We performed sequence-based allele mining for the genes predicted for sclerotinia and aphid resistance.

Sclerotinia Resistance: We investigated a total of 27 genes, including *SBT4.4, SBT4.9, SBT4.12, AT1G65850, AT3G04220, AT5G11250, GSTT2, GSTT3, AT5G16590, BSK4, BSK7, BSK8, AOC3, AOC2, AOC1, AOC4, LCR73, PDF2.1, PDF2.3, PDF2.5, LCR71, LCR76, SBT3.3, PBL7, AT1G65850,* and *AT3G04220*. Details regarding the discovery of these genes are available in Section 10.3.3.2 of this chapter. A total of 51 SNPs and 6 INDELs were reported for these genes. Ten SNPs and four INDELs were identified at positions 6274763–994 on chromosome A03 in loci *AT1G65850, AT3G04220*, and *AT5G11250*. As shown in Table 10.5, a large number of SNPs were also observed for *BSK4* (9), *BSK7* (9), *BSK8* (9), *PBL7* (8), and *SBT3.3* (7) on various chromosomes.

TABLE 10.4
Summary of Allele Mining for Nitrogen Use Efficiency (NUE) in *B. juncea*

Trait	Chr	Position(s)	Gene	AT ID	Reference	SNPs	INDELs
NUE	A03	30554135-9	*NRT1.1*	*AT1G12110*	Gupta et al. (2021)	1	0
	A05	29290258-532	*AT3G09430*	*AT3G09430*		2	0
	A08	14124984-5145	*NRT1.8*	*AT4G21680*		3	2
	A09	14976303-453	*AAP1*	*AT1G58360*		2	0
	A09	34062230-85	*RSL4*	*AT1G27740*		2	2
	A10	14342041-233	*ATG5*	*AT5G17290*		8	0
	B02	57937940	*AAH*	*AT4G20070*		1	0
	B03	1789526	*BFN1*	*AT1G11190*		1	0
	B03	50251730	*RHD6*	*AT1G66470*		1	0
	B03	52982944	*NRT 3.1*	*AT5G50200*		1	0
	B05	3270436-66	*DJ1C*	*AT4G34020*		2	0
	B05	16757342	*RLF*	*AT5G09680*		1	0
	B05	17204057-154	*ATG3*	*AT5G61500*		1	1
	B05	24283528	*GS1*	*AT5G16570*		1	0
	B05	54719294-29595	*NPF3.1*	*AT1G68570*		9	3
	B07	30294832	*AT1G47670*	*AT1G47670*		1	0
	B08	24046120	*CAT9*	*AT1G05940*		1	0

Aphid Resistance: Details regarding the discovery of the genes putatively associated with aphid resistance are available in Section 10.3.3.4. Sequence analysis allowed us to detect 177 SNPs and 194 INDELs in *PER45, ESR/TASTY, PER69, PER71, GTR1, PER29, COI1, BSK7, ROM-like protein, SOT18, SDI2, 4CLL2, ARID3, CAT3, STP1, TMT1, GATL6, AGB1, ADF3, FAMT, PER64, SLI1, ERF1, ERF016, CAF1B, NSP1, CYP81F2, CNGC20* and *OXS3*. *NSP1* and *CYP81F2*, both located on chromosome B08, had the maximum number of SNPs (35) and INDELs (45) between positions 13959010 and 140200. Many SNPs were also identified in *ERF016* (11), *ROM-like protein* (10), and *SDI2* (10), among others. Likewise, a large number of INDELs were also detected in *SDI2* (43), *ROM-like protein* (35), and other genes (Table 10.5).

TABLE 10.5
Summary of Allele Mining for Biotic Stresses in *B. juncea*

Trait	Chr	Position(s)	Gene	AT ID	Reference	SNPs	INDELs
Sclerotinia Resistance	A03	6235895-6020	*SBT4.4, SBT4.9, SBT4.12*	*AT5G59090*	Rana et al. (2019)	2	0
	A03	6274763-994	*AT1G65850, AT3G04220, AT5G11250*	*AT1G65850, AT3G04220, AT5G11250*		10	4
	A03	6337002-047	*GSTT2, GSTT3*	*AT5G41240, AT5G41220*		2	0
	A03	6390210-381	*AT5G16590*	*AT5G16590*		4	0
	A06	14196946-7052	*BSK4, BSK7, BSK8*	*AT1G01740, AT1G63500, AT5G41260*		9	0
	A06	26227171-7735	*AOC3, AOC2, AOC1, AOC4*	*AT3G25780, AT3G25770, AT3G25760, AT1G13280*		4	1

(Continued)

TABLE 10.5 (*Continued*)
Summary of Allele Mining for Biotic Stresses in *B. juncea*

Trait	Chr	Position(s)	Gene	AT ID	Reference	SNPs	INDELs
	A06	27175971-6193	*LCR73, PDF2.1, PDF2.3, PDF2.5, LCR71, LCR76*	*AT2G02147, AT2G02120, AT2G02130, AT5G63660, AT2G02135, AT2G31953*		4	0
	B03	924478-5003	*SBT3.3*	*AT1G32960*		7	1
	B03	998752-898	*PBL7*	—		8	0
	B04	1922364	*AT1G65850, AT3G04220, AT5G11250*	*AT1G65850, AT3G04220, AT5G11250*		1	0
Aphid Resistance	A01	3662846-55	*PER45*	*AT4G30170*	Atri et al. (2023), in preparation	2	0
	A01	20385850	*ESR/TASTY*	*AT1G54040*		1	0
	A02	36898456-564	*PER69, PER71*	*AT5G64100, AT5G64120*		3	0
	A03, B05	29644674-748, 52333008-7944	*GTR1*	*AT3G47960*		2, 9	0,0
	A05	23020649-735	*PER29*	*AT3G17070*		4	0
	A07, B08	34183265-70, 22042465-738	*COI1*	*AT2G39940*		2, 8	0, 6
	B01	29629278	*BSK7*	*AT1G63500*		1	1
	B01	45105527-23116	*ROM-like protein*	*AT3G07910*		10	35
	B03	303164-429	*SOT18*	*AT1G74090*		3	5
	B03	381772-420991	*SDI2*	*AT1G04770*		10	43
	B03	16459279	*4CLL2, ARID3, CAT3, STP1, TMT1*	*AT1G20480, AT1G20910, AT1G20620, AT4G02050, AT1G20840*		1	0
	B03	17968228	*GATL6*	*AT4G02130*		1	0
	B05	2870247	*AGB1*	*AT4G34460*		1	1
	B05	28890648	*ADF3*	*AT5G59880*		1	0
	B05	37542566-88	*FAMT*	*AT3G44860*		3	0
	B06	27599943-600140	*PER64*	*AT5G42180*		6	2
	B07	2904790	*SLI1*	*AT3G10680*		1	0
	B07	24909394	*ERF1*	*AT4G17500*		1	0
	B08	12263168-565	*ERF016*	*AT5G21960*		11	2
	B08	12400503-648	*CAF1B*	*AT5G22250*		3	0
	B08	13959010-14020030	*NSP1*	*AT3G16400*		35	45
	B08	13959010–14020030, 26317676-30452	*CYP81F2*	*AT5G57220*		35, 7	45, 3
	B08	14368391-70892	*CNGC20*	*AT3G17700*		8	3
	B08	14368391-70892	*OXS3*	*AT5G56550*		8	3

ACKNOWLEDGMENTS

Genome sequencing data of mustard genotypes and ILs used for sequence-based allele mining was supported by grants received under the National Agricultural Science Fund aided project "Creating a fully characterized genetic resources pipeline for mustard improvement" and the Department of Biotechnology, Government of India, in the form of Centre of Excellence and Innovation in Biotechnology "Germplasm enhancement for crop architecture and defensive traits in *Brassica juncea* (L.) Czern. and Coss" SSB is also grateful for salary support in the form of Raja Ramana Fellowship, Department of Atomic Energy, Government of India.

REFERENCES

Agrawal N, Gupta M, Atri C, Akhatar J, Kumar S, Heslop-Harrison PJS, Banga SS (2021) Anchoring alien chromosome segment substitutions bearing gene(s) for resistance to mustard aphid in *Brassica juncea - B. fruticulosa* introgression lines and their possible disruption through gamma irradiation. *Theor Appl Genet* 134(10):3209–3224. https://doi.org/10.1007/s00122-021-03886-z

Akhatar J, Banga SS (2015) Genome-wide association mapping for grain yield components and root traits in *Brassica juncea* (L.) Czern & Coss. *Mol Breed* 35:1–14. https://doi.org/10.1007/s11032-015-0230-8

Akhatar J, Goyal A, Kaur N, Atri C, Mittal M, Singh MP, Kaur R, Rialch I, Banga SS (2021) Genome wide association analyses to understand genetic basis of flowering and plant height under three levels of nitrogen application in *Brassica juncea* (L.) Czern & Coss. *Sci Rep* 11(1):1–14. https://doi.org/10.1038/s41598-021-83689-w

Akhatar J, Singh MP, Sharma A, Kaur H, Kaur N, Sharma S, Bharti B, Sardana VK, Banga SS (2020) Association mapping of seed quality traits under varying conditions of nitrogen application in *Brassica juncea* L. Czern & Coss. *Front Genet* 11:744. https://doi.org/10.3389/fgene.2020.00744

Arora H, Padmaja KL, Paritosh K, Mukhi N, Tewari AK, Mukhopadhyay A, Gupta V, Pradhan AK, Pental D (2019) *BjuWRR1*, a *CC-NB-LRR* gene identified in *Brassica juncea*, confers resistance to white rust caused by *Albugo candida*. *Theor Appl Genet* 132(8):2223–2236. https://doi.org/10.1007/S00122-019-03350-Z/METRICS

Atri C, Akhatar J, Gupta M, Gupta N, Goyal A, Rana K, Kaur R, Mittal M, Sharma A, Singh MP, Sandhu PS, Barbetti MJ, Banga SS (2019) Molecular and genetic analysis of defensive responses of Brassica *juncea - B. fruticulosa* introgression lines to Sclerotinia infection. *Sci Rep* 9(1). https://doi.org/10.1038/s41598-019-53444-3

Atri C, Kumar B, Kumar H, Kumar S, Sharma S, Banga SS (2012) Development and characterization of *Brassica juncea - B. fruticulosa* introgression lines exhibiting resistance to mustard aphid (*Lipaphis erysimi* Kalt). *BMC Genet* 13:104. https://doi.org/10.1186/1471-2156-13-104

Augustine R, Majee M, Gershenzon J, Bisht NC (2013) Four genes encoding *MYB28*, a major transcriptional regulator of the aliphatic glucosinolate pathway, are differentially expressed in the allopolyploid *Brassica juncea*. *J Exp Bot* 64(16):4907–4921. https://doi.org/10.1093/jxb/ert280

Axelsson T, Bowman CM, Sharpe AG, Lydiate DJ, Lagercrantz U (2000) Amphidiploid *Brassica juncea* contains conserved progenitor genomes. *Genome* 43(4):679–688. https://doi.org/10.1139/g00-026

Banga SK, Kumar P, Bhajan R, Singh D, Banga SS (2015) Genetics and breeding. In: Kumar A, Banga SS, Meena PD, Kumar PR (eds) *Brassica Oilseeds: Breeding and Management*. CABI, London, UK, pp. 11–41.

Banga SS (1996) Breeding designer rapeseed-mustard. *Crop Improv* 23:151–153.

Banga SS, Banga SK (2009) Crop improvement strategies in rapeseed mustard. In: Hedge DM (ed) *Vegetable Oils Scenario: Approaches to Meet the Growing Demands*. ISOR, Hyderabad, pp. 13–35.

Banga SS, Banga SK (2016) Genetic diversity and germplasm patterns in *Brassica juncea*. In: Rajpal VR, Rao SR, Raina SN (eds) *Sustainable Development and Biodiversity*, Vol. 10. Springer, Cham, pp. 163–186.

Banga SS, Banga SK, Labana KS (1983) Nucleo-cytoplasmic interactions in *Brassica*. In: *Proceedings of 6th International Rapeseed Congress*. Paris, France, pp. 602–606.

Barbetti MJ, Banga SK, Fu TD, Li YC, Singh D, Liu SY, Ge XT, Banga SS (2014) Comparative genotype reactions to *Sclerotinia sclerotiorum* within breeding populations of *Brassica napus* and *B.* juncea from India and China. *Euphytica* 197(1):47–59. https://doi.org/10.1007/S10681-013-1049-1/METRICS

Barker GC, Larson TR, Graham I a, Lynn JR, King GJ (2007) Novel insights into seed fatty acid synthesis and modification pathways from genetic diversity and quantitative trait Loci analysis of the *Brassica* C genome. *Plant Physiol* 144(4):1827–1842. https://doi.org/10.1104/pp.107.096172

Becker HC, Léon J (1988) Stability analysis in plant breeding. *Plant Breed* 101:1–23.

Bhandari SR, Jo JS, Lee JG (2015) Comparison of glucosinolate profiles in different tissues of nine *Brassica* crops. *Molecules* 20(9):15827–15841. https://doi.org/10.3390/molecules200915827

Bhat MA, Gupta ML, Banga SK, Raheja RK, Banga SS (2002) Erucic acid heredity in *Brassica juncea* - Some additional information. *Plant Breed* 121(5):456–458. https://doi.org/10.1046/j.1439-0523.2002.731805.x

Bhayana L, Paritosh K, Arora H, Yadava SK, Singh P, Nandan D, Mukhopadhyay A, Gupta V, Pradhan AK, Pental D (2020) A mapped locus on LG A6 of *Brassica juncea* line tumida conferring resistance to white rust contains a CNL type R gene. *Front Plant Sci* 10:1690. https://doi.org/10.3389/fpls.2019.01690

Bisht NC, Gupta V, Ramchiary N, Sodhi YS, Mukhopadhyay A, Arumugam N, Pental D, Pradhan AK (2009) Fine mapping of loci involved with glucosinolate biosynthesis in oilseed mustard (*Brassica juncea*) using genomic information from allied species. *Theor Appl Genet* 118(3):413–421. https://doi.org/10.1007/s00122-008-0907-z

Brader G, Tas É, Palva ET (2001) Jasmonate-dependent induction of indole glucosinolates in Arabidopsis by culture filtrates of the nonspecific pathogen *Erwinia* carotovora. *Plant Physiol* 126(2):849–860. https://doi.org/10.1104/pp.126.2.849

Burkill IH (1930) The Chinese mustards in the Malay peninsula. *Gardners Bull* 5:99–117.

Burns MJ, Barnes SR, Bowman JG, Clarke MHE, Werner CP, Kearsey MJ (2003) QTL analysis of an intervarietal set of substitution lines in *Brassica napus*: (I) seed oil content and fatty acid composition. *Heredity (Edinb)* 90(1):39–48. https://doi.org/10.1038/sj.hdy.6800176

Burton WA, Ripley VL, Potts DA, Salisbury PA (2004) Assessment of genetic diversity in selected breeding lines and cultivars of canola quality *Brassica juncea* and their implications for canola breeding. *Euphytica* 136(2):181–192. https://doi.org/10.1023/B:EUPH.0000030668.62653.0d

Ceciliani F, Bortolotti F, Menegatti E, Ronchi S, Ascenzi P, Palmieri S (1994) Purification, inhibitory properties, amino acid sequence and identification of the reactive site of a new serine proteinase inhibitor from oil-rape (*Brassica napus*) seed. *FEBS Lett* 342(2):221–224. https://doi.org/10.1016/0014-5793(94)80505-9

Chen S, Wan Z, Nelson MN, Chauhan JS, Redden R, Burton WA, Lin P, Salisbury PA, Fu T, Cowling WA (2013) Evidence from genome-wide simple sequence repeat markers for a polyphyletic origin and secondary centers of genetic diversity of *Brassica juncea* in China and India. *J Hered* 104(3):416–427. https://doi.org/10.1093/jhered/est015

Cheung WY, Gugel RK, Landry BS (1998a) Identification of RFLP markers linked to the white rust resistance gene (Acr) in mustard (*Brassica juncea* (L.) Czern. and Coss.). *Genome* 41(4):626–628. https://doi.org/10.1139/G98-043

Cheung WY, Landry BS, Raney P, Rakow GFW (1998b) Molecular mapping of seed quality traits in *Brassica* juncea L. Czern. and Coss. *Acta Hortic* 459:139–147.

Dhaka N, Mukhopadhyay A, Paritosh K, Gupta V, Pental D, Pradhan AK (2017a) Identification of genic SSRs and construction of a SSR-based linkage map in *Brassica juncea. Euphytica* 213(1):15. https://doi.org/10.1007/s10681-016-1814-z

Dhaka N, Rout K, Yadava SK, Sodhi YS, Gupta V, Pental D, Pradhan AK (2017b) Genetic dissection of seed weight by QTL analysis and detection of allelic variation in Indian and east European gene pool lines of *Brassica juncea. Theor Appl Genet* 130(2):293–307. https://doi.org/10.1007/s00122-016-2811-2

Ecke W, Uzunova M, Weißleder K (1995) Mapping the genome of rapeseed (*Brassica napus* L.). II. Localization of genes controlling erucic acid synthesis and seed oil content. *Theor Appl Genet* 91(6–7):972–977. https://doi.org/10.1007/BF00223908

Erickson LR, Straus NA, Beversdorf WD (1983) Restriction patterns reveal origins of chloroplast genomes in *Brassica* amphiploids. *Theor Appl Genet* 65(3):201–206. https://doi.org/10.1007/BF00308066

Garg H, Kohn LM, Andrew M, Li H, Sivasithamparam K, Barbetti MJ (2010) Pathogenicity of morphologically different isolates of *Sclerotinia sclerotiorum* with *Brassica napus* and *B. juncea* genotypes. *Eur J Plant Pathol* 126(3):305–315. https://doi.org/10.1007/S10658-009-9547-7/METRICS

Gill KS, Kaur G, Kaur G, Kaur J, Sra SK, Kaur K, Gurpreet K, Sharma M, Bansal M, Chhuneja P, Banga SS (2021) Development and validation of kompetitive allele-specific PCR assays for erucic acid content in Indian mustard [*Brassica juncea* (L.) Czern and Coss.]. *Front Plant Sci* 12:1–9. https://doi.org/10.3389/fpls.2021.738805

Grami B, Stefansson BR (1977) Paternal and maternal effects on protein and oil content in summer rape. *Can J Plant Sci* 57(3):945–949. https://doi.org/10.4141/cjps77-135

Gunasekera CP, Martin LD, Siddique KHM, Walton GH (2006) Genotype by environment interactions of Indian mustard (*Brassica* juncea L.) and canola (B. napus L.) in Mediterranean-type environments. 1. Crop growth and seed yield. *Eur J Agron* 25(1):1–12. https://doi.org/10.1016/j.eja.2005.08.002

Gupta N, Gupta M, Akhatar J, Goyal A, Kaur R, Sharma S, Goyal P, Mukta A, Kaur N, Mittal M, Singh MP, Bharti B, Sardana VK, Banga SS (2021) Association genetics of the parameters related to nitrogen use efficiency in *Brassica juncea* L. *Plant Mol Biol* 105(1–2). https://doi.org/10.1007/s11103-020-01076-x

Gupta V, Mukhopadhyay A, Arumugam N, Sodhi YS, Pental D, Pradhan AK (2004) Molecular tagging of erucic acid trait in oilseed mustard (*Brassica juncea*) by QTL mapping and single nucleotide polymorphisms in *FAE1* gene. *Theor Appl Genet* 108(4):743–749. https://doi.org/10.1007/s00122-003-1481-z

Hirani AH, Geng J, Zhang J, Zelmer CD, McVetty PBE, Daayf F, Li G (2016) Quantitative trait loci mapping and candidate gene identification for seed glucosinolates in *Brassica rapa* L. *Crop Sci* 56(3):942–956. https://doi.org/10.2135/cropsci2014.12.0837

Huang Z, Ban Y, Yang L, Zhang Y, Li H, Xiao E, Xu A, Zhang D (2012) Fine mapping of the yellow seed locus in *Brassica juncea* L. *Genome* 55(1):8–14. https://doi.org/10.1139/G11-072

Huang Z, Liu L, Lu H, Lang L, Zhao N, Ding J, Xu A (2016) Development of IP and SCAR markers linked to the yellow seed color gene in *Brassica juncea* L. *Breed Sci* 66(2):175–180. https://doi.org/10.1270/JSBBS.66.175

Huang Z, Zhang Y, Qiang HL, Yang L, Yuan BAN Y, Xia AX, Shi XIAO E (2011) AFLP and SSR markers linked to the yellow seed colour gene in *Brassica juncea* L. *Czech J Genet Plant Breed* 47(4):149–155.

James DW, Dooner HK (1990) Isolation of EMS-induced mutants in Arabidopsis altered in seed fatty acid composition. *Theor Appl Genet* 80(2):241–245. https://doi.org/10.1007/BF00224393

Kang L, Qian L, Zheng M, Chen L, Chen H, Yang L, You L, Yang B, Yan M, Gu Y, Wang T, Schiessl SV, An H, Blischak P, Liu X, Lu H, Zhang D, Rao Y, Jia D, Zhou D, Xiao H, Wang Y, Xiong X, Mason AS, Chris Pires J, Snowdon RJ, Hua W, Liu Z (2021) Genomic insights into the origin, domestication and diversification of *Brassica juncea*. *Nat Genet* 53(9):1392–1402. https://doi.org/10.1038/s41588-021-00922-y

Kaur H, Wang L, Stawniak N, Sloan R, van Erp H, Eastmond P, Bancroft I (2020a) The impact of reducing fatty acid desaturation on the composition and thermal stability of rapeseed oil. *Plant Biotechnol J* 18(4):983–991. https://doi.org/10.1111/pbi.13263

Kaur J, Akhatar J, Goyal A, Kaur N, Kaur S, Mittal M, Kumar N, Sharma H, Banga S, Banga SS (2020b) Genome wide association mapping and candidate gene analysis for pod shatter resistance in *Brassica juncea* and its progenitor species. *Mol Biol Rep* 47(4):2963–2974. https://doi.org/10.1007/s11033-020-05384-9

Kaur P, Banga S, Kumar N, Gupta S, Akhatar J, Banga SS (2014) Polyphyletic origin of *Brassica juncea* with *B. rapa* and *B. nigra* (Brassicaceae) participating as cytoplasm donor parents in independent hybridization events. *Am J Bot* 101(7):1157–1166. https://doi.org/10.3732/ajb.1400232

Kaur S, Atri C, Akhatar J, Mittal M, Kaur R, Banga SS (2021) Genetics of days to flowering, maturity and plant height in natural and derived forms of *Brassica rapa* L. *Theor Appl Genet* 134(2):473–487. https://doi.org/10.1007/s00122-020-03707-9

Khan HMWA, Chattha AA, Munir M, Zia A (2009) Evaluation of resistance in sugarcane promising lines against whip smut. *Pak J Phytopathol* 21(1):92–93.

Kim HR, Choi SR, Bae J, Hong CP, Lee SY, Hossain MJ, Van Nguyen D, Jin M, Park BS, Bang JW, Bancroft I, Lim YP (2009) Sequenced BAC anchored reference genetic map that reconciles the ten individual chromosomes of *Brassica rapa*. *BMC Genomics* 10(1):432. https://doi.org/10.1186/1471-2164-10-432

Kirk JT, Hurlstone C (1983) Variation and inheritance of erucic acid content in *Brassica juncea*. *J Plant Breed* 90(4):331–338.

Ku KM, Becker TM, Juvik JA (2016) Transcriptome and metabolome analyses of glucosinolates in two broccoli cultivars following jasmonate treatment for the induction of glucosinolate defense to Trichoplusia ni (Hübner). *Int J Mol Sci* 17(7):1135. https://doi.org/10.3390/ijms17071135

Kumar R, Saini DK, Kumar M, Priyanka V, Akhatar J, Kaushik D, Sharma A, Dhanda PS, Kaushik P (2022) Revealing the genetic architecture of yield-related and quality traits in Indian Mustard [*Brassica juncea* (L.) Czern. and Coss.] using meta-QTL analysis. *Agronomy* 12(10):2442. https://doi.org/10.3390/agronomy12102442

Kumar S, Atri C, Sangha MK, Banga SS (2011) Screening of wild crucifers for resistance to mustard aphid, *Lipaphis erysimi* (Kaltenbach) and attempt at introgression of resistance gene(s) from *Brassica fruticulosa* to *Brassica juncea*. *Euphytica* 179(3):461–470. https://doi.org/10.1007/s10681-011-0351-z

Kumar S, Banga SS (2017) Breeding for aphid resistance in rapeseed mustard. In: Arora R, Sandhu S (eds) *Breeding Insect Resistant Crops for Sustainable Agriculture*. Springer, Singapore, pp 171–199.

Kumari P, Bhat SR (2019) Allohexaploid (H2) (IC0626000; INGR18031), an Indian mustard (*Brassica juncea*) germplasm with heat tolerance. Resistant to *Alternaria brassicae*. *Indian J Plant Genet Resourc* 32:439.

Kumari P, Bisht DS, Bhat SR (2018) Stable, fertile somatic hybrids between *Sinapis alba* and *Brassica juncea* show resistance to *Alternaria brassicae* and heat stress. *Plant Cell Tiss Org Cult* 133(1):77–86. https://doi.org/10.1007/S11240-017-1362-9/METRICS

Latté KP, Appel KE, Lampen A (2011) Health benefits and possible risks of broccoli - An overview. *Food Chem Toxicol* 49:3287–3309.

Lepiniec L, Debeaujon I, Routaboul JM, Baudry A, Pourcel L, Nesi N, Caboche M (2006) Genetics and biochemistry of seed flavonoids. *Annu Rev Plant Biol* 57:405–430.

Li J, Zou X, Zhang L, Cao L, Chen L (2016) Linkage map construction using SSR markers and QTL analyses of stem expansion traits in *Brassica juncea*. *Sci Hortic (Amsterdam)* 209:67–72. https://doi.org/10.1016/j.scienta.2016.06.017

Lionneton E, Aubert G, Ochatt S, Merah O (2004) Genetic analysis of agronomic and quality traits in mustard (*Brassica juncea*). *Theor Appl Genet* 109(4):792–799. https://doi.org/10.1007/s00122-004-1682-0

Lionneton E, Ravera S, Sanchez L, Aubert G, Delourme R, Ochatt S (2002) Development of an AFLP-based linkage map and localization of QTLs for seed fatty acid content in condiment mustard (*Brassica juncea*). *Genome* 45(6):1203–1215. https://doi.org/10.1139/G02-095

Mahmood T, Ekuere U, Yeh F, Good AG, Stringam GR (2003) RFLP linkage analysis and mapping genes controlling the fatty acid profile of *Brassica juncea* using reciprocal DH populations. *Theor Appl Genet* 107(2):283–290. https://doi.org/10.1007/s00122-003-1244-x

Mahmood T, Rahman MH, Stringam GR, Raney JP, Good AG (2005a) Molecular markers for seed colour in *Brassica juncea. Genome* 48(4):755–760. https://doi.org/10.1139/G04-122

Mahmood T, Rahman MH, Stringam GR, Yeh F, Good A (2005b) Molecular markers for yield components in *Brassica juncea* - Do these assist in breeding for high seed yield? *Euphytica* 144(1–2):157–167. https://doi.org/10.1007/s10681-005-5339-0

Mahmood T, Rahman MH, Stringam GR, Yeh F, Good AG (2006) Identification of quantitative trait loci (QTL) for oil and protein contents and their relationships with other seed quality traits in *Brassica juncea. Theor Appl Genet* 113(7):1211–1220. https://doi.org/10.1007/s00122-006-0376-1

Mathur S, Paritosh K, Tandon R, Pental D, Pradhan AK (2022) Comparative analysis of seed transcriptome and coexpression analysis reveal candidate genes for enhancing seed size/weight in *Brassica juncea. Front Genet* 13:814486. https://doi.org/10.3389/fgene.2022.814486

Mei J, Qian L, Disi JO, Yang X, Li Q, Li J, Frauen M, Cai D, Qian W (2011) Identification of resistant sources against *Sclerotinia sclerotiorum* in *Brassica* species with emphasis on *B. oleracea*. *Euphytica* 177(3):393–399. https://doi.org/10.1007/S10681-010-0274-0/METRICS

Mohd Saad NS, Severn-Ellis AA, Pradhan A, Edwards D, Batley J (2021) Genomics armed with diversity leads the way in *Brassica* improvement in a changing global environment. *Front Genet* 12:1–28. https://doi.org/10.3389/fgene.2021.600789

Negi MS, Devic M, Delseny M, Lakshmikumaran M (2000) Identification of AFLP fragments linked to seed coat colour in *Brassica juncea* and conversion to a SCAR marker for rapid selection. *Theor Appl Genet* 101(1-2):146–152. https://doi.org/10.1007/S001220051463/METRICS

Neumann GM, Condron R, Polya GM (1996) Purification and sequencing of napin-like protein small and large chains from *Momordica charantia* and *Ricinus communis* seeds and determination of sites phosphorylated by plant Ca2+-dependent protein kinase. *Biochim Biophys Acta - Protein Struct Mol Enzymol* 1298(2):223–240. https://doi.org/10.1016/S0167-4838(96)00133-1

Olsson G (1960) Species crosses within the genus Brassica: II. artificial *Brassica napus* L. *Hereditas* 46(3–4):351–386. https://doi.org/10.1111/j.1601-5223.1960.tb03092.x Padmaja LK, Agarwal P, Gupta V, Mukhopadhyay A, Sodhi YS, Pental D, Pradhan AK (2014) Natural mutations in two homoeologous *TT8* genes control yellow seed coat trait in allotetraploid *Brassica juncea* (AABB). *Theor Appl Genet* 127(2):339–347. https://doi.org/10.1007/S00122-013-2222-6

Padmaja LK, Arumugam N, Gupta V, Mukhopadhyay A, Sodhi YS, Pental D, Pradhan AK (2005) Mapping and tagging of seed coat colour and the identification of microsatellite markers for marker-assisted manipulation of the trait in *Brassica juncea*. *Theor Appl Genet* 111(1):8–14. https://doi.org/10.1007/s00122-005-1933-8

Palmer JD, Shields CR, Cohen DB, Orton TJ (1983) Chloroplast DNA evolution and the origin of amphidiploid *Brassica* species. *Theor Appl Genet* 65(3):181–189. https://doi.org/10.1007/BF00308062

Panjabi P, Jagannath A, Bisht NC, Lakshmi KL, Sharma S, Gupta V, Pradhan AK, Pental D (2008) Comparative mapping of *Brassica juncea* and *Arabidopsis thaliana* using Intron Polymorphism (IP) markers: Homoeologous relationships, diversification and evolution of the A, B and C *Brassica* genomes. *BMC Genomics* 9(1):1–19. https://doi.org/10.1186/1471-2164-9-113/FIGURES/8

Panjabi-Massand P, Yadava SK, Sharma P, Kaur A, Kumar A, Arumugam N, Sodhi YS, Mukhopadhyay A, Gupta V, Pradhan AK, Pental D (2010) Molecular mapping reveals two independent loci conferring resistance to Albugo *Candida* in the East European germplasm of oilseed mustard *Brassica juncea. Theor Appl Genet* 121(1):137–145. https://doi.org/10.1007/S00122-010-1297-6/METRICS

Paritosh K, Gupta V, Yadava SK, Singh P, Pradhan AK, Pental D (2014) RNA-seq based SNPs for mapping in *Brassica juncea* (AABB): Synteny analysis between the two constituent genomes A (from B. rapa) and B (from B. nigra) shows highly divergent gene block arrangement and unique block fragmentation patterns. *BMC Genomics* 15(1):396. https://doi.org/10.1186/1471-2164-15-396

Paritosh K, Yadava SK, Singh P, Bhayana L, Mukhopadhyay A, Gupta V, Bisht NC, Zhang J, Kudrna DA, Copetti D, Wing RA, Reddy Lachagari VB, Pradhan AK, Pental D (2021) A chromosome-scale assembly of allotetraploid *Brassica juncea* (AABB) elucidates comparative architecture of the A and B genomes. *Plant Biotechnol J* 19(3):602–614. https://doi.org/10.1111/PBI.13492

Patel MK, Chaudhary R, Taak Y, Pardeshi P, Nanjundan J, Vinod KK, Saini N, Vasudev S, Yadava DK (2022) Seed coat colour of Indian mustard [*Brassica juncea* (L.) Czern. and Coss.] is associated with *Bju. TT8* homologs identifiable by targeted functional markers. *Front Plant Sci* 13:1012368. https://doi.org/10.3389/fpls.2022.1012368

Petit RJ, Mousadik AEL, Pons O (1998) Identifying populations for conservation on the basis of genetic markers. *Conserv Biol* 12(4):844–855. https://doi.org/10.1046/j.1523-1739.1998.96489.x

Prabhu K V., Somers DJ, Rakow G, Gugel RK (1998) Molecular markers linked to white rust resistance in mustard *Brassica juncea*. *Theor Appl Genet* 97(5-6):865–870. https://doi.org/10.1007/S001220050966/METRICS

Pradhan AK, Gupta V, Mukhopadhyay A, Arumugam N, Sodhi YS, Pental D (2003) A high-density linkage map in *Brassica juncea* (Indian mustard) using AFLP and RFLP markers. *Theor Appl Genet* 106(4):607–614. https://doi.org/10.1007/S00122-002-1083-1

Pradhan AK, Prakash S, Mukhopadhyay A, Pental D (1992) Phytogeny of *Brassica* and allied genera based on variation in chloroplast and mitochondrial DNA patterns: Molecular and taxonomic classifications are incongruous. *Theor Appl Genet* 85(2–3):331–340. https://doi.org/10.1007/BF00222878

Prakash S, Hinata K (1980) Taxonomy, cytogenetics and origin of crop Brassicas, a review. *Opera Bot* (55):50–57.

Pushpa HD, Yadava DK, Singh N, Vasudev S, Saini N, Muthusamy V, Prabhu KV (2016) Validation of molecular markers linked to low glucosinolate QTLs for marker assisted selection in Indian mustard (*Brassica juncea L. Czern & Coss). Indian J Genet* 76(1):64–68. https://doi.org/10.5958/0975-6906.2016.00009.2

Rabbani MA, Iwabuchi A, Murakami Y, Suzuki T, Takayanagi K (1998) Phenotypic variation and the relationships among mustard (*Brassica juncea* L.) germplasm from Pakistan. *Euphytica* 101(3):357–366. https://doi.org/10.1023/A:1018305201279/METRICS

Rahman M, McVetty PBE (2011) A review of *Brassica* seed color. *Can J Plant Sci* 91(3):437–446. https://doi.org/10.4141/CJPS10124/ASSET/IMAGES/LARGE/CJPS10124F1.JPEG

Ramchiary N, Bisht NC, Gupta V, Mukhopadhyay A, Arumugam N, Sodhi YS, Pental D, Pradhan AK (2007a) QTL analysis reveals context-dependent loci for seed glucosinolate trait in the oilseed *Brassica juncea*: Importance of recurrent selection backcross scheme for the identification of "true" QTL. *Theor Appl Genet* 116(1):77–85. https://doi.org/10.1007/s00122-007-0648-4

Ramchiary N, Padmaja KL, Sharma S, Gupta V, Sodhi YS, Mukhopadhyay A, Arumugam N, Pental D, Pradhan AK (2007b) Mapping of yield influencing QTL in *Brassica juncea*: Implications for breeding of a major oilseed crop of dryland areas. *Theor Appl Genet* 115(6):807–817. https://doi.org/10.1007/s00122-007-0610-5

Rana K, Atri C, Akhatar J, Kaur R, Goyal A, Singh MP, Kumar N, Sharma A, Sandhu PS, Kaur G, Barbetti MJ, Banga SS (2019) Detection of first marker trait associations for resistance against sclerotinia sclerotiorum in *Brassica juncea - Erucastrum cardaminoides* introgression lines. *Front Plant Sci* 10:1015. https://doi.org/10.3389/fpls.2019.01015

Rana K, Atri C, Gupta M, Akhatar J, Sandhu PS, Kumar N, Jaswal R, Barbetti MJ, Banga SS (2017) Mapping resistance responses to Sclerotinia infestation in introgression lines of *Brassica juncea* carrying genomic segments from wild Brassicaceae *B. fruticulosa*. *Sci Rep* 7(1):1–12. https://doi.org/10.1038/s41598-017-05992-9

Ripley VL, Roslinsky V (2005) Identification of an ISSR marker for 2-propenyl glucosinolate content in *Brassica juncea* L. and conversion to a SCAR marker. *Mol Breed* 16:57–66. https://doi.org/10.1007/s11032-005-3572-9

Rout K, Yadav BG, Yadava SK, Mukhopadhyay A, Gupta V, Pental D, Pradhan AK (2018) QTL landscape for oil content in *Brassica juncea*: Analysis in multiple bi-parental populations in high and "0" erucic background. *Front Plant Sci* 871:1448. https://doi.org/10.3389/fpls.2018.01448

Sachan JN, Kolt SJ, Singh B (2000) Inheritance of resistance of white rust (*Albugo candida* race 2) *Brassica juncea*. *Indian Phytopath* 53(2):206–209.

Saini N, Singh N, Kumar A, Vihan N, Yadav S, Vasudev S, Yadava DK (2016) Development and validation of functional CAPS markers for the *FAE* genes in *Brassica juncea* and their use in marker-assisted selection. *Breed Sci* 66(5):831–837. https://doi.org/10.1270/jsbbs.16132

Saini N, Yashpal, Koramutla MK, Singh N, Singh S, Singh R, Yadav S, Bhattacharya R, Vasudev S, Yadava DK (2019) Promoter polymorphism in *FAE1.1* and *FAE1.2* genes associated with erucic acid content in *Brassica juncea*. *Mol Breed* 39(5):75. https://doi.org/10.1007/s11032-019-0971-x

Sandhu SK, Pal L, Kaur J, Bhatia D (2019) Genome wide association studies for yield and its component traits under terminal heat stress in Indian mustard (*Brassica juncea* L.). *Euphytica* 215(11):188. https://doi.org/10.1007/s10681-019-2489-z

Sharma D, Nanjundan J, Singh L, Parmar N, Singh KH, Verma KS, Thakur AK (2022) Genetic diversity and population structure analysis in indian mustard germplasm using phenotypic traits and SSR markers. *Plant Mol Biol Report* 40(3):579–594. https://doi.org/10.1007/s11105-022-01339-5

Sharma M, Mukhopadhyay A, Gupta V, Pental D, Pradhan AK (2016) *BjuB.CYP79F1* regulates synthesis of propyl fraction of aliphatic glucosinolates in oilseed mustard *Brassica juncea*: Functional validation through genetic and transgenic approaches. *PLoS One* 11(2):e0150060. https://doi.org/10.1371/JOURNAL.PONE.0150060

Shockey JM, Fulda MS, Browse JA (2002) Arabidopsis contains nine long-chain acyl-coenzyme A synthetase genes that participate in fatty acid and glycerolipid metabolism. *Plant Physiol* 129(4):1710–1722. https://doi.org/10.1104/pp.003269

Singh KH, Singh L, Parmar N, Kumar S, Nanjundan J, Singh G, Thakur AK (2022) Molecular characterization and genetic diversity analysis in Indian mustard (*Brassica* juncea L. Czern & Coss.) varieties using SSR markers. *PLoS One* 17(8):e0272914. https://doi.org/10.1371/journal.pone.0272914

Singh KP, Kumari P, Rai PK (2021) Current status of the disease-resistant gene(s)/QTLs, and strategies for improvement in *Brassica juncea*. *Front Plant Sci* 12:295.

Singh S, Mohapatra T, Singh R, Hussain Z (2013) Mapping of QTLs for oil content and fatty acid composition in Indian mustard [*Brassica juncea* (L.) Czern. and Coss.]. *J Plant Biochem Biotechnol* 22(1):80–89. https://doi.org/10.1007/s13562-012-0113-6

Somers DJ, Rakow G, Rimmer SR (2002) *Brassica napus* DNA markers linked to white rust resistance in *Brassica juncea*. *Theor Appl Genet* 104(6–7):1121–1124. https://doi.org/10.1007/S00122-001-0812-1/METRICS

Sonderby IE, Burow M, Rowe HC, Kliebenstein DJ, Halkier BA, Sønderby IE, Burow M, Rowe HC, Kliebenstein DJ, Halkier BA (2010) A complex interplay of three R2R3 MYB transcription factors determines the profile of aliphatic glucosinolates in Arabidopsis. *Plant Physiol* 153(1):348–363. https://doi.org/10.1104/pp.109.149286

Song KM, Osborn TC, Williams PH (1988) *Brassica* taxonomy based on nuclear restriction fragment length polymorphisms (RFLPs) -1. Genome evolution of diploid and amphidiploid species. *Theor Appl Genet* 75(5):784–794. https://doi.org/10.1007/BF00265606

Sundaresan V (2005) Control of seed size in plants. *Proc Natl Acad Sci USA* 102:17887–17888.

Svensson J, Palva ET, Welin B (2000) Purification of recombinant *Arabidopsis thaliana* dehydrins by metal ion affinity chromatography. *Protein Expr Purif* 20(2):169–178. https://doi.org/10.1006/prep.2000.1297

Tiwari AS, Petrie GA, Downey RK (1988) Inheritance of resistance to *Albugo candida* race 2 in mustard (*Brassica juncea* (L.) Czern.). *Can J Plant Sci* 68(2):297–300. https://doi.org/10.4141/CJPS88-039

U N (1935) Genome-analysis in *Brassica* with special reference to the experimental formation of *B. napus* and peculiar mode of fertilization. *Jpn J Bot* 7:389–452 Uchimiya H, Wildman SG (1978) Evolutin of fraction I protein in relation to origin of amphidiploid *Brassica* species and other members of the Cruciferae. *J Hered* 69(5):299–303. https://doi.org/10.1093/oxfordjournals.jhered.a108954

Varshney A, Mohapatra T, Sharma RP (2004) Development and validation of CAPS and AFLP markers for white rust resistance gene in *Brassica juncea*. *Theor Appl Genet* 109(1):153–159. https://doi.org/10.1007/S00122-004-1607-Y/METRICS

Vaughan JG, Hemingway JS, Schofield HJ (1963) Contributions to a study of variation in *Brassica juncea* Coss. & Czern. *J Linn Soc London Bot* 58(374):435–447. https://doi.org/10.1111/j.1095-8339.1963.tb00913.x

Vavilov NI (1951) The origin, variation, immunity and breeding of cultivated plants. *Soil Sci* 72(6):482. https://doi.org/10.1097/00010694-195112000-00018

Vera CL, Woods DL (1982) Isolation of independent gene pairs at two loci for seed coat color in *Brassica juncea*. *Can J Plant Sci* 62(1):47–50. https://doi.org/10.4141/CJPS82-006

Vera CL, Woods DL, Downney RK (1979) Inheritance of seed coat color in *Brassica juncea*. *Can J Plant Sci* 59(3):635–637. https://doi.org/10.4141/CJPS79-100

Walker KC, Booth EJ (2001) Agricultural aspects of rape and other *Brassica* products. *Eur J Lipid Sci Technol* 103(7):441–446. https://doi.org/10.1002/1438-9312(200107)103:7

Wanasundara JP (2011) Proteins of Brassicaceae oilseeds and their potential as a plant protein source. *Crit Rev Food Sci Nutr* 51(7):635–677. https://doi.org/10.1080/10408391003749942

Wang N, Shi L, Tian F, Ning H, Wu X, Long Y, Meng J (2010) Assessment of *FAE1* polymorphisms in three *Brassica* species using EcoTILLING and their association with differences in seed erucic acid contents. *BMC Plant Biol* 10:137. https://doi.org/10.1186/1471-2229-10-137

Warwick SI, Black LD (1991) Molecular systematics of *Brassica* and allied genera (Subtribe Brassicinae, Brassiceae) -chloroplast genome and cytodeme congruence. *Theor Appl Genet* 82(1):81–92. https://doi.org/10.1007/BF00231281

Weselake RJ, Taylor DC, Rahman MH, Shah S, Laroche A, McVetty PBE, Harwood JL (2009) Increasing the flow of carbon into seed oil. *Biotechnol Adv* 27:866–878.

Woods DL (1980) The association of yellow seed coat with other characters in mustard *B. juncea*. *Crucif Newsl* 5:23–24.

Xu A, Huang Z, Ma C, Xiao E, Tian G, Zhang X, Tu J, Fu T, Zhang G (2010a) Inheritance of seed color and molecular markers linked to the seed color gene in *Brassica juncea*. *Mol Breed* 25(1):57–65. https://doi.org/10.1007/s11032-009-9308-5

Xu J, Qian X, Wang X, Li R, Cheng X, Yang Y, Fu J, Zhang S, King GJ, Wu J, Liu K (2010b) Construction of an integrated genetic linkage map for the A genome of *Brassica napus* using SSR markers derived from sequenced BACs in *B. rapa*. *BMC Genomics* 11(1):594. https://doi.org/10.1186/1471-2164-11-594

Yadav M, Rana JS (2018) Quantitative analysis of Sinigrin in *Brassica juncea*. *J pharmacogn phytochem* 7(2):948–954.

Yadava SK, Arumugam N, Mukhopadhyay A, Sodhi YS, Gupta V, Pental D, Pradhan AK (2012) QTL mapping of yield-associated traits in *Brassica juncea*: Meta-analysis and epistatic interactions using two different crosses between east European and Indian gene pool lines. *Theor Appl Genet* 125(7):1553–1564. https://doi.org/10.1007/s00122-012-1934-3

Yadava SK, Ramchiary N (2022) Advanced molecular breeding for yield. In: Kole C, Mohapatra T (eds) *The Brassica juncea Genome*. Springer International Publishing, Cham, pp 399–412.

Yan M, Liu Z, Guan C, Chen S, Yuan M, Liu X (2009) Inheritance and molecular markers for the seed coat color in *Brassica juncea*. *Front Agric China* 3(1):1–6. https://doi.org/10.1007/S11703-009-0019-9/METRICS

Yang J, Liu D, Wang X, Ji C, Cheng F, Liu B, Hu Z, Chen S, Pental D, Ju Y, Yao P, Li X, Xie K, Zhang J, Wang J, Liu F, Ma W, Shopan J, Zheng H, Mackenzie SA, Zhang M (2016) The genome sequence of allopolyploid *Brassica juncea* and analysis of differential homoeolog gene expression influencing selection. *Nat Genet* 48(10):1225–1232. https://doi.org/10.1038/ng.3657

Yang J, Wang J, Li Z, Li X, He Z, Zhang L, Sha T, Lyu X, Chen S, Gu Y, Li Z, Hu Z, He H, Bancroft I, Zhang M (2021) Genomic signatures of vegetable and oilseed allopolyploid *Brassica juncea* and genetic loci controlling the accumulation of glucosinolates. *Plant Biotechnol J* 19(12):2619–2628. https://doi.org/10.1111/pbi.13687

Yang Y, Zhu K, Li H, Han S, Meng Q, Khan SU, Fan C, Xie K, Zhou Y (2018) Precise editing of *CLAVATA* genes in *Brassica napus* L. regulates multilocular silique development. *Plant Biotechnol J* 16(7):1322–1335. https://doi.org/10.1111/pbi.12872

Zhang D, Hamauzu Y (2004) Phenolics, ascorbic acid, carotenoids and antioxidant activity of broccoli and their changes during conventional and microwave cooking. *Food Chem* 88(4):503–509. https://doi.org/10.1016/j.foodchem.2004.01.065

Zhao N, Liu C, Meng Y, Hu Z, Zhang M, Yang J (2019) Identification of flowering regulatory genes in allopolyploid *Brassica juncea*. *Hortic Plant J* 5(3):109–119. https://doi.org/10.1016/j.hpj.2019.04.003

11 Prospects of Allele Mining for Improvement of Qualitative and Quantitative Traits in Safflower (*Carthamus tinctorius* L.)

Sarita Pandey, Shailendra Goel, Mulpuri Sujatha, Abhishek Anand, and Chittaranjan Kole

11.1 INTRODUCTION

Carthamus tinctorius L., commonly known as "safflower," has seed oil that contains significantly higher levels of nutritionally required unsaturated fatty acids compared to other oilseed crops (Ambreen et al. 2018; Mani et al. 2020; Jegadeeswaran et al. 2021). It is diploid (2n = 24), a highly self-pollinating dicot with an estimated haploid genome size of 1.4 GB (Ambreen et al. 2015). Safflower is also used for the extraction of dye, medicinal applications, and as a plant factory for pharmaceutical production (Flider 2013; Carlsson et al. 2014; Zhou et al. 2014b). Transgenic safflower has been used as a plant factory to produce critical pharmaceuticals of human interest, such as insulin and apo lipoprotein (Saberianfar and Menassa 2018). Safflower is reported to have anticoagulant, antioxidant, and neuroprotective properties (Wang et al. 2007). The discovery of safflower oil as a rich source of linoleic (polyunsaturated) and oleic (monounsaturated) fatty acids has increased interest in safflower cultivation. Furthermore, its ability to grow in low rainfall and high salinity environments would give it a competitive advantage over other oilseed crops as a climate-resilient crop that can adapt to a wide range of growing conditions (Kaya et al. 2011; Bahrami et al. 2014; Yeilaghi et al. 2015; Ebrahimi et al. 2016; Emongor and Emongor 2023). Nonetheless, safflower has failed to establish itself as a major oilseed crop. The key factors limiting its cultivation are low seed yield and oil content, susceptibility to several biotic stresses, the presence of spines, and a lack of seed dormancy Nimbkar (2008). Genetic improvement of safflower is therefore critical for increasing its acceptability and utility as a globally important oilseed crop (Ambreen et al. 2018). The International Plant Genetic Resources Institute (IPGRI) and the German Agency for Technical Cooperation (GTZ) identified safflower as one of the 25 underutilized crops that should be the focus of future development in the mid-1990s (Jegadeeswaran et al. 2021). This interest in safflower was motivated by its local and regional importance, the potential for global socio-economic and agricultural development, adaptation to areas with limited surface moisture, and the risk of genetic erosion within the crop gene pool (Emongor and Emongor 2023). However, safflower is considered as an "orphan" crop in the genomics revolution (Varshney et al. 2012) and breeding efforts have not been complemented with the latest molecular tools that would otherwise allow for more rapid genetic improvement. However, things have begun to change, and few research groups are developing appropriate molecular tools, techniques, and resources (Pearl and Burke 2014; Wu et al. 2019, 2021).

DOI: 10.1201/9781003385035-11

11.1.1 Importance of the Crop

Originally, safflower was grown for the rich red pigments (carthamine) in its florets, which were employed as a source of dye for a variety of cultural applications. Flower extracts were used as a food additive and are prized for their alleged therapeutic benefits Weiss (2000). In Asian countries, the entire safflower plant is used as medicine – roots, leaves, inflorescences, seeds, in the most diverse forms of use: decoction, tea, coloring, ointments, and encapsulated oil, among others, for various purposes (Zhou et al. 2014b) (Figure 11.1). Carotene, riboflavin, and vitamins B, C, D, and E are abundant in safflower seeds (Kinupp and Lorenzi 2014). Safflower oil, a type of industrial oil, has a wide range of possible applications, including the production of paints, enamels, and soaps (Ambreen et al. 2018; Mani et al. 2020; Jegadeeswaran et al. 2021). According to Jegadeeswaran et al. (2021), the excellent stability and biodegradability of safflower oil, which contains an average of 72% oleic acid, have attracted the plastics, resins, inks, printing pigments, solvents, cosmetics, and lubricants industries, among others, to replace petroleum derivatives. Due to its popularity, the Institutes of China and India developed a sweetened composition that is rich in amino acids, minerals, and vitamins of complex B, C, D, and E (Nimbkar 2017). In China, safflower is predominantly cultivated to obtain inflorescences, which are used in treating diseases in the form of tonic tea (Emongor 2010; Nimbkar 2017). The principal active ingredient in medicine from safflower is the water-soluble yellow coloring dye; however, another ingredient, i.e., the extracts of alcohol, has been used in the preparation of a medication for menstruation issues, cardiovascular disorders, pain and swelling related to trauma, purgative, antithermal, antioxidant, anti-inflammatory, analgesic, and anticonvulsant, among other conditions, as suggested by several clinical and laboratory studies (Asgarpanah and Kazemivash 2013; Zhou et al. 2014b). Safflower inflorescences have ornamental value due to their beauty, rusticity, and adaptability and can be utilized as fresh or dried-cut flowers as they have post-harvest durability of between 15 days and more than two months (Khalil et al. 2013). Safflower is used as supplemental feed for cattle, goats, sheep, pigs, and poultry in the form of grazing forage, silage, and seed cakes (Arantes 2011). Safflower seeds are also used for small animals like guinea pigs, hamsters, chinchillas, and rabbits, as well as for pet birds like parrots, parakeets, and canaries (Ambreen et al. 2018; Mani et al. 2020; Jegadeeswaran et al. 2021).

FIGURE 11.1 Illustrations to depict the different stages of safflower and their uses.

11.1.2 Production, Productivity, and Area under Safflower Cultivation

A record 36,000 tons of oilseeds are expected to be produced nationwide in 2020–21 (Figures 11.2a and 11.3). Nonetheless, despite large changes in area under cultivation, India remained the crop's top average producer from 1994 to 2021 (Figures 11.2b, c, 11.3, and 11.4). The major producers of safflower are Kazakhstan (223,895 tons), the Russian Federation (151,384 tons), the USA (61,310 tons), Mexico (52,553 tons), and India (36,000 tons), accounting for 83% of total production (FAOSTAT 2023) (Figures 11.4–11.6). Asia will share 52% of the production of safflower seed by region in 2021 (Figure 11.5).

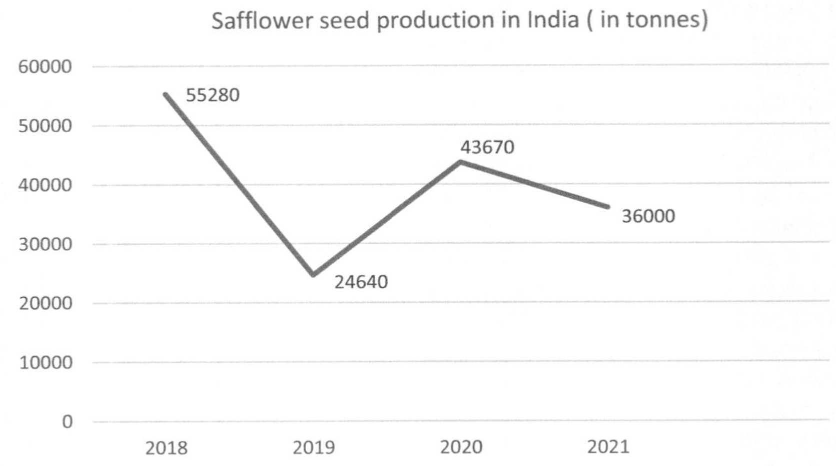

FIGURE 11.2A Safflower seed yield in India between 2018 and 2021 as on 19th March, 2023, (*Source*: https://www.fao.org/faostat/en).

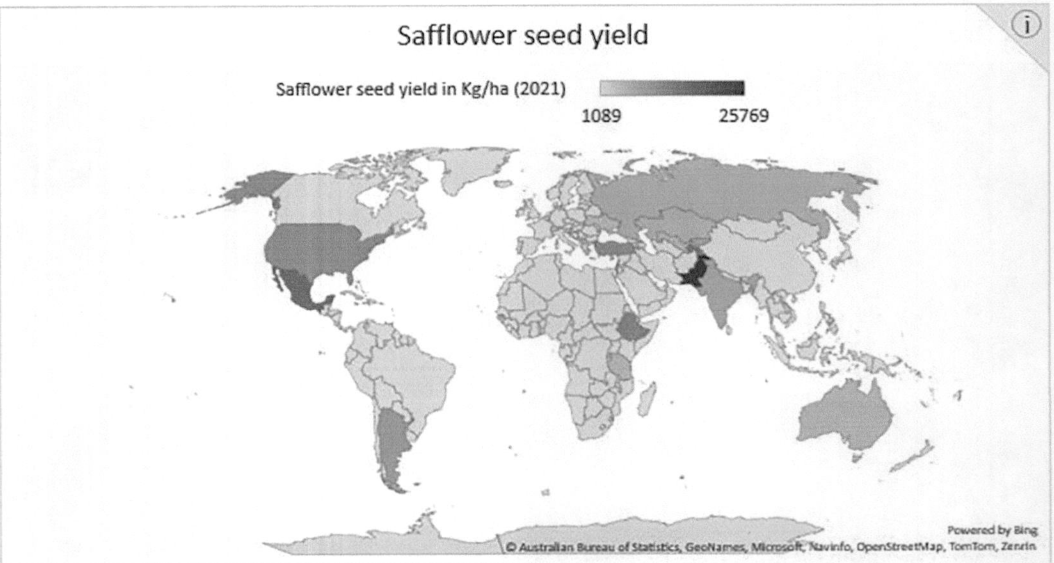

FIGURE 11.2B Safflower seed yield in 2021 as on 4th March 5, 2023 (*Source*: https://www.fao.org/faostat/en).

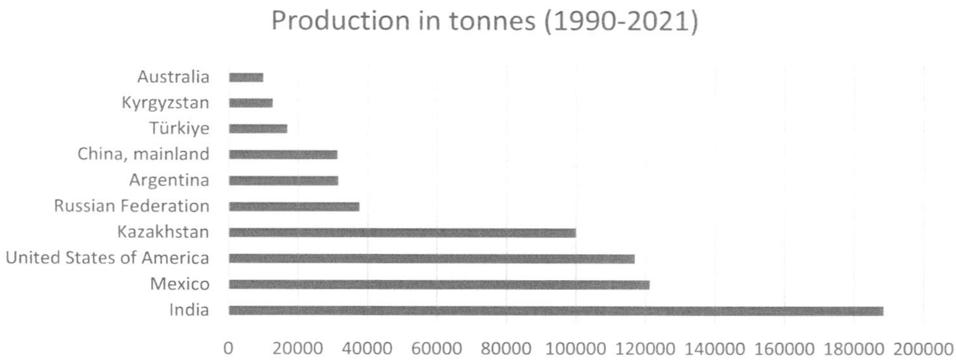

FIGURE 11.2C Average production of safflower seed from 1990 to 2021 (*Source*: https://www.fao.org/faostat/ April 1, 2023).

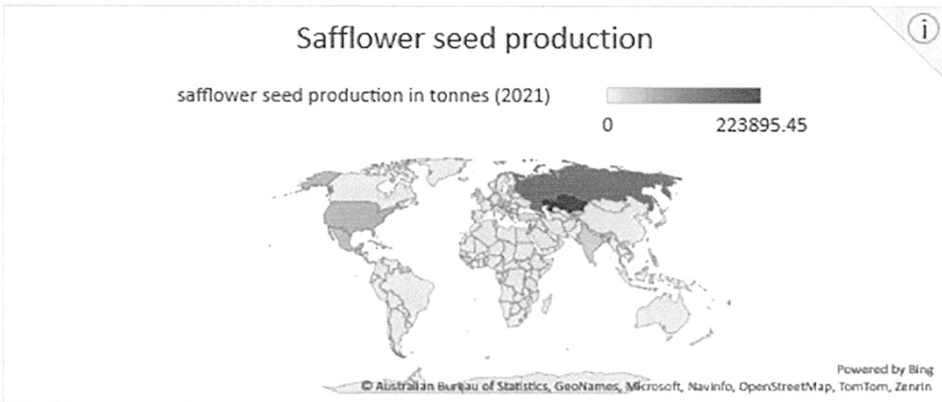

FIGURE 11.3 Safflower seed production in 2021 as on 4th March 5, 2023 (*Source*: https://www.fao.org/faostat/en).

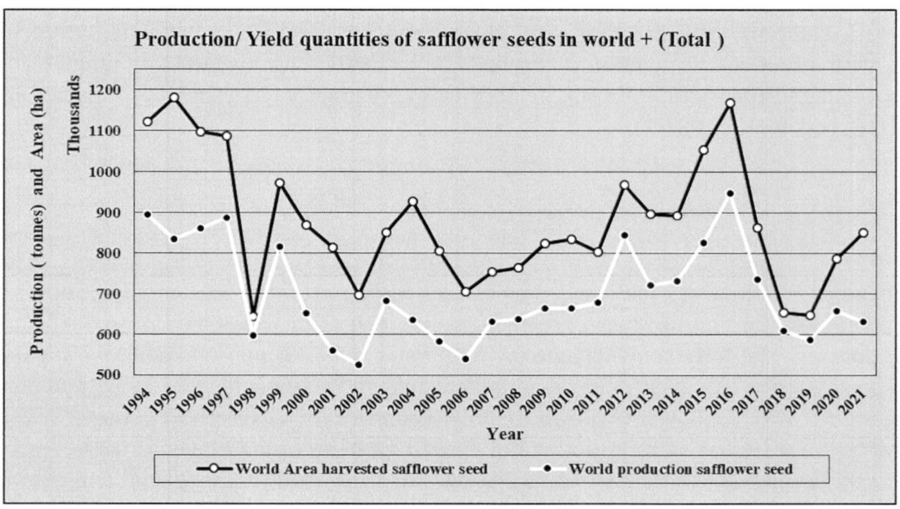

FIGURE 11.4 The Production/Yield quantities of safflower seed in world between 1994 and 2021 as on 18th March 5, 2023 (*Source*: https://www.fao.org/faostat/en).

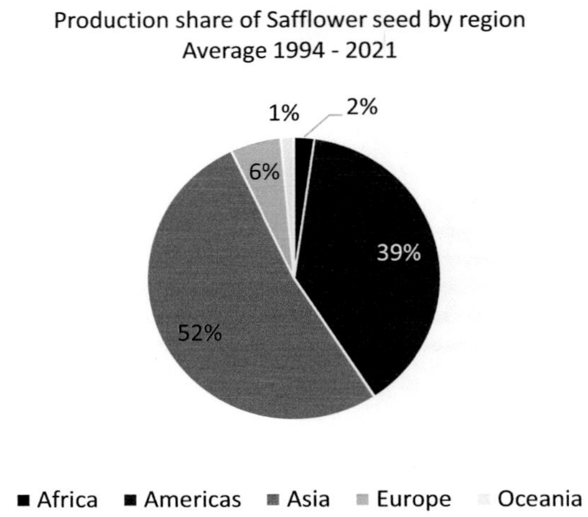

FIGURE 11.5 Production share of safflower seed by region in 2021 as on 4th March 5, 2023 (*Source*: https://www.fao.org/faostat/en).

11.1.3 Botanical and Morphological Characteristics

The name *Carthamus* comes from the Arabic word "quartum or gurtum," which refers to the color of the flower petals. Safflower has secondary and tertiary branching patterns, with each ending in a capitulum. Spiny leaves cover the branches and stem. The seeds are known as achenes, and they are usually white and shiny (Emongor 2010). Safflower is an annual thistle-like plant with many spines on the leaves and bracts that is mostly farmed in dry, hot temperature conditions. Its axillary flowers grow in the leaf axils and can grow to a height of 0.3–2.1 m. Flowers begin in orange and then turn red. It has a robust tap root that can reach depths of 2–3 m, allowing it to survive in dry regions (McPherson et al. 2004; Emongor 2010). The capitulum is 6–8 cm long, obviate or pear-shaped at the base, and has protruding long ribs. The heads with upper leaves are up to 4 by 3 cm long. The *Carthamus* species has a thin fusiform root, and its stem is erect, simple, or branched at the top into stiff, glabrous, whitish-yellow, and glossy branches. The leaves are long, rather soft, and glabrous, with a thorny-serrate margin and tip. The leaf's dimensions typically range from 2.5 to 5.0 cm in width and 10–15 cm in length, but they can vary greatly from species to species and plant to plant (Delshad et al. 2018).

11.1.4 Chemical Composition of Safflower Oil

More than 200 different types of chemicals are found in safflower, including phenols, glycosides, phenylethanoics, coumarins, flavonoids, omega-6, polyunsaturated and monounsaturated fatty acids, antioxidants, carthamin, beta-carotene, lignins, polysaccharides, vitamins B, C, D, and E, and essential oils. The fatty acids include 0.1% myristic acid, 6% palmitic acid, 0.2% palmitoleic acid; 2% stearic acid, 72% oleic acid, 16% linoleic acid, 1.4% linolenic acid, 0.3% arachidic acid, and 2% gadoleic acid. Safflower is grown in many countries for its flavonoid content. These flavonoids have been utilized as medications and/or dyes in a variety of industries. Safflower has yielded more than 60 flavonoids. These flavonoids are classified into two types: special and common flavonoids, both of which are active medicinal substances that are used to treat cardiovascular and cerebrovascular illnesses. The production of flavonoids in safflower has been examined using gene functions (Ren et al. 2022; Sajid et al. 2023).

11.1.5 Safflower Improvement

Plant breeding progress in terms of generating superior and high-yielding agricultural crop varieties has been made possible by the accumulation of favorable alleles from vast plant genetic resources available around the world. Yet, many of these better genes were not successfully deployed because they were lost during domestication. This latent genetic variation in crop plants, wild relatives, and landraces could be profitably utilized for the development of agronomically superior cultivars. Prior research has shown that certain alleles and their combinations have the capacity to drastically modify trait expression when introduced into a genetically diverse background (Lombardi et al. 2014). Hence, there is a need to revisit the large germplasm resources in search of new alleles to further improve the crop varieties for various agronomic attributes. Rapid strides have been made in depositing a large amount of sequence data into GenBank, which is increasing exponentially (Wu et al. 2021). The accumulation of massive sequence and expression data in various genomic databases should accelerate the discovery and annotation of new genes, allowing the construction of allele-specific markers (Wu et al. 2019, 2021). Genes and genome sequences enable direct access to key alleles that confer tolerance to biotic and abiotic stresses, higher nutrient utilization efficiency, yield, and quality. Using contemporary genomic approaches, similar alleles responsible for a specific trait and their variants in different genotypes can be identified. This is referred to as 'dissection of naturally occurring variation at candidate genes/loci' or simply 'allele mining'. Identification of allelic variation in germplasm collections not only provides new germplasm for delivering novel alleles to targeted trait development, but also categorizes germplasm entries for conservation. This chapter elucidates the concepts, approaches, and applications of allele mining, as well as the challenges associated with them, emphasizing the need for more refined 'mining' strategies to accelerate the process of allele discovery for exploitation through molecular breeding in safflower.

11.1.6 Taxonomy

Safflower belongs to the tribe Cardueae (thistles) and subtribe Centaureinae of the family Asteraceae (Compositae) as per the classification given below (Bervillé et al. 2005). The number of species in the genus has ranged from 4 to 25, and subspecies are split into five sections. The sections were based on the five chromosomal groupings n = 10, 11, 12, 22, and 32 found by Ashri and Knowles (1960).

Kingdom: Plantae
Clade: Tracheophytes
Clade: Angiosperms
Clade: Eudicots
Clade: Asterids
Order: Asterales
Family: Asteraceae
Genus: *Carthamus*
Species: *C. tinctoris*

Safflower is a member of the *Carduncellus-Carthamus* complex, and it has not been possible to divide it into distinct sections or genera using morphological or cytological traits alone. Species have been shifted between the genera *Carthamus* and *Carduncellus* depending on the taxonomist and the emphasis on specific physical traits (McPherson et al. 2004). Because of the low amounts of genetic variation that arise when there are obvious phenotypic differences, determining species relationships is considerably more challenging (Mayerhofer et al. 2011). The three sections of the species—the Sections *Carthamus* (n = 12), *Odonthagnathis* (n = 10 or 11), *Atractylis* (n = 22 or 32), and two species with questionable placement—have been further separated based on chromosomal counts. It was formerly believed that *C. tinctorius*' progenitor species were *Carthamus oxyacanthus* frequently referred to as *C. oxyacantha* and *C. persicus* (Ashri and Knowles 1960; Lombardi et al. 2014; Gendall, and Hayden 2017). *C. palaestinus* is the wild ancestor of safflower, totally cross-compatible with cultivated safflower, and originated in the Middle East, close to Israel, according to genetic studies and geographic evidence (Pearl and Burke 2014).

11.1.7 Wild Relatives and Phylogenetics

Safflower was domesticated from its putative wild progenitor, *C. palaestinus* Eig, about 4,500 years ago in the Fertile Crescent region (Mani et al. 2020). Safflower cultivation spread to the Middle East, northern Africa, India, and the Far East once it was domesticated. Safflower was brought to North America in the late 1890s, and commercial production started in the 1950s. Among 18 species of the genus *Carthamus*, only one (*C. tinctorius* L.) is domesticated. Many agronomically advantageous characteristics of the wild species of *Carthamus* are known to exist; however, they were lost during the domestication of safflower (Ambreen 2015). One of the significant wild species, *C. oxyacanthus* ($2n = 2x = 24$), can be found from western Iraq, Iran, and Afghanistan to northern India. Moreover, *C. oxyacanthus* oil grade and content are comparable to those of the genotypes grown in cultivation (Sabzalian et al. 2009). *C. lanatus* ($2n = 2x = 44$) is a weed that grows along roadsides and in open spaces (Knowles and Ashri 1958). It is believed that *C. oxyacanthus* and *C. lanatus* may be the ancestors of cultivated safflower (Chavan 1961). The improvement of novel safflower genotypes with beneficial features, such as resistance to certain diseases, pests, cold, and drought, depends heavily on these two species (Sabzalian et al. 2010; Majidi et al. 2011). In the Mediterranean and Asia, where they are sympatric, hybridization with a number of wild species of *Carthamus* may have contributed to the development of *C. tinctorius* (Ashri and Knowles 1960). Iran, Pakistan, and India all have records of natural hybrids of *C. tinctorius* and *C. oxyacanthus* (Knowles and Ashri 1995). When either species is employed as the recipient female in a cross between *C. tinctorius* and *C. lanatus*, hybrids are produced (Heaton and Klisiewicz 1981).

11.1.8 Cytogenetics of Safflower

According to cytogenetical research conducted by Imrie and Knowles, (1970) and Khidir and Knowles (1970), the self-compatible wild species *C. palaestinus*, which is only found in the deserts of southern Israel and western Iraq, is the progenitor of the weedy species *C. oxyacanthus*, which includes both self-compatible and self-incompatible types, as well as *C. persicus* Wild. These species are sequentially regarded as the parents of the domesticated species *C. tinctorius* L. (Ashri and Knowles 1960). The genome designated as "BB" with chromosomal number $2n = 24$ is represented by the four identified safflower species. They may be crossed in all possible ways to create fertile hybrids (Knowles 1958). There was complete chromosomal pairing in hybrids, which is uncommon when the parental species have different numbers of chromosomes, because these species have similar numbers of chromosomes (Ashri and Knowles 1960). To enhance the cultivation of safflower, weedy and cultivated species should be introduced (Zeven and Zhukovsky 1975). Throughout the regions where safflower is produced, the weedy ancestors of farmed safflower are widespread. Branched, extremely spiny, and an annual weed, *C. oxyacanthus*, is a serious common weed in Pakistan and Northwest India (Ashri and Knowles 1960). Seeds of some of the wild species can be used as fuel for lighting and cooking because they contain about 28% oil (Weiss 1983). Most seeds are tiny, black, and lack pappus. *C. persicus* is additionally a serious weed in Syria and Lebanon. Kumar et al. (1981) carried out detailed karyotype analysis to identify translocation homozygotes, and the result showed the identification of ten translocation homozygotes out of 42 translation heterozygotes. Flourescent *in situ* hybridization (FISH) analysis in safflower using pCt kpnI-1 and pCt kpnI-2 repeated sequences simultaneously discovered that the pCt kpnI-1 sequence was solely localized at sub-telomeric regions on most of the chromosomes, but the pCt kpnI-2 sequence was distributed on two nucleolar and one non-nucleolar chromosome pair (Raina et al. 2005).

11.2 GERMPLASM AND GENETIC DIVERSITY

An essential initial step in protecting and utilizing the resource base of underutilized crops is the establishment and genetic characterization of germplasm collections (Bavec et al. 2018; Thoday-Kennedy et al. 2023). These germplasm collections frequently contain cultivated materials gathered from all around the world and may also contain closely related wild species. Hence, for current plant breeding efforts, these collections constitute a potentially significant source of genetic variation (Singh et al. 2022). Unimproved landraces and wild germplasm could be very beneficial sources of previously unknown genes for crop plants' response to environmental constraints (Thoday-Kennedy et al. 2023). Safflower, a highly self-pollinating plant that is cultivated in a variety of agro-climatic zones worldwide, exhibits geographic patterning (Thoday-Kennedy et al. 2023; Ali et al. 2019). In a recent study by Thoday-Kennedy et al. (2023), field trials comprising a globally diverse collection of 350 safflower genotypes were conducted during 2017–2019 with the use of unmanned aerial vehicle (UAV) multispectral data for estimating vegetation indices. Resources from germplasm serve as a storehouse for trait variability and are crucial for crop improvement. However, their size and varied structure limit their usability and potential applications (Thoday-Kennedy et al. 2023). Frankel (1984) established the idea of "core collection" for the efficient administration and utilization of these resources. A core collection is a representative sample made up of the fewest possible non-redundant individuals, which best captures the range of variability present throughout the whole germplasm collection. The evaluation of morphological features and geographic distribution has been the foundation of prior initiatives to combine safflower genetic resources into core collections (Hassani et al. 2020b). Johnson et al. (1993) evaluated a germplasm collection of 2,042 accessions from about 50 countries and created the first core collection of 210 accessions for safflower. A second core collection of 570 accessions was created by Dwivedi et al. (2005) from a total collection of 5,522 safflower accessions from 38 different countries. Several procedures and sampling techniques were proposed by Balakrishnan and Suresh (2001a, b); they found core collections using geographic origins and 28 descriptors on 3,250 safflower accessions from 32 countries. Safflower germplasm is conserved in 25,179 accessions across 22 gene banks in 15 different nations (http://www.ipgri.cgiar.org/regions/apo/safflower.html). About 7,000 safflower accessions from 40 different nations were classified for a few morphological and agronomic descriptors by the Safflower Project Coordination Center in Solapur, Maharashtra, India (Dwivedi et al. 2005). Because there was a paucity of allelic information, the morphological and geographic information employed to build the original core collections in safflower would have underrepresented the genetic diversity available in the crop. For safflower researchers to build a more useful and robust core collection, efforts must be made to incorporate genetic diversity based on molecular markers. The study by Kumar et al. (2016) discusses the phenotypic assessment of a global collection representing 531 safflower accessions and the construction of a strong core collection in safflower utilizing a maximizing technique. These core collections will be helpful for analyzing the molecular causes of trait variability and for more effectively constructing safflower improvement initiatives.

11.2.1 Conventional Breeding and Selection

Genetic variability for characteristics such as plant height, oil content, test weight of seeds, days to blossom, and bolts were discovered. However, the diversity was limited for traits like floral color and fatty acid profile in the study by Rampure et al. (2014). The findings of Akbari et al. (2020) showed that, in drought-stricken conditions, melatonin-seed priming boosted seed production, harvest index (HI), and oil yield of safflower. The findings of this study have shown that the sowing time and variety may have a significant impact on the safflower seed's oil quality and content. According to Coşge et al. (2007) sowing throughout the winter is advised given the climate in Ankara. In the study by La Bella et al. (2019), hierarchical cluster analysis was used to separate the four main groups of safflower accessions based on their fatty acid profiles. The primary fatty acids in the accessions were linoleic, oleic, and palmitic acids. Alleles at the *st* (stearic acid) and *ol* (oleic acid) loci that impact the fatty acid composition of safflower seed oil separated independently. Increases in stearic acid percentage were accompanied by drops in oleic acid, linoleic acid, or both of those acids' percentages (Ladd and Knowles 1971). Over the years, various varieties with desirable agronomic traits have been developed through conventional breeding. The few notable safflower varieties developed by selection in India and abroad are NARI 57, NARI 96, DSH-1 (India 2017), SF 301 (India 2015), Centennial (USA, 2020), Megha (India, 2016), Oresca (Spain, 2017), Saffire Gold (Australia, 2019), Sironaria (Greece, 2021), Centennial (USA, 2020), Shwetank (India, 2021), Annigeri-2020 (ANG-17–102) (India, 2021), DSAF-1 (ANG-18-02) (India, 2021), IGKV Kusum (RSS 2016-03) (India, 2021), Phule Gold (SSF-15-65) (India, 2021), Phule Kiran (SSF-16-02) (India, 2021), RVSAF 14-1 (India, 2021), RVSAF 18-1 (India, 2021), ISH 402 (India, 2023), PBNS-154 (India, 2023), Rajvijay Safflower 18-3 (RVSAFF18-3) (India, 2023) (source: www.seednet.gov.in) (Table 11.1).

TABLE 11.1
List of Notified Varieties/Hybrids of Safflower with Characters

Variety	Year of Release	Released Centre	Yield Potential (Kg/ha)	Oil Content (%)	Recommended States/ Regions Situations	Salient Features/Traits
SSF-658	2009	MPKV, Solapur	1,430	28	All India	Tolerant to Wilt and Aphids
SSF-708	2012	MPAU, Solapur	1,300–2,200	29	Western Maharashtra	Moderately tolerant to aphid
PKV-Pink (AKS-311)	2013	SRS, Oilseeds, PDKV, Akola	2,200–2,500	33	Vidharbha region of Maharashtra	Moderately tolerant to wilt
NARI-57	2015	AICRP, Nari, Phaltan	1,500	29–30	Maharashtra, KNK, MP, Raj, Punjab, UP, J&K, WB	Resistant to Wilt, Recommended for irrigated Area
NARI-96	2018	AICRP, Nari, Phaltan	2,020	33	Maharashtra, Telangana, AP, MP, Chhattisgarh & Rajasthan.	Recommended for irrigated Area
DSH-185	2018	IIOR Hyderabad	1,600–2,250	28–29	All India	Resistant to Fussarium Wilt
Pride (ISF 1)	2020	IIOR Hyderabad	1,236 (Rainfed) 1864 (Irrigated)	30.5	Telengana, AP, MP, Chhattisgarh, Karnataka & Maharashtra	-
Lakshmi Priya (ISF 764)	2020	IIOR Hyderabad	1,583 (Rainfed) 2,274 (Irrigated)	30.6	Telengana, AP, MP, Chatisgarh, Karnataka & Maharashtra, Gujarat, UP, Bihar, WB, and Odisha	-
			Hybrid			
NARI-H-23	2013	NARI, Phaltan	1,710	31	Maharashtra, Karnataka, MP, Chhattisgarh, Rajasthan, WB	Suited for Irrigated and assured rainfall, Tolerant to Wilt
DSH-185	2013	IIOR, Hyderabad	1,430 (rainfed) -1,740 (irrigated)	28–29	All India (Maharashtra, Karnataka, Andhra Pradesh, Telangana, Madhya Pradesh, Chhattisgarh)	Resistant to Fusarium wilt

Source: Seed Division, GOI Website (seednet.gov.in) & ICAR Institutes.

Many crop species have shown significant improvement due to the use of molecular markers in crop breeding. Strong molecular marker-trait associations are necessary for the successful application of molecular breeding to crops. Safflower enhancement requires a holistic approach to enhance productivity. Yet, the genetics and genomics of safflower are relatively less known, although they have started to change in recent years, and the availability of reference genome sequences in the crop will pave the way for rapid generation of genomic resources, which will help in enhancement through molecular breeding programs.

11.2.2 Morphological Diversity

To establish successful crop improvement techniques, a thorough assessment of the widespread genetic and phenotypic variation among the crop's worldwide germplasm is needed. Initially, morphological criteria and/or geographic distribution were used to build core collections (Upadhyaya and Ortiz 2001; Upadhyaya et al. 2009). Therefore, it has been made possible to create more robust core collections using molecular markers alone (Zhang et al. 2009) or in combination with phenotypic data in a variety of crop species due to the availability of molecular markers and their increased effectiveness in illuminating genetic diversity (Liu et al. 2015).

11.2.3 Phenotyping

Safflower germplasm has shown substantial variability for plant height in earlier investigations (Yeilaghi et al. 2015; Kumar et al. 2016). Early maturing cultivars are preferred, and CartAP analysis revealed several early flowering accessions (Kumar et al. 2016). The CartAP collection's accessions contained between 16 and 203 capitula (Kumar et al. 2016). According to Kumar et al. (2016), there was a direct association between the number of primary branches and the capitula per plant. According to Dajue and Mündel (1996), safflower seed oil content can be greatly increased by lowering the amount of hull present. The germplasm has been assessed to employ key lines for breeding programs and conservation of genetic resources (Mahmood et al. 2010). The degree of genetic diversity and heritability of a character are the major factors that determine the effectiveness of a selection program (Falconer and Mackay 1996). Although a variety of approaches are used to quantify the genetic diversity, morphological features, such as leaves, stems, branches, and flowers, are the main traits used because of their shape, size, and variability. Safflower's flower color and spininess are the main sources of diversity (Bradley and Lang 1999). Amazing genetic variation can be found in safflowers in many parts of the world (Knowles 1989). However, only limited research has been carried out to evaluate the genetic diversity of safflower using agro-morphological features (Jaradat and Shahid 2006). A study by Shinwari et al. (2014a) involved 122 safflower germplasm samples from diverse agro-climatic regions and nations, in which the principal component analysis revealed significant variation. These results indicated that the main characteristics that set the germplasm apart are capsule diameter, capsules/plant, seeds/capsule, days to maturity, plant height, and time of flowering. Five main clusters were identified through cluster analysis (Shinwari et al. 2014a). Ali et al. (2020c) assessed the morpho-agronomic performance of 94 safflower accessions from 26 different nations. Considering the large amount of germplasm available, more efforts are required to characterize the existing safflower germplasm based on morphological parameters.

11.2.4 Inheritance Studies in Safflower

Joglekar and Deshmukh (1956) revealed a recessive monogenic control of white vs. yellow florets in safflower. Classen (1952) discovered that four distinct main genes regulate blossom color. Moreover, some researchers have discovered that several distinct main genes, some of which have epistatic effects, regulate safflower floret color (Narkhede and Deokar 1986). Spiny vs. spineless character was similarly shown to be under monogenic control, with the spiny trait being fully or partially dominant

(Deshmukh 1958). According to Classen (1952), spininess was affected by an undetermined number of genes. Also, it was shown that four genes with epistatic effects are implicated and that in safflower, spines almost outnumber spinelessness (Narkhede and Deokar 1990). Two epistatic loci, each with two alleles, determined the color of the flowers, with an allele ratio of 13:3 in the segregating F_2 population for plants with orange and yellow flowers. Moreover, duplicate dominant and duplicate recessive kinds of epistasis were seen to affect flower color in other crosses, resulting in ratios of 7:9 and 15:1, respectively, for plants with orange and yellow flowers (Pahlavani et al. 2004).

11.2.5 Biochemical Diversity

A set of 116 different accessions of the safflower germplasm were examined through sodium dodecyl sulfate polyacrylamide gel electrophoresis (SDS-PAGE) of seed storage proteins exhibiting wide diversity (Shinwari et al. 2014b). In another study, 97 safflower accessions and lines were examined using agro-morphological characterization and SDS-PAGE analysis of total seed storage proteins, grouping them into heterotic groups and divulging the degree of genetic distance between the groups, disclosing genetic potential for seed yield and related features (Ibrar et al. 2015).

11.2.6 Mutation Breeding

The total tocopherol content and profile of 77 mutants from bulk seed samples of six species were examined. An altered tocopherol profile in a safflower mutant was identified and genetically characterized (Velasco et al. 2005). In another study, four distinct safflower genotypes were optimized for the GR50 dosage of gamma rays based on information about seedling height at 42 days after sowing (Shrivastava and Shrivastava 2012). Kumar and Srivastava (2010) isolated a few notable cytological mutants, such as those that produce chains and multivalent compounds during metaphase and cytoplasmic bridges during anaphase. The results showed that laser-treated sets had a larger percentage of chromosomal abnormalities as compared to gamma-ray-treated sets, indicating that gamma rays might be useful for successfully increasing the genetic diversity of safflower (Kumar and Srivastava 2010). A study was conducted to increase the diversity of isolating high-oleic acid mutants of the Indian variety "Bhima" using induced mutagenesis (Rampure et al. 2017). An ethylmethane sulphonate (EMS) mutagenized M_2 population of 15,403 plants contained three mutants with high oleic acid levels (Rampure et al. 2015, 2017).

11.2.7 Hybrid Breeding

The basic aim of hybridization is to combine the desirable characteristics of two or more kinds into a single cultivar. Additionally, inter- and intra-specific hybridization in safflower breeding, along with molecular characterization, may aid in gene tracking and successful gene introgression in cultivated germplasm (Sujatha et al. 2008). The importance of both additive and non-additive gene actions has been reported in safflower for important traits like seed yield (Mani et al. 2020). Breeding techniques like biparental mating followed by reciprocal recurrent selection and diallel selective mating may be used in safflower to maximize gene action and increase the likelihood of generating desired recombinants (Wang et al. 2020). While seed weight has been shown to be negatively correlated with oil content, seed production per plant, number of capitula per plant, and oil output per plant showed positive associations (Singh et al. 2004). To increase safflower seed yield, oil content, and other desirable qualities, different approaches have been frequently followed, such as the pedigree method, the single-seed descent method, the backcross method, etc. (Mani et al. 2020). Safflower is a good option for exploitation of hybrid vigor in the crop due to the presence of several features of economic value and the presence of both genetic male sterility (GMS) and cytoplasmic male sterility (CMS) systems. The development and release of spiny safflower hybrids DSH-129 and MKH-11, the first non-spiny hybrid safflower NARI-NH-1 in 2001, and the spiny hybrid NARI-H-15 followed the finding of genetic male sterility sources in safflower (Singh et al. 2003). The hybrid DSH-185 was developed through a cross between CMS-A-133 and 1705-p22 by the ICAR-Indian Institute of Oilseeds Research, Hyderabad and released in 2013. These cytoplasmic male sterile lines (A-lines) were developed using oxy-cytoplasm for the conservation of several spiny, non-spiny, and sparsely spiny lines. The wild species *Carthamus oxyacanthus* was identified as the source of male sterile cytoplasm (www.seednet.org/, accessed March 2023) (Table 11.1).

11.3 MOLECULAR MARKERS

The lack of effective genomic resources and inadequate understanding of the genetic diversity within *C. tinctorius* have hindered breeding efforts to improve agro-economic features. Several research teams have realized that the safflower genome needs to be molecularly characterized and that genetic markers must be developed (Khadijeh et al. 2020; Wu et al. 2021). Finding and describing appropriate genetic markers is one of the most crucial steps in developing a molecular breeding system that works well. Initially, development of genomic and molecular resources was limited in safflower as compared to other members of the Asteraceae family, and commonly, simple sequence repeat (SSR), random amplified polymorphic DNA (RAPD), and amplified fragment length polymorphism (AFLP) markers were used for evaluation of landraces, germplasm, and genetic diversity and to assess the outcrossing of safflower with weedy relatives (Sujatha et al. 2008). Genetic diversity in safflower has been analyzed using several molecular markers (Johnson et al. 2007; Amini et al. 2008; Khan et al. 2009; Sehgal et al. 2009; Chapman et al. 2009; Lee et al. 2014; Pearl and Burke 2014; Ambreen et al. 2015; Kumar et al. 2015) (Table 11.2). Several studies on the assessment of genetic diversity are based on SSR-markers in safflower (Naresh et al. 2009; Chapman et al. 2009; Hamdan et al. 2011; Mayerhofer et al. 2011; Derakhshan et al. 2014; Lee et al. 2014; Luo et al. 2015; Ambreen et al. 2015). The genetic diversity in safflower using RAPD markers in a diverse set of accessions was also employed by many investigators (Maali-Amiri et al. 2001; Amini et al. 2008; Rahimi et al. 2011) (Table 11.2).

TABLE 11.2
List of Markers Used for Marker Assisted Selection in Safflower

Molecular Marker	Trait	No. of Makers	Reference
AFLP	Agronomic traits	12	Johnson et al. (2007)
	Diversity analysis	10	Kumar. et al. (2015)
	Drought	30	Poodineh et al. (2021)
ISSR	Seed yield and number of seeds per head.	40	Majidi and Zadhoush (2014)
	Drought tolerance		Mirzahashemi et al. (2015)
	Assessment of genetic diversity	12	Ali et al. (2019)
	Phenotypic traits	13	Houmanat et al. (2021)
	Agronomic traits	24SCoT and 10 ISSR	Rahimi et al. (2011)
iPBS	13 iPBS-retrotransposon markers	13	Ali et al. (2019)
RAPD	Seed trait	8	Hacioglu et al. (2013)
	Agro-morphological (plant height and days to flowering), biochemical traits (oleic and linoleic acid)	8	Khan et al. (2009)
SNPs	Diversity analysis	133	Pearl and Burke (2014)
SSRs	Number of capitula per plant, number of seeds per plant, 1,000-seed weight, and seed yield	10	Hassani et al. (2020a)
	Drought tolerance	145	Mirzahashemi et al. (2015)
	Variation for seed yield-related traits	48	Kiran et al. (2017)
	Assessment of genetic diversity	48	Singh et al. (2022)

(*Continued*)

TABLE 11.2 (*Continued*)
List of Markers Used for Marker Assisted Selection in Safflower

Molecular Marker	Trait	No. of Makers	Reference
SRAP	Number of capitula per plant, number of seeds per plant, 1,000-seed weight, and seed yield	10	Hassani et al. (2020a)
DArTseq markers	Agronomic traits	12,232	Ali et al. (2020a)

Sequence-related amplified polymorphism (SRAP) markers;Single nucleotide polymorphisms (SNPs); Inter-primer binding site (iPBS) retrotransposon markers; Inter simple sequence repeats (ISSR); simple sequence repeats (SSRs); Random amplified polymorphic DNA (RAPD); Amplified fragment length polymorphism (AFLP); Diversity arrays technology (DArT).

Genetic variation among populations of wild safflower, *C. oxyacanthus*, was analyzed by agro-morphological traits and inter-simple sequence repeat (ISSR) markers by several researchers (Sabzalian et al. 2009; Golkar et al. 2011; Ali et al. 2020c). Genetic diversity using SSR markers has been investigated in world safflower germplasm (Ambreen et al. 2018; Mani et al. 2020; Jegadeeswaran et al. 2021), *C. lanatus* populations, and *C. oxyacanthus*. These studies aimed to construct linkage maps and tag agronomic parameters. Ebrahimi et al. (2016) have explored marker-trait associations in safflower through association mapping using AFLP markers. A total of 190 safflower accessions were genotyped using 133 single nucleotide polymorphism (SNP) markers (Pearl and Burke 2014).

The investigations on the enhancement of linkage maps and tagging of agronomic traits applying various markers were initiated in safflower (Mayerhofer et al. 2011; Hamdan et al. 2012; Pearl and Burke 2014; Jegadeeswaran et al. 2021). The initial genetic linkage map in safflower was provided by Mayerhofer et al. (2011). Hamdan et al. (2011) reported mapping of the genes monitoring high oleic acid content in safflower. They constructed a genetic linkage map of the F_2 population of CL-1×CR-9 crosses, comprising 15 linkage groups (LGs) and 116 SSR markers. Pearl and Burke (2014) mapped 61 quantitative trait loci (QTLs) underlying 24 domestication-related traits. Jegadeeswaran et al. (2021) reported QTLs associated with tolerance to aphids in the recombinant inbred line (RIL) population derived from CO-1×EC-523368–2. In a study by Poodineh et al. (2021), 30 QTLs were detected, including 17 major QTLs with additive effects and 13 determined QTLs in the multi-environment. Here, the five QTL positions, including qNc-3– 1, qTSW-3– 4, qHI-4– 3, qNB4– 4, and qOY-4– 4, were altered in multi-environment analysis, whereas three QTLs, qNB-11– 1, qGY-4– 4, and qDTT-5– 1, were found to be stable QTLs.

11.3.1 Association Mapping and Marker Trait Association

Using association analysis, markers associated with each of the eight agronomic characteristics of safflower were found (Ambreen et al. 2018). Lack of known marker-trait associations (MTAs) is a significant barrier to the creation of effective marker-assisted breeding programs in safflower. The appropriateness of a safflower panel (CartAP) consisting of 124 accessions from two core collections was evaluated, and 11 MTAs with strong correlation coefficients were found to be linked with oil content (Ambreen et al. 2018). A genome-wide association study (GWAS) was conducted using a total of 12,232 diversity array technology (DArT)-seq markers to identify the MTAs for important agronomic traits in an international safflower panel derived from 26 different countries around the world (Ali et al. 2020a). GWAS was carried out with a globally diverse safflower collection for six agronomic traits in four field trials; a high number of significant MTAs were identified for seed oil content (OL), of which 18 were shared across sites and 12 were site-specific, indicating the complex genetic nature of this trait (Zhao et al. 2022a).

11.3.2 Genomics-Assisted Breeding

Since the publication of *Arabidopsis thaliana's* genome assembly in 2000—the first for any land plant—a number of plant genomes have been sequenced, assembled, and made publically accessible on GenBank and other genomic data repositories. The genomes of just about 0.16% of the 350,000 living land plants have been sequenced, and these efforts have not been spread equally among clades. In safflower, the first high-quality genome assembly (contig N50 of 21.23 Mb) for the 12 pseudo-chromosomes of safflower using single-molecule real-time sequencing, Hi-C mapping technologies and a genetic linkage map is published (Wu et al. 2021). Bowers et al. (2016) performed low-coverage, whole-genome shotgun sequencing on 96 F_6 RILs of a cross between cultivated safflower and its wild progenitor (*C. palaestinus* Eig), providing a powerful tool for the assembly of a low-cost draft genome of safflower. After sequencing a single, short insert library to 21 depths, the research team created a draft genome assembly of *C. tinctorius* that covered 866 million bp (two-thirds) of the expected 1.35 Gbp genome (Bowers et al. 2016). The safflower cultivar 'Anhui1' was used for *de novo* assembly to create a chromosome-level reference genome. The chloroplast (cp) genome was 152,963 bp in length and was made up of two inverted repeats (25,128 bp) separated by a large single-copy region (84,124 bp) and a small single-copy region (18,583 bp) (Wu et al. 2019).

11.3.3 Functional Genomics

Advances in sequencing technologies, particularly long-read sequencing, have resulted in better assembly of genomes in many crops (Table 11.3). Safflower population genetics have also advanced significantly in the last decade. Many safflower accessions have been genotyped to investigate the breeding footprints, and this has provided valuable insights into the molecular mechanisms that govern safflower growth, development, and response to environmental stresses. Several functional genomics approaches, such as transcriptomics, proteomics, and metabolomics, have been employed to study various aspects of safflower biology.

11.4.1 Transcriptomics

Li et al. (2012) carried out transcriptome analysis to establish a relation between fatty acid production and gene expression at various phases of seed development that shed light on the underlying lipid metabolism and served as a crucial building block for the genetic engineering of safflower

TABLE 11.3
Status of Allele Mining in Oilseed Crops

Crop	Allele/Locus	Trait/Name of the Protein	Reference
Soybean	Gm DGAT	Kennedy pathway	Wang et al. (2006)
Soybean	SKTI	Soybean Kunitz trypsin Inhibitor	Wang et al. (2008)
Brassica napus	Genetic loci,	Inter-subgenomic heterosis	Zou et al. (2010)
Soybean	Satt343_3, Sat_295_6 and Satt598_7	Disease resistant	Nguyen et al. (2012)
Soybean	qSS-02 and qSS-16	Big grain protein	Sun et al. (2021)
Soybean	qGI10–1	Drought stress	Sun et al. (2022)
Soybean	*GmNAC12* gene	drought stress tolerance	Yang et al. (2022)
Soybean	Rag1	Disease resistant	Kim et al. (2010)
Soybean	rbs1, rbs3	Brown stem root resistance	Eathington et al. (1995)
Soybean	Rhg1, Rhg4	SCN resistance	Meksem et al. (2001)
Soybean	Rfs1, Rfs2	SDS resistance	Meksem et al. (2001)
Groundnut	SSRs	Drought tolerance	Ravi et al. (2011)
Ground nut	SSRs	Drought tolerance	Gautami et al. (2012)

types. A *de novo* transcriptome made up of 153,769 unigenes was created using Solexa-based deep sequencing on safflower seed, leaf, and petal samples. The *de novo* transcriptome analysis of the distinctive transcripts gave candidate gene resources for researching genes involved in flavonoid production and metabolism in safflower (Li et al. 2012). In another study, large datasets were created by the 454 pyrosequencing of floral RNA at various developmental phases to explore genes involved in safflower yellow biosynthesis. A total of 51,591 unigenes with a mean length of 679 bp and a maximum length of 5,109 bp were created (Liu et al. 2015). The first report on a comprehensive and systematic analysis of the safflower MADS-box transcription factor family was conducted by Wang et al. (2023), where a total of 77 members of the MADS-box transcription factor family were identified in the safflower genome. Kouhi et al. (2020a,b) identified and reported 107 miRNAs in safflower, where they first compared all safflower whole genome sequences with plant miRNAs in the mirBase database to identify potential miRNA homologs in safflower. In a later study by Kouhi et al. (2020), the expression profiles of seven conserved miRNAs related to drought, salinity, heat, and Cd stress in the leaf and root organs using qRT-PCR were studied.

11.4.2 Metabolomics

A total of 272 CtMYBs were discovered and divided into 12 subgroups, and these transcription factors were shown to play a variety of roles during plant growth, the regulation of secondary metabolites, and various abiotic stress responses (Hong et al. 2022). In a study by Chen et al. (2018), the full-length flavonoid genes were identified, along with their evolutionary relationships and patterns of expression. A total of 212 flavonoid metabolites, including carthamin, anthocyanins, and hydroxysafflor yellow A, were found, which demonstrated significant differences (Ren et al. 2022). A total of 788 possible differential metabolites were identified by comparing the entire metabolomics database between the transgenic and control groups using PCA and PLS-DA, and most of them showed up-regulated patterns. In addition, transgenic safflower plants accumulated certain identified secondary metabolites, such as rutin, kaempferol-3-O-rutinoside, and dihydrokaempferol (Guo et al. 2019). In another study by Waki et al. (2021), the team identified the genes coding for the enzyme responsible for the formation of carthamin from precarthamin PC, termed 'carthamin synthase' (CarS), using enzyme purification and transcriptome analysis. Kim et al. (2020) investigated the metabolite accumulation in the florets of three wild species (C. *lanatus*, C. *palaestinus*, and C. *turkestanicus*) and one cultivated species (C. *tinctorius*) of safflower at three developmental stages.

11.4.3 Proteomics

A total of 72 proteins involved in photosynthesis, as well as the metabolism of carbohydrates, proteins, energy, and defense, were differentially accumulated (Çulha et al. 2021). Oil bodies and membrane lipids were likely degraded by the enzymes involved in glycerolipid metabolism and fatty acid degradation, which ultimately destroy the structure and reduce seed vigor during natural seed aging (Chen et al. 2022). The stress-responsive proteins that were discovered may help safflower acquire salt tolerance and maintain its productivity in saline environments (Shaki et al. 2020a,b). Oleic acid desaturation in *Synechococcus sp.* strain PCC7942 was confirmed following the cloning of the safflower plastidial oleashakite desaturase gene, CtFAD6. The findings demonstrated that oleate desaturation was tissue-specifically controlled by temperature at the transcriptional level (Li et al. 2023).

11.4.4 Genetic Transformation and Gene Editing

By introducing certain genes from safflower and other sources, safflower has recently been genetically enhanced for agronomic, nutritional, and other properties. The creation of a transformation technique, gene(s) of interest, and *in vitro* propagation strategy are requirements for a successful transformation. To successfully transfer genes in safflower, the *Agrobacterium*-mediated transformation technique has so far proved successful in enhancing γ-linolenic acid (GLA), α-linolenic acid (ALA), oleic acid, bioactive peptides, and bioactive flavonoids, as well as resistance to the fungus *Alternaria carthami* (Belide et al. 2011). Genetically modified (GM) safflower is currently being grown commercially in Australia, Canada, and the United States (Nitnaware et al. 2021). The first comprehensive approach to genetic transformation of safflower was described by Belide et al. (2011), with a transformation efficiency of 4.8% for high oleic acid content and 3.1% for high linoleic acid content. A transgenic safflower plant was also developed by the pollen-tube method by Guo et al. (2017).

Genome editing technology enabled the researchers to obtain safflower with high oleic acid. Two guide RNA sequences were designed to target fatty acid desaturase 2 (*FAD2-1*) genes, which were targeted within the coding region and at a distance of 640 bp from each other. The guide sequences along with the codon-optimized Cas9 gene were cloned in the T-DNA region of the *Agrobacterium* construct and transferred to the safflower by the *in-planta* transformation method. The resulting seeds were cultivated and showed that the amount of oleic acid in the seeds of one of the lines reached 53.14% on average (Neycee et al. 2022).

11.5 PROSPECTS OF ALLELE MINING IN SAFFLOWER

11.5.1 Need for Allele Mining in Safflower

Safflower breeding has received less research attention than other major crops due to its lower prominence in agricultural systems. However, the available germplasm's diverse genetic background can be an important resource for safflower germplasm improvement and a valuable training set for allele mining (Zhao et al. 2022a). Allele mining is a promising strategy for analyzing naturally occurring allelic variation at potential candidate genes affecting important agronomic parameters like seed production in safflowers that could be employed in crop development projects. Several international agricultural research institutes are working on allele mining initiatives in safflower (Zhao et al. 2022a; Pearl and Burke 2014; Ali et al. 2020b). To accelerate the discovery and application of alleles in safflower molecular breeding, more advanced "mining" techniques are required (Zhao et al. 2022a).

11.5.2 Allele Mining

Initially, allele mining studies concentrated solely on identifying SNPs/InDels (insertion/deletions) in coding sequences or exons of the gene because these variations were predicted to alter the encoded protein structure and/or function. Mining for sequence variation, particularly in regulatory regions of genes, has gained importance in the context of gene expression in recent years. In order to account for the majority of the functional variations of importance in the genes, "true" allele mining should additionally analyze the non-coding and regulatory portions of the candidate genes in addition to the sequence differences in the coding regions of the key genes for agronomical attributes (Bowers et al. 2016). Till date, the safflower crop suffers from a lack of well-documented allele mining reports, except for a few sequence-based allele mining reports (Zhao et al. 2022a; Pearl and Burke 2014; Ali et al. 2020). The status of allele mining in different oilseed crops is given in Table 11.3. Through the subtractive genomic library and compositae data base, 285 nucleotide sequences and 41,000 expressed sequence tags (ESTs) have been reported in safflower as of right now. However, development in biotechnology and functional genomics for safflower may help to develop improved varieties (Mani et al. 2020).

11.5.3 Promoter Mining

The promoters of genes have been selected as a potential target for allele mining, also known as "promoter mining" (Figure 11.6). This type of study of the promoter sequences also reveals the conservation of regulatory motifs that are unique to distinct types of differentiated cells and tissues. To recognize and classify cis-regulatory elements and to comprehend their function in gene regulation, methods like *in vitro* and *in silico* promoter mining must be incorporated into regular allele mining programs. This is because the number of global gene expression experiments to identify sets of genes with similar expression patterns is rapidly increasing (Figures 11.7 and 11.8).

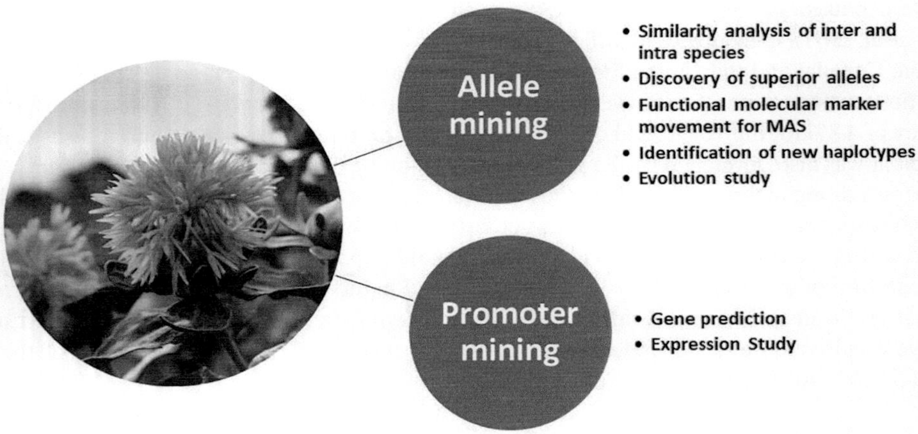

FIGURE 11.6 Roadmap of application of allele mining in safflower.

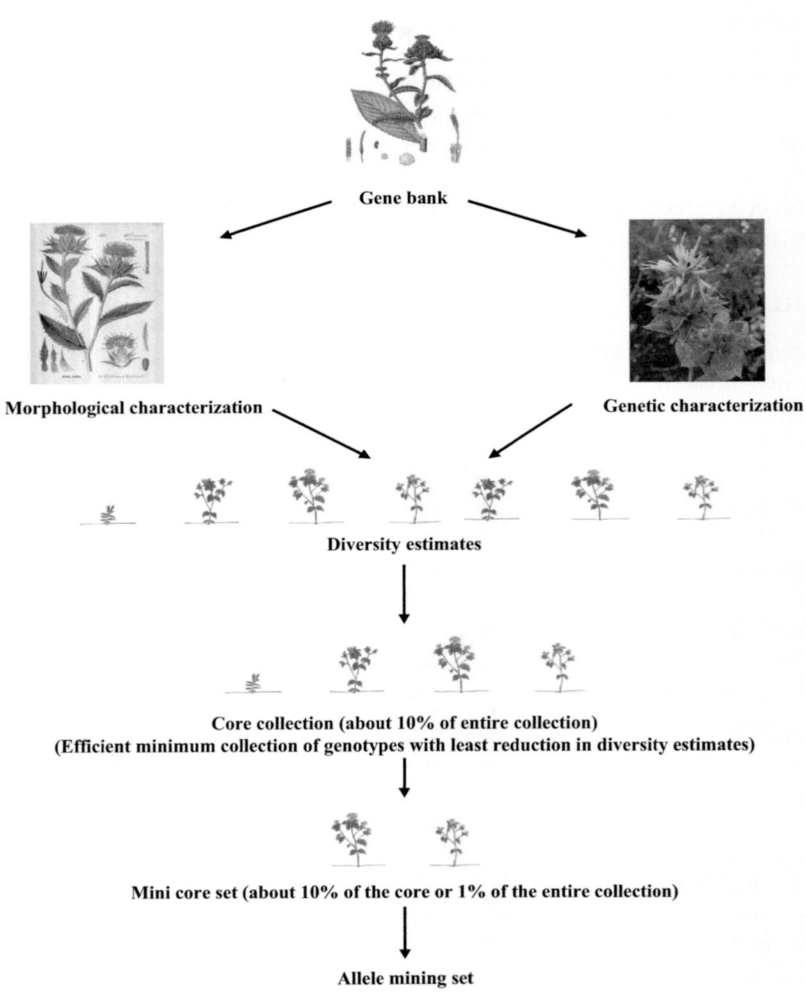

FIGURE 11.7 Potential strategy for creating allele mining set in safflower.

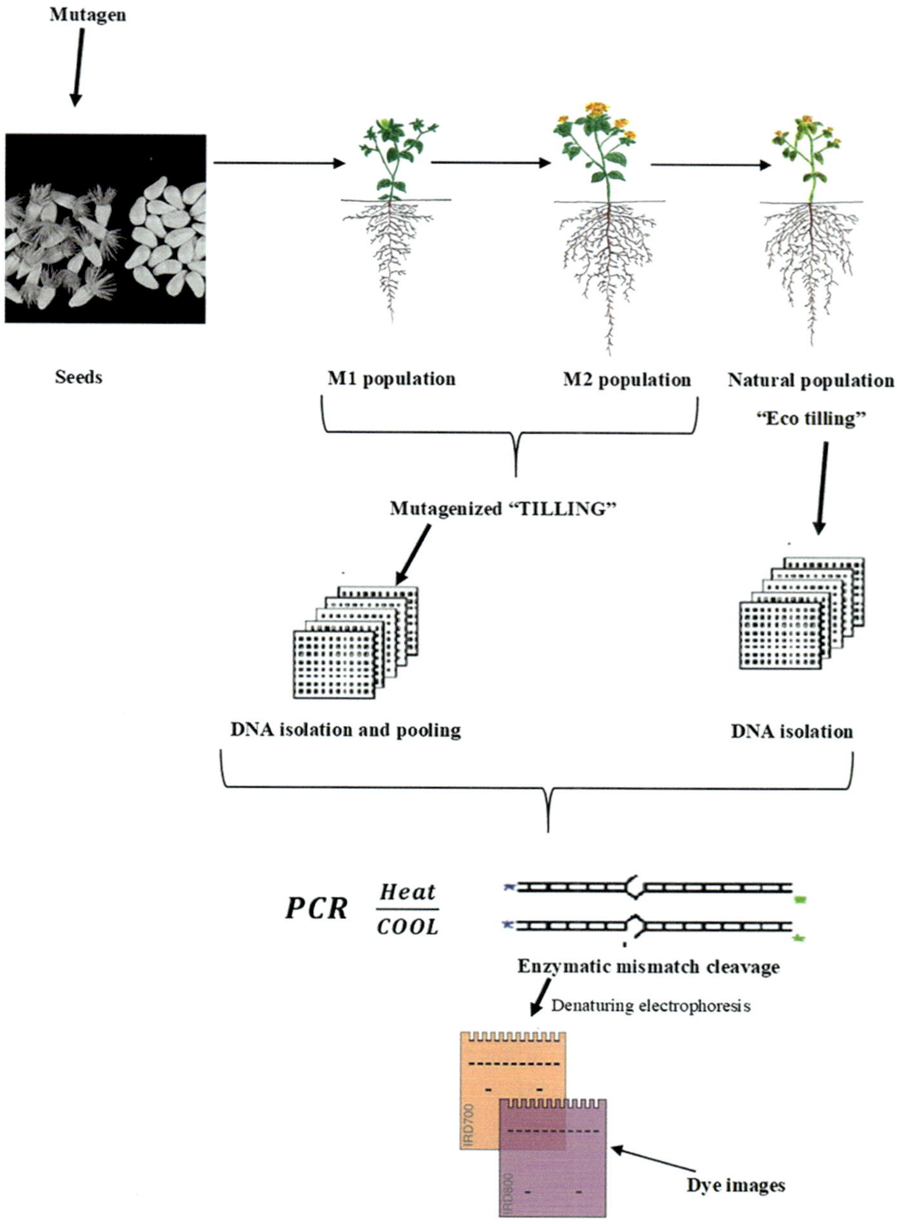

FIGURE 11.8 Outline of the basic steps for typical TILLING and Eco-TILLING assays.

11.5.4 Approaches for Allele Mining

Sequence polymorphisms for a specific gene can be found in naturally occurring populations using one of two main methods. These are (i) the Eco-TILLING approach, a modified version of TILLING (targeting induced local lesions in genomes) and (ii) allele mining based on sequencing.

11.5.4.1 Eco-TILLING

One of the widely applicable plant genetics methods known as Eco-TILLING (Targeting Induced Local Lesions IN Genomes) is used to identify mutations in specific genes of interest. In order to identify natural genetic variation in a population of plants, this high-throughput technique combines conventional mutagenesis with DNA sequencing technologies (Malik 2023). The technique entails assembling a collection of mutant plants, each with a different genetic modification in the desired gene. Then, using DNA sequencing technology, the library is checked for plants with mutations that affect the target gene's function. Eco-TILLING is an effective technique for genetic research as it enables us to find and select plants that have desired characteristics, such as enhanced yield or disease resistance, without the need for genetic modification (Malik 2023).

11.5.4.2 Sequencing-Based Allele Mining

In this method, alleles from various genotypes are amplified using PCR, and then DNA sequencing is used to identify nucleotide variation. To analyze individuals for haplotype structure and diversity and infer genetic association studies in safflower, sequencing-based allele mining might be helpful. Sequencing-based allele mining, in contrast to Eco-TILLING, involves sequencing costs rather than specialized equipment or time-consuming methods. A low-coverage, whole-genome shotgun sequencing on 96 F_6 RILs of a cross between safflower and its wild progenitor *C. palaestinus Eig* produced a draft genome assembly of *C. tinctorius* covering 866 million bp after sequencing a single, short insert library to ~21 × depth (Bowers et al. 2016). Hassani et al. (2020b), reported the first application of DArTseq technology: an efficient genotyping-by-sequencing analysis of the genetic diversity and population structure of 89 safflower accessions from worldwide origin, with a total of 19,639 DArTseq markers, 10,130 SilicoDArTs, and 9,509 SNPs generated through DArTseq genotyping.

11.5.5 Applications of Allele Mining in Safflower

Genetic diversity analyses have typically been performed using several types of sparse molecular markers (Zhao et al. 2022a). More recently, a high-density genetic linkage map was created from a bi-parental cross using lowcoverage whole-genome shotgun sequencing to discover millions of SNPs (Bowers et al. 2016; Zhao et al. 2022b). Such reports provide a valuable genomic resource to support more detailed allele mining. While safflowers have been grown worldwide, breeding efforts for crop improvement have typically been modest. Allele mining can be utilized efficiently to uncover superior alleles by 'mining' the gene of interest from various genetic resources. It can also shed light on the molecular basis of unique phenotypic variations and find nucleotide sequence changes linked to superior alleles in this orphan crop.

REFERENCES

Akbari GA, Heshmati S, Soltani E, Amini Dehaghi M (2020) Influence of seed priming on seed yield, oil content and fatty acid composition of safflower (*Carthamus tinctorius* L.) grown under water deficit. *Int J Plant Product* 14: 245–258.

Ali F, Nadeem MA, Barut M, Habyarimana E, Chaudhary HJ et al (2020a) Genetic diversity, population structure and marker-trait association for 100-seed weight in international safflower panel using SilicoDArT marker information. *Plants* 9(5): 652.

Ali F, Nadeem MA, Habyarimana E, Yılmaz A, Nawaz MA et al (2020b) Molecular characterization of genetic diversity and similarity centers of safflower accessions with ISSR markers. *Braz J Bot* 43: 109–121.

Ali F, Yilmaz A, Chaudhary HJ, Nadeem MA, Rabbani MA et al (2020c) Investigation of morphoagronomic performance and selection indices in the international safflower panel for breeding perspectives *Turk J Agric For* 44(2): 103–120.

Ali F, Yılmaz A, Nadeem MA, Habyarimana E, Subaşı I et al (2019) Mobile genomic element diversity in world collection of safflower (*Carthamus tinctorius* L.) panel using iPBS-retrotransposon markers. *PLoS One* 14(2): 0211985.

Ambreen H, Kumar S, Kumar A, Agarwal M, Jagannath A et al (2018) Association mapping for important agronomic traits in safflower (*Carthamus tinctorius* L.) core collection using microsatellite markers. *Front Plant Sci* 9: 402.

Ambreen H, Kumar S, Variath MT, Joshi G, Bali S et al (2015) Development of genomic microsatellite markers in *Carthamus tinctorius* L. (safflower) using next generation sequencing and assessment of their cross-species transferability and utility for diversity analysis. *PLoS One* 10(8): 0135443.

Amini F, Saeidi G, Arzani A (2008) Study of genetic diversity in safflower genotypes using agro-morphological traits and RAPD markers. *Euphytica* 163: 21–30.

Arantes AM, (2011) Cártamo (Carthamus tinctorium L.) produção de biomassa, grãos, óleo e avaliação nutritiva da silagem. Alcides Meneghelli Arantes. Dissertação. Nova Odessa-SP.

Asgarpanah J, Kazemivash N (2013) Phytochemistry, pharmacology and medicinal properties of (*Carthamus tinctorius* L.) *Chin J Integr Med* 19: 153–159.

Ashri A, Knowles PF (1960) Cytogenetics of safflower (*Carthamus tinctorius* L.) species and their hybrids 1. *Agron J* 52(1): 11–17.

Bahrami F, Arzani A, Karimi V (2014) Evaluation of yield-based drought tolerance indices for screening safflower genotypes. *Agron J* 106: 1219–1224.

Balakrishnan R, Suresh KK (2001a) Strategies for developing core collections of safflower (*Carthamus tinctorius* L.) germplasm - Part II. Using an information measure for obtaining a core sample with predetermined diversity levels for several descriptors simultaneously *Indian J Plant Genet Resour* 14: 32–42.

Balakrishnan R, Suresh KK (2001b) Strategies for developing core collections of safflower (*Carthamus tinctorius* L.) germplasm - Part III. Obtaining diversity groups based on an information. *Indian J Plant Genet Resour* 14: 342–349.

Belide S, Hac L, Singh, SP, Green AG, Wood CC (2011) Agrobacterium-mediated transformation of safflower and the efficient recovery of transgenic plants via grafting. *Plant Methods*, 7: 1–13.

Bervillé A, Muller MH, Poinso B, Serieys H (2005) Ferality-risks of gene flow between sunflower and other Helianthus species. *Crop Ferality and Volunteerism*, 209–230.

Bowers JE, Pearl SA, Burke JM (2016) Genetic mapping of millions of SNPs in safflower (*Carthamus tinctorius* L.) via whole-genome resequencing. *G3 Genes Genomes Genet* 6(7): 2203–2211.

Bradley MM, Lang PJ (1999) Affective norms for English words (ANEW): Instruction manual and affective ratings. Technical report C-1, the center for research in psychophysiology, University of Florida. 30(1): 25–36.

Carlsson AS, Zhu LH, Andersson M, Hofvander P (2014) Platform crops amenable to genetic engineering-a requirement for successful production of bio-industrial oils through genetic engineering. *Biocatal Agric Biotechnol* 3(1): 58–64.

Chapman MA, Hvala J, Strever J, Matvienko M, Kozik A et al (2009) Development, polymorphism, and cross-taxon utility of EST-SSR markers from safflower (*Carthamus tinctorius* L.). *Theor Appl Genet* 120(1): 85–91.

Chavan VM (1961) *Niger and Safflower*. Indian Central Oilseeds Committee, Hyderabad, India. pp. 57–150.

Chen C, Wang R, Dong S, Wang J, Ren CX et al (2022) Integrated proteome and lipidome analysis of naturally aged safflower seeds varying in vitality. *Plant Biol* 24(2): 266–277.

Chen J, Tang X, Ren C, Wei B, Wu Y et al (2018) Full-length transcriptome sequences and the identification of putative genes for flavonoid biosynthesis in safflower. *BMC Genom* 19(1): 1–13.

Classen CE (1952) Inheritance of sterility, flower color, spinlessness, attached pappus and rust resistance in safflower (*Carthamus tinctorius* L.) *Agron J* 42: 381.

Coşge B, Gürbüz b, Kiralan M (2007) Oil content and fatty acid composition of some safflower (*Carthamus tinctorius* L.) varieties sown in spring and winter. *Int J Nat Eng Sci* 1(3): 11–15.

Çulha EŞ, Eyidoğan F, Ekmekçi Y (2021) Comparative physiological and proteomic analysis of cultivated and wild safflower response to drought stress and re-watering. *Physiol Mol Biol Plants* 27(2): 281–295.

Dajue L, Mündel HH (1996) Safflower, *Carthamus tinctorius* L. Bioversity International, Rome, Italy.

Delshad E, Yousefi M, Sasannezhad P, Rakhshandeh H, Ayati Z (2018) Medical uses of *Carthamus tinctorius* L. (Safflower): A comprehensive review from traditional medicine to modern medicine. *Elect Physician* 2018 10(1): 6672–6681.

Derakhshan E, Majidi MM, Sharafi Y, Mirlohi A (2014) Discrimination and genetic diversity of cultivated and wild safflowers (*Carthamus spp.*) using EST-microsatellites markers. *Biochem Syst Ecol* 54: 130–136.

Deshmukh PL (1958) Steroid therapy in apoplexy. *Bri Med J* 1(5077): 1001.

Dwivedi SL, Upadhyaya HD, Hegde DM (2005) Development of core collection using geographic information and morphological descriptors in safflower (*Carthamus tinctorius* L.) germplasm. *Genet Resour Crop E* 52(1): 821–830.

Eathington SR, Nickell CD, Gray LE (1995) Inheritance of brown stem rot resistance in soybean cultivar BSR 101. *J Heredity*, 86(1): 55–60.

Ebrahimi F, Majidi MM, Arzani A, Mohammadi-Nejad G (2016) Oil and seed yield stability in a worldwide collection of safflower under arid environments of Iran. *Euphytica* 212(1): 131–144.

Emongor VE (2010) Safflower (*Carthamus tinctorius L.)* the underutilized and neglected crop: A review. *Asian J Plant Sci* 9(6): 299–306.

Emongor VE, Emongor RA (2023) Safflower (*Carthamus tinctorius* L.). In: Muhammad F (ed) Neglected *and* Underutilized Crops, Kadambot HMS Academic Press, Cambridge, Massachusetts, pp. 683–731. https://doi.org/10.1016/B978-0-323-90537-4.00024-7.

Falconer DS, Mackay TFC (1996) Introduction to Quantitative Genetics, Edn 4. Longmans Green, Harlow, Essex, UK.

Flider FJ (2013) Development and commercialization of GLA safflower oil. *Lipid Technol* 25: 227–229.

Food and Agriculture Organization of the United Nations (2023) FAOSTAT Statistical Database FAO, Rome.

Frankel EN (1984) Lipid oxidation: Mechanisms, products and biological significance. *J Amer Oil Chem Soc* 61(12): 1908–1917.

Gautami B, Fonćeka D, Pandey MK, Moretzsohn MC, Sujay V et al (2012) An international reference consensus genetic map with 897 marker loci based on 11 mapping populations for tetraploid groundnut (*Arachis hypogaea* L.). *PLoS One* 7(7): 41213.

Gendall AR, Hayden MJ (2017) Genetic diversity, population structure and ancestral origin of Australian wheat. *Front Plant Sci* 8(1): 2115.

Golkar P, Arzani A, Rezaei AM (2011) Inheritance of flower colour and spinelessness in safflower (*Carthamus tinctorius* L.). *J Genet* 89(1): 259–262.

Guo D, Gao Y, Liu F, He B, Jia X et al (2019) Integrating molecular characterization and metabolites profile revealed CtCHI1's significant role in *Carthamus tinctorius* L. *BMC Plant Biol* 19(1): 1–13.

Guo D, Xue Y, Li D, He B, Jia X et al. (2017) Overexpression of CtCHS1 increases accumulation of quinochalcone in safflower. *Front Plant Sci* 8:1409

Hacioglu BT, Yaman HÜ, M. E. Y. R. A, Arslan, Y, Katar D (2013) Investigation of molecular diversity of Asian safflower (*Carthamus tinctorius* L.) accessions by RAPD markers for using in hybridization programme. *Res Crops* 14(1): 169–174.

Hamdan YAS, García-Moreno MJ, Redondo-Nevado J, Velasco L, Pérez-Vich B (2011) Development and characterization of genomic microsatellite markers in safflower (*Carthamus tinctorius* L.) *Plant Breed* 130(2): 237–241.

Hamdan YA, García-Moreno MJ, Fernández-Martínez JM, Velasco L, Perez-Vich B (2012) Mapping of major and modifying genes for high oleic acid content in safflower. *Molecular Breeding* 30: 1279–1293.

Hassani SM, Talebi R, Pourdad SS, Naji AM., Fayaz F (2020a) Morphological description, genetic diversity and population structure of safflower (*Carthamus tinctorius* L.) mini core collection using SRAP and SSR markers. *Biotechnol Biotechnol Equip* 34(1): 1043–1055.

Hassani SMR, Talebi R, Pourdad SS, Naji AM, Fayaz F (2020b) In-depth genome diversity, population structure and linkage disequilibrium analysis of worldwide diverse safflower (*Carthamus tinctorius* L.) accessions using NGS data generated by DArTseq technology. *Mol Biol Rep* 47(3): 2123–2135.

Heaton TC, Klisiewicz JM (1981) A disease-resistant safflower alloploid from *Carthamus tinctorius* L. to C. lanatus L. *Can J Plant Sci* 61(1): 219–224.

Hong Y, Zhang J, Lv Y, Yao, N, amp, Liu X (2022) The safflower bHLH transcription factor CtbHLH41 negatively regulates SA-induced leaf senescence through interaction with CtCP1. *Environ Exp Bot* 199(1): 104883.

Houmanat K, Douaik A, Charafi J, Hssaini L, El Fechtali M et al (2021) Appropriate statistical methods for analysis of safflower genetic diversity using agglomerative hierarchical cluster analysis through combination of phenotypic traits and molecular markers. *Crop Sci* 61(6): 4164–4180.

Ibrar A, Ahmed W, Saeed A (2015) Synthetic approaches, functionalization and therapeutic potential of quinazoline and quinazolinone skeletons: The advances continue. *Eur J Med Chem* 90(1): 124–169.

Imrie BC, Knowles PF (1970) Inheritance studies in interspecific hybrids between *Carthamus flavescens* and C. tinctorius L. *Crop Sci*, 10(4): 349–352.

Jaradat AA, Shahid M (2006) Patterns of phenotypic variation in a germplasm collection of *Carthamus tinctorius* L. from the Middle East. *Genet Resour Crop E* 53(1): 225–244.

Jegadeeswaran M, Kadirvel P, Srinivas PS, Senthilvel S, Selvaraj VM (2021) Genetic mapping reveals a major QTL associated with tolerance to the aphid, *Uroleucon compositae* (Theobald) in safflower (*Carthamus tinctorius* L). *Plant Breed* 140(2): 320–330.

Joglekar RG, Deshmukh NY (1956) Inheritance of florets color in safflower (*Carthamus tinctorius* L.) *Proc Bihar Acad Agric Sci* 5(1): 90–116.

Johnson RC, Kisha TJ, MA Evans (2007) Characterizing safflower germplasm with AFLP molecular markers. *Crop Sci* 47(1): 1728–1736.

Johnson RC, Stout DM, and Bradley VL (1993) The US collection: A rich source of safflower germplasm. In: Dajue L, Yuanzhou H (eds) Proceedings of the Third International Safflower Conference, Institute of Botany, Chinese Academy of Sciences, Beijing Botanical Garden, Beijing, pp. 202–208.

Kaya MD, Bayramin S, Kaya G, Uzun O (2011) Seed vigor and ion toxicity in safflower (*Carthamus tinctorius* L.) seedlings produced by various seed sizes under NaCl stress. *Arch Biol Sci* 63(1): 723–729.

Khadijeh AY, Ghodratollah S, Mohammad RS (2020) Drought stress improves the composition of secondary metabolites in safflower flower at the expense of reduction in seed yield and oil content Industrial crops and products. *Ind Crops Prod* 154: 112496.

Khalil KA, Fouad H, Elsarnagawy T, Almajhdi, FN (2013) Preparation and characterization of electrospun PLGA/silver composite nanofibers for biomedical applications. *Int J ElectSci* 8(3): 3483–3493.

Khan MA, von Witzke-Ehbrecht S, Maass BL, Becker HC (2009) Relationships among different geographical groups, agro-morphology, fatty acid composition and RAPD marker diversity in safflower (*Carthamus tinctorius* L.). *Genet Resour Crop E* 56(1): 19–30.

Khidir MO, Knowles PF (1970). Cytogenetic studies of Carthamus species (Compositae) with 32 pairs of chromosomes. II. Intersectional hybridization. *Can J Genet Cytol* 12(1): 90–99.

Kim J, Assefa AD, Song J, Mani V, Park S et al (2020) Assessment of metabolic profiles in florets of *Carthamus* species using ultra-performance liquid chromatography-mass spectrometry. *Metabolites* 10(11): 440.

Kim KS, Bellendir S, Hudson KA, Hill CB, Hartman GL, Hyten DL, Diers BW (2010) Fine mapping the soybean aphid resistance gene Rag1 in soybean. *Theor Appl Genet*, 120: 1063–1071.

Kinuup VF, Lorenzi H (2014) *Non-Conventional Food Plants (PAN'C) in Brazil: Identification Guide, Nutritional Aspects and Illustrated Recipes*. New Odessa: Ed. Plantarum, Instituto Plantarum, São Paulo, pp. 768–774.

Kiran BU, Mukta N, Kadirvel P, Alivelu K, Senthilvel S (2017) Genetic diversity of safflower (*Carthamus tinctorius* L.) germplasm as revealed by SSR markers. *Plant Genet Resour* 15(1): 1–11.

Knowles PF (1958) Safflower. *Adv Agron* 10(1): 289–323.

Knowles PF (1989) Safflower. In: Röbbelen G, Downey RK, Ashri A (eds) *Oil Crops of the World*. McGraw Hill, New York, pp. 363–374.

Knowles PF, Ashri A (1995) Safflower: *Carthamus tinctorius* (Compositae). In: Smartt J, Simmonds NW (eds) *Evolution of Crop Plants*. 2nd edn. Longman, Harlow, UK, pp. 47–50.

Knowles P, Ashri A (1958) Wild safflower in California: Improvement of cultivated safflower through plant-breeding program to obtain desirable characteristics of wild species. *California Agri*, 12(4): 4–5.

Kouhi F, Sorkheh K, Ercisli S (2020a) Identification of conserved miRNAs and their putative target genes in safflower (*Carthamus tinctorius* L). *Modern Genet* 2018 13(1): 363–379.

Kouhi F, Sorkheh K, Ercisli S (2020b) MicroRNA expression patterns unveil differential expression of conserved miRNAs and target genes against abiotic stress in safflower. *PLoS One* 15(2): 0228850.

Kumar G, Srivastava P (2010) Comparative radio-cytological effect of gamma rays and laser rays on safflower. *Rom J Biol Plant Biol* 55(1): 105–111.

Kumar S, Ambreen H, Murali TV, Bali S, Agarwal M (2015) Assessment of genetic diversity and population structure in a global reference collection of 531 accessions of *Carthamus tinctorius* L. (Safflower) using AFLP markers. *Plant Mol Biol Rep* 33(5): 1299–1313.

Kumar S, Ambreen H, Variath MT, Rao AR, Agarwal M (2016) Utilization of molecular, phenotypic, and geographical diversity to develop compact composite core collection in the oilseed crop, safflower (*Carthamus tinctorius* L.) through maximization strategy. *Front Plant Sci* 7(1): 1554.

Kumar VGMD, German MD, Shih CF (1981) Engineering Approach for Elastic-plastic Fracture Analysis (No. EPRI-NP–1931). General Electric Co.

La Bella S, Tuttolomondo T, Lazzeri L, Matteo R, Leto C, Licata M (2019) An agronomic evaluation of new safflower (*Carthamus tinctorius* L.) germplasm for seed and oil yields under Mediterranean climate conditions. *Agronomy* 9(8): 468.

Ladd SL, Knowles PF (1971) Interactions of alleles at two loci regulating fatty acid composition of the seed oil of safflower (*Carthamus tinctorius* L.) 1. *Crop Sci* 11(5): 681–684.

Lee GA, Sung JS, Lee SY, Chung JW, Yi JY (2014) Genetic assessment of safflower (*Carthamus tinctorius* L.) collection with microsatellite markers acquired via pyrosequencing method. *Mol Ecol Resour* 14(1): 69–78.

Li D, Li K, Zhou G, He S (2023) Effects of temperature and salt stress on the expression of delta-12 fatty acid desaturase genes and fatty acid compositions in safflower. *Int J Mol Sci* 24(3): 2765.

Li H, Dong Y, Yang J, Liu X, Wang Y (2012) De novo transcriptome of safflower and the identification of putative genes for oleosin and the biosynthesis of flavonoids. *PLoS One* 7(2): 30987.

Liu W, Shahid MQ, Bai L, Lu Z, Chen Y et al (2015) Evaluation of genetic diversity and development of a core collection of wild rice (Oryza rufipogon Griff.) populations in China. *PLoS One* 10: e0145990.

Lombardi M, Materne M, Cogan NOI, Rodda, M, Daetwyler HD, Slater AT, Kaur S (2014) Assessment of genetic variation within a global collection of lentil (Lens culinaris Medik.) cultivars and landraces using SNP markers. *BMC Genet* 15(1): 150.

Luo Z, Zeng H, Ye Y, Liu L, Li S (2015) Safflower polysaccharide inhibits the proliferation and metastasis of MCF-7 breast cancer cell. Mol Med Rep 11(6): 4611–4616.

Maali-Amiri RM, Yazdi-Samadi B, Ghannadha MR, Abd-Mishani C (2001) Detection of DNA polymorphism in landrace populations of safflower in Iran using RAPDPCR technique. *Iran J Agric Sci* 32(1): 737–745.

Mahmood T, Muhammad S, Shinwari ZK (2010) Molecular and morphological characterization of *Caralluma* species. *Pak J Bot* 42(1): 1163–1171.

Majidi MM, Tavakoli V, Mirlohi A, Sabzalian MR (2011) Wild safflower species (*Carthamus oxyacanthus* Bieb.): A possible source of drought tolerance for arid environments. *Aust J Crop Sci* 5(8): 1055–1063.

Majidi MM, Zadhoush S (2014) Molecular and morphological variation in a world-wide collection of safflower. *Crop Sci* 54(5): 2109–2119.

Malik AH (2023) TILLING and EcoTILLING: Tools of trade for crop improvement in changing climate. In: Bhat TA, Hakeem, KR (eds.). *Biotechnologies and Genetics in Plant Mutation Breeding: Volume 1: Mutagenesis and Crop Improvement* (1st ed.). Apple Academic Press,, pp. 249–266. https://doi.org/10.1201/9781003305064

Mani V, Lee SK, Yeo Y, Hahn BS (2020) A metabolic perspective and opportunities in pharmacologically important safflower. *Metabolites* 10(1): 253.

Mayerhofer M, Mayerhofer R, Topinka D, Christianson J, Good AG (2011) Introgression potential between safflower (*Carthamus tinctorius*) and wild relatives of the genus *Carthamus*. *BMC Plant Biol* 11(1): 1–10.

McPherson MA, Good AG, Keith C, Topinka A, Hall LM (2004) Theoretical hybridization potential of transgenic safflower (*Carthamus tinctorius* L.) with weedy relatives in the New World. *Can J Plant Sci* 84(3): 923–934.

Meksem K, Pantazopoulos P, Njiti VN, Hyten LD, Arelli PR et al (2001) Forrest'resistance to the soybean cyst nematode is bigenic: Saturation mapping of the Rhg1and Rhg4 loci. *Theor Appl Genet* 103(1): 710–717.

Mirzahashemi M, Mohammadi-Nejad G, Golkar P (2015) A QTL linkage map of safflower for yield under drought stress at reproductive stage. *Irani J Genet Plant Breed* 4(2): 20–27.

Naresh V, Yamini KN, Rajendrakumar P, Dinesh Kumar V (2009) EST-SSR marker-based assay for the genetic purity assessment of safflower hybrids. *Euphytica* 170(3): 347–353.

Narkhede BN, Deokar AB (1986) Inheritance of corolla color in safflower. *J Maharashtra Agric Univ* 11: 278–281.

Narkhede BN, Patil AM, Deokar AB (1990) Combining ability for yield and its components in safflower. *J MAU* 5(1)

Neycee MA, Mohsenpour M, Rahnama H (2022) Investigating the increase of oleic acid in safflower plant by changes in Fatty Acid Desaturase-2 gene after using CRISPR-Cas9 method. *Crop Biotechnol* 12(39): 69–83.

Nguyen VT, Vuong TD, VanToai T, Lee JD, Wu, X et al (2012) Mapping of quantitative trait loci associated with resistance to *Phytophthora sojae* and flooding tolerance in soybean. *Crop Sci* 52(6): 2481–2493.

Nimbkar N (2008) Issues in safflower production in India In: Knights SE, Potter TD (eds) Safflower: Unexploited Potential and World Adaptability, Proceedings of the Seventh International Safflower Conference, Wagga Wagga, NSW, Australia.

Nimbkar N (2017) Safflower rediscovered though safflower flowers find mention in of ayurveda and in European and japanese pharmacopoeias, the interest in this crop has been rekindled in the last few years. *Times Agri J*. 2017.

Nitnaware KM, Naikawadi VB, Chavan SS, Shelk DB, Barmukh RB, Naik AA, Nikam TD. (2021). Genetic Engineering in Safflower (Carthamus tinctorius L.): Retrospect and Prospect. *Genetically Modified Crops: Current Status, Prospects and Challenges Volume 1*, 201–226.

Pahlavani M, Mirlohi A, Saeidi G (2004) Inheritance of flower color and spininess in safflower (*Carthamus tinctorius* L.). *J Hered* 95(3): 265–267.

Pearl SA, Burke JM (2014) Genetic diversity in (*Carthamus tinctorius* L.) (Asteraceae; safflower), an underutilized oilseed crop. *Amer J Bot* 101(10): 1640–1650.

Poodineh, M, Nezhad NM, Mohammadi-Nejad G, Fakheri BA, Ebrahimi F (2021) Identification of safflower (*Carthamus tinctorius* L.) QTL under drought stress and normal conditions. *Ind Crops Prod* 171(1): 113889.

Rahimi M, Marashi SH, Farsi M, Ghorbanzadeh M (2011) Detection of DNA polymorphism of safflower by using of RAPD markers. *J Agric Sci Sustain Prod* 21(3): 13–23.

Raina P, O'Donnell M, Rosenbaum P, Brehaut, J, Walter SD et al (2005) The health and well-being of caregivers of children with cerebral palsy. *Pediatrics* 115(6): 626–636.

Rampure NH, Choudhary AD, Jambhulkar SJ, Badere RS (2015) Ethyl methanesulphonate-induced high oleic acid mutants in safflower (*Carthamus tinctorius* L.) *J Crop Improv* 29(6): 720–727.

Rampure NH, Choudhary AD, Jambhulkar SJ, Badere RS (2017) Isolation of desirable mutants in safflower for crop improvement. *Indian J Genet* 77(01): 134–144.

Rampure NH, Majumdar PN, Badere RS (2014) Genetic variability for morphological and biochemical characters in safflower (*Carthamus tinctorius* L.) *Indian J Genet* 74(03): 353–361.

Rao M (1943) Inheritance of characters in safflower- (*Carthamus tinctorius* L). *Madras Agric J* 31(1): 141–148.

Ravi K, Vadez V, Isobe S, Mir RR, Guo Y, Nigam SN, Varshney RK (2011) Identification of several small main-effect QTLs and a large number of epistatic QTLs for drought tolerance related traits in groundnut (*Arachis hypogaea* L.). *TAG* 122: 1119–1132.

Ren, C, Chen, C, Dong S, Wang R, Xian B et al (2022) Integrated metabolomics and transcriptome analysis on flavonoid biosynthesis in flowers of safflower (*Carthamus tinctorius* L.) during colour-transition. *Peer J* 10(1): 13591.

Saberianfar R, Menassa R (2018) Strategies to increase expression and accumulation of recombinant proteins. In: *Molecular Pharming: Applications, Challenges, and Emerging Areas* (pp. 119–135).

Sabzalian MR, Mirlohi A, Saeidi G, Rabbani MT (2009) Genetic variation among populations of wild safflower, Carthamus oxyacanthus analyzed by agro-morphological traits and ISSR markers. *Genet Resour Crop Evol* 56: 1057–1064.

Sabzalian MR, Saeidi G, Mirlohi A, Hatami B (2010) Wild safflower species (*Carthamus oxyacanthus*): A possible source of resistance to the safflower fly (*Acanthiophilus helianthi*). *Crop Protec* 29(1): 550–555.

Sajid M, Munir H, Khaliq A et al (2023) Unveiling safflower yield, oil content, water use efficiency, and membrane stability under differential irrigation regimes. *Arab J Geosci* 16: 249.

Sehgal D, Raina SN, Devarumath M, Sasanuma T, T. Sasakuma (2009) Nuclear DNA assay in solving issues related to ancestry of the domesticated diploid safflower (*Carthamus tinctorius* L.) and the polyploid (Carthamus) taxa, and phylogenetic and genomic relationships in the genus *Carthamus* L. (Asteraceae). *Mol Phylogenet E* 53: 631–644.

Shaki F, Maboud HE, Niknam V (2020a) Differential proteomics: Effect of growth regulators on salt stress responses in safflower seedlings. *Pesticide Biochem Physiol* 164: 149–155.

Shaki F, Niknam V, Ebrahimzadeh Maboud H (2020b) Effects of penconazole on hormonal crosstalk and fatty acids from salt-stressed safflower. *Iran J Plant Physiol* 10(3): 3213–3221.

Shinwari ZK, Rehman H, Rabbani MA (2014a) Morphological traits based genetic diversity in safflower (Carthamus tinctorius L.) *Pak J Bot* 46(4): 1389–1395.

Shinwari ZK, Rehman HINA, Rabbani MA (2014b) SDS-Page based genetic divergence in safflower (Carthamus Tinctorius L.). *Pak J Bot* 46(3): 811–815.

Shrivastava SR, Shrivastava PS (2012) Evaluation of trained Accredited Social Health Activist (ASHA) workers regarding their knowledge, attitude and practices about child health. *Rural Remote Health* 12(4): 1–7.

Singh RK, Sreenivasulu N, Prasad M (2022) Potential of underutilized crops to introduce the nutritional diversity and achieve zero hunger. *Funct Integr Genom* 22(1): 1459–1465.

Singh S, Sinha S, Saxena R, Pandey K, Bhatt K (2004) Translocation of metals and its effects in the tomato plants grown on various amendments of tannery waste: Evidence for involvement of antioxidants. *Chemosphere* 57(2): 91–99.

Singh V, Deshpande MB, Nimbkar N (2003) NARI-NH-1: The first non-spiny hybrid safflower released in India. *Sesame Safflower News* 118: 77–79.

Sujatha M, Geetha A, Sivakumar P, Palanisamy N (2008) Biotechnological interventions for genetic improvement of safflower. In Proceedings of VIIth International Safflower Conference (Vol. 3, No. 6). Wagga Wagga, NSW, Australia, 3–6 Nov 2008 (1–5) Agri-MC Marketing and Communication.

Sun M, Sun Q, Zhao C, Huang Y, Jiang J et al (2022) Degradation of diclofenac sodium with low concentration from aqueous milieu through polydopamine-chitosan modified magnetic adsorbent-assisted photo-Fenton process. *Separ Purif Technol* 289: 120771.

Sun Y, Li R, Wu X, Yu Y (2021) AdipoRon promotes the osseointegration of dental implants in mice with type 2 diabetes mellitus. *Front Physiol* 12(1): 697738.

Thoday-Kennedy E, Banerjee B, Panozzo J, Maharjan P, Hudson D et al (2023) Dissecting physiological and agronomic diversity in safflower populations using proximal phenotyping. *Agriculture* 13(3): 620.

Upadhyaya HD, Ortiz R (2001) A mini core subset for capturing diversity and promoting utilization of chickpea genetic resources in crop improvement. *Theor Appl Genet* 102(1): 1292–1298.

Upadhyaya NM, Zhu QH, Zhou XR, Eamens AL, Hoque MS et al (2009) Dissociation (Ds) constructs, mapped Ds launch pads and a transiently expressed transposase system suitable for localized insertional mutagenesis in rice. *Theor Appl Genet* 112: 1326–1341.

Varshney RK, Ribaut JM, ES Buckler, R Tuberosa, JA Rafalski, P Langridge (2012) Can genomics boost productivity of orphan crops *Nat Biotechnol* 30: 1172– 1176.

Velasco L, Pérez-Vich B, Fernández-Martínez, JM (2005) Identification and genetic characterization of a safflower mutant with a modified tocopherol profile. *Plant Breed* 124(5): 459–463.

Waki T, Terashita M, Fujita N, Fukuda K, Kato M et al (2021) Identification of the genes coding for carthamin synthase, peroxidase homologs that catalyze the final enzymatic step of red pigmentation in safflower (*Carthamus tinctorius* L.) *Plant Cell Physiol* 62(10): 1528–1541.

Wang C, Gao Y, Zhang Z, Chi Q, Liu Y et al (2020) Safflower yellow alleviates osteoarthritis and prevents inflammation by inhibiting PGE2 release and regulating NF-κB/SIRT1/AMPK signaling pathways. *Phytomedicine* 78: 153305.

Wang C, Zhang D, Li G, Liu J, Tian J et al (2007) Neuroprotective effects of safflor yellow B on brain ischemic injury. *Exp Brain Res* 177: 533–539.

Wang D, Wang LJ, Zhu FX, Zhu JY, Chen XD et al (2008) In vitro and in vivo studies on the antioxidant activities of the aqueous extracts of Douchi (a traditional Chinese salt-fermented soybean food). *Food Chem* 107(4): 1421–1428.

Wang L, Guan Y, Guan R, Li Y, Ma, Y et al (2006) Establishment of Chinese soybean Glycine max core collections with agronomic traits and SSR markers. *Euphytica* 151: 215–223.

Wang Y, Ge H, Ahmad N, Li J, Wang Y et al (2023) Genome-wide identification of MADS-box family genes in safflower (*Carthamus tinctorius* L.) and functional analysis of CtMADS24 during flowering. *Int J Mol Sci* 24(2): 1026.

Weiss EA (1983) *Oilseed Crops*. Longman Group Ltd.

Weiss EA (2000) Safflower. In: Weiss EA (ed) *Oilseed Crops*. Blackwell Science, London, UK, pp. 93–129.

Wu Z, Liu H, Zhan W, Yu Z, Qin E et al (2021) The chromosome-scale reference genome of safflower (*Carthamus tinctorius* L.) provides insights into linoleic acid and flavonoid biosynthesis. *Plant Biotechnol J* 19 (9): 1725–1742.

Wu Z, Wang M, Yang S, Chen S, Chen X et al (2019) A global coexpression network of soybean genes gives insight into the evolution of nodulation in non-legumes and legumes. *New Phytol* 223: 2104–2119.

Yang C, Huang Y, Lv P, Antwi-Boasiako, A, Begum N, Zhao T, Zhao J (2022) NAC transcription factor GmNAC12 improved drought stress tolerance in soybean. *Int J Mol Sci* 23(19): 12029.

Yeilaghi H, Arzani A, Ghaderian M (2015) Evaluating the contribution of ionic and agronomic components toward salinity tolerance in safflower. *Agron. J* 107: 2205–2212.

Zeven AC, Zhukovsky PM (1975) Dictionary of cultivated plants and their centres of diversity, excluding ornamentals, forest trees and lower plants. *Center for Agricultural Publishing and Documentation* (pp. 1–219), Wageningen. ISBN 978-9022005491.

Zhang CY, Chen XS, Zhang YM, Yuan ZH, Liu ZC et al (2009) A method for constructing core collection of *Malus sieversii* using molecular markers. *Sci Agric Sin* 42: 597–604.

Zhao H, Savin KW, Li Y, Breen EJ, Maharjan P et al (2022a) Genome-wide association studies dissect the G×E interaction for agronomic traits in a worldwide collection of safflowers (*Carthamus tinctorius* L.) *Mol Breed* 42: 24.

Zhao L, Ren C, Xue X, Lu H, Wang K (2022b) Safflomin A: A novel chemical marker for (*Carthamus tinctorius* L.) safflower monofloral honey. *Food Chem* 366: 130584.

Zhou J, Shen B, Zhang W, Wang J, Yang J et al (2014a) One-step generation of different immunodeficient mice with multiple gene modifications by CRISPR/Cas9 mediated genome engineering. *Int J Biochem Cell Biol* 46: 49–55.

Zhou X, Tang L, Xu Y, Zhou G, Wang Z (2014b) Towards a better understanding of medicinal uses of *Carthamus tinctorius* L. In traditional Chinese medicine: A phytochemical and pharmacological review. *J Ethnopharmacol* 151: 27–43.

Zou J, Jiang C, Cao Z, Li R, Long Y et al (2010) Association mapping of seed oil content in Brassica napus and comparison with quantitative trait loci identified from linkage mapping. *Genome* 53(11): 908–916.

For Product Safety Concerns and Information please contact our EU representative GPSR@taylorandfrancis.com Taylor & Francis Verlag GmbH, Kaufingerstraße 24, 80331 München, Germany